D. Hellwig · B. L. Bauer (Eds.)

Minimally Invasive Techniques for Neurosurgery

Springer-Verlag Berlin Heidelberg GmbH

D. Hellwig · B. L. Bauer (Eds.)

Minimally Invasive Techniques for Neurosurgery

Current Status and Future Perspectives

With 161 Figures

Springer

PD Dr. med. Dieter Hellwig
Professor Dr. med. Bernhard L. Bauer
Klinikum der Universität Marburg
Zentrum für Operative Medizin I
Baldingerstraße 35

35043 Marburg

Cover Design with approval by Prinz & Partner, Marburg

ISBN 978-3-642-63701-8

Library of Congress Cataloging-in-Publication Data
Minimally invasive techniques for neurosurgery : current statuts and future perspectives
/ [edited by] D. Hellwig, B. L. Bauer.   p.   cm.
Includes bibliographical references and index.
ISBN 978-3-642-63701-8   ISBN 978-3-642-58731-3 (eBook)
DOI 10.1007/978-3-642-58731-3
1. Nervous system – Endoscopic surgery. 2. Nervous system – Surgery. I. Hellwig, D. (Die-
ter), 1956- . II. Bauer, B. L. (Bernhard Ludwig), 1929- . [DNLM: 1. Neurosurgery – methods,
2. Surgery, Minimally Invasive – methods. 3. Endoscopy – methods.
WL 368 M6654 1998]   RD593.M546   1998   617.4'8 – dc21

Cover Design: E. Kirchner, Heidelberg
Production: ProduServ GmbH Verlagsservice, Berlin
Typesetting: Fotosatz-Service Köhler OHG, Würzburg
SPIN: 10634615   81/3020 – 5 4 3 2 1 0 – Gedruckt auf säurefreiem Papier

Preface

Seven years have passed since the First Congress of Minimally Invasive Neurosurgery and 5 years since the second one was held in Marburg. Much has happened during that time. These concerning the availability of sophisticated equipment and also the broad and worldwide acceptance of neuroendoscopic techniques for specific neurosurgical indications. Many of the newly developed neurosurgical, diagnostic, and therapeutic technologies and methods are minimally invasive. Therefore our goal was to provide an overview of all facets of minimally invasive approaches in neurosurgery.

In addition to neuroendoscopy, the major topics of this volume include neuronavigation, functional neurosurgery, neurotransplantation, and "molecular neurosurgery". None of these is discussed in its full variety, but "the state of the art" is presented for each to demonstrate the future prospects of minimally invasive neurosurgery. A panel of international experts contribute to this aim.

It is our pleasure to present this for reaching overview of minimally invasive neurosurgery today and potential developments in the future.

Dieter Hellwig   Bernhard L. Bauer

Contents

Part 7   MIEN: Prospects for the Future

Part 8   Neurotransplantation

Part 9   Molecular Neurosurgery

Authors

Alexander Vasilevich Afonasiev
Department of Neurosurgery and Neurology
Main Military Clinical Hospital
ul. Gospitalna 18
252016 Kiev
Ukraine

F. K. Albert
Department of Neurosurgery
Paracelsus Klinik
Am Natruper Holz 69
49076 Osnabrück
Germany

O. Alberti
Gough-Cooper Department of Neurological Surgery
The National Hospital for Neurology
and Neurosurgery
London WCW 3 BG
UK

G. Arango
Department of Neurosurgery
Nordstadt Hospital
Haltenhoffstraße 41
30167 Hannover
Germany

Hans Arnold
Department of Neurosurgery
Medical University of Lübeck
Ratzeburger Allee 160
23538 Lübeck
Germany

Dorothea Auer
MPI für Psychiatrie
Kraepelinstraße 10
80804 München
Germany

Ludwig M. Auer
Neurochirurgische Klinik
Abteilung für spez. Neurochirurgie
Oscar Orth-Straße
66421 Homburg
Germany

Volker Bartel
Erbe Elektromedizin GmbH
Waldhörnlestraße 17
72072 Tübingen
Germany

Bernhard L. Bauer
Department of Neurosurgery
Philipps University
Baldingerstraße
35043 Marburg
Germany

R. Becker
Department of Neurosurgery
Philipps University
Baldingerstraße
35043 Marburg
Germany

A. Behnke
Department of Neurosurgery
Christian Albrecht University
Weimarer Straße 8
24106 Kiel
Germany

L. Benes
Department of Neurosurgery
Philipps University
Baldingerstraße
35043 Marburg
Germany

H. Bertalanffy
Department of Neurosurgery
Philipps University
Baldingerstraße
35043 Marburg
Germany

S. Bien
Department of Neuroradiology
Rudolf-Bultmann-Straße 8
35033 Marburg
Germany

M. M. Bonsanto
University of Heidelberg
Im Neuenheimer Feld 400
69120 Heidelberg
Germany

J. Boschert
Department of Neurosurgery
University of Heidelberg
Theodor-Kutzer-Ufer 1–3
68167 Mannheim
Germany

T. Brinker
Department of Neurosurgery
Nordstadt Hospital
Haltenhoffstraße 41
30167 Hannover
Germany

Giovanni Broggi
Department of Neurosurgery
National Neurological Institute "C Besta"
Via Celoria 11
20133 Milano
Italy

P. Brundin
Department of Neurology
University of Lund
Sweden

J. Brunner
Department of Neurosurgery
University of Heidelberg
Theodor-Kutzer-Ufer 1–3
68167 Mannheim
Germany

P. Brüser
Department of Hand, Plastic,
and Reconstructive Surgery
Malteser Hospital
von Hompesch-Straße 1
53123 Bonn
Germany

W. Burkert
Department of Neurosurgery
Faculty of Medicine
Martin Luther University
Magdeburger Straße
06112 Halle-Wittenberg
Germany

C. Busch
Department of Neurosurgery
Dr. Horst-Schmidt-Kliniken GmbH
Ludwig-Erhard-Straße 100
65199 Wiesbaden
Germany

J. Buurman
Integrated Clinical Solutions
Philips Medical Systems Nederland BV
Best
The Netherlands

V. A. Coenen
Department of Neuroanatomy
Technical University RWTH
Pawelstraße 30
52057 Aachen
Germany

Y. S. Cordoliani
Department of Radiology
Val-de-Grâce Hospital
74 Boulevard de Port Royal
75230 Paris CEDEX 05
France

Alexander Geoirgievich Danchin
Department of Neurosurgery and Neurology
Main Military Clinical Hospital
ul. Gospitalna 18
252016 Kiev
Ukraine

Andrei Alexandrovich Danchin
Department of Neurosurgery and Neurology
Main Military Clinical Hospital
ul. Gospitalna 18
252016 Kiev
Ukraine

W. Dauber
Department of Anatomy for Dentists
University of Tübingen
Hoppe-Seyler-Straße 3
72076 Tübingen
Germany

J. Debus
Universitäts-Strahlenklinik
Im Neuenheimer Feld 400
69120 Heidelberg

M. Desgeorges
Department of Neurosurgery and Neurosciences
Val-de-Grâce Hospital
74 Boulevard de Port Royal
75230 Paris CEDEX 05
France

A. Dijkstra
Integrated Clinical Solutions
Philips Medical Systems Nederland BV
Best
The Netherlands

I. Dones
Department of Neurosurgery
National Neurological Institute "C Besta"
Via Celoria 11
20133 Milano
Italy

N. L. Dorward
The National Hospital for Neurology
and Neurosurgery
Gough-Cooper Department of Neurological Surgery
Queen Square
London WCW 3 BG
UK

F. Duffner
Department of Neurosurgery
University of Tübingen
Hoppe-Seyler-Straße 3
72076 Tübingen
Germany

Christopher Earl
Department of Neurology
Philipps University
Rudolf-Bultmann-Straße 8
35033 Marburg
Germany

Frank Eggers
Department of Neurosurgery
Philipps University
Baldingerstraße
35043 Marburg
Germany

T. Eichhorn
Department of Neurology
Philipps University
Baldingerstraße
35043 Marburg
Germany

R. Engenhardt-Cabilic
Klinik für Strahlentherapie
Philipps Universität Marburg
Baldingerstraße
35043 Marburg
Germany

R. Filippi
Department of Neurosurgery
Johannes Gutenberg University
Langenbeckstraße 1
55131 Mainz
Germany

M. Fornari
Department of Neurosurgery
National Neurological Institute "C Besta"
Via Celoria 11
20133 Milan
Italy

Angelo Franzini
Department of Neurosurgery
National Neurological Institute "C Besta"
Via Celoria 11
20133 Milan
Italy

D. Freudenstein
Department of Neurosurgery
University of Tübingen
Hoppe-Seyler-Straße 3
72076 Tübingen
Germany

Georg Fries
Department of Neurosurgery
Johannes Gutenberg University
Langenbeckstraße 1
55131 Mainz
Germany

Michael J. Fritsch
Pediatric Neurosurgery
Phoenix Children's Hospital
909 East Brill Street
Phoenix, AR 85006
USA

Michael R. Gaab
Department of Neurosurgery
Ernst Moritz Arndt University
Sauerbruchstraße
17487 Greifswald
Germany

F. A. Gerritsen
Integrated Clinical Solutions
Philips Medical Systems Nederland BV
Best
The Netherlands

J. M. Gilsbach
Department of Neurosurgery
Technical University RWTH
Pawelstraße 30
52057 Aachen
Germany

S. Giombini
Department of Neurosurgery
National Neurological Institute "C Besta"
Via Celoria 11
20133 Milan
Italy

Jan Gliemroth
Department of Neurosurgery
Medical University of Lübeck
Ratzeburger Allee 160
23538 Lübeck
Germany

M. Grisoli
Department of Neuroradiology
National Neurological Institute "C Besta"
Via Celoria 11
20133 Milan
Italy

E. H. Grote
Department of Neurosurgery
University of Tübingen
Hoppe-Seyler-Straße 3
72076 Tübingen
Germany

J. A. Grotenhuis
Radboud oost
Reinier Postlaan 4
Postbus 9101
6500 HB Nijmegen
The Netherlands

A. E. Guber
Forschungszentrum Karlsruhe GmbH
Institut für Mikrostrukturtechnik
Postfach 3640
76021 Karlsruhe
Germany

H. K. Gumprecht
Department of Neurosurgery
University Hospital
Englschalkinger Straße 77
81925 München-Bogenhausen
Germany

Reiner Haag
Erbe Elektromedizin GmbH
Waldhörmlestraße 17
72072 Tübingen
Germany

P. Hagell
Department of Neurosurgery
University of Lund
Sweden

J. Hamer
Department of Neurosurgery
Paracelsus Klinik
Am Natruper Holz 69
49076 Osnabrück
Germany

M. Hardenack
Department of Neurosurgery
Ruhr University of Bochum
Knappschaftskrankenhaus
In der Schornau 23 – 25
44892 Bochum
Germany

A. Harders
Department of Neurosurgery
Ruhr University of Bochum
Knappschaftskrankenhaus
In der Schornau 23 – 25
44892 Bochum
Germany

A. Heinemann
Department of Neurosurgery
Philipps University
Baldingerstraße
35043 Marburg
Germany

Dieter Hellwig
Department of Neurosurgery
Philipps University
Baldingerstraße
35043 Marburg
Germany

H. J. Hoff
Department of Neurosurgery
Krankenanstalten Gilead
Burgsteig 4
33617 Bielefeld
Germany

Nikolai J. Hopf
Department of Neurosurgery
Johannes Gutenberg University
Langenbeckstraße 1
55131 Mainz
Germany

F. Hor
Department of Neurosurgery and Neurosciences
Val-de-Grâce Hospital
74 Boulevard de Port Royal
75230 Paris CEDEX 05
France

B. O. Hütter
Department of Neurosurgery
Technical University RWTH
Pawelstraße 30
52057 Aachen
Germany

R. Jennemann
Department of Neurosurgery
Philipps University
Baldingerstraße
35043 Marburg
Germany

Ali Kafadar
Department of Neurosurgery
Johannes Gutenberg University
Langenbeckstraße 1
55131 Mainz
Germany

J. Kaminsky
Department of Neurosurgery
Nordstadt Hospital
Haltenhoffstraße 41
30167 Hannover
Germany

Uwe Kehler
Department of Neurosurgery
Medical University of Lübeck
Ratzeburger Allee 160
23538 Lübeck
Germany

R. T. Kelley
Department of Otolaryngology
SUNY-HSC University Hospital
Syracuse, NY 13210
USA

R.M. Kellman
Department of Otolaryngology
SUNY-HSC University Hospital
Syracuse, NY 13210
USA

D. Graf von Keyserlingk
Department of Neuroanatomy
Technical University RWTH
Pawelstraße 30
52057 Aachen
Germany

N. Kitchen
Gough-Cooper Department of Neurological Surgery
The National Hospital
for Neurology and Neurosurgery
Queen Square
London WCW 3 BG
UK

H. Klinge
Department of Neurosurgery
Christian Albrecht University
Weimarer Straße 8
24106 Kiel
Germany

M. Knauth
Department of Radiology
University of Heidelberg
Heidelberg
Germany

Ulrich Knopp
Department of Neurosurgery
Medical University of Lübeck
Ratzeburger Allee 160
23538 Lübeck
Germany

L. Krasznai
Department of Neurosurgery
University of Tübingen
Hoppe-Seyler-Straße 3
72076 Tübingen
Germany

G. A. Krombach
Department of Neurosurgery
Technical University RWTH
Pawelstraße 30
52057 Aachen
Germany

W. Küker
Department of Neuroradiology
Technical University RWTH
Pawelstraße 30
52057 Aachen
Germany

Thomas J. Kuhn
Department of Neurosurgery
Philipps University
Baldingerstraße
35043 Marburg
Germany

S. Kunze
University of Heidelberg
Im Neuenheimer Feld 400
69120 Heidelberg
Germany

A. Kupsch
Department of Neurology
Rudolf-Virchow-Krankenhaus
der Humboldt-Universität zu Berlin
Augustenburger Platz 1
13353 Berlin
Germany

G. Larkin
Department of Hand, Plastic, and Reconstructive
Surgery
Malteser Hospital
von Hompesch-Straße 1
53123 Bonn
Germany

O. Lindvall
Department of Neurology
University of Lund
Sweden

C. B. Lumenta
Department of Neurosurgery
University Hospital
Englschalkinger Straße 77
81925 München-Bogenhausen
Germany

Kim H. Manwaring
Pediatric Neurosurgery
Phoenix Children's Hospital
909 East Brill Street
Phoenix, AR 85006
USA

L. Mayfrank
Department of Neurosurgery
Technical University RWTH
Pawelstraße 30
52057 Aachen
Germany

M. Mazanek
Department of Neurology
Zentrum für Nervenheilkunde
Philipps University
Rudolf-Bultmann-Straße 8
35033 Marburg
Germany

H. M. Mehdorn
Department of Neurosurgery
Christian Albrecht University
Weimarer Straße 8
24106 Kiel
Germany

Matthias Meinhardt
Department of Neurosurgery
Philipps University
Baldingerstraße
35043 Marburg
Germany

Hans-Dieter Mennel
Department of Neuropathology
University of Marburg
Baldingerstraße
35043 Marburg
Germany

S. Mense
Institute of Anatomy and Cell Biology
University of Heidelberg
Heidelberg
Germany

H. Mewes
Department of Neurosurgery
Philipps University
Baldingerstraße
35043 Marburg
Germany

Bernhard Meyer
Department of Neurosurgery
University of Bonn
Sigmund-Freud-Straße 25
53127 Bonn
Germany

A. Nabavi
Department of Neurosurgery
Christian Albrecht University
Weimarer Straße 8
24106 Kiel
Germany

K. Niemann
Department of Neuroanatomy
Technical University RWTH
Wendlingweg 2
52074 Aachen
Germany

G. Nikkhah
Neurochirurgie
Krankenhaus Nordstadt
Haltenhoffstraße 41
30167 Hannover

P. Odin
Department of Neurosurgery
University of Lund
Sweden

Wolfgang Oertel
Department of Neurology
Philipps University
Rudolf-Bultmann-Straße 8
35033 Marburg
Germany

Falk Oppel
Department of Neurosurgery
Krankenanstalten Gilead
Burgsteig 4
33617 Bielefeld
Germany

H. W. Pannek
Department of Neurosurgery
Krankenanstalten Gilead
Burgsteig 4
33617 Bielefeld
Germany

K. Pietz
Department of Neurosurgery
University of Lund
Sweden

Axel Perneczky
Department of Neurosurgery
Johannes Gutenberg University
Langenbeckstraße 1
55131 Mainz
Germany

B. Petersen
Department of Neurosurgery
Christian Albrecht University
Weimarer Straße 8
24106 Kiel
Germany

Karl H. Plate
Department of Neuropathology
Albert Ludwig Unversity
Breisacherstraße 64
79106 Freiburg
Germany

O. Pogarell
Department of Neurology
Philipps University
Rudolf-Bultmann-Straße 8
35033 Marburg
Germany

N. G. Rainov
Department of Neurosurgery
Faculty of Medicine
Martin Luther University
Halle-Wittenberg
Germany

M. H. Reinges
Department of Neurosurgery
Technical University RWTH
Pawelstraße 30
52057 Aachen
Germany

Robert Reisch
Department of Neurosurgery
Johannes Gutenberg University
Langenbeckstraße 1
55131 Mainz
Germany

K. D. M. Resch
Department of Neurosurgery
Johannes Gutenberg University
Langenbeckstraße 1
55131 Mainz
Germany

T. Riegel
Department of Neurosurgery
Philipps University
Baldingerstraße
35043 Marburg
Germany

A. Rieger
Department of Neurosurgery
Faculty of Medicine
Martin Luther University
Halle-Wittenberg
Germany

Kurt Ringel
Department of Neurosurgery
Johannes Gutenberg University
Langenbeckstraße 1
55131 Mainz
Germany

G. S. Rodziewicz
Department of Neurosurgery
SUNY-HSC University Hospital
Syracuse, New York 13210
USA

V. Rohde
Department of Neurosurgery
Technical University RWTH
Pawelstraße 30
52057 Aachen
Germany

J. Rohlfs
Department of Neurosurgery
Philipps University
Baldingerstraße
35043 Marburg
Germany

Dirk van Roost
Department of Neurosurgery
University of Bonn
Sigmund-Freud-Straße 25,
53127 Bonn
Germany

M. Samii
Department of Neurosurgery
and Medical School
Nordstadt Hospital
Haltenhoffstraße 41
30167 Hannover
Germany

L. Sanchin
Department of Neurosurgery
Faculty of Medicine
Martin Luther University
Halle-Wittenberg
Germany

Jürgen Sautter
Department of Neurology
Humbold University, Charité
Augustenburger Platz 1
13353 Berlin
Germany

Carl Schaller
Department of Neurosurgery
University of Bonn
Sigmund-Freud-Straße 25
53127 Bonn
Germany

Jürgen Schlegel
Department of Neuropathology
University of Marburg
Baldingerstraße
35043 Marburg
Germany

P. Schmiedek
Department of Neurosurgery
Mannheim
Theodor-Kutzer-Ufer 1–3
68167 Mannheim
Germany

K. Schmieder
Department of Neurosurgery
Ruhr University of Bochum
Knappschaftskrankenhaus
In der Schornau 23–25
44892 Bochum
Germany

A. Schneider
Department of Neurosurgery
Philipps University
Baldingerstraße
35043 Marburg
Germany

M. Scholz
Department of Neurosurgery
Ruhr University of Bochum
Knappschaftskrankenhaus
In der Schornau 23–25
44892 Bochum
Germany

R. Schönmayr
Department of Neurosurgery
Dr. Horst-Schmidt-Kliniken GmbH
Ludwig-Erhard-Straße 100
65199 Wiesbaden
Germany

G. Schöpp
St. Elisabeth Hospital
Department of Radiology
Halle
Germany

Johannes Schramm
Department of Neurosurgery
University of Bonn
Sigmund-Freud-Straße 25
53127 Bonn
Germany

Henry W. S. Schroeder
Department of Neurosurgery
Ernst Moritz Arndt University
Sauerbruchstraße
17487 Greifswald
Germany

R. Schubert
Department of Neurosurgery
Johannes Gutenberg University
Langenbeckstraße 1
55131 Mainz
Germany

H. Senyurt
Neurosurgical Clinic
Altstadtstraße 25
44534 Lünen
Germany

D. Servello
Department of Neurosurgery
National Neurological Institute "C Besta"
Via Celoria 11
20133 Milano
Italy

M. Skalej
Department of Neuroradiology
University of Tübingen
Hoppe-Seyler-Straße 3
72076 Tübingen
Germany

I. Slansky
Department of Neurosurgery
Technical University RWTH
Pawelstraße 30
52057 Aachen
Germany

M. V. Smith
Department of Neurosurgery
SUNY-HSC University Hospital
Syracuse, NY 13210
USA

U. Spetzger
Department of Neurosurgery
Technical University RWTH
Pawelstraße 30
52057 Aachen
Germany

A. Staubert
Department of Neurosurgery
Paracelsus Klinik
Am Natruper Holz 69
49076 Osnabrück
Germany

Gabi Stumm
Department of Neuropathology
University of Marburg
Baldingerstraße
35043 Marburg
Germany

C. Teo
Division of Pediatrics Neurosurgery
Arkansas Children's Hospital
Little Rock
Arkansas
USA

H.-U. Thal
Neurosurgical Clinic
Altstadtstraße 25
44534 Lünen
Germany

D. G. T. Thomas
Gough-Cooper Department of Neurological Surgery
The National Hospital
for Neurology and Neurosurgery
Queen Square
London WCW 3 BG
UK

U. Thorns
Department of Anatomy
Medical School
Hannover
Germany

I. Tijsma
Department of Neurosurgery
Dr.Horst-Schmidt-Kliniken GmbH
Ludwig-Erhard-Straße 100
65199 Wiesbaden
Germany

M. Traina
Department of Neurosurgery and Neurosciences
Val-de-Grâce Hospital
74 Boulevard de Port Royal
75230 Paris CEDEX 05
France

V. M. Tronnier
Department of Neurosurgery
University of Heidelberg
Im Neuenheimer Feld 400
69120 Heidelberg
Germany

V. Urban
Department of Neurosurgery
Dr. Horst-Schmidt-Kliniken GmbH
Ludwig-Erhard-Straße 100
65199 Wiesbaden
Germany

B. Velani
Gough-Cooper Department of Neurological Surgery
The National Hospital
for Neurology and Neurosurgery
Queen Square
London WCW 3 BG
UK

J. Wahrburg
Institute of Control Engineering,
and Center for Sensory Systems
57068 Siegen
Germany

D. C. Widenka
Department of Neurosurgery
University Hospital
Englschalkinger Straße 77
81925 München-Bogenhausen
Germany

H. Widner
Department of Neurosurgery
University of Lund
Sweden

H. Wiegandt
Physiologisch-Chemisches Institut
Philipps University
Baldingerstraße
35043 Marburg
Germany

P. Wieneke
Aesculap AG
Postfach 40
78532 Tuttlingen
Germany

U. Wildförster
Department of Neurosurgery
Ruhr University of Bochum
Knappschaftskrankenhaus
In der Schornau 23 – 25
44892 Bochum
Germany

Grigoriy Vladimirovich Zcvigun
Department of Neurosurgery and Neurology
Main Military Clinical Hospital
ul. Gospitalna 18
252016 Kiev
Ukraine

C.-R. Wirtz
Department of Neurosurgery
University of Heidelberg
Im Neuenheimer Feld 400
69120 Heidelberg
Germany

G. V. Zcvigun
Department of Neurosurgery and Neurology
Main Military Clinical Hospital
ul. Gospitalna 18
252016 Kiev
Ukraine

**Part 1
New Techniques and Technologies
for MIEN**

Biportal Endomicrosurgery in the Intracranial Subarachnoid Space

G. Fries, R. Reisch, A. Kafadar, and A. Perneczky

Summary

While bi- or multiportal approaches to circumscribed indications have been adopted in all fields of surgery including spine surgery, the uniportal access to the skull is still a traditional principle in neurosurgery. Although sporadic microneurosurgical procedures using biportal endoscope-assisted techniques have been described, biportal endomicrosurgical dissection in the intracranial subarachnoid space has never been tested with respect to feasibility, utility, and safety in a preclinical study.

In 25 adult human cadavers, a total of 33 biportal endomicrosurgical dissections into and within the basal cisterns were carried out. Lensscopes of 0°, 30° and 70° with outer diameters of 4.2 mm and trochars with outer diameters of 5.0–6.5 mm were used for endomicrosurgical dissection.

Six different endoscopic routes to the basal cisterns and a total of ten different combinations of these approaches for biportal endoneurosurgery could be described, but it was found that not all of them were useful and safe.

Biportal endomicrosurgical subarachnoid preparations were effective and safe in the olfactory groove, the prechiasmatic cistern, the region of the optic chiasm, the entire suprasellar area, parts of the parasellar area, the pre- and perimesencephalic cisterns, and the prepontine cistern.

Introduction

As a traditional principle based on the phrase "one hole in the skull is enough", uniportal approaches are commonly used for brain surgery. Unless the craniotomy is very large, which "per se" may cause adverse side effects, the traditional uniportal microsurgical or endoscopical dissection of intracranial structures – especially at the base of the brain – is afflicted with the necessity of almost coaxial control of the tip of the microinstruments. Frequently, these instruments must be used through narrow preformed anatomical windows framed by sensitive nerval or vascular structures that do not tolerate extensive retraction [4, 6, 7]. Besides, many microinstruments at least partially obscure the visual control of their tips and of parts of the anatomical situation. This problem that is important during microsurgery for aneurysms and tumors at the base of the brain, may not completely be solved by amelioration of the design of microinstruments [5] or by more detailed preoperative planning [4].

However, with the availability of highquality lensscopes and allied modern technical equipment, it may be possible to solve the above problem with the method of biportal endomicrosurgical dissection of the intracranial subarachnoid space [2].

The aim of this preclinical cadaver study was to create a biportal endomicrosurgical technique suitable for dissection in the subarachnoid space at the base of the brain and to evaluate this new, uncommon technique with respect to feasibility, utility, and safety. For this purpose, data on possible intraoperative positioning of the head, the location of skin incisions and trephinations, the extent of brain retraction and brain selfretraction due to the impact of gravity, the frequency and severity of tissue traumatization, and microanatomical landmarks of the intracranial subarachnoid space were collected.

Methods and Material

In 25 adult human cadavers, a total of 33 biportal endomicrosurgical dissections into and within the basal cisterns was performed.

Following skin incisions of 15–25 mm in length burr-hole trephinations extended to diameters of 12–22 mm were applied in the following locations: supraorbital with the linea temporalis in the center of the trephination for the subfrontal approach, frontal paramedian about 4–4.5 cm above the nasion and 0.5 cm from the midline for the anterior interhemispheric approach, at the coronary suture about 2–2.5 cm from the midline for the transventricular approach, into the squamous part of the temporal bone above the zygomatic arch for the anterior subtemporal approach, and about 10 mm above the incisura mastoidea for the posterior subtemporal approach. The subfrontal approach was carried out either intradurally or epidurally.

After opening of the dura mater and the arachnoid, the cerebrospinal fluid was removed from the subarachnoid cisterns in order to perform the biportal endomicrosurgical dissection in an air environment and induce the self-retraction of the brain.

Lensscopes of 0°, 30°, and 70° (Aesculap AG, Tuttlingen, Germany) with 4.2-mm outer diameter and trochars of 5.0- to 6.5-mm outer diameter were used. Under endoscopic visual control, the trochars were carefully advanced into the chosen subarachnoid cistern and then fixed in the desired intracranial position by retractor arms attached to the head holder. Thus, in each specimen, two ports and two guiding channels for microinstruments and endoscopes were created.

Each step of the dissections was documented as a video sequence and as a color photograph using endoscope video camera equipment (Aesculap, Tuttlingen, Germany) and Nikon and Olympus photo cameras with endoscope ocular adapters. Fujichrome Provia 1600 ASA daylight color reversal films and Kodak EPJ 1600 ASA daylight color reversal films were used for photographic documentation.

Results

According to the dissections performed, a total of six different endomicrosurgical approaches and a total of ten different biportal combinations of these approaches to the intracranial subarachnoid cisterns could be described.

Out of the six endomicrosurgical approaches, the transventricular access through the floor of the third ventricle to the prepontine cistern, although desirable for certain indications, was evaluated as not safe enough for the situation in a real operating room when combined with other approaches in a biportal fashion. In all transventricular dissections of the prepontine cistern, the endoscope trochar had to pass the interventricular foramen of Monro and the stoma that was created at the floor of the third ventricle in the area of the tuber cinereum. Irrespective of the existence of a hydrocephalic configuration of the ventricles and the interventricular foramen, a relaxation and caudal shift of the brain was regularly noted. This dislocation of the brain in relation to the endoscope trochar consistently caused traumatization to the trochar of the fornix, the hypothalamus, and in some cases the pituitary stalk. Thus, biportal endomicrosurgical approach combinations involving

the transventricular route cannot be recommended for intraoperative use.

In contrast, the following endomicrosurgical dissections could be performed easily, reproducibly, safely, and without traumatization of cerebral tissue: (a) supraorbital subfrontal epidural, (b) supraorbital subfrontal intradural, (c) anterior subtemporal, (d) posterior subtemporal, and (e) frontal interhemispheric. The supraorbital subfrontal epidural dissection was carried out under endoscopic control down to the falciform ligament and the anterior clinoid process, where the dura was opened with a hook knife. Usually this maneuver provided an excellent overview of the intracranial course of the optic nerve and the supraclinoid part of the internal carotid artery.

Various biportal combinations of the above approaches, either intradural or epidural, were judged to be useful and safe enough for application in the operating room.

1. Supraorbital subfrontal bilateral
2. Supraorbital subfrontal combined with anterior subtemporal ipsilateral
3. Supraorbital subfrontal combined with anterior subtemporal contralateral
4. Supraorbital subfrontal combined with posterior subtemporal ipsilateral
5. Supraorbital subfrontal combined with frontal interhemispheric ipsilateral
6. Supraorbital subfrontal combined with frontal interhemispheric contralateral
7. Anterior subtemporal combined with frontal interhemispheric ipsilateral
8. Anterior subtemporal combined with frontal interhemispheric contralateral

Following adequate positioning of the head, most of the endomicrosurgical dissections could be performed without retraction of cerebral tissue. By making use of the gravityinduced selfretraction of the brain, the subarachnoid space of the basal cisterns and the interhemispheric fissure could be easily opened for insertion of endoscope trochars and microinstruments. With the exception of a few anterior subtemporal preparations in cases with a low temporal lobe, the use of brain retractors was not necessary for biportal endomicrosurgery.

In 23 out of 25 specimens, a rather solid Lillequist's membrane was encountered, closing the premesencephalic and prepontine cisterns. Using hook knives or pistolshaped microscissors for transsphenoidal surgery to precisely open the mesencephalic and diencephalic layers of Lillequist's membrane, special attention was paid to small perforating arteries which could be easily controlled in a biportal endoscopic fashion. In the remaining two specimens, Lillequist's membrane consisted only of tiny arachnoid trabeculae which could be passed by the trochars using blunt dissection.

When using the biportal technique, the angle of vision towards the microinstruments was 60°–180° depending on the viewing angle of the lens-scopes, the location and combination of the burr-hole trephinations, and the individual shape of the head.

Discussion

The use of biportal endomicrosurgical strategies allowed for an effective and safe dissection of structures within the subarachnoid space at the base of the brain. Compared to uniportal endoscope-assisted microsurgical techniques [1, 3], the angle of vision towards the tip of the microinstruments and the visual control of neighboring anatomical structures were much better and more convenient with biportal endomicrosurgical approaches.

With the application of the biportal endomicrosurgical technique to dissect neurosurgically relevant intracranial structures within the subarachnoid space at the base of the brain, effective and safe surgical approaches to the following regions were feasible: the olfactory groove and the cribriform plate, the prechiasmatic cistern, the entire suprasellar region including the optic chiasm, parts of the parasellar region, the pre- and perimesencephalic cisterns, the prepontine cistern, and the entire arterial circle of Willis.

Recently, other surgical fields such as abdominal surgery, thoracic surgery, vascular surgery, orthopedic surgery, gynecology, and urology have replaced uniportal "large" surgical approaches by multiportal minimally invasive endoscopic techniques with better results for the treatment of well-defined disease entities. According to the results of the study presented here and following the establishment of clearly defined indications, it might also be desirable for neurological surgery to abandon the traditional principle of uniportal "large" craniotomies in favor of biportal minimally invasive endomicrosurgical approaches to intracranial structures in the subarachnoid space. Future indications for such minimally invasive endomicrosurgical biportal techniques might be elective procedures for the surgical repair of incidental cerebral aneurysms, microvascular decompression of cranial nerves, fenestration of arachnoid cysts, repair of cerebrospinal fluid fistulas involving the anterior cranial fossa, selective amygdalohippocampectomy, and removal of small tumors involving the pre-, supra-, and retrosellar area.

References

1. Cohen AR, Pernecky A, Rodziewicz GS, Gingold SI (1995) Endoscope-assisted craniotomy: approach to the rostral brainstem. Neurosurgery 36: 1128–1130
2. Fries G, Reisch R (1996) Biportal neuroendoscopic microsurgical approaches to the subarachnoid cisterns. A cadaver study. Minim Invasive Neurosurg 39: 99–104
3. Grotenhuis JA (1996) Endoscope-assisted craniotomy. In: Techniques in neurosurgery, vol 1, no 3. Lippincott-Raven, Philadelphia, pp 201–212
4. Perneczky A (1992) Planning strategies for the suprasellar region. Philosophy of approaches. Neurosurgeons 11: 343–348
5. Perneczky A, Fries G (1995) Use of a new aneurysm clip with inverted-spring mechanism to facilitate visual control during clip application. J Neurosurg 82: 898–899
6. Perneczky A, Tschabitscher M, Resch KDM (1993) Endoscopic anatomy for neurosurgery. Thieme, Stuttgart
7. Van Lindert E, Perneczky A, Fries G, Pierangeli E (1997) The supraorbital keyhole approach to supratentorial aneurysms. Concept and technique. Surg Neurol (in press)

A New Endoscopic System for Neurosurgical Procedures

F. Duffner, W. Dauber, D. Freudenstein, L. Krasznai, M. Skalej, and E. H. Grote

Summary

A new endoscopic tool for the CRW (Cosman-Roberts-Wells) system is presented. The endoscope was designed and manufactured for neurosurgical needs. It has a special optical bundle with an image close to rod-lens quality without forfeiting the flexibility of a fiberscope. The camera can be positioned to the nonsterile area in the operating theatre. The optical and light bundles leave the operation field within the same cable. The outer diameter is 4.5 mm and small enough for the endoscope to traverse the foramen of Monro at occlusive hydrocephalus. Further tools such as multifunctional instruments are presented.

Introduction

Endoscopy was integrated in to neurosurgical procedures in the first decades of the twentieth century. The most famous pioneers of neuroendoscopy were W. E. Dandy and W. J. Mixter. Dandy removed parts of the choroid plexus to reduce CSF production by means of a cystoscope [2,3]. The fundamental idea for endoscopic treatment of occlusive hydrocephalus came from W. J. Mixter. He recommended perforation of the floor of the third ventricle between the hypophyseal recess and the corpora mammilaria [5].

Due to the size and unsuitability of the early instruments, endoscopically guided operations for hydrocephalus did not become routine. During recent years, minimally invasive endoscopic neurosurgery (MIEN) [1] was fostered by innovations from other fields such as general surgery or gastroenterology.

Minimizing approaches using preexisting spaces in the neurocranium became benefits available to modern neuroendoscopy. In spite of new developments there are still various problems to resolve. There are for instance, the unusual anatomical view, new and different planning strategies, problems with the instruments, and operation standards, that need to be established.

Combining an endoscope and a stereotactic frame, disorientation in the brain can be prevented by calculating different passage points of the endoscope. Due to the complex demands, a 3-dimensional planning system becomes increasingly necessary.

After testing a number of endoscopes available from different manufacturers on cadavers, the following aspects appeared in need of improvement:

1. The majority of instruments were not adaptable to a stereotactic system.
2. Several endoscopes did not fit to anatomical structures in the brain, for example, an 8-mm-diameter endoscope would not pass gently through the foramen of Monro with a diameter of less than 6 mm [4].
3. Connection of the video camera to the endoscope led to a disadvantageous center of gravity behind the investigator's hand; the problem was exacerpated by the optical and light bundles leaving the table.
4. Steerable fiberscopes were not sufficient because of the reduced optical qualities resulting from the use of small optical bundles to enable a bending of the tip.
5. Due to the prospective orientation of the fiberscopes, bending of the tip entails the risk of undetected traumatisation of tissue, particularly while retracting the instrument.

Bending of the endoscope or instruments sometimes becomes necessary during an operation. In such a case, we prefer a steerable catheter running through the working channel of the endoscope (mother-baby principle).

Because of these shortcomings, we decided to develop a new instrument. The idea was to create a rigid endoscope which could be used in combination with a stereotactic system. It needed to be easy and safe to handle for neurosurgeons inexperienced in neuroendoscopy.

Technical Report

A new endoscopy system was designed and manufactured in cooperation with Schölly Co. (Denzlingen,

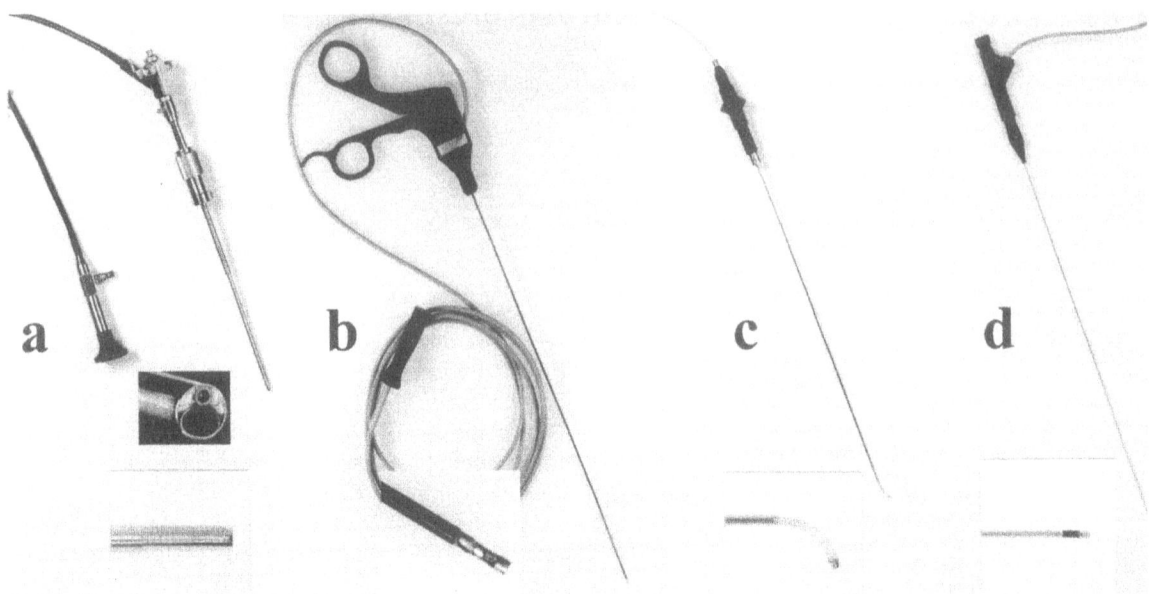

Fig. 1. a The endoscope has a microdrive which allows telescopic setting of each possible turn. The working channel is oval and easy to clean. The outer diameter of the guiding cannula is 4.5 mm. **b** The bipolar punch (prototype) enables the surgeon to grasp and induce coagulation in a single step. **c** The multifunctional catheter makes instruments such as a ballon catheter, a baby endoscope, or the Erbe 0.9-mm bipolar probe controllable. **d** The coagulation probe is furnished with a movable wire at the center of the tip which allows cutting or coagulation according to the position of the wire

Germany); (Fig. 1). The principal component is an endoscope of a length of 2.30 m. Using a special optical bundle with 20000 fibers that merged (Imagefiber), the quality of the image comes very close to that of the rod-lens system and presents an improvement over the typical use of optical bundles in fiberscopes. Because of the chosen length of the endoscope (2.30 m), the camera can be positioned in the nonsterile area of the operating theater (Fig. 1). The optical and the light bundles leave the operative field in a single cable. Furthermore, the problematic center of gravity resulting from connecting the camera behind the investigator's hand is avoided. The system is designed with standard camera mounting. Therefore the same camera can be used for other purposes. The camera is used with the microscope, either.

Furthermore, this two-in-one bundle technique could be a spacesaving solution for endoscopically guided procedures with open MRI. The whole endoscope has to be sterilized with ethylene oxide. Sterile covering is no longer necessary.

The rigid part of the endoscope has a length of 31 cm with an external diameter of 4.0 mm. The diameter of the guiding cannula measures 4.5 mm for the first 17 cm from the tip and 6.35 mm for the next 7 cm (according to the Radionics guide block of the stereotactic system).

The working channel is diagonally oval and its smallest diameter is 2 mm. The oval form of the working channel allows rinsing along the instruments and easy cleaning. The nontraumatic tip of the endoscope includes the lens with an 8° view in the direction of the working channel which shows the passing instruments very quickly (Fig. 1a).

A microdrive was adapted to the distal part of the guiding cannula. It has a range of 3 cm in the forward direction with a 1 mm telescopic setting following each possible turn. The view does not change while moving. The 3 cm feed is normally enough to overcome the distance between the foramen of Monro and the floor of the third ventricle [4]. All parts of the system (endoscope, guiding cannula, microdrive) have scales showing the depth of the instrument in centimeters for easy orientation. The stereotactic arc can be removed without changing the position of the tip of the guiding cannula. Meanwhile, our instrumentation has been completed by further components.

A multifunctional catheter has been developed in cooperation with Schölly Co. This instrument (2 mm diameter) fits through the working channel of the endoscope. It has a 1 cm flexible section at the tip and a 0.9 mm-diameter working channel within itself (Fig. 1c). The steerable catheter functions as a hole-in-hole system in the working channel of the endoscope. The bendable section allows control of flexible instruments, such as ballon catheters or coagulation probes, while they pass through the working channel.

It is possible to use the catheter as a guide for a baby endoscope. This multifunctional catheter combines the advantages of a controllable fiberscope with rigid endoscopes and makes endoscopy safer by permanent visual control of the instruments while guiding them to the target. The multifunctional catheter is completed by a 0.9 mm bipolar coagulation probe which is manufactured by Erbe (Tübingen, Germany). Combining these two instruments, the neurosurgeon can coagulate vessels is able to in the visual field without changing the position of the endoscope. In addition, a rigid bipolar coagulation probe with a 1.5-mm diameter has been designed by Erbe Co. (Fig. 1d). Both coagulation probes are furnished with a movable wire at the center of the tip which allows cutting and coagulation by changing the position of the wire with a screw. The coagulation probes are designed for single use. Another new development is a bipolar punch allowing coagulation and grasping of tissue (e.g., a cyst wall) in a single step (Fig. 1b). Other instruments such as forceps and scissors are standard.

Conclusions

We present a new endoscopic tool for the CRW stereotactic system. It has been designed and manufactured for neurosurgical needs and safe procedures. The camera in the nonsterile area, the single cable leaving the operation field, the telescopic forward movement, and the multifunctional catheter are innovations. The endoscopic system has been tested and our expectations were met.

References

1. Bauer BL, Hellwig D (1994) Minimally invasive endoscopic neurosurgery – a survey. In: Bauer BL (eds) Minimally invasive neurosurgery II. Acta Neurochir Suppl (Wien) 61:1–12
2. Dandy WE (1922) Cerebral ventriculoscopy. Bull Johns Hopkins 33:189
3. Dandy WE (1918) Extirpation of the choroid plexus of the lateral ventricles in communicating hydrocephalus. Ann Surg 68:569–579
4. Lang J (1992) Topographic anatomy of preformed intracranial spaces. In: Bauer BL (eds) Minimally invasive neurosurgery I. Acta Neurochir Suppl (Wien) 54:1–10
5. Mixter WJ (1923) Ventriculoscopy and puncture of the third ventricle, preliminary report of a case. Boston Med Surg 188:277–278

Innovative Instruments for Endoscopic Neurosurgery

A. E. Guber and P. Wieneke

Summary

Novel instruments are required for the performance of monoportal or biportal interventions in future endoscopic neurosurgery. As part of the MINOP joint project funded by the BMBF, novel trocar systems and the corresponding instruments have been developed. For the first time, joint-free microinstruments of nickel-titanium alloys have been produced for use in endoscopic neurosurgery, e.g., deflectable suction units, laser applicators, and extremely thin microforceps and microscissors which are made of nickel-titanium wires only 0.63 mm thick. These novel instruments allow surgery to be performed "around the corner" and can be applied in extremely small surgical areas.

Introduction

As commonly known, a variety of special instruments is available for neurosurgical operations. The maximum dimensions of the instruments are determined by the anatomy of the brain or vertebral column. Mostly, instruments with relatively small outer diameters are needed. They are produced by some medical engineering firms by means of high-precision surgery mechanics. Future minimally invasive interventions, however, will require much smaller instruments or at least strongly miniaturized partial instruments which will have to be brought to the site of operation by suitable guiding systems.

Development and manufacturing of very small components, so-called microstructural bodies, began in the early 1980s at several research institutions. At Karlsruhe Research Center (formerly "Karlsruhe Nuclear Research Center"), two entirely new techniques for microshaping metals and plastics have been developed since then, namely, the LIGA technique and mechanical microengineering [6]. By means of these techniques, the most minute components can be manufactured with the dimensions of 10 μm (0.01 mm), i.e., of cell size.

The design and manufacturing of future neuro-endoscopic instrument generations will presumably result in an increasing use of these microengineering techniques. Within the context of the MINOP joint project (microsystems technology for use in minimally invasive neurosurgical operation techniques; grant no. 13 MV 0323) funded by the BMBF, Aesculap AG, and Karlsruhe Research Center, we therefore started to manufacture very small operation instruments that can be deflected at the distal end and have a minimum outer diameter of only 0.63 mm. These instruments can be brought to the site of operation by various trocar systems.

Methods and Materials

A possible operation scenario is represented schematically in Fig. 1. Contrary to conventional neurosurgical interventions, the MINOP concept does not

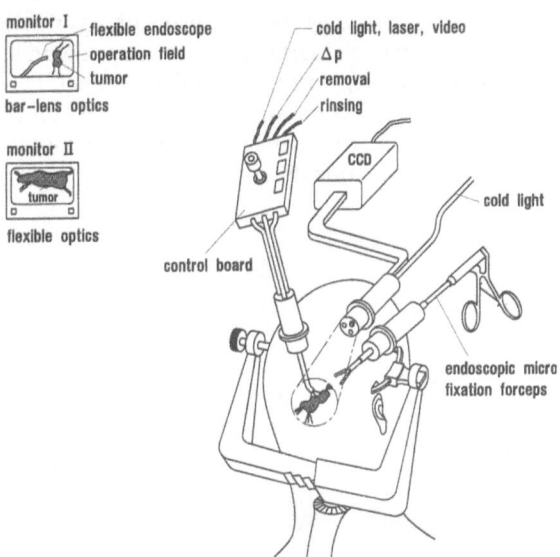

Fig. 1. Schematic representation of a possible operation scenario. Besides monoportal interventions, biportal interventions are also conceivable in principle. By observing instruments from both directions, view of the operator is improved during the endoscopic operation

only allow monoportal operation [4, 7]. Biportal interventions are conceivable under certain boundary conditions. The operation system shown here consists of three separate units which may be applied individually or combined. The first unit consists of a trocar that is equipped with the endoscope optics. In addition, it contains at least one channel for rinsing or sucking or for instruments approaching the site of operation. The second unit consists of a thin trocar through which a smaller flexible endoscope can be brought to the site of operation. Under permanent optical control of the first unit, the flexible endoscope may then be safely driven out of the trocar. The flexible endoscope consists of a four-lumen plastic hose with an external armor. It is equipped with fiber bundles for illumination, a microobjective and image fiber, and a working channel for sucking and rinsing or for guiding of the laser fiber. The fourth lumen is equipped with an integrated microfluidic control system, by means of which the distal end of the flexible endoscope can be deflected by a certain angle. If necessary, small microforceps or microscissors could be introduced into a further trocar for gripping or cutting tissue. The distal end of the instruments is deflectable and thus allows to operate "around the corner", if required. It is the advantage of the bi- or triportal intervention that the site of operation can permanently be observed by the surgeon from at least two different directions. Thus, control of the operation instruments applied is improved.

Figure 2 shows seven different trocars that have been developed within the MINOP project. Their outer diameters vary and they are equipped with a variable number of channels serving as working channel, endoscope channel, rinsing, or outlet channel. Obturators are provided for atraumatic introduction of the trocars. These trocars may be used for the mono- or biportal intervention described in Fig. 1.

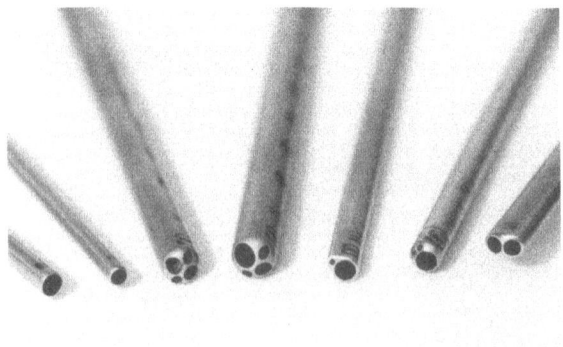

Fig. 2. The MINOP trocar family consists of various trocars with varying outer diameters. The channels can be equipped with optics or instruments according to the specific needs

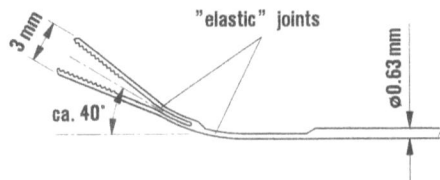

Fig. 3. The microinstruments are engineered directly from a NiTi wire 0.63 mm thick. By making use of the superelasticity effect of nickel-titanium alloys, the commonly applied mechanical joints are no longer required

Monopolar or bipolar electrodes or most minute flexible scissors and forceps may be introduced via the channels of the trocars. The distal end of the instruments is usually equipped with very small and rather filigree mechanical joints to allow a cutting or gripping movement. In addition to available flexible instruments with an outer diameter of about 2.1 mm, first prototypes of much smaller gripping and cutting instruments were produced within the MINOP project. The distal end of the instruments is smaller with a factor of 3, and the maximum outer diameter of the functional unit is only 0.63 mm.

The design of this new generation of microinstruments is rather simple. When using nickel-titanium alloys instead of the commonly applied surgical stainless steel, mechanical joints are not necessary due to the superelasticity effect (Fig. 3) [1, 2]. The forceps profile is directly engineered into a thin NiTi wire from one side by means of the µEDM technique (microelectrical discharge manufacturing) [5]. By removing some wire material from behind the actual forceps profile, thus leaving a recess, and by a subsequent thermal treatment of the forceps fixed in a bending facility, the open and deflected type is obtained. Opening and closing of the forceps are accomplished by the longitudinal movement of a thin flexible polytetrafluoroethylene hose. For closing the bits of the forceps the hose is moved forwards, for opening it is pulled back slightly (Fig. 4). By moving a stiff metal deflection pipe over the polytetrafluoroethylene hose, the direction of the deflected forceps is straightened. The outer protective pipe serves to protect the entire system. Hence, guiding of the microforceps via the trocars referred to above is accomplished quite easily. The forceps are operated by means of a special handle (Fig. 1).

Four different types of forceps have been produced within the MINOP project. Scanning electron microscope images (SEMs) of an opened and closed pair of microforceps are presented in Fig. 5. The forceps were engineered directly from a NiTi wire of only 0.63 mm thickness. Over a length of about 4 mm, the

Fig. 4.
Setup and functioning principle of microforceps. The bits are opened and closed by moving the flexible hose backwards and forwards. The deflected forceps can be straightened by moving a stiff metal pipe forwards

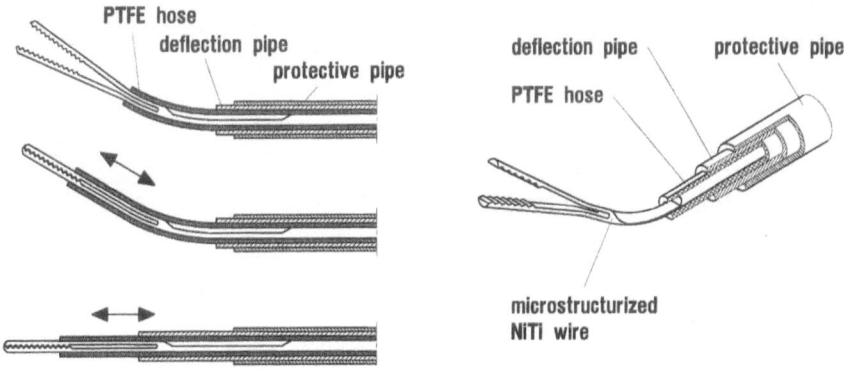

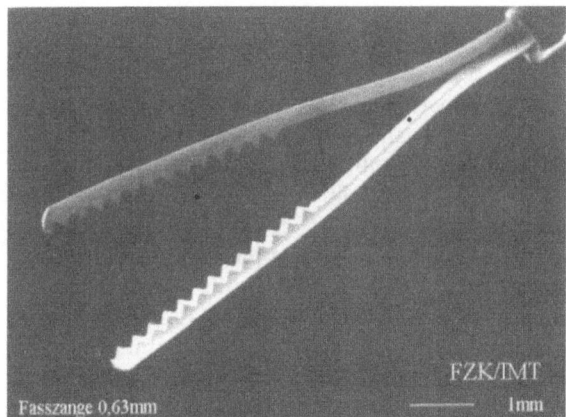

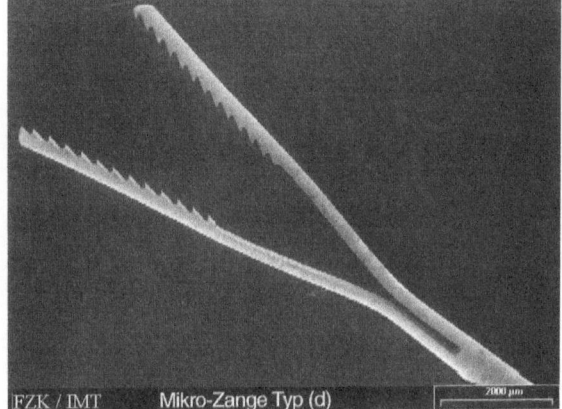

Fig. 6. SEM of an open pair of microforceps with a much more aggressive type of bit profile. The small barbs allow a better grip of tissue sections

Fig. 5. SEMs of an open (*top*) and closed (*bottom*) pair of microforceps with a simple zig-zag profile

bits are provided with a zig-zag profile. Other profiles are possible, e. g., small teeth directed inside or small barbs (Fig. 6), which increase the grip of the forceps. This type may preferentially be used as biopsy forceps. The slightly rough surface between the bit sections results from fabrication technology and

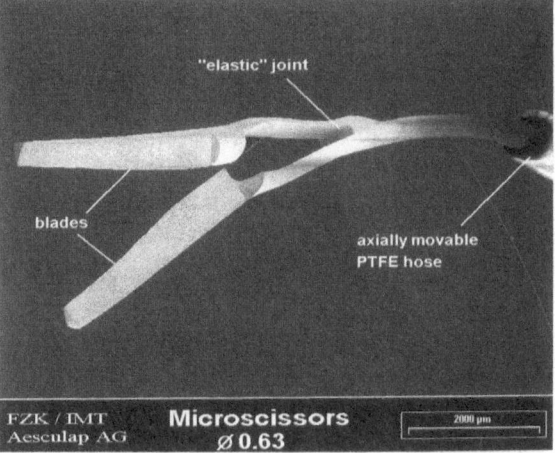

Fig. 7. SEM of open microscissors which were directly engineered from a NiTi wire of only 0.63 mm thickness. The blades are about 5 mm long and are closed by the forward movement of the flexible PTFE hose

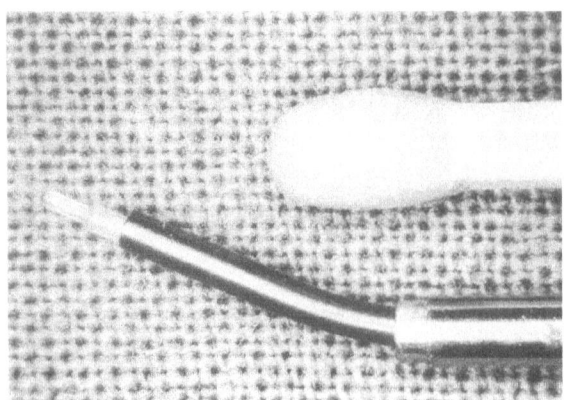

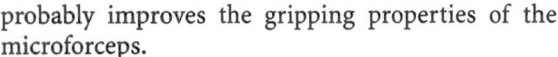

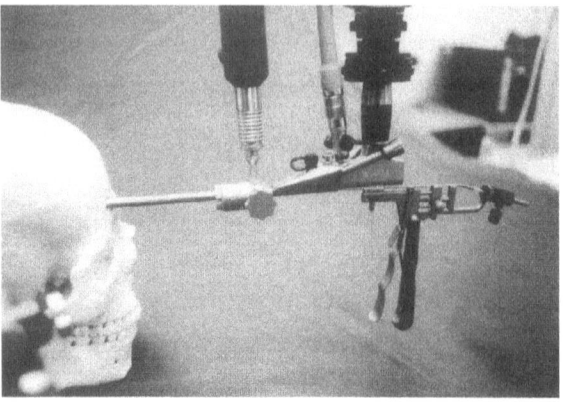

Fig. 8. Deflectable applicator on the basis of a thin NiTi tube. It may be applied for guiding laser fibers

Fig. 9. Ventriculoscope trocar equipped with a microinstrument and endoscope optics. For the first time, the deflectable instruments at the distal end of the trocar allow to work "around the corner"

probably improves the gripping properties of the microforceps.

Besides the gripping and taking of tissue sections by microforceps, reliable cutting of tissue plays an important role. Using the movement principle of the above microforceps and replacing the profiled bit components by minute blades, microscissors can be generated that are joint-free and deflectable. In Fig. 7, open microscissors engineered from a NiTi wire of only 0.63 mm thickness are shown. The length of the blades is about 5 mm. By moving the PTFE hose forwards, the scissors are closed. In order to obtain good closing, the blades are braced against each other. When closing the scissors, the cutting point moves from the back to the front. By means of the stiff deflection pipe, the microscissors can be straightened.

Furthermore, a sucking instrument and an applicator for laser fibers are necessary for endoscopic interventions. They can also be produced easily on the basis of thin NiTi tubes. A photograph of a deflectable laser applicator for guiding a 400 µm thick laser is shown in Fig. 8. Straigthening of the laser applicator is achieved by a stiff deflection pipe [3].

Results

Within the context of the MINOP project, several new instruments have been designed and produced as prototypes. First functional test were performed as part of the project at different places. In terms of functionality, the trocars are precisely adjusted and allow the introduction of optics and various instruments (Fig. 9). The entire system and the individual components are designed such that they can be disassembled easily and hence subjected to

autoclave treatment. Working or expendable parts can be easily exchanged.

The microinstruments made of the 0.63 mm-thin NiTi wires were subjected to a number of test series. By means of a specially developed manipulator, long-term test were performed. The deflectable sucking instrument and the laser applicator proved to withstand about 10 000 load cycles; the microforceps and microscissors even survived about 30 000 load cycles before fracturing at the deflection points occurred. Consequently, sufficient redundancy is guaranteed. Moreover, the microforceps and microscissors were subjected to various gripping and cutting tests. For example, reliable cutting of the wing of a common housefly was demonstrated.

Discussion

For future neuroendoscopic interventions, a new operation system is being developed within the context of the joint MINOP project. The instruments needed, such as sucking systems, laser applicators, microforceps, and scissors, can be brought to the site of operation via various trocars. If necessary, even biportal approach of the site of operation is possible, resulting in improved surgical safety with the instruments under permanent visual control by the operator. The design of some instruments allows a certain deflection in the range of 0° – 40° after they have been driven out of the trocar end. To this purpose, it is not necessary to change the position of a trocar once it has been fixed. By rotation of the instruments around their longitudinal axis and translatory movement in the trocar, a new "freedom of movement" is obtained in the area of operation below. For the first time, the neurosurgeon is able to operate "around the corner". Joint-free and 0.63-mm-thin microforceps and microscissors of biocompatible nickel-titanium alloys will allow future operative interventions at much finer biological structures.

References

1. Giordano N, Lutze T, Weißhaupt D, Wieneke P (1996) Chirurgisches Rohrschaftinstrument – Abwinkelbare Mikrozange. German patent P 44 42 439.6
2. Giordano N, Lutze T, Weißhaupt D, Wieneke P (1996) Scherenförmiges, chirurgisches Werkzeug und Verfahren für dessen Herstellung. German patent P 195 12 559.2
3. Giordano N, Dötzkirchner V, Guber AE (1997) Die MINOP-Instrumenten- und Trokarfamilie. Reihe Innovationen in der Mikrosystemtechnik, vol 50. Final report of the MINOP project, pp 126–149
4. Guber AE, Wieneke P (1996) MINOP – development of a miniaturized endoscopic operation system for neurosurgery. SPIE 2676:2–13
5. Guber AE, Giordano N, Schüssler A, Baldinus O, Loser M, Wieneke P (1996) Nitinol-based instruments for endoscopic neurosurgery. Actuator 96 : 375–378
6. Menz W, Bley P (1993) Mikrosystemtechnik für Ingenieure. VCH, Weinheim, pp 189–268
7. Wieneke P (1997) Das MINOP-Projekt. Reihe Innovationen in der Mikrosystemtechnik, vol 50. Final report of the MINOP project, pp 18–25

Investigation of Approaches to Localize and Control Flexible Neuroendoscopes Automatically

J. Wahrburg

Summary

The use of flexible endoscopes in neurosurgery is concentrated on those indications which cannot be reached by a straight access or require a "look around the corner". The benefits of using flexible endoscopes could be further increased if the surgeon knew the exact position of the tip of the instrument in order to correlate it with the image information obtained by scanning procedures. However, devices do not yet offer this possibility.

This contribution illustrates our approaches to investigating suitable techniques and procedures for automatic localization and control of flexible endoscopes. The work is essentially twofold. On the one hand, a testbed is constructed which enables the motor-driven, force-controlled actuation of the endoscope handle. The resulting 3D position can be measured by a very precise mechanical coordinate measuring system. On the other hand, research focuses on the investigation of suitable sensing techniques to determine the position of the endoscope tip during the operation with a resolution of up to 1 mm.

Introduction

New approaches for minimally invasive surgery are being widely evaluated and discussed. Among the various techniques under consideration endoscopic procedures attract particular interest both in research and development and in clinical application. The endoscopic equipment already available is characterized by a broad variety of instruments and tools depending on the application area of the body and the associated requirements. In neurosurgery the navigation of an endoscope is a specific aspect of crucial importance as there are no "natural paths" to follow. The basic problem can be solved by stereotactic planning methods and by mounting a rigid endoscope on a stereotactic frame or, in newer and future applications, on a frameless mechanical system [1, 2].

However, there are still some important restrictions that prevent a wider application of endoscopic equipment in neurosurgical applications. These limitations apply not only for the neuroendoscopes themselves but even more for the tool systems which may be guided through one or more working channels of the endoscope to perform a certain intervention. Improvements are required essentially in two main areas: The instruments must become more flexible in combination with increased functionality, and they must be better controllable by the surgeon. Due to the very small size of miniaturized instruments novel technologies are required which are not yet available. Sophisticated new designs are still in the research phase. The present paper contributes to this work by investigating suitable techniques to localize and control the tip of a flexible endoscope.

Objectives of the Research Project

The motivation behind our work starts from the question: Is it possible to localize the tip of a flexible endoscope precisely with a resolution of about 1 mm? If this can be achieved, significant benefits will result:

- The position of the endoscope tip can be correlated with preoperative scanning data from CT or MR.
- A corresponding symbol can be inserted into the image of scan data on a computer monitor.
- In this way the surgeon knows exactly where he is navigating with the endoscope.

The technology of a new flexible endoscope with a tip which is precisely localizable can be exploited in two ways. First, it will improve the diagnosis and therapy of indications such as anatomical or pathological hollow spaces and lesions of the ventricular system (occlusion, cyst). Secondly, it may form the basis to extend research investigations to the working instruments for endoscopic procedures. Localizable and controllable tools will greatly enhance the possible actions that a neurosurgeon can perform, independently of whether he uses a flexible or a rigid endoscope.

The latter task is of course even more challenging due to the very small dimensions of such instruments (diameter around 1 mm). On the other hand, there is presently a large gap between the good diagnostic facilities offered by modern endoscopes and the restricted therapeutic actions which can be performed by endoscopic instruments. A long-term goal of our work therefore consists in contributing to improve the functionality of such instruments. The investigations to localize and control a flexible endoscope may be regarded as a first step in this direction.

Technical Background and Roadmap of the Project

Presently there is no equipment available to measure and monitor the position of an endoscope tip. Technical solutions might be based on emitter-receiver principles, with one component (e.g., a transmitting device) attached to or integrated in the tip, and the complementary one (e.g., one or more receivers) located outside the skull in the operating theatre. However, due to strong demands on miniaturization, attainable resolution, robustness and medical compatibility no commercial version of such a system has been reported or offered so far.

Another technical solution may emerge from the new technology of real-time scanning devices, such as open MR scanners. Unfortunately, many existing instruments, particularly those containing magnetic materials, are not compatible with the new scanners, and if they are developed at all, it will take some time until they are available.

With regard to this background and to the complexity of the task we have decided to restrict our investigations initially to a few key issues, the solution of which is essential for future research. The roadmap of our research project to localize and control flexible instruments therefore comprises the following milestones.

1. Reduction of the problem. Research will focus initially on the bending section of flexible endoscopes, presuming that the other part may be regarded to be stiff, for example, by guiding it through a fixed and rigid mandrel.
2. Investigation of the characteristics of currently available endoscopes. The construction of a suitable testbed facilitates the automatic actuation of the endoscope and precise external measurement of the resulting 3 D tip position.
3. Search for and evaluation of suitable techniques to detect the bending angle by integrated sensing elements.
4. Deduction of improvements for the next generation of flexible endoscopes. It is expected that the

results of points 2 and 3 will lead to suggestions on how to improve the design of new endoscopes with regard to better localization and controllability.

The present paper essentially describes the work which has been carried out under points 1 and 2.

Investigation of Available Flexible Endoscopes

Construction of a Testbed

A suitable test setup is required to facilitate quantitative and reproducible measurements of the properties of the bending section of flexible endoscopes. The design of our testbed has been guided by the objective to determine the following important features.

- Motion range of the bending section
- Necessary input signals (angle and force) at the endoscope handle to move the tip to a certain position
- Hysteresis and reproducibility of the devices
- Effect of instruments pushed through the working channel

A block diagram of the prototype system is shown in Fig. 1. It has been designed to be adaptable to existing endoscopes by use of adjustable elements. Some manufacturers of flexible endoscopes have kindly put a specimen of their devices at our disposal, which gave us the opportunity to investigate various commercially available products. The main system components of the testbed consist of:

- A motorized actuation system, comprising a stepper motor and sensing elements to register the rotation angle of the endoscope handle and the forces exerted on it.

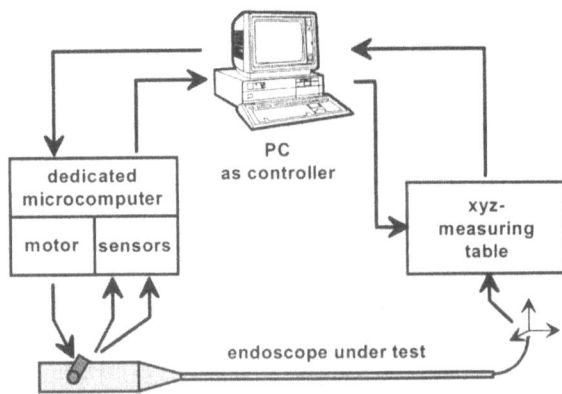

Fig. 1. Main components of the test setup

Fig. 2. Motorized actuation the endoscope handle

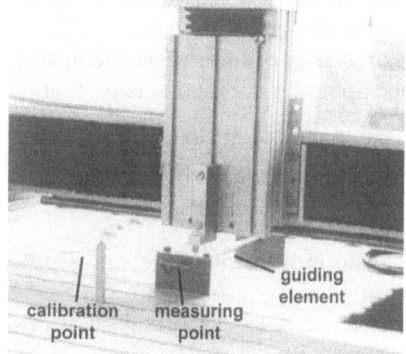

Fig. 3. Measurement of the position of the endoscope tip

- A measuring point which is attached to the movable head of an xyz coordinate table. By moving this point to the tip of the endoscope and recording the coordinates, the three-dimensional position of the tip can be located.
- A PC system to control the actuation and the registration process and to log and visualize the measuring data.

Motorized Actuation of the Endoscope Handle

The endoscope is mounted on the xyz coordinate table by a suitable fixation device, as illustrated in Fig. 2. The endoscope handle is moved by an adjustable gripper which is connected to the axis of a steppermotor. Force sensors in the gripper detect the actuation forces, while a high-precision potentiometer attached to the opposite side of the motor axes measures the angle of the handle. A small microcomputer system links sensors and motor to the control PC. This preprocesses the sensor data and generates a microstepping actuation of the motor, enabling a high resolution of 0.36° per step.

Measuring the 3D Position of the Endoscope Tip

The actuation of the handle generates a change of the position of the endoscope tip which can be recorded by moving the carriage of the xyz table to the new position. To facilitate a fast and convenient motion of the carriage the surgeon can use a joystick connected to the control PC. The PC software generates the necessary input signals for the axes motors of the coordinate table and records the xyz coordinates when a new position is located. A resolution of 0.03 mm for the axis movements, the exact definition of a reference point at the endoscope tip, and the use of a thin measuring needle attached to the carriage

facilitate precise measurements. Figure 3 illustrates the registration of a new position.

By stepping through the motion range of the endoscope handle a characteristic plot of the endoscope can be generated, giving the 3D position of its tip in terms of the angle of the handle. This plot may be regarded as a specific calibration curve for the endoscope under test, which facilitates the setting of a desired position of the endoscope tip by adjusting the handle at the corresponding angle. Figure 4 presents an example plot which clearly shows a typical hysteresis in the forward-backward movement. Simultaneous recording of the actuating forces at the endoscope handle together with the angle enables a clear discrimination which part of the curve is actually valid, and determines the offset which must be added if the direction of motion changes.

The presence of endoscopic instruments in the working channel(s) generally changes the results which have been obtained without instruments. It is therefore necessary to record additional curves for these cases. The calibration procedure in the testbed finally leads to a field of characteristic plots each of

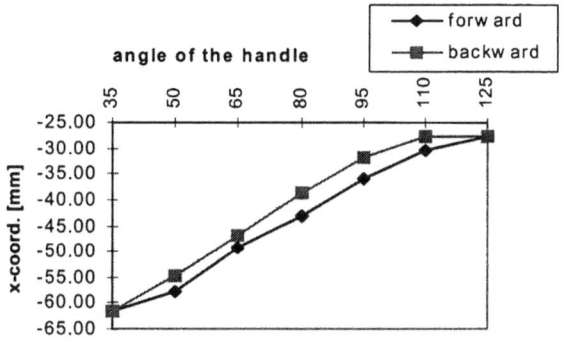

Fig. 4. Calibration curve for the positioning the endoscope tip

which represents a specific use and working environment of the evaluated endoscope.

The characteristic plots deliver the position of the endoscope tip under the prerequisites that:

- The flexible part of the endoscope, except for the bending section at its distal end, is guided through rigid elements which are fixed in a well-known position.
- There is no movement of the endoscope in axial direction.
- There is no torsion of the endoscope around its axis.

Of course these assumptions present a certain restriction. However, the primary goal of the present investigation is to evaluate the characteristics of the bending section of the endoscope as the most important part of the endoscope to change the location of its tip. Furthermore the laboratory results appear to be transferable to neurosurgical applications in the operating theater if a rigid guiding system similar to that of the test setup is attached to a stereotactic frame.

Conclusion and Outlook

It turns out that the test setup is well suited to determine the characteristics of the bending section of a flexible endoscope. Evaluation of the measuring data

facilitates the generation of a field of characteristic plots for the endoscope under test, showing the 3D position of its tip versus the angle of the handle and the actuation force. These calibration plots enable a defined setting of the endoscope tip by the handle when the flexible part, except of the bending section, is guided through a rigid system.

This prerequisite may be overcome at least partly if the position of the tip can be determined by sensors whose measuring principle is not directly related to the movement of the handle. Our present research concentrates on the investigation of suitable technologies for sensing elements small enough to be integrated into or near the endoscope tip. The installed test setup is an ideal platform to evaluate the performance of suitable new devices.

Acknowledgement. We express our thanks to Prof. Bauer and Dr. Hellwig, Philipps University, Marburg, for encouraging and supporting this project.

References

1. Bauer BL, Hellwig D (1994) Minimally invasive endoscopic neurosurgery. Acta Neurochir Suppl 61 (Wien):1–12
2. Grunert P, Perneczky A, Resch K (1994) Endoscopic procedures through the foramen interventriculare under stereotactical conditions. Minim Invasive Neurosurg 37:2–9

Endo-Neuro-Sonography: Basics and Current Use

K. D. M. Resch and A. Perneczky

Summary

In order to evaluate the usefulness of transendo-scopic sonography, we studied two new sono-graphic probes of 6-F diameter in 15 fresh specimens. We saw precise imaging of well-known anatomical structures and an additional dimension in endoscopy, as the sonograph adds to the endoscopic view a transversal scan like, as if it were a mini CT at the tip of the probe. Thus, we also experienced the navigation character-istics of this imaging technique, both in real time and on-line. Some three-dimensional reconstruc-tions of the foramen of Monro region are examined. First clinical use has been carried out, and adaptation of the equipment to neuro-surgery is necessary.

Introduction

The step from microneurosurgery to endoneuro-surgery not only means a step to minimally invasive technique but also a decrease in safety, which leads to a reduction of applicability. The goal for further developments must therefore be to make neuro-endoscopy safer. One concept is to establish an on-line guiding system which has been undertaken with MR or CT with an enormous financial and technical effort.

A very simple and inexpensive concept is to equip the scope with a sono-guiding system. A sonographic probe (Aloka/B & K medical) is introduced into the working canal and a radial sonographic view serves as a "mini CT" of 1- to 5-cm diameter in which the tip of the probe itself can be seen and guided on line. As this sonographic system presents very typical but dif-ferent anatomical aspects of well-known structures, the first step is to present the typical anatomical features of endo-neuro-sonography.

The use of 3 D endo-neuro-sonography is worth being examined as well as clinical experience and reflections on a preliminary indication list.

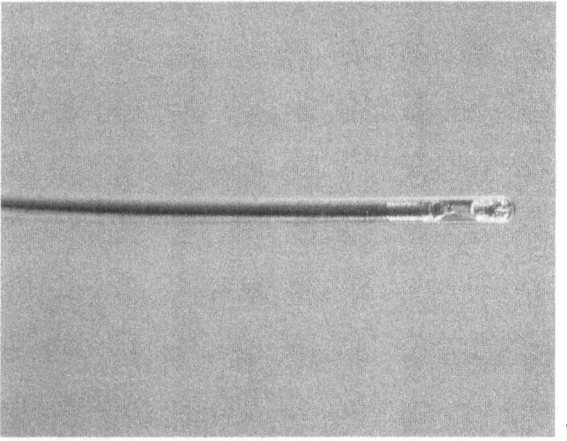

Fig. 1. a Video tower, b Sono probe (Aloka)

Material and Method

In 15 specimens, the sonographic anatomical aspects of the whole CNS have been worked out and documented by prints, photography and parallel sono- and endoscopy video recording. The examinations were performed in fresh specimens in situ as this offers the model best comparable to surgery and to sonographic echo characteristics of tissue.

Two different sonographic catheters (Aloka, B & K Medical) with a diameter of 1.9 mm (6 F) were used and introduced into the working canal of an endoscope. The specifications (Aloka, B & K Medical) were: frequency 20/12.5 MHz; diameter 6/6.2 F; length of the probe cabel 192/110 cm; display magnification 9 – 124 mm in 24 steps; frame rate maximally 15/s; image adjustment as gain, STC, contrast, and received-frequencies bandwidth; image rotation as 360°; measurement function as distance with unit of 0.1 mm; display of gray scales 256 levels, and video signals as TV standard video out and in (BNC)/S-VHS(Y/C). The sonographic pictures are displayed on a monitor, on which some parameters can be varied and were also studied to obtain the best view of different anatomical structures (Fig. 1). Two 3 D reconstructions were examined using a post-processing computer (Tomtec). In a patient with aqueduct stenosis, the imaging quality during ventriculocisternostomy was evaluated.

Results

Endo-neuro-sonography is a technique that adds to the anterior view of the scope an axial view of the probe tip position which itself is visible in the scope and on the sonographic view. This axial vision is like an axial "mini CT" of the tip plane on which the position of the tip can be localized in relation to this axial overview. The zoom-function allows to adapt

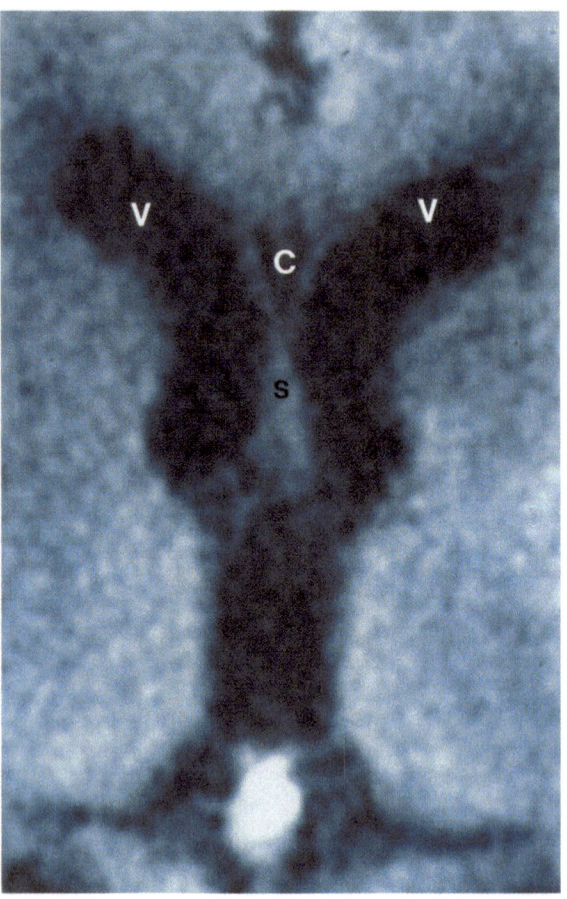

a

Fig. 2 a, b.
The comparison of **a** CT and **b** sonographic photographs shows the good visualization of anatomical structures by endosonography. The cavum of pellucid septum (C), lateral ventricles (V), and septum pellucidum (S) are well visible with comparable quality in both methods. Position of the sonographic probe (P) is localized and can be navigated in real-time towards the plexus (p) to the foramen of Monro

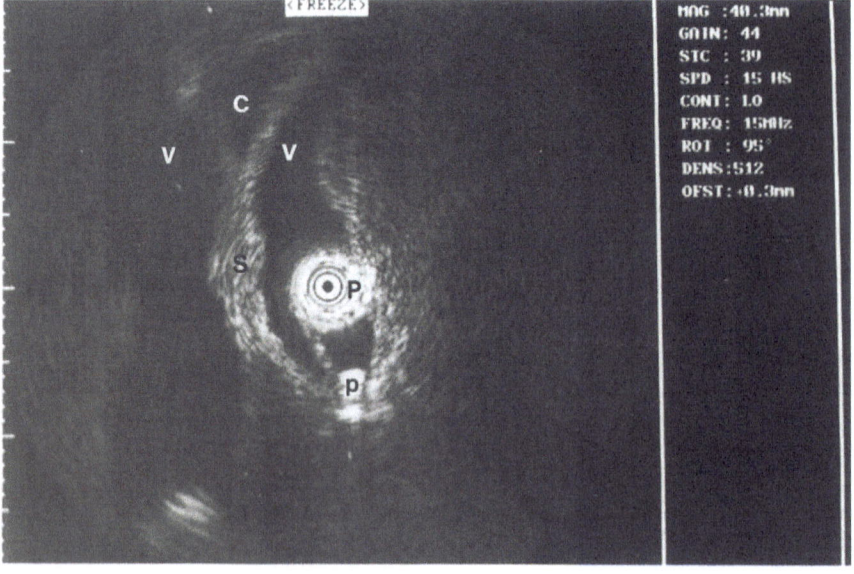

b

the view in relation to the size of structures. It is possible to see an overview or only one small cistern. Furthermore it allows to anticipate aspects that will come into endoscopics view next by "looking through the parenchyma" and overviewing a larger area as the scope does. The "on-line mode" of the sonographic view allows to observe changes of structures such as size of ventricles or pulsations. In connection with the endographic view, it allows to guide the scope on-line and safely in real time.

The resolution (Fig. 2) for structures within 3 cm is comparable to that of CT with respect to problems such as guiding of the scope. A septum cyst, for example, is visible in the sonograph as well as in the CT. The lateral ventricles (Fig. 3) invaded by the endo-sonograph are shown in an overview of both lateral ventricles while the scope only sees the right one. This shows the difference between the optical characteristics. The plane angle of the sonographic scan depends on the approach angle in the case of lens scopes: for example, in a supraorbital approach, the supracellar area (Fig. 4) is seen in the scope viewing anteriorly and showing the hypophyseal stalk and gland in between the optic nerves (Fig. 4a), while the sonographic view (Fig. 4b) presents the anatomy in a semi-sagittal plane showing the hypothalamus with an infundibular recess with lumen of the third ventricle. Decreasing the zoom, the sonograph shows an overview of the whole sella with pituitary gland and left optic nerve (Fig. 4b).

Fig. 3a, b.
The cella media of the right lateral ventricle with septum pellucidum *(S)* and the choroid plexus *(p)* is approached using a frontal burrhole approach frontal burr-hole and the sono probe *(P)* is visible. In the sonograph, both ventricles *(V)*, the septum and sonographic probe contacting the plexus can be well visualized

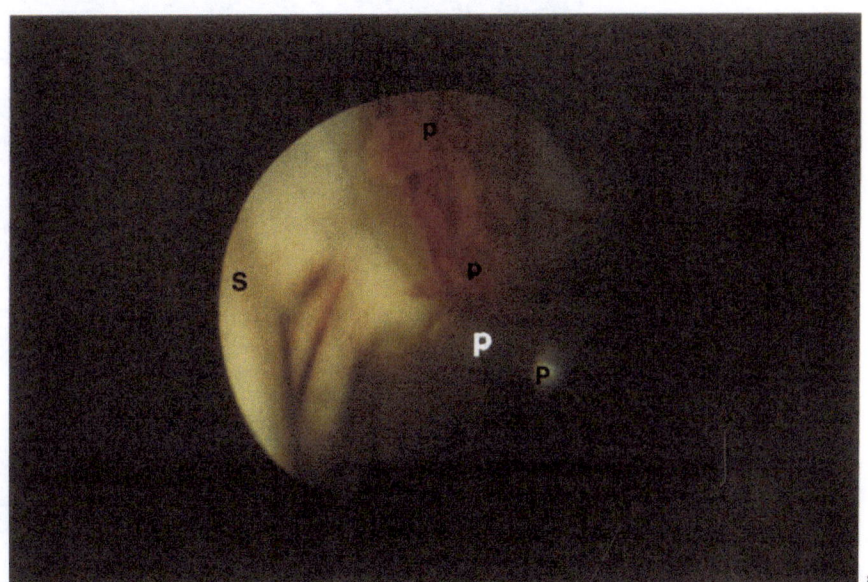

a

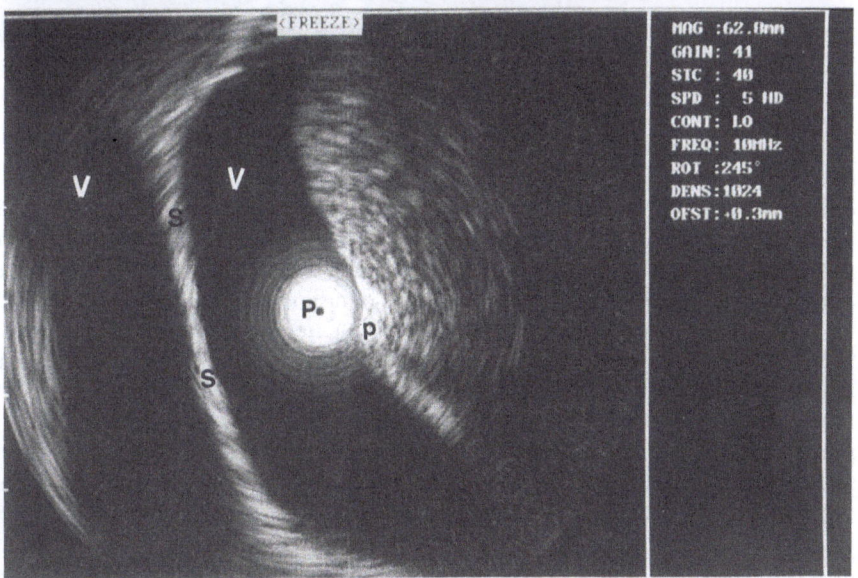

b

Fig. 4.
a From a right supraorbital approach, the suprasellar area is visible in the scope showing both optic nerves *(O)*, the hypophyseal gland *(H)* with the stalk *(St)* and, under visual control the sonographic probe *(P)*. b Additional, the endosonography presents a semisagittal scan with good anatomical correlation with the endographic view: parenchyma of hypophyseal gland *(H)* with stalk, left optical nerve *(O)*, the lumen of anterior third ventricle *(V)*, and hypothalamus *(h)*. The sono probe can be seen in the suprasellar cistern

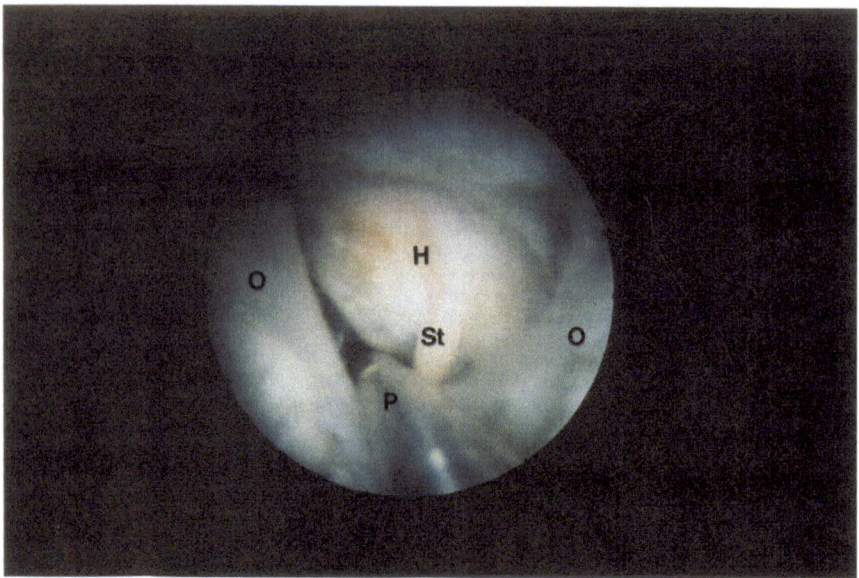

a

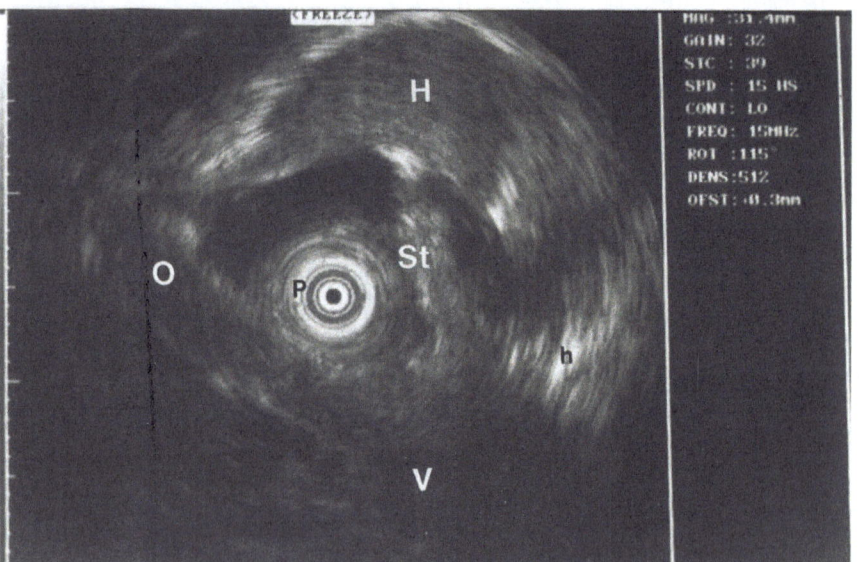

b

Running ventrally to the brain stem in caudal direction, the cerebellopontine angle (CPA) with both its cisterns and cranial nerve 7/8 accompanied by the AICA loop with labyrinthine artery are well seen in the scope (Fig. 5a). The sonograph shows in the scan of pons with basilar artery and clivus and the 7/8 bundle an overview (Fig. 5b). Endoscopically, it is possible in same cases to reach and see ventral to the brain stem the foramen magnum area with the medulla and cervical nerves C1 and C2 (Fig. 6a). Moreover the sonograph can "see" the entire axial shape of the medulla and cervical nerves C1 on both sides (Fig. 6b). Pushing the sonographic probe carefully along the spinal canal (Fig. 7), we see the typical scan of different levels such as cervical (Fig. 7a) or thoracal (Fig. 7b) ones.

The first results of 3D post-processing (Fig. 8) in two cases showed the foramen of Monro area and the tissue around the sonographic catheter as presenting a volume with the penetration depth as radius and the catheter's route as length. On the display, this volume moves in a selected manner of rotation angle and frequency. We reconstructed one freehand move of the catheter and one automatic move (Fig. 8a–c). For 3D reconstruction, it is necessary to move the catheter equably.

The first clinical case (Fig. 9) was a patient presenting a third-ventricular hydrocephalus due to aqueduct stenosis. The floor of the third ventricle was transparent and all structures could be seen and controlled endoscopically, ensuring that each sono-

Fig. 5.
a From a prepontine position the left CPA is endoscopically seen with bundle 7/8 as well as the AICA loop (*a*) with labyrinthine artery (*L*) near cranial base with the sono probe (*P*) under visual control. **b** The endosonograph shows in a semi-axial scan the parenchyma of the pons (*p*), the basilar artery (*B*), the cranial-base bone (*C*), the bundle 7/8, and the AICA loop (*a*). The sonographic probe is seen in the pontocerebellar cistern (*c*)

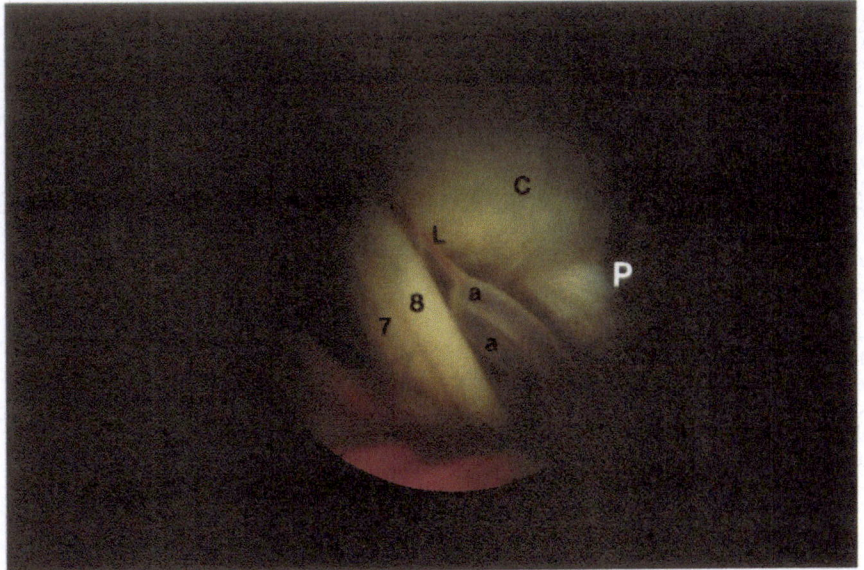

a

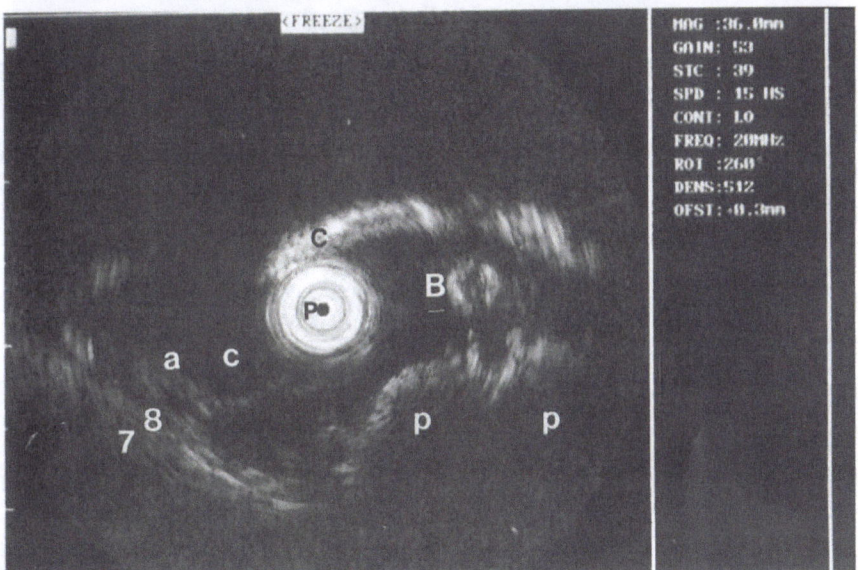

b

graphic probe position could be controlled and each sonographic scan correlated to an endoscopic view. The intraoperatively imaging quality was better in contrast enhancement than in specimen examination. Moreover, of course, all movements were visible such as pulsation of vessels or CSF flow. In particular all vessels were much bigger in size and gave an excellent contrast.

Imaging characteristics and first limited clinical experience gives an idea of a working list of possible indications in future.

Discussion

The basic principles of endo-neuro-sonography and the anatomy of the ventricular system and the subarachnoidal cisterns as determined by endo-neuro-sonography have already been described [35, 36, 38] as well as 3D endo-neuro-sonograpy [39]. The sonographic catheter has been used in other disciplines such as cardiology [7], angiology [1, 2, 16, 21, 28], or gastroenterology and urology [9, 10, 19]. In neurosurgery, sonography was previously used to assist in open microsurgery [3, 4, 20, 24, 32, 46] and stereotactic procedures [17, 23, 26, 44, 47, 48]. The first routine use in neurosurgery was

Fig. 6.
a The premedullary view through the endoscope shows the ventral foramen magnum *(M)*, the medulla oblongata and medulla spinalis *(m)* with cervical motor root C1 and C2 *(s)*, and the endosonographic probe *(P)* entering the spinal canal. **b** The endosonographic scan presents the whole transversal parenchyma shape of the medulla, the ventral and dorsal cervical roots on both sides, and the entire border of the foramen magnum. The endosonographic probe is well visible and can be guided in real time

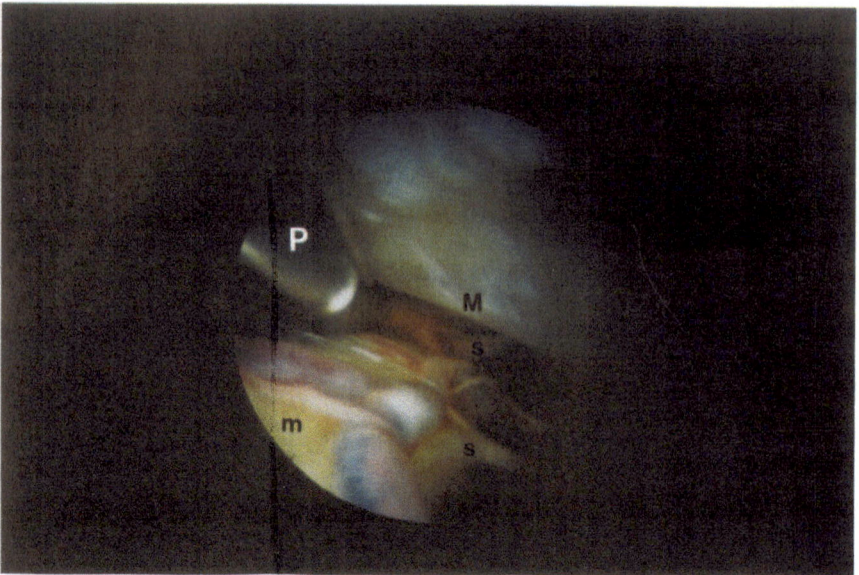

a

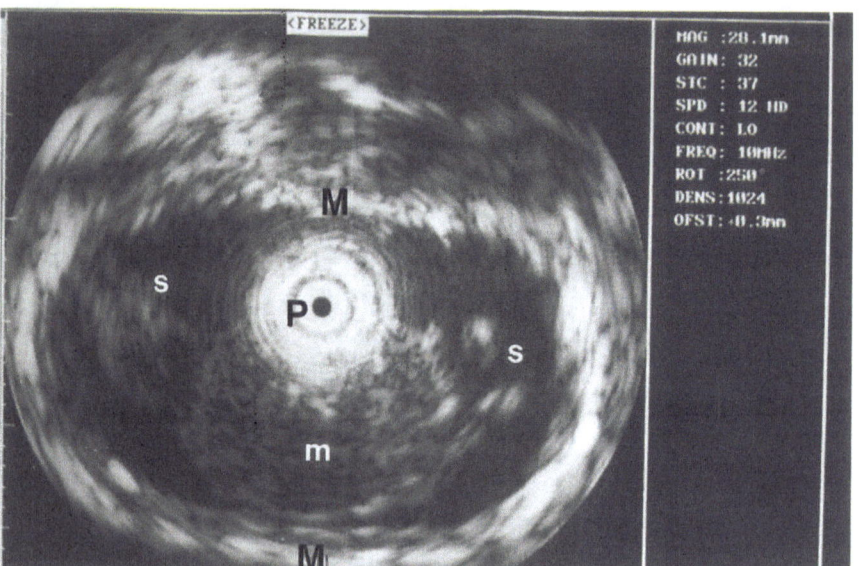

b

in pediatric neurosurgery [6, 15, 50]. In two cases, stereotactic biopsies were assisted by the sonographic catheter in an interdisciplinary group [11]. One pediatric neurosurgical department has also started clinical use [45]. It was always our concept that anatomical studies should precede surgery [31, 33, 34, 37], when we first began laboratory investigations before doing our first clinical case [35, 36,38].

The guidance of an endoscope is performed mainly by endoscopic anatomy [31, 33, 34, 37] and methods of stereotaxy [13]. Orientation, navigation, and safety equipment became more complicated when neurosurgery evolved from micro- to minimal-

ly invasive techniques. This kind of surgery can be summarized under the heading "image-guided neurosurgery" and the idea is to present as much information as possible on a single display [12,14,22]. These techniques contain an extraordinary potential for the future. However, currently they are characterized by the fact that a technical solution is looking for its application. Moreover, they are extremely expensive and immobile and the course of development is quite unclear. In this context endo-neuro-sonography seems to be a very interesting, cheap, and elegant technique.

In contrast to other disciplines [1, 2, 7, 9, 10, 16, 19, 21, 28] in neurosurgery, the transendoscopic cathe-

Fig. 7.
a Through the small subdural (epiarachnoidal) spinal gap, the endosono probe can be carefully advanced, and a typical sonographic scan was obtained at a cervical level with the medulla spinalis *(m)* the vertebral arch *(s)* with processus spinosus, and the sonographic probe in between *(P)*. b At a thoracic level, we see the small transversal shape of the medulla, the entire bony border of the spinal canal, and some strong subarachnoidal fibers *(f)* near the sonographic probe

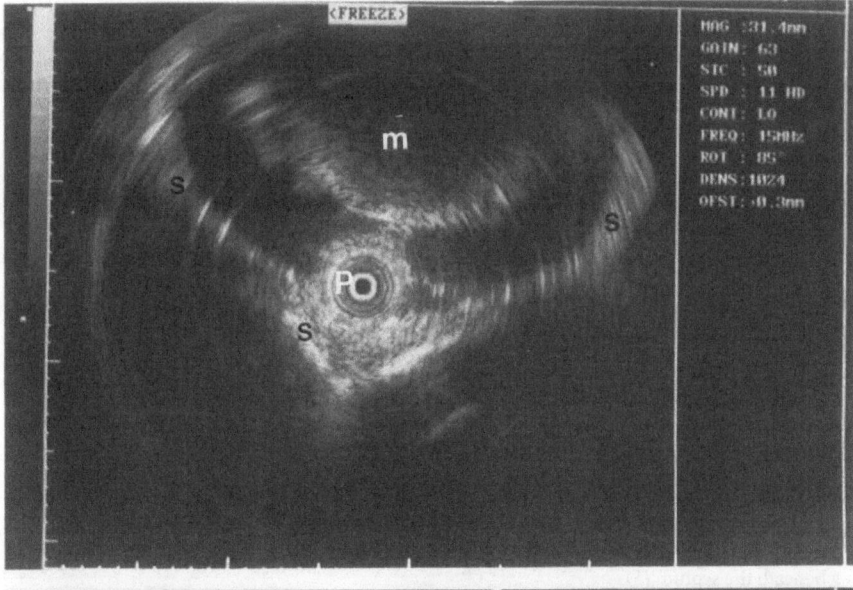

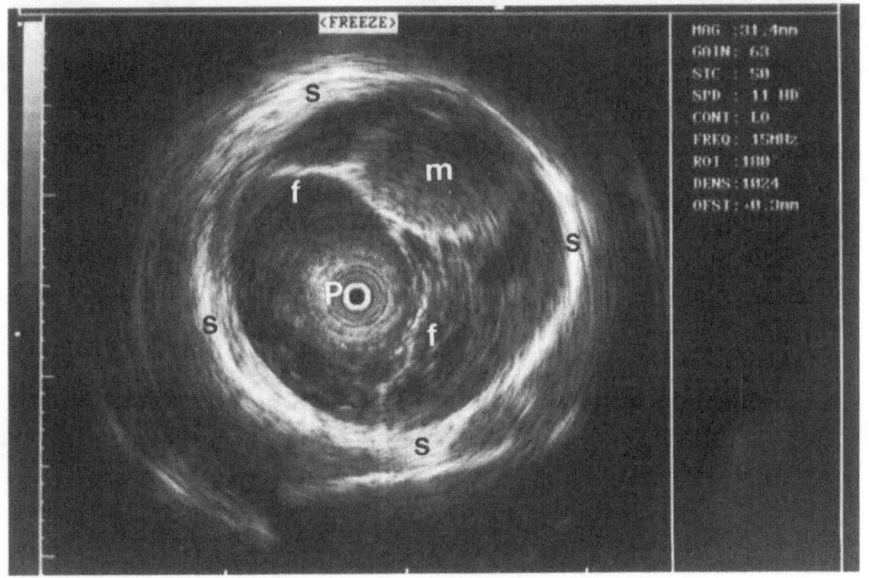

ter exhibits a navigation characteristic if the echo is strong enough and navigates the endoscope on-line and in real-time [35, 36, 38]. From the cisterns, it is possible to "listen" into the parenchyma and get an anatomical overview as with a mini-CT in the plane of the tip of the scope, thus acquiescing additional information on the endoscopic view. The additional imaging information can (1) compensate visual deficits of the scope in case of bleeding, undetectable space, lens hidden by tissue, (2) compensate the blind angle of the endoscope, (3) give an axial topographical overview around the endoscope tip, and (4) give a real-time imaging for navigation and an on-line monitoring for all changes around the target area. The advantage of the combination of both techniques results in a safer imaging than with endoscopy alone [35, 36, 38].

Three-dimensional reconstruction is still somewhat distant from real-time characteristics and might be interesting for documentation, learning and research [6, 8, 25, 27, 30, 39 – 43]. For clinical use, endoneuro-sonography is very simple, requiring no change in the OR. For neurosurgery, it has to be adapted to more practical handling. Pure electronic and non-mechanical catheters will be a step making endoscopes safer.

Fig. 8a–c.
3D endo-neuro-sonography is performed by a post-processing computer (Tomtec). It is possible to see different scans **a** coronar or **b** axial. The axial level is localized by a (yellow) line in the coronar view. In the 2D view, the volume to be reconstructed in 3D can be individually defined. **c** The 3D sonographic view presents a volume along the catheter route and has a cylinder form. The animation will be presented with a chosen angle and frequency of movement to present a 3D imagination. In the foramen Monro area, we can follow the pathway of the choroid plexus *(p)* coming from the left ventricle *(V, white)* and running beneath the septum *(S)* to the right ventricle *(black)*. The structures surround the surface of the catheter *(P)*

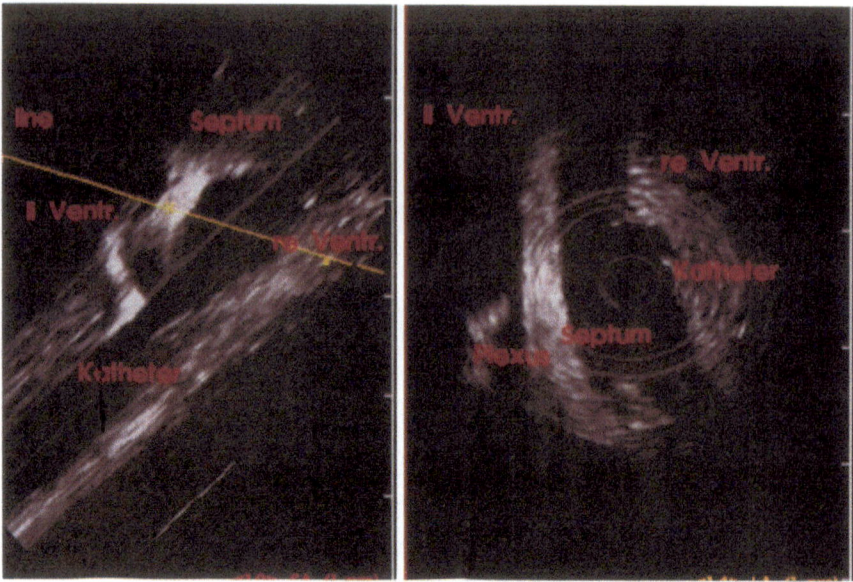

a

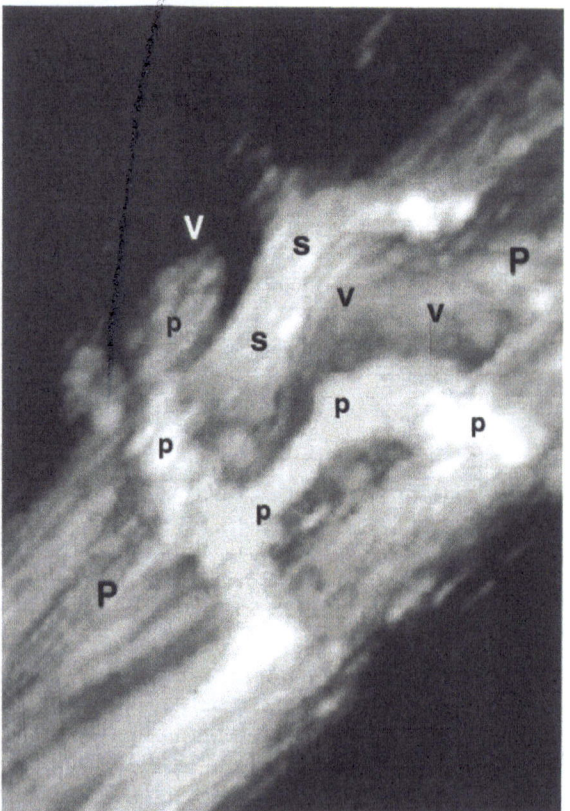

b

This development and some adaptations to endoneurosurgery are necessary, and it is especially important to additionally sound in the direction the scope is moved to detect structures in front of the lens in case of loosing orientation for various reasons. Finally of course, the scope must be equipped more intelligently to become as safe as the microscope. Endo-neuro-sonography seems worth further studies and clinical use.

Fig. 9a–c.
This clinical case of hydrocephalus due to aqueduct stenosis was a situation well suited to show the quality of imagination intraoperatively and experience the increase in safety (which can only be demonstrated by video). **a** All important structures are well seen in a good overview through the right foramen of Monro with the fornix (*f*) dorsal to the dorsum sellae (*d*) and ventral to the mammillary bodies (*m*). The head of basilar artery (*b*) is visible through the transparent premamillary membrane. The choroid plexus (*p*) hides the complete view to the interthalamic adhesion (*a*). (The sonographic probe is retracted back into the working canal). **b** The semi-axial sonographic scan of the third ventricle (*V*) presents near the sonographic probe (*S*) the right part of the interthalamic adhesion (*a*), the choroid plexus (*P*), and, not visible in the endoscopic view, the contralateral foramen of Monro (*M*), and the subependymal parenchym of the thalamus (*T*). **c** Retrosellarly the sonographic scan shows the sonographic probe (*S*) in the position of the Fogarty catheter with the ventriculocisternostomy in between the dorsum of sellae (*ol*) and the basilar artery (*b*) near the interpeduncular fossa (*F*). On the screen of course, the pulsation and the flow of irrigation can be seen. (Photographs taken from the monitor)

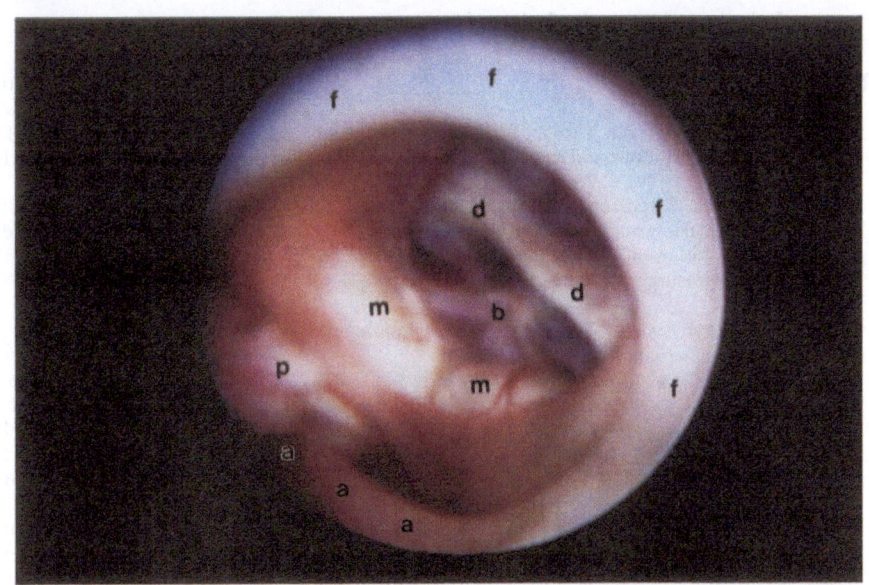

a

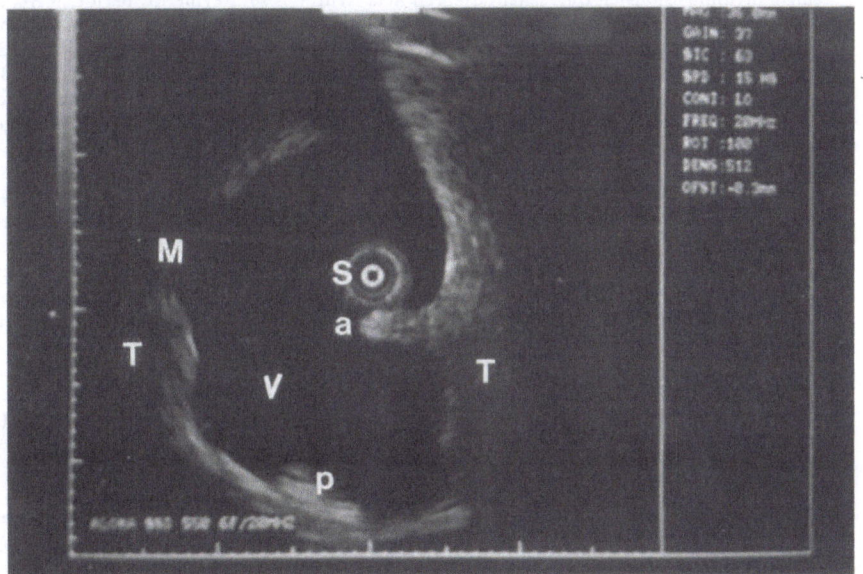

b

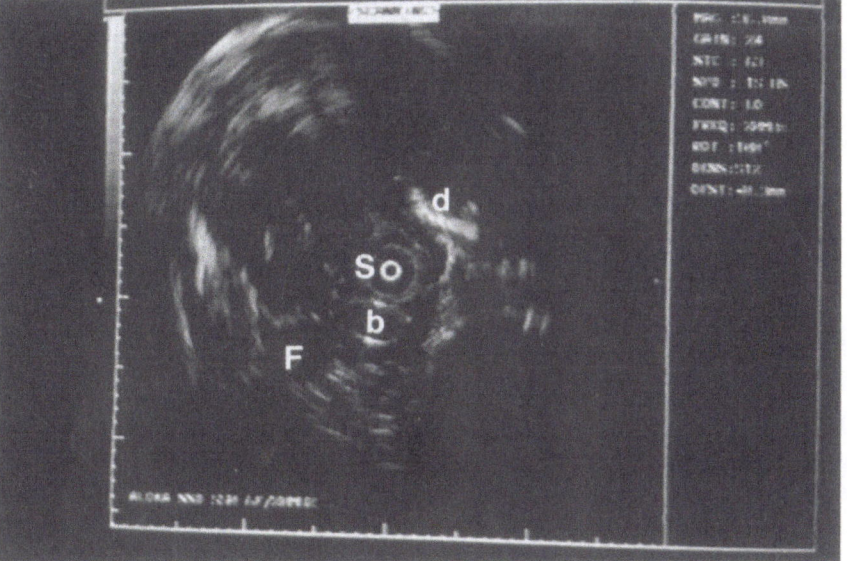

c

References

1. Aschermann M, Fergusson JJ (1992) Present possibilities of use of intravascular ultrasound examinations. Cas-Lek-Cesk 131:516–520

2. Aschermann M, Fergusson JJ, Raymond-Martimbeau-P (1992) Endovascular echography. J Mal Vasc 17:123–126

3. Auer LM, van Velthoven V (1990) Intraoperative ultrasound imaging in neurosurgery. Springer, Berlin Heidelberg New York

4. Auer LM, Holzer P, Ascher PW, Heppner F (1988) Endoscopic neurosurgery. Acta Neurochir (Wien) 90:1–14

5. Cavaye DM, Tabbara MR, Kopchok GE, Laas TE, White RA (1991) Three-dimensional vascular ultrasound imaging. Am Surg 57:751–755

6. Chadduck WM (1989) Perioperative sonography. J Child Neurol 4:91–100

7. Coy KM, Maurer G, Siegel RJ (1991) Intravascular ultrasound imaging: a current perspective. J Am Coll Cardiol 18:1811–1823

8. Delcker A, Diener C (1994) Die quantitative Erfassung arteriosklerotischer Wandveränderungen der Karotiden mit einem dreidimensionalen Ultraschallverfahren Acta Neurol 21:20–27

9. Frank N, Grieshammer G, Zimmermann W (1994) A new miniature ultrasonic probe for gastrointestinal scanning: feasibility and preliminary results. Endoscopy 26:603–608

10. Frank N, Holzapfel P, Wenk A (1994) Experience with a new endosonographic mini probe. Endoskopie Heute 3:238–244

11. Froelich J, Bien S, Hoppe M, Eggers F, Klose KJ (1996) An intracerebral sonographic catheter as an adjunct to stereotactic guided endoscopic procedure. Minim Invasive Neurosurg 39:9

12. Grönemeyer DHW, Seibel RMM, Schmidt A, Tschabitscher M, Alesch F, Wenz K (1994) Safety by CT/EBT or MRI for endoscopic instrument guidance. Minim Invasive Therapy 3: Suppl. 1 p 1

13. Grunert P, Perneczky A, Resch KDM (1995) Endoscopic procedures through the foramen interventriculare of Monro under stereotactic conditions. Minim Invasive Neurosurg 38:2–8

14. Hayashi N, Endo S, Kurimoto M, Nishio H, Ono T, Takaku A (1995) Functional image-guided neurosurgical simulation system using computerized three-dimensional graphics and diploë tracing. Neurosurgery 37:694–702

15. Horwitz AE, Sorensen N (1990) Intraoperativer Ultraschall in der pädiatrischen Neurochirurgie. Röntgenblätter 43:220–223

16. Isner JM, Rosenfield K, Losordo DW, et al. (1990) Percutaneous intravascular US as adjunct to catheter-based interventions: preliminary experience in patients with peripheral vascular disease. Radiology 175:61–70

17. Kanazawa I, Shiraishi K, Kamitani H, Sato J, Masuzawa H (1986) Intraoperative ultrasonography through a burr hole: clinical trail of ultrasound-guided sterotactic surgery. No Shinkei Geka 14:295–300

18. Key A, Retzius G (1875) Studien in der Anatomie des Nervensystems und des Bindegewebes, vol 1. Samson and Wallin, Stockholm, pp 1–155

19. Köstering B (1991) Endosonographie – technische Grundlagen und klinische Anwendung. In: Zimmermann W, Nitzsche H, Schentke U, Sessner H (eds) Grenzen und Möglichkeiten der Endoskopie. Sonographie in der Gastroenterologie. Dresdener Kongreßbericht, vol 2. Dustri

20. Koivukangas J, Louhisalmi Y, Alakuijala J, Oikarinen J (1993) Ultrasound-controlled neuronavigator-guided brain surgery. J Neurosurg 79:36–42

21. Ludwig M, Wetzig H, Sauer A, Vetter H (1995) Experiences with use of an intravascular 6 French endosonography catheter in vivo. Klin Wochenschr 68:570–575

22. Maciunas RJ (1993) Interactive image-guided neurosurgery. AANS Publication Committee

23. Masuzawa H, Kanazawa I, Kamitani H, Sato J (1985) Intraoperative ultrasonography through a burr-hole. Acta Neurochir (Wien) 77:41–45

24. Mayfrank L, Bertalanffy H, Spetzger U, Klein HM, Gilsbach JM (1994) Ultrasound-guided craniotomy for minimally invasive exposure of cerebral convexity lesions. Acta Neurochir (Wien) 131:270–273

25. Mintz GS, Pichard AD, Satler LF, Popma JJ, Kent KM, Leon MB (1993) Three-dimensional intravascular ultrasonography: reconstruction of endovascular stents in vitro and in vivo. J Clin Ultrasound 21:609–615

26. Moringlane JR, Voges M (1995) Real-time ultrasound imaging of cerebral lesions during "target point" stereotactic procedures through a burr hole. Technical note. Homburg/Saar, Federal Republic of Germany. Acta Neurochir (Wien) 132:134–137

27. Müller S, Bartel T, Baumann G, Ebel R (1996) Preliminary report: evaluation of three-dimensional echocardiographic volumetry by simultaneous thermal dilution in coronary heart disease. Cardiology 87:552–559

28. Neville RF, Bartorelli AL, Sidawy AN, Almagor Y, Potkin B, Leon MB (1989) An in vivo feasibility study of intravascular ultrasound imaging. Am J Surg 158:142–145

29. O'Malley CD, Saunders JB de CM (1983) Leonardo on the human body. Dover, New York

30. Pandian NG, Sugeng L, Vogel M, Marx G (1993) Three-dimensional echocardiography: the future in cardiac imaging. Learning Center Highlights 6, Boston, Mass 6–12

31. Perneczky A, Tschabitscher M, Resch KDM (1993) Endoscopic anatomy for neurosurgery (atlas/video). Thieme, Stuttgart

32. Reich J, Onik GM, Maroon J (1988) Intracerebral biopsy hemorrhage: monitoring and intervention guided by intraoperative sonography. AJNR Am J Neuroradiol 9:1240–1241

33. Resch KDM, Perneczky A (1993) Endoscopic approaches to the suprasellar region: anatomy and current clinical applications. Adv Neurosurg 22:225–230

34. Resch KDM, Perneczky A (1994) Endoneurosurgery: anatomical basics. In: Samii M (ed) Skull base surgery. Karger, Basel

35. Resch KDM, Perneczky A (1997) Endo-neuro-sonography: anatomical aspects of the basal cisterns. MITAT

36. Resch KDM, Reisch R (1997) Endo-neuro-sonography: anatomical aspects of the ventricles. Minim Invasive Neurosurg 1

37. Resch KDM, Perneczky A, Tschabitscher M, Kindel S (1994) Endoscopic anatomy of the ventricles. In: Bauer BL, Hellwig D (eds) Minimally invasive neurosurgery II. Acta Neurochir Suppl (Wien) 61

38. Resch KDM, Reisch R, Hertel F, Perneczky A (1996) Endo-Neuro-Sonographie: eine neue Bildgebung in der Neurochirurgie. Endoskopie Heute 2, 3:152–157

39. Resch KDM, Perneczky A, Schwarz M, Voth D (1997) Principles and 3D technique. Child Nerv Syst (in press)

40. Roelandt JRTC, ten Cater FJ, Brunning N, Salustri A, Vletter WB, Mumm B, v d Putten N (1993/1994) Transoesophageal rotoplane echo-CT. A novel approach to dynamic three-dimensional echocardiography. Thorax Centre 6 (1):4–8

41. Roelandt JRTC, ten Cater FJ, Vletter WB et al. (1994) Ultrasonic dynamic three-dimensional visualization of the heart with a multiplane transoesophageal imaging transducer. J Am Soc Echocardiography 7:217–229

42. Rosenfield K, Losordo DW, Ramaswamy K et al. (1991) Three-dimensional reconstruction of human coronary and peripheral arteries from images recorded during two-dimensional intravascular ultrasound examination circulation. 84:1938–1956

43. Schwartz S, Cao Q, Azevedo J, Pandian NG (1994) Simulation of intraoperative visualization of cardiac structures and study of dynamic surgical anatomy with real-time three-dimensional echocardiography. Am J Cardiol 73:501–507

44. Slovis TL, Canady A, Touchette A, Goldstein A (1991) Transcranial sonography through the burr hole for detection of ventriculomegaly. A preliminary report. J Ultrasound Med 10:195–200

45. Soerensen N (1997) Personal communication. Neuropädiatrische Arbeitstagung. Würzburg 1997

46. Sutcliffe JC (1991) The value of intraoperative ultrasound in neurosurgery. Br J Neurosurg 5:169–178

47. Tsutsumi Y, Andoh Y, Sakaguchi J (1989) A new ultrasound-guided brain biopsy technique through a burr hole. Technical note. Acta Neurochir (Wien) 96:72–75

48. Yamakawa K, Kondo K, Yoshioka M, Takakura K (1994) Ultrasound guided endoscopic neurosurgery – new surgical instrument and technique. Acta Neurochir Suppl (Wien) 61:46–48

49. Yasargil MG (1984) Microneurosurgery, vol 1. Thieme, Stuttgart

50. Zorzic C, Angonese I (1989) Subependymal pseudocysts in the neonate. Eur J Pediatr 148:462–464

New Possibilities and Future Developments of Electrosurgery in MIEN

R. Haag, V. Bartel, D. Hellwig, B. L. Bauer and F. Eggers

Summary

Complex demands and new indications in neurosurgery have led to the need for new electrosurgical generators (ESUs). For the first time, newly developed, automatically controlled ESUs make it possible to use extremely small and flexible instruments safely and to achieve reproducible microcuts and microcoagulations.

Introduction

High-frequency (HF) surgery has been used in medicine for approximately 70 years. Together with technical development and increasingly frequent use, there has been a demand for more precise instruments and generators. The greatest problem was the influence of cutting depth, cutting speed, electrode size, and differing tissues on the cutting quality. In practice, this often meant that an insufficiently low power setting would deliver a less than desirable cut and rather a mechanical "pulling up" of the tissue. This led to using a higher output power in order to ensure a safe cut. When a superficial cut was then made, the electrical voltage was too high, i.e., the applied power was too high for the cut. The consequence was either carbonization during incision or after the cut, or the tip of fine instruments could burn out.

An insufficiently low power setting for coagulation could result in insufficient hemostasis, a too high power setting the electrode's sticking to the tissue, which also could lead to bleeding when the electrode was removed. Another unintended effect was that carbonization could occur.

These problems were solved by the development of an automatically controlled ESU and precise instruments made of novel materials. Thus, for the first time it was possible to achieve reproducible cutting and coagulation qualities even using differing instruments.

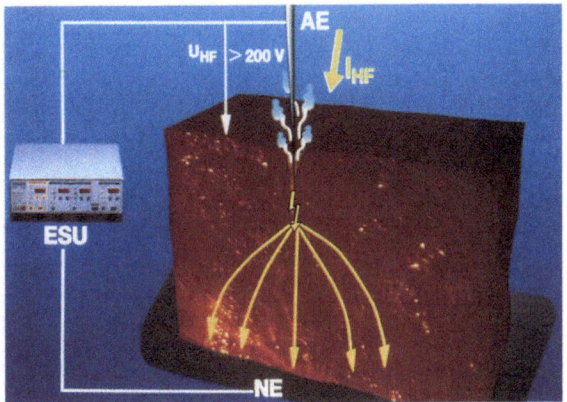

Fig. 1. Monopolar cut. Spark and high current density at the distal end of the electrode

Technical Report

Recent research has shown that high-frequency electrosurgical incision is only possible when electric voltage between active electrode and tissue is at least large enough for electric arcs to initiate. This is the case if the voltage is at least $200\,V_p$. The HF current is thus concentrated on one point (high current density) of the tissue. High temperature levels are reached very rapidly at the points where the electric arcs strike the tissue like microscopic flashes of lightning, so that the struck tissue is immediately vaporized or burnt away. If the active electrode is moved through tissue, then electric arcs spark over stochastically wherever the distance between the active electrode and the tissue is small enough, creating an electrical cut (Fig. 1).

Under the same conditions, the following applies: if the voltage is raised further, then the spark intensity rises proportionally, a higher spark intensity leading in turn to a greater depth of coagulation during cutting.

Figure 2 shows the principle course of output voltage during a cut with a conventional ESU. The output voltage fluctuates considerably between a maximum value and a minimum value depending on following variables: (a) size and shape of the cutting

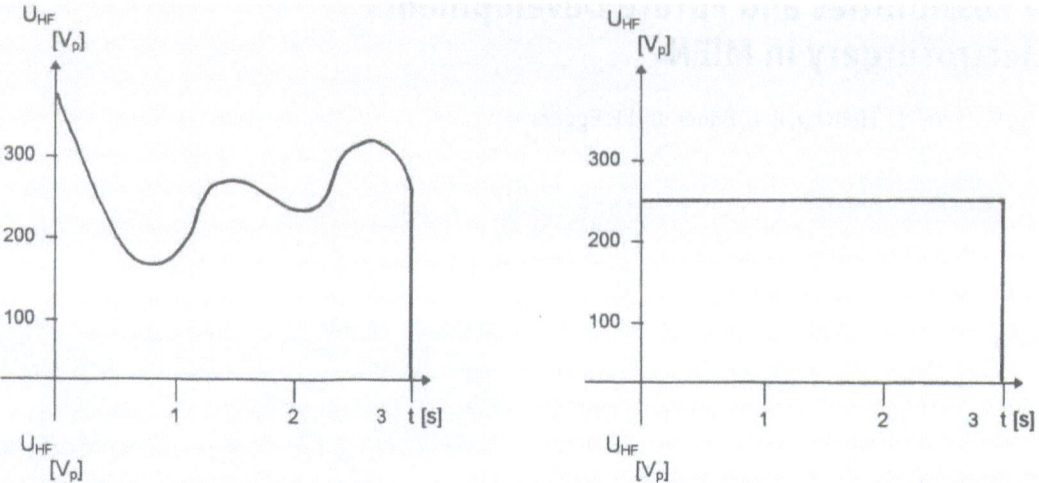

Fig. 2. Voltage of a cut made with conventional ESU

Fig. 3. Voltage of a cut made with an automatically controlled ESU

electrodes, (b) depth and speed of the cut, and (c) differing tissue properties. Here the maximum output voltage level can increase so much that carbonization can occur. The minimum value of the output voltage can decrease to such an extent that electrical cutting is no longer possible. If we regard the strongly fluctuating voltage level during a cut with a conventional ESU and the influence of output voltage on cutting quality, we recognize that it is physically impossible to realize reproducible cutting qualities with a conventional ESU. This knowledge was consequently used – an automatically controlled generator was designed where output voltage and spark intensity remain constant, forming the basis for a reproducible cut. According to the latest research, voltage regulation in neurosurgery is more relevant than spark regulation. Therefore, voltage regulation is described in more detail.

Figure 3 shows that the output voltage remains constant, i.e., unaffected by existing influence variables as mentioned above. Voltages can be preselected within a useful range, i.e., no voltages below 200 V_p can be set which would make it physically impossible to cut, and no excessively high voltages (>600 V_p) can be set, or else excessively strong carbonization and metallic erosion of the active electrodes can occur.

Once the operator has preselected his individual setting, the preselected effects will automatically be held constant. By means of such regulation, the cutting quality remains reproducible, independent of the described influence variables (Fig. 4).

Another advantage of regulated constant output voltage is that the current adapts to the influence variables in dependence on the various resistances of the influence variables. Thus, the existing output

Fig. 4. Comparison of cutting qualities

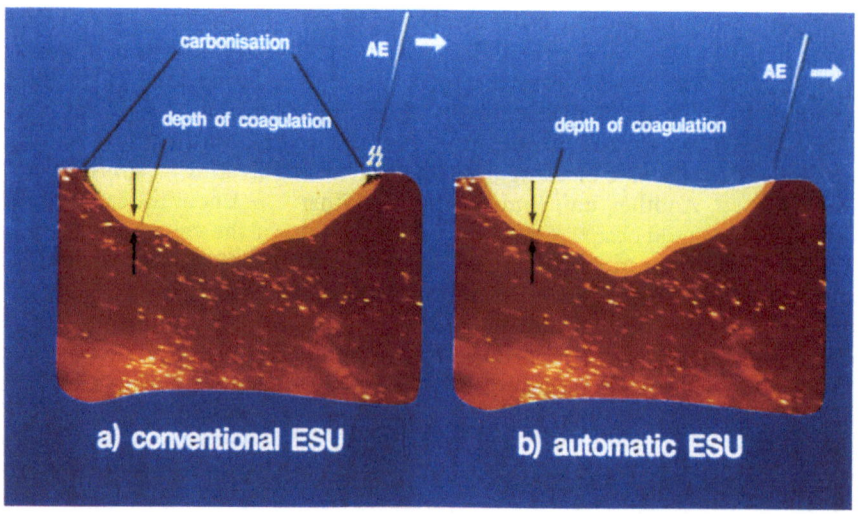

power is regulated to be as low as possible and as high as necessary, resulting in a substantially higher safety level.

Due to the voltage regulation, it is possible to achieve good cutting qualities with a relatively low voltage level. This is the prerequisite to be able to use finest instruments as described below.

Three principal modes of coagulation are distinguished, of which soft coagulation is an extremely gentle coagulation mode. The electric voltage is relatively low ($<190\ V_p$), so that there is no electric arc between active electrode and tissue. This results in the following advantages: no unintended cutting effect, no carbonization, and significantly reduced sticking effect between active electrode and tissue. These advantages are particularly useful in neurosurgery.

In addition to "soft coagulation", there is the function "soft coagulation with auto stop". In this mode, a sensor measures the optimum coagulation point occuring when intracellular and extracellular liquids are converted to vapor. The generator switches off automatically before carbonization or sticking of electrodes to the tissue can occur.

To complete the picture, two further coagulation modes are mentioned which are of no relevance to MIEN at the present time.

Forced coagulation fulfills all demands placed on standard coagulation and allows the surgeon to work effectively and swiftly. However, in forced coagulation, the voltages are so high that electric arcs may be generated between active electrode and tissue, so that an unintended cut can occur. In addition, carbonization can occur.

In *spray coagulation*, the output voltages are so high that long electric arcs can arise between active electrode and tissue. Spray coagulation thus allows coagulation without direct contact between active electrode and tissue. Such non-contact coagulation is generally expedient where superficial coagulation is required.

Due to the technological advantages offered by automatically controlled ESUs, it is possible to use finest monopolar and bipolar instruments at a much higher safety level. A good example of this is the completely newly developed bipolar needle for cutting and coagulation (Fig. 5). This single-use bipolar needle is very flexible, with an outer diameter of 0.9 mm and a length of 80 cm.

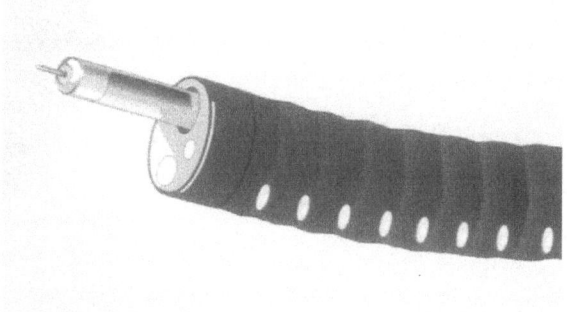

Fig. 5. Flexible bipolar needle

The probe itself has a Teflon tube for ease of movement in the endoscope without much mechanical resistance. The distal end has a special anti-sticking coating to further reduce the sticking effect. The length of the cut wire can be adjusted. Therefore, it is possible to preadjust the cutting depth up to a maximum of 2 mm. The probe has been sucessfully used in about 100 cases third ventriculostomy and ventriculo-septostomy where at first the tissue was coagulated and then cut with the same instrument.

Conclusion

Only an automatically controlled ESU makes it possible to obtain reproducible cutting and coagulation effects. Especially in neurosurgery, the voltage regulation and soft-coagulation functions are very beneficial and offer new operating possibilities with completely new instruments. These monopolar and bipolar instruments with small outer dimensions can be safely used, for example, in third ventriculostomy and ventriculo-septostomy.

References

1. Farin G (1995) Hochfrequenzchirurgie in der minimal invasiven Chirurgie. In: Pier A, Schippers E (eds) Minimal invasive Chirurgie: Grundlagen, Technik, Ergebnisse, Trends, Thieme, Stuttgart, pp 81–88
2. Haag R, Cuschierei A (1993) Review: recent advances in high-frequency electrosurgery: development of automated systems. J R Coll Surg Edinb 38 : 354–364

Three-Dimensional Endoscopy – A Useful Tool for Neuroendoscopy? An Anatomical Study Using MRI and Three-Dimensional Endoscopy

M. Meinhardt, T. Riegel, D. Hellwig, B. L. Bauer

Summary

The ventricular system of 22 formol-saline fixed cadaver brain hemispheres was investigated using four different approaches in a free hand endoscopy mode based on MRI reconstructions. MRI and 3D endoscopy were used to perform 198 measurements of extra- and intraventricular distances. Comparison of 184 matched trajectories showed a mean deviation of MRI from 3D endoscopy of –1 mm, but no correlation between length of measurements and deviation of the two methods could be found. Additionally we analysed the applicability of different 3D endoscopes in neurosurgery. The advantages and disadvantages of the 3D prototypes in comparison to common 2D endoscopes are discussed. Suggestions of improvements to 3D endoscopes are made.

Introduction

Minimally invasive techniques using endoscopes for image-guided diagnosis and therapy become more and more important in neurosurgery [1, 6]. Today, the availability of ultrathin 3D endoscopes enables neurosurgeons for the first time to have a 3D visualization of the human ventricular system. Since August 1994, we have used a 3D endoscopy workstation and 3D endoscope prototypes for anatomical and correlative studies. Neuroendoscopic interventions are at present performed either by fluoroscopic or CT guidance [6]. For the future it seems to be certain that the outstanding contrast resolution of soft tissue and the excellent 3D localization of MRI will lead to its widespread use in guiding procedures, but at the moment there is an uncertainty concerning the use of MRI in guiding procedures.

The two major aims of our study were to find out whether a reliable coherence in MRI measurements and endoscopic distances exists and to evaluate the applicability of 3D endoscopy in neurosurgery.

Methods and Material

The ventricular system of 22 formol-saline fixed cadaver brain hemispheres was investigated using four different approaches in a free hand endoscopy mode based on MRI reconstructions. For each approach the extraventricular distance from the cortex to the lateral ventricle and intraventricular distances to defined landmarks were determined. Altogether 198 measurements were performed with MRI and endoscopy.

The study consisted of a radiological and an endoscopical part. The first part included the MRI examination of the cadaver brains in a 3D FISP technique (TR 40 ms, TE 10 ms, flip-angle 20°, slicethickness 1 mm, 192 × 256 matrix) using a Magnetom Impact at 1.0 Tesla (Siemens).

MRI Reconstructions

A special software program, the so-called MPR (multiplanar reconstruction mode) was used to reconstruct the lateral ventricle in transverse, sagittal, coronal and oblique scans. To obtain reproducible data for the investigation, a standardized reconstruction protocol was developed. Therefore transverse scans were created parallel to a reference line through the anterior and posterior commissure (AC-PC). Sagittal and coronal scans were set orthogonal to this AC-PC scan [3, 9, 21].

Approaches to the Ventricular System

Four operative approaches to the lateral ventricle were simulated in the multiplanar reconstruction mode (MPR). The frontal horn was reached by the well-known precoronal (puncture is similar to Kocher's point [12], 2 cm lateral to the midline and 2.5 cm anterior to the coronal suture) and the transfrontal-sagittal approach [12]. A transparietal-vertical approach and an occipital-sagittal approach provided a view into the atrium ventriculi. The latter was performed from a point which is generally reknown as Dandy's point [12].

Fig. 1 a, b.
Trajectories of the approaches
to the anterior horn (**a**) and
the atrium ventriculi (**b**)

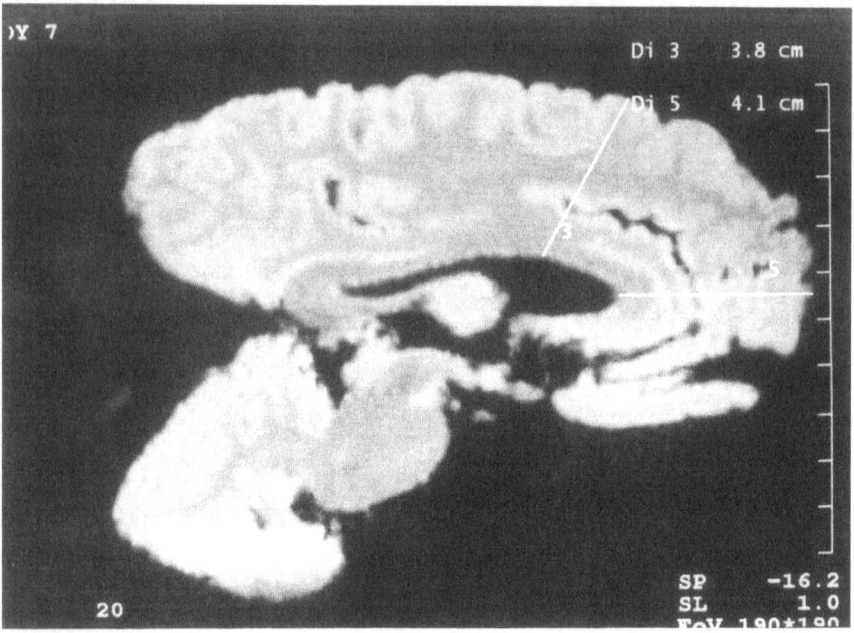

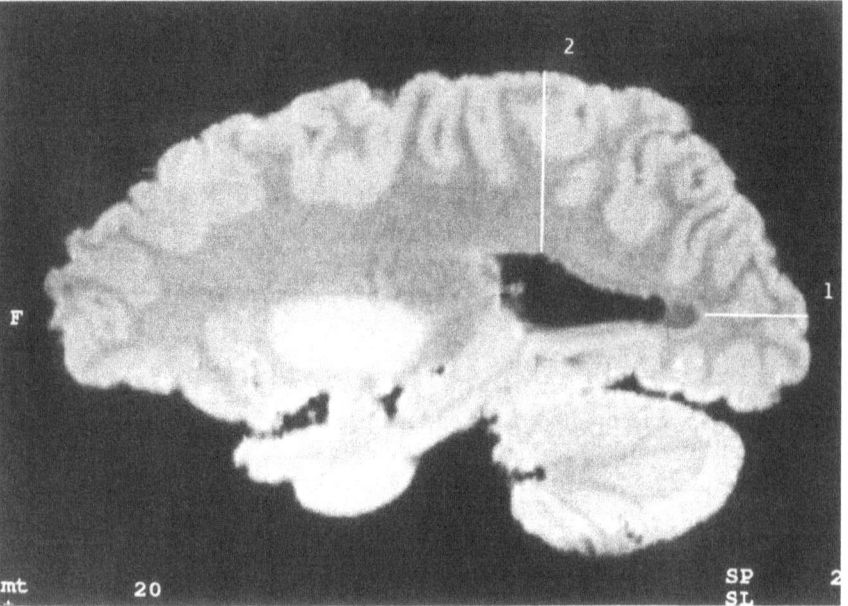

Fig. 1 a, b. Trajectories of the approaches to the anterior horn (**a**) and the atrium ventriculi (**b**)

Additionally, multiple planning scans had to be reconstructed to allow precise calculation of the puncture trajectories. Due to the absence of outer skull landmarks to determine the exact trajectories, these MRI reconstructions became necessary.

Endoscopical Technique

During the second part of the study, 3D endoscopy was performed according to the MRI calculated trajectories. We used a 3D endoscope prototype (Richard Wolf) with 0° optic, an outer diameter of

4 mm, one optical channel, and a prism for image doubling, to determine the distances. Hence a metal probe in a sliding cannula was attached to the endoscope shaft.

After fixing and positioning of the brains in a special device, the ventricular puncture with a guiding cannula was performed, aided by the previously calculated planning scans. Thereafter the probe was positioned to the defined landmarks under endoscopical control.

For inspection of the ventricular system we also tested 3D endoscope prototypes with an outer diameter of 5 mm, two optical channels, and 25°/70° optics.

Fig. 2.
Tips of the different endoscopes from left to right: 0°,
25°, and 70° optic

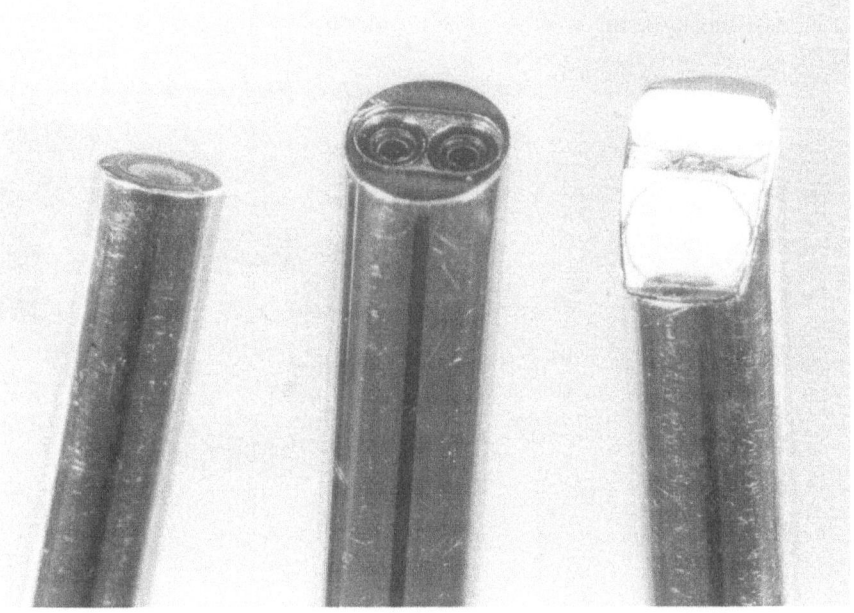

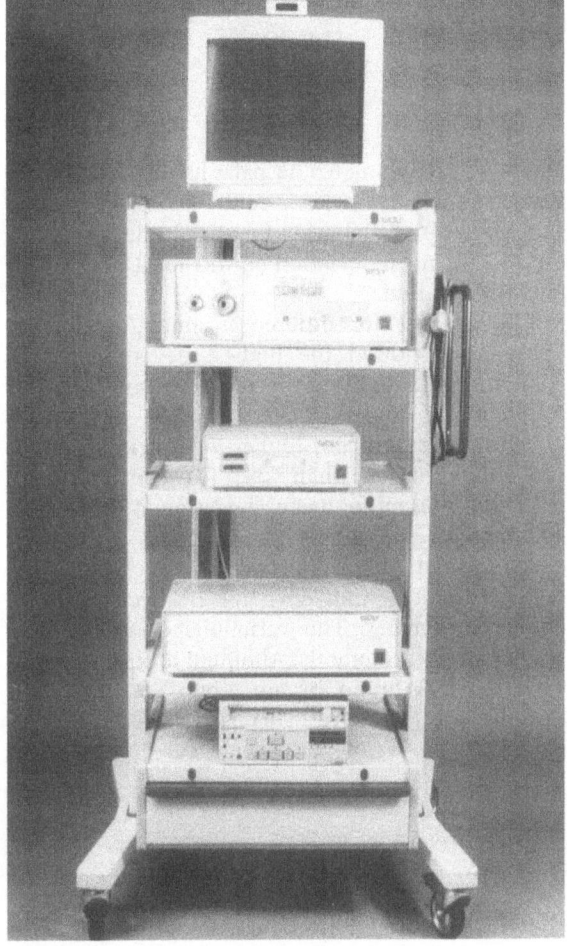

Fig. 3. The three-dimensional endoscopy workstation consists of (from top to bottom): an infrared emitter or liquid crystal display shutter (100 Hz), which controls the LCD glasses; a monitor; a light source; a double camera system; a 3 D converter (100 Hz) that projects the two camera pictures of the same object over each other on the screen; and an S-VHS video recorder for documentation

Working Principles of the 3 D Workstation

The main problem of 3 D endoscopy is to produce a three-dimensional picture on a two-dimensional screen. During the physiological process of vision the cerebral cortex takes the picture of each eye and calculates, during a highly complicated process called fusion, a 3 D image. This process has to be imitated by a technical process in real-time 3 D endoscopy.

According to human eyes, with their anthropomorphic interpupillary distance and convergence, the 3 D endoscopes present a view of an object from different points of view. Two images of the same object are taken simultaneously and displayed on the monitor. The two charge-coupled device (CCD) camera pictures have to be computerized by the 3 D converter. This 3 D converter is able to display the images in a 100 Hz mode. Each picture is projected onto the monitor with 50 Hz frequency. This frequency is necessary to obtain a flicker-free image. At a lower frequency, below 45 Hz, the visual cortex would split up the 3 D picture into its two single images [2, 17, 18, 23].

Fig. 4.
Liquid crystal display (LCD)
glasses

The transmission of the left and right partial image to the corresponding eye is achieved in a sequential display mode. Actively controlled shutter glasses (liquid crystal display or LCD glasses) display the right image to the right eye and the left image to the left eye. At the same moment when the left LCD glass is transparent the left partial image appears on the screen while the right LCD glass is blind. Because of the high frequency (50 Hz), the cerebral cortex is able to merge the pictures into a virtual 3D impression. The 3D effect exists only as a result of the cortical computerization of two simultaneously perceived pictures [17, 18, 23].

Statistics

- The difference between endoscopic and MRI values was defined to compare the measurements of both methods (Δd = endoscopical value – MRI value).
- The Shapiro-Wilks Test was used to determine whether the differences were distributed normally (Gaussian distribution).
- We used the parametric Student-Newman-Keuls Test to analyze the differences in accuracy amongst the values of the various approaches.
- The same test examined the correlation between length of measurements and deviation between endoscopy and MRI.

Results

Measurements

It was possible to evaluate 184 pairs of measurements. Due to a missing match, 14 pairs had to be excluded from the study.

A Gaussian distribution for all Δd values could be found. The mean deviation of MRI to 3D endoscopy amounts to 1 mm.

The analysis for differences amongst the four approaches showed the highest deviation in the transfrontal-sagittal approach. The mean value was – 2.6 mm.

This is a significant higher value with an error probability of less than 5%. The remaining approaches present a mean deviation of less than – 1.0 mm.

No correlation could be found between the length of measurements and deviation between MRI and endoscopic values. The variations of values were similar in both methods. Although a high standard variation was found in the group of the longest distance, no statistical correlation could be calculated.

Fig. 5.
Distribution of deviations
of MRI from endoscopy
measurements

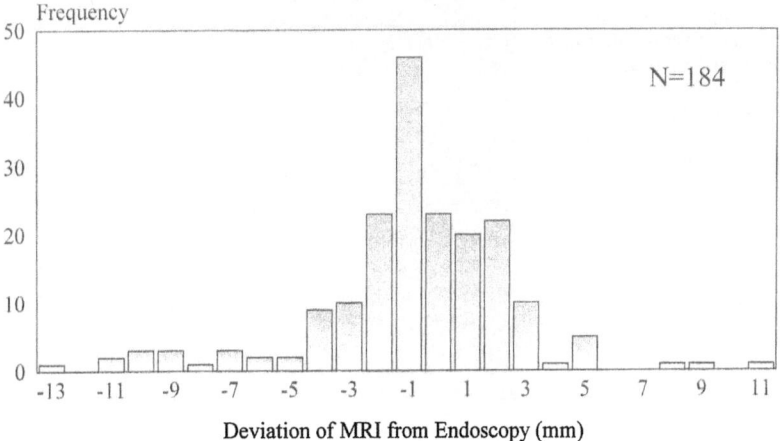

Distribution of deviations of MRI from
Endoscopy measurements

Fig. 6.
Deviation of endoscopy and
MRI measurements in
relation to approaches

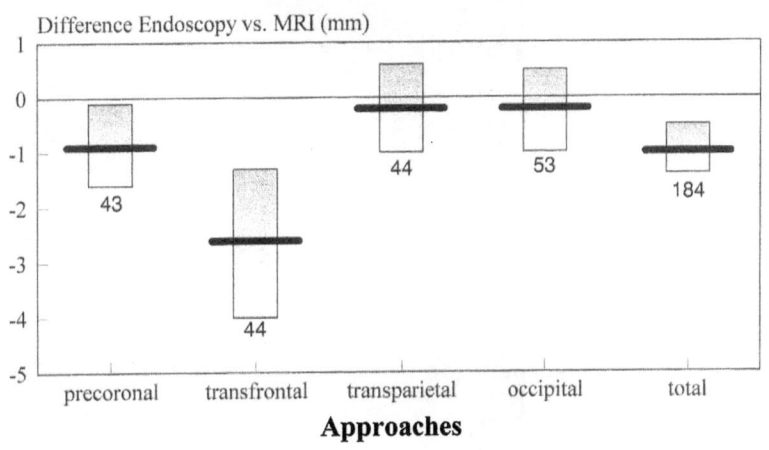

Deviation of endoscopy and MRI measurements
in relation to approaches

Fig. 7.
Influence of length of
measurements to differences
between endoscopy and MRI
measurements

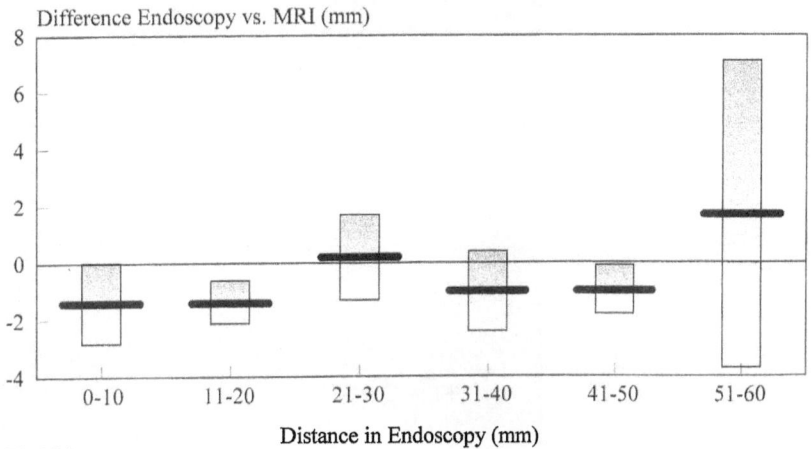

Influence of length of measurements to differences
between Endoscopy and MRI measurements

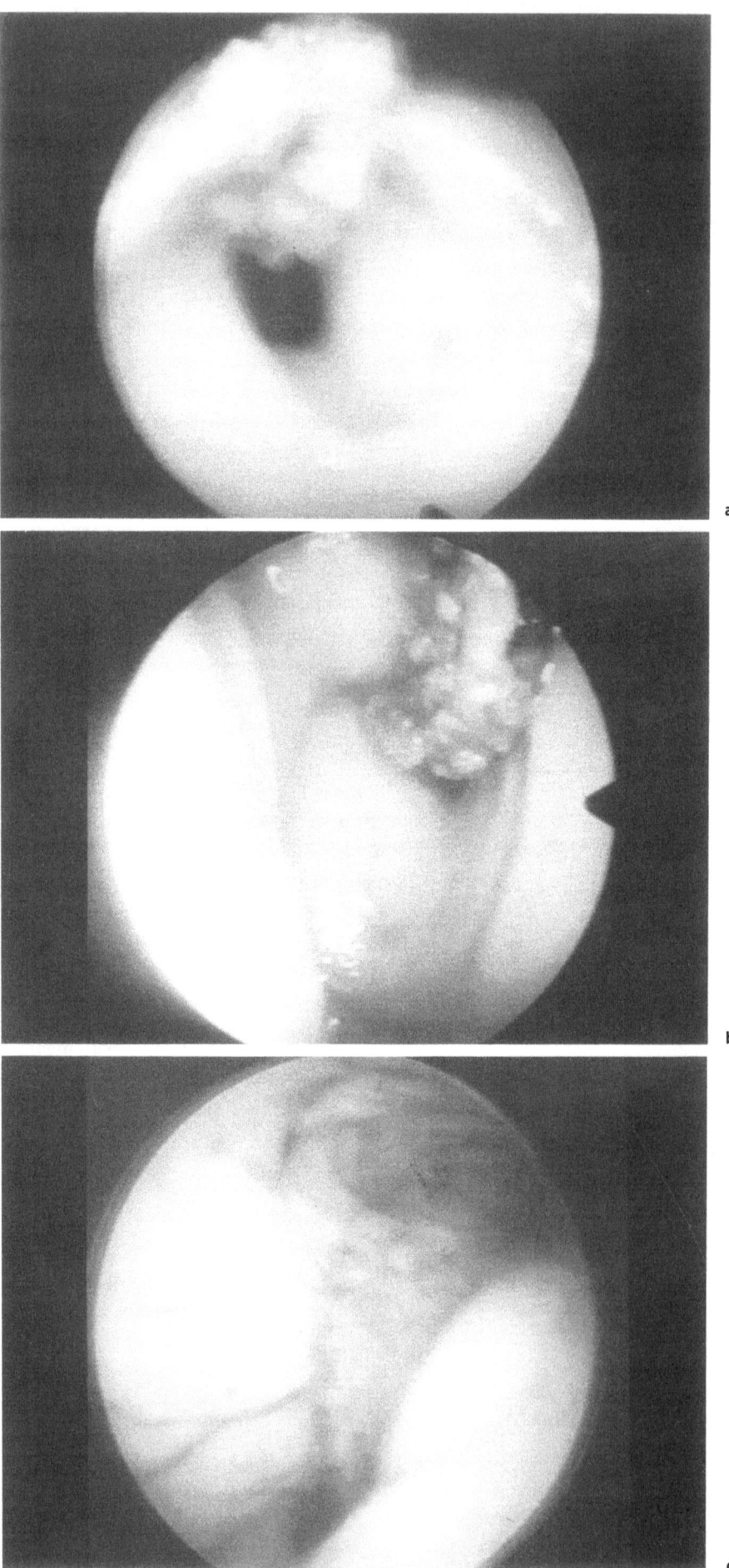

Fig. 8.
a Anatomy of the foramen of Monro as seen from the coronal approach. b Structures of the atrium ventriculi from a transparietal-vertical approach. c Typical view presented by the occipital-sagittal approach through the posterior horn

Anatomy

The approaches to the anterior horn present the anatomy of the anterior horn and the anterior part of the cella media from different points of view.

The important landmarks of this part of the lateral ventricle are

- The typical Y-shape of the foramen of Monro, consisting of choroid plexus, thalamostriate vein, and confluence of septal veins [11, 13].
- The anterior and lateral limits of the foramen are formed by the fornix [11, 13].

The precoronal approach offers an excellent view of these two landmarks.

- The medial septal vein serves as a guideline to the foramen of Monro during the transfrontal-sagittal approach.
- Laterally, the head of the caudate nucleus accompanies the endoscopist on the way to the foramen. The descent of the choroid plexus from the higher leveled cella media to the frontal horn is easily visualized [20].

The approaches to the atrium or so-called trigonum were performed from two different anatomical regions:

- The transparietal-vertical approach allows a view from the roof of the atrium to the anatomical structures. The survey of the anatomy presents the glomus choroideum, the dorsal part of the thalamus, the commissure of the fornix passing into the hippocampus, and the entrance into the posterior horn [3, 7, 8]. (A backward view by a 70° optic shows the anatomy of the posterior horn.)
- The atrium can also be reached through the posterior horn – if present – by the occipital-sagittal approach. The survey shows the descent of the choroid plexus from the cella media to the inferior horn, medially accompanied by the pulvinar and laterally by the crus fornicis [3, 7, 8].

3D Endoscopy Compared to 2D endoscopy

The assessment of the applicability of 3D endoscopes in neurosurgery requires one to make an important distinction between the two different optical principles of the prototypes. Prototype A consists of one optical channel and needs a prism for image doubling. Prototypes B and C are equipped with two integrated optical channels. A CCD camera is attached to each of them.

During our experiments with 3D endoscope prototypes we discovered that the double-channel 3D endoscopes present a higher quality of field of depth and therefore a more impressive perception of the third dimension [10]. But both methods allow the endoscopist to precisely estimate distances within preformed spaces [10]. The clear and high-resolution images are another major advantage compared to common 2D endoscopes.

But we also noticed the typical pitfalls of these endoscopes. Due to their constructional principle, double pictures can disturb the investigation. If one optical channel is blinded by tissue or if the area of optical convergence (5–25 mm) is exceeded, double pictures will occur [18, 22]. It can be very confusing to try to pass the endoscope through two foramina of Monro with neither being the real one.

The fact that the endoscopes do not have working or rinsing channels and lack a smaller diameter compared to fiberscopes is simply due to their stage of development: they are prototypes. If a neuroendoscopist is used to a fiberscope, he will also miss the high maneuverability of the bending section.

The advantages and disadvantages of 3D endoscopes compared to common "2D" endoscopes are listed in Table 1.

Tabelle 1.
Advantages and disadvantages of 3D endoscopy versus 2D endoscopy

	Advantages	Disadvantages
3D endoscopy	Clear, high-resolution images	Limited manoeuverability (rigid lens scope)
	Precise estimation of distance Highly accurate in endoscopic procedures	Necessity of LCD glasses Absence of working channels double pictures (in special conditions)
2D endoscopy (fiberscope)	Lower outer diameter	Low-resolution images due to limited numer of fibers
	High manoeuverability – bending section of fiberscopes	No precise estimation of distances

Discussion

Measurements

The fact that the MRI measurements are on average 1.0 mm larger than 3D measurements seems to be rather a systematic mistake than an error of variance or one dependent on investigator's accuracy. The deviation can be explained by a systematic dispersion of spatial point-to-point accuracy in MRI. The so-called pincushion distortion ranges from 1.0 mm in the coil center to 2–3 mm in the periphery of the coil [5, 16]. The two basic potential causes of geometric distortion in MRI are gradient field nonlinearities and magnetic field inhomogeneities [16, 19]. In our study we tried to minimize the pincushion effect by the use of a 3D FISP sequence with 1 mm slices and positioning of the brains in the coil center.

The second main source of deviations between 3D endoscopic and MRI measurements is the free hand mode during endoscopy. Despite detailed and precise planning, slight deviations of the punctures could not be avoided in all cases. Additionally, minor angulations of the endoscope inside the guiding cannula could also result in deviations from the calculated trajectories.

The highest significant deviation was found in the transfrontal-sagittal approach. This might be due to the more practical difficulties entailed by this specific puncture.

We think that MRI provides reliable data for neuroendoscopic interventions. Of course, a higher accuracy can be achieved with the use of stereotactical devices than with our free-hand procedures. But this relatively simple guiding method already displays a highly reliable level of accuracy. MRI-guided neuroendoscopy will enter new dimensions in neurosurgery.

3D Endoscopy

Our first experiences with three-dimensional endoscopy are promising. In reference to the above mentioned disadvantages for the prototypes, we think that some further developments have to be made. Only then will the advantages of 3D endoscopy be able to surpass the high quality of "2D" endoscopes used at the moment [23].

The following improvements need to be made:

- Reduction of outer diameters.
- Working and rinsing channels.
- Integration of anthropomorphic convergence and accommodation to avoid the optical limitation of the 3D convergence area [23].

The established 2D endoscopic systems provide neuroendoscopists with a high technical standard of vision. So, why use 3D endoscopy?

A comparative study of endoscopical two-dimension and three-dimensional vision systems has shown that inexperienced as well as experienced surgeons have solved complex operation tasks in a significantly shorter period of time and with fewer mistakes using the 3D system [10].

If we want to work in a minimally invasive manner, we should aim to work with the highest possible safety. Most of the future-orientated concepts for neurosurgery take a 3D vision for granted [4, 14, 15]. Living in a three-dimensional world, it is desirable to work with three-dimensional vision.

References

1. Bauer BL, Hellwig D (1994) Minimally invasive endoscopic neurosurgery – a survey. Acta Neurochir Suppl (Wien) 61:1–12
2. Becker H, Melzer A, Schurr O, Buess G (1993) 3-D video techniques in endoscopic surgery. Endosc Surg Allied Technol 1 (1):40–46
3. Duvernoy H (1991) The human brain – surface, three dimensional sectional anatomy and MRI. Springer, Berlin Heidelberg New York
4. Guthrie BL (1994) The medical videoscope: neurosurgery into the 21st century. Microsurgery 15 (8):547–554
5. Groenemeyer HW et al (1995) Image-guided access techniques. Endosc Surg (3):69–75
6. Hellwig D, Bauer BL (1991) Endoscopic procedures in stereotactic neurosurgery. Acta Neurochir Suppl (Wien) 52:30–32
7. Hussein S, Woischneck D (1990) Topographie des Atrium ventriculi und ihre mikrochirurgische Bedeutung. Neurochirurgia 33:8–10
8. Kretschmann HJ, Weinrich W (1991) Klinische Neuroanatomie und kranielle Bilddiagnostik CT und MR, 2nd edn. Thieme, Stuttgart, pp 304–307
9. Kretschmann HJ, Weinrich W (1996) Hirnstrukturen und Koordinatensysteme. In: Kretschmann HJ, Weinrich W (eds) Dreidimensionale Computergraphik neurofunktioneller Systeme. Thieme, Stuttgart
10. Kunert W, Van Bergen P, Schurr MO, Buess G (1996) Comparative study of endoscopic two-dimensional and three-dimensional vision systems. Endoskopie heute 4:345–351
11. Lang J (1992) Topographic anatomy of preformed intracranial spaces. Acta Neurochir Suppl (Wien) 54:1–10
12. Mapstone TB, Ratcheson RA (1996) Techniques of ventricular puncture. In: Wilkins RH, Rengachary SS (eds) Neurosurgery, 2nd edn. McGraw-Hill, New York, pp 151–152
13. Riegel T, Hellwig D, Bauer BL, Mennel HD (1994) Endoscopic anatomy of the third ventricle. Acta Neurochir Suppl (Wien) 61:54–56
14. Rininsland HH (1993) Basics of robotics and manipulators in endoscopic surgery. Endosc Surg Allied Technol 1:154–159
15. Satava RM (1993) 3-D vision technology applied to advanced minimally invasive surgery systems. Surg Endosc 7:429–431
16. Schad LR, Ehricke HH, Wowra B et al. (1992) Correction of spatial distortion in magnetic resonance angiography for radiosurgical treatment planning of cerebral arteriovenous malformations. Magn Reson Imaging 10:609–621

17. Schragl E, Bigenzahn W, Donner A et al. (1994) Laryngeal surgery by three-dimensional (3D) endoscopy via jet-laryngoscope using superimposed high frequency jet ventilation (SHFJV) [German]. Laryngorhinootologie 73:412–416

18. Strutz J (1993) Die 3D Endoskopie. HNO 41:128–130

19. Sumanaveera TS et al. (1994) Characterization of spatial distortion in MRI and its implications for stereotactic surgery. Neurosurgery 35 (4):696–703

20. Torkildsen A (1934) The gross anatomy of the lateral ventricles. J Anat 68:480–491

21. Villemure JG, Marchaud E, Peters T et al. (1987) MRI stereotaxy: recognition and utilization of the commissures. Appl Neurophysiol 50:57–62

22. Wenzl R, Lehner R, Vry U et al. (1994) Three-Dimensional video-endoscopy: clinical use in gynecological laparoscopy. Lancet 344:1621–1622

23. Zobel J (1993) Basics of three-dimensional endoscopic vision. Endosc Surg Allied Technol; 1 (1):36–39

Spinal Endoscopy in Fixed Human Cadavers: Preliminary Findings with Possible Implications for Implanting Bladder Stimulators in Paraplegic Patients

J. Boschert, J. Brunner, S. Mense, and P. Schmiedek

Summary

In paraplegic patients after spinal cord injury voluntary urination can be simulated by implanting an electronic bladder stimulator. For proper functioning the spasticity of the detrusor muscle must be eliminated by dorsal rhizotomy of the sacral roots S2 to S4 bilaterally. Additionally, voiding can be improved if the motor rootlets innervating the sphincter muscle (S2) are also dissected on both sides. The commonly used approach for these procedures at present is a laminectomy of at least two vertebral arches at the level of the medullary cone. The aim of our study is to develop a minimally invasive method applying endoscopic techniques. Using fixed human cadavers we tested a flexible and a rigid endoscope. Based on our first experience the rigid instrument was preferred. A bilateral-biportal approach was most apt. Even interlaminary spaces at upper lumbar/lower thoracic levels could easily be punctured percutaneously with trocars of 5 mm in diameter. However, such a percutaneous puncture is not under visual control and thus may be perilous for the intrathecal structures. We suggest a "microsurgically assisted" endoscopic procedure: (a) A unilateral flavectomy at the level of the medullary cone and opening of the dura are performed under microsurgical conditions. (b) A rigid endoscope is introduced allowing a percutaneous puncture of the contralateral interlaminary space under visual control. (c) Rhizotomy is then possible interchanging both approaches as working versus optical channels.

Introduction

In patients with spinal cord lesions cranial to the medullary level S2 inhibitory influences on the sacral micturition center are disrupted, resulting in a bladder spasticity with hyperreflexy of both the detrusor and the sphincter muscle. Consequently, a reflex incontinence develops severely impairing the patients's quality of life. Additionally, the spasticity of the urinary bladder endangers the patient by damage to the upper urinary tract and chronic renal insufficiency and hence reduces life expectancy. Therefore several attempts have been made to restore an almost physiological bladder function [4, 24]. Voluntary urination can be simulated by implantation of a bladder stimulator for electro-stimulation of sacral anterior roots S2 to S4. The stimulator is implanted subcutaneously in the lower abdominal wall and connected to the motor roots innervating the bladder either by intradural [5] or extradural [17, 19] placement of electrodes. The spasticity of the detrusor muscle is eliminated by dorsal rhizotomy of the sacral roots S2 – S4 bilaterally [22, 12]. The commonly used approach for these procedures is a laminectromy of at least two vertebral arches at the level of the lower lumbar spine [20, 17]. To our knowledge only few attempts have been made so far to minimize the operative trauma necessary for dorsal selective rhizotomy and electrode implantation [1]. We therefore performed explorative studies in fixed human cadavers to evaluate the potential benefits of neuro-endoscopic techniques.

Methods and Material

For spinal endoscopy we applied a rigid endoscope (outer diameter of 2.7 mm, 30° deflection of the optical axis, no working channel; (Aesculap endoscope PE 203 A) and a flexible endoscope (outer diameter: 1.9 mm, no angulation of the optical path, one rinsing channel; (Aesculap syringoscope PF 612 Fleximed). A halogen endoscopic illuminator (Aesculap light source 300) served as a light source. The endoscopic image was transferred to a monitor by means of a CCD video camera and recorded on tape.

In three fixed male human bodies the vertebral levels L5/S1 up to T10/T11 were examined. At each level a left-sided flavectomy was performed. The contralateral interlaminary spaces and the dura underneath were percutaneously punctured by means of

Touhy needles and trocars of increasing diameters from 2 to 5 mm under endoscopic control, the rigid instrument being introduced through the flavectomy.

Our explorative experiments addressed on the following issues:

1. What kind of approach (monoportal vs biportal, percutaneous vs open) offers most advantages?
2. At which spinal level should the approach be made?
3. What kind of endoscope (flexible vs rigid, 30° vs 0° angulation of the optical axis) is to be favoured for selective posterior rhizotomy and minimally invasive implantation of electronic bladder controllers?
4. Are surgical manipulations such as dissection of the arachnoid membrane, identification of rootlets or posterior rhizotomy possible under endoscopic control?

Results

In all three fixed cadavers interlaminar spaces could be punctured percutaneously even at the upper levels (T10/T11) without severe obstacles. This was the case although we had no fluoroscope available and in one cadaver the clinical signs of polyarthritis were obvious. The entry point at the skin was located 1–2 cm paramedian at the tip of the processus spinosus of the upper vertebra. The Touhy neddle and trocars were directed about 5° medially from the sagittal plane and about 10°–15° cranially with respect to the horizontal plane. First, a Touhy needle was introduced which also was used to slit the dura mater by tilting the needle in a pendulating manner in a craniocaudal direction. Next, the puncture channel was enlarged using trocars of increasing diameters of up to 5 mm. After withdrawal of the obturator the last trocar was used for introducing the endoscope. With blind percutaneous punctures we observed injuries to rootlets and the ventral dura. No such lesions occurred using a combined approach (flavectomy and contralateral percutaneous puncture under endoscopic control).

At identical spinal levels the anatomical aspects as endoscopically visualized were similar in all three cadavers. At lower lumbar levels (L5/S1) the terminal intraspinal root of roots L5 and S1 could be seen entering their dural sheaths and running towards the intervertebral foramina. The lower rootlets (S2 to C1), however, could not be seen. At middle lumbar levels (L3/L4 and L2/L3) the bundle of the cauda equina could be seen clearly, but no individual or groups of rootlets could be identified. Occasionally, the filum terminale could be found in the ventral portion of the whole bundle. In the upper lumbar region (L2/L3 and L1/L2) the filum terminale could rapidly be identified and continuously followed toward the tip of the medullary cone. Left- and right-sided bundles of rootlets could be identified without difficulties. In the lower thoracic spine (T12/L1, T11/T12 and T10/T11) the medullary cone and its continuity towards either medulla spinalis or the filum could be visualized. Left-and right-sided rootlets could be well identified. Differentiation between anterior (motor) and posterior (sensory) radicules was not possible.

In order to introduce the fiberscope through a flavectomy a skin incision of at least 10 cm in length and dissection of the erector trunci muscle were necessary. Alternatively, it could be forwarded through a trocar which then had to be introduced almost parallel to the skin. At the ventral dura the instrument always bent up- or downward to follow the spinal channel along its ventral wall. It was impossible to steer the flexible endoscope towards a certain point. In all cases the lens was located ventral to the neural structures. With the rigid instrument location of its tip and direction of view were well defined and controlled, allowing orientation and identification of anatomical structures. After some training manipulations under endoscopic vision were easy and safe. The straight-aiming lens had a limited visual field which could not be enlarged very much by tilting the instrument. Structures appeared highly magnified, yielding almost no perception of depth. With the 30° endoscope at least one spinal level upward and downward from the entry point could be visualized. Shadows and fish-eye effect were strong, with both together resembling three-dimensional perception. Without injuring rootlets or adjacent structures the following operative manipulations were performed under endoscopic vision through the flavectomy: dissection of the arachnoid membrane and, at L1/L2 and at T12/L1, clipping of the filum terminale with a vessel clip.

Discussion

With respect to the approach we initially tried to use a strictly percutaneous design. Surprisingly, it was not too difficult to puncture interlaminar spaces and the dura underneath with trocars of up to 5 mm diameter percutaneously even at upper lumbar/lower thoracic levels. However, such a percutaneous puncture was perilous for the intrathecal structures due to the lack of visual control. We therefore switched to a combined endoscopic-microsurgical procedure using a bilateral-biportal approach. Through a unilateral flavectomy the dura was opened under microscopical control. The contralateral interlaminar space was punctured under visual control through the endo-scope which was introduced through the flavectomy.

For establishing the most suitable vertebral level of approach this combined technique was performed in all interlaminary spaces from L5/S1 up to T10/T11. Due to the lumbar lordoses the radicules of the cauda equina are stretched in a cranioventral and caudo-dorsal direction. As a result, at lower lumbar levels (L5/S1 and L4/L5) the more caudal rootlets (S2–C1) cling to the mediodorsal dura. Consequently, during penetration of the dura the instrument separates these lower rootlets at its entry point and then shuffles the radicules aside which therefore never could be visualized. Hence, in this region the endoscope seems to dive into an anatomically "empty space". Accordingly, no manipulations could be performed on rootlets S2 to S4. Thus, the lower two lumbar levels cannot be regarded as a proper approach for sacral deafferentation. In the middle of the lumbar spine the bundle of the cauda equina could be seen clearly. However, no identification of rootlets was possible according to anatomical criteria; even the side of their origin from the medullary cone could not be determined. At upper lumbar and lower thoracic segments at least some anatomical orientation is provided by the medullary cone and the filum termi-nale. For an endoscopically controlled implantation of electrodes as well as sacral deafferentation these levels might be most suitable. This is in line with previous findings of Toczek et al [21] who reported a more precise identification of anterior roots at the level of the medullary cone than with rhizotomies at the sacral level [22].

For our purposes we could not find any beneficial use for a flexible endoscope. First, with the fibrescope we had to ensure a sufficiently smooth slope into the spinal canal. Otherwise the endoscope's shaft would, due to its rigidity and the small dimensions of the anatomical space being entered, not bend up- or downward to follow the spinal canal. Such a smooth slope only can be achieved either through a flavec-tomy with a rather long and therefore traumatic incision of the covering soft tissue or through a percutaneous puncture at an almost imprecable angle within the sagittal plane. Second, while being introduced through a trocar or a flavectomy the instrument passed through the rootlets of the cauda equina ventrally and was deflected at the ventral dura in a cranial or caudal direction. Therefore the lens was always placed ventral to the neural structures.

Additionally, even after a training period, the tip of the endoscope could not be steered towards a certain anatomical structure; the instrument sought its own way while being forwarded in the spinal canal. Due to the fact that the endoscope could not be steered and that there was some uncontrollable torsion of the

shaft the interpretation of the endoscopic image was intolerably difficult. Finally, the quality of the endo-scopic image was very blurred as compared with that of a rigid instrument. This experience of the limited use of a flexible instrument for spinal endoscopy agrees with reports from the literature [3, 7, 8, 14, 15]. Karakhan et al. [10], however, reported on precise stereotopography at lumbar and cauda equina levels in fresh human cadavers using a flexiscope with a bending tip. Using a rigid endoscope in our experi-ments, we were able to view up to one interspace in both cranial and caudal directions, as described by other investigators [2, 16].

A comparison of the endoscopes with respect to their optical characteristics (0° vs 30° deflection of the optical path) revealed the superiority of the 30° lens. With a straight instrument the field of view is limited. Since the endoscope is introduced nearly perpendicular to the axis of the spinal canal the working depth only measures 1.5 cm due to the ana-tomical dimensions. Consequently, the structures always lie close to the tip of the instrument, resulting in a strong magnification. Therefore, the image seems to be almost completely flat, giving no impression of depth. With the 30° optic the viewing direction is brought in parallel with the axis of the spinal canal. Thus, by simply rotation the instrument around its mechanical axis the anatomical structures can visual-ly be followed at least one vertebral level cranially and caudally from the approach. With this very extended field of view the 30° instrument is characterized by a strong fish-eye effect and strong shadows. Both factors result in a 3D-like impression which makes manipulations under endoscopic control much easier. Due to the fish-eye effect the structures straight in front of a 30° instrument are also visu-alized, thus ensuring visual control of the instru-ments's tip during introduction. Other investigators also preferred a rigid instrument with an outer dia-meter of 2.7 mm and 30° deflection of the optical axis for spinal endoscopy in fresh human cadavers [9].

We were repetitively able to clip the filum termi-nale under endoscopic control without injuring neural tissue. This performance shows that endo-scopically controlled surgical procedures within the cauda equina are possible and sufficiently safe for neighbouring neural structures.

We therefore suggest the following "microsurgically assisted" endoscopic procedure and will further test it on fixed human bodies. After MRI localization of the patients' individual spinal level of the medullary cone a unilateral flavectomy is performed followed by perforation of the dura applying microsurgical techniques. A rigid endoscope is introduced and the medullary cone is visualized. Through a percuta-neous puncture of the contralateral interlaminary space the dura underneath is perforated by means of a trocar under visual control through the endoscope. After withdrawing the obturator the trocar is used as a working channel for performing rhizotomy contralateral to the flavectomy. Dissection of rootlets on the other side may be performed either microsurgically or under endoscopic control in which case the trocar is used for introducing the endoscope.

Besides the dimensions of the currently used electrodes the major problem is the identification of rootlets [18]. This can most probably be achieved endoscopically on the basis of morphological criteria [6, 11, 23] in combination with electrostimulation [13], as is used in the case of macrosurgery.

Acknowledgment. This study was supported by the Federal Ministry of Education and Science, program priority area "Neurotraumatology and Neuropsychological Rehabilitation".

References

1. Barolat G (1991) Dorsal selective rhizotomy through a limited exposure of the cauda equina at L-1. J Neurosurg 75 : 804 – 807
2. Blomberg RG (1994) Epiduroscopy and spinaloscopy: endoscopic studies of lumbar spinal spaces. Acta Neurochir (Wien) 61 [Suppl]: 106 – 107
3. Blomberg RG (1995) Fibrous structures in the subarachnoid space: a study with spinaloscopy in autopsy subjects. Anesth Analg 80 : 875 – 879
4. Brindley GS (1994) The first 500 patients with sacral anterior root stimulator implants: general description. Paraplegia 32 : 795 – 805
5. Brindley GS, Polkey CE, Rushton DN (1982) Sacral anterior root stimulators for bladder control in paraplegia. Paraplegia 20 : 365 – 381
6. D'Avella D, Mingrino S (1979) Microsurgical anatomy of lumbosacral spinal roots. J Neurosurg 51 : 819 – 823
7. Döhring S, Ooi Y, Schulitz KP, Satoh Y (1984) Myeloskopische Befunde im Bereich der unteren lumbalen Wirbelsäule. Beitr Orthop Traumatol 31 : 120 – 126
8. Fukushima T, Schramm J (1975) Klinischer Versuch der Endoskopie des Spinalkanals: Kurzmitteilung. Neurochirurgia 18 : 199 – 203
9. Hertz H, Schabus R, Wunderlich M (1985) Endoskopie des Spinalkanals – experimentelle Untersuchungen an Leichen. Unfallchirurgie 11 : 275 – 277
10. Karakhan VB, Filimonov BA, Grigoryan YA, Mitropolsky VB (1994) Operative spinal endoscopy: stereotopography and surgical possibilities. Acta Neurochir (Wien) 61 [Suppl]: 108 – 114
11. Lang J, Geisel U (1983) Über den lumbosakralen Teil des Duralsackes und die Topographie seines Inhalts. Morphol Med 3 : 27 – 46
12. MacDonagh RP, Forster DMC, Thomas DG (1990) Urinary continence in spinal injury patients following complete sacral posterior rhizotomy. Br J Urol 66 : 618 – 622
13. Mersdorf A, Schmidt RA, Tanagho EA (1993) Topographic-anatomical basis of sacral neurostimulation: Neuroanatomical variations. J Urol 149 : 345 – 349
14. Ooi Y, Mita F, Satoh Y (1990) Myeloscopic study on lumbar spinal canal stenosis with special reference to intermittent claudication. Spine 15 : 544 – 549
15. Ooi Y, Satoh Y, Inoue K, Mikanagi K, Morisaki N (1981) Myeloscopy, with special reference to blood flow changes in the cauda equina during Lasègue's test. Int Orthop 4 : 307 – 311
16. Peek RD, Thomas JC, Wiltse LL (1993) Diagnosis of lumbar arachnoiditis by myeloscopy. Spine 18 : 2286 – 2289
17. Sauerwein D, Ingunza W, Fischer J, Madersbacher H, Polkey CE, Brindley GS, Colombel P, Teddy PJ (1990) Extradural implantation of sacral anterior root stimulators. J Neurol Neurosurg Psychiatr 50 : 681 – 684
18. Schalow G (1985) The problem of cauda equina nerve root identification. Zentralbl Neurochir 46 : 322 – 330
19. Tanagho EA, Schmidt RA (1998) Electrical stimulation in the clinical management of the neurogenic bladder. J Urol 140 : 1331 – 1339
20. Tanagho EA, Schmidt RA, Orvis BR (1989) Neural stimulation for control of voiding dysfunction: a preliminary report in 22 patients with serious neuropathic voiding disorders. J Urol 142 : 340 – 345
21. Toczek SK, McCullough DC, Boggs JS (1978) Sacral rootlet rhizotomy at the conus medullaris for hypertonic neurogenic bladder. J Neurosurg 48 : 193 – 196
22. Toczek SK, McCullough DC, Gargour GW, Kachman R, Baker R, Luessenhop AJ (1975) Selective sacral rootlet rhizotomy for hypertonic neurogenic bladder. J Neurosurg 42 : 567 – 574
23. Wall EJ, Cohen MS, Abitbol JJ, Garfin SR (1990) Organization of intrathecal nerve roots at the level of the medullary cone. J Bone Joint Surg Am 72 : 1495 – 1499
24. Wipfler G, Jünemann KP (1995) Funktionelle Wiederherstellung der Blasenfunktion bei spastischer Querschnittslähmung mittels Elektrostimulation und sakraler Deafferentation. Aktuelle Urol 26 : 14 – 26

Minimally Invasive Techniques in Monitoring of Brain Tissue Oxygenation and Intracranial Pressure in Neurosurgical Intensive Care

T. J. Kuhn, L. Benes, A. Schneider, H. Mewes, B. L. Bauer, and H. Bertalanffy

Summary

Monitoring of partial pressure of brain tissue oxygenation P(ti)O$_2$, brain tissue ph (brph), partial pressure of brain tissue CO$_2$ [P(ti)CO$_2$] and intracranial pressure (ICP) by invasive methods, calculation of cerebral perfusion pressure (CPP), and hemodynamic monitoring of middle arterial blood pressure (MABP) are the main strategies in multimodal neuromonitoring. Using this multimodal approach the therapy of raised intracranial pressure with diffuse brain edema has undergone fundamental changes during recent years [2, 3, 7, 9, 19, 24, 25, 40, 42].

These parameters are of great importance in view for the therapeutical strategies used in raised ICP. Raised ICP and the consecutive reduction of CPP and P(ti)O$_2$ are the mayor cause for secondary ischemic events as well as for morbidity and mortality following severe head injury (SHI), subarachnoid hemorrhage (SAH high degree), and diffuse brain edema of other origin.

Our results indicate that a reduction of raised ICP (rICP) and a consequent increase of CPP and P(ti)O$_2$ serve to improve the survival rate, better neurological symptoms, and also serve as prevention of secondary ischemic brain injury [7, 21]. Similar findings were made in other situations with raised ICP such as intracerebral bleeding, stroke with edema, encephalitis, and global an- or hypoxia [20, 28].

Introduction

Invasive monitoring of raised ICP has been a routine method in neurosurgical intensive care for nearly 20 years. Different methods have been developed, like epidural monitoring and intraventricular or intraparenchymal techniques. The intraparenchymal monitoring and treatment of raised ICP is generally accepted and represents the state of the art in the treatment of patients with brain edema [5, 13, 14, 18, 23, 31, 41].

In order to prevent secondary ischemic brain damage by hypoxia and hypotension, the monitoring of CPP, P(ti)O$_2$, P(ti)CO$_2$ and pHbr will become more and more important in neurosurgical intensive care medicine. The calculated parameter CPP (MABP – ICP = CPP) has become a guideline in managing patients with severe brain injury [21]. If cerebral blood flow decreases in situations of raised ICP and autoregulatory dysfunction, the CPP and also the P(ti)O$_2$ will decrease and the patient will suffer from hypoxia and secondary ischemic lesions.

There are two main causes for ischemia: extracerebral circulation and intracerebral blood flow. Preliminary animal studies by Mass, Fleckenstein et al. [24] have been followed by clinical investigations [28]. In these studies, normal values and the varying behaviors elicited by changes in P(ti)O$_2$ values have been studied. The investigators found different therapy adjusted effects, like CO$_2$-reactivity, hyperoxygenation, in- and decreasing MABP, effects of sorbitol and mannitol, increasing ICP, and at last of all, hypoxia and brain death.

The monitoring of P(ti)O$_2$ is possible in brain tissue and in CSF with the same probe [24]. The insertion of these probes can be performed in the intensive care unit (ICU).

Patients and Monitoring Methods

Patients and Including Criteria

In this clinical investigation we included patients with SHI, an initial Glasgow Come Seale below 8, a Marshall classification [27] grade II or more in the initial CT scan, and patients with SAH degrees IV or V. During an 18-month period from July 1995 to December 1996 we monitored 20 patients with SHI and eight patients with SAH. Our patients were monitored for 4.4 days on average.

Monitoring Methods

ICP Monitoring

For ICP monitoring we prefer the intraparenchymal system, designed by Camino (San Diego, USA) [11, 14,

Table 1. Patients – suffering from severe head injury (SHI) and subarachnoid hemorrhage (SAH) degrees IV and V – studied

	No.	M	F	Mean age	Good recovery	Defect syndrome	Death	%)
SHI	20	13	7	30.3	11	3	3	17.6
SAH	8	3	5	40.2	1	1	6	85.6
Total	28	16	13		12		9	37.5

Fig. 1.
Computed tomography scan of severe brain injury and brain edema

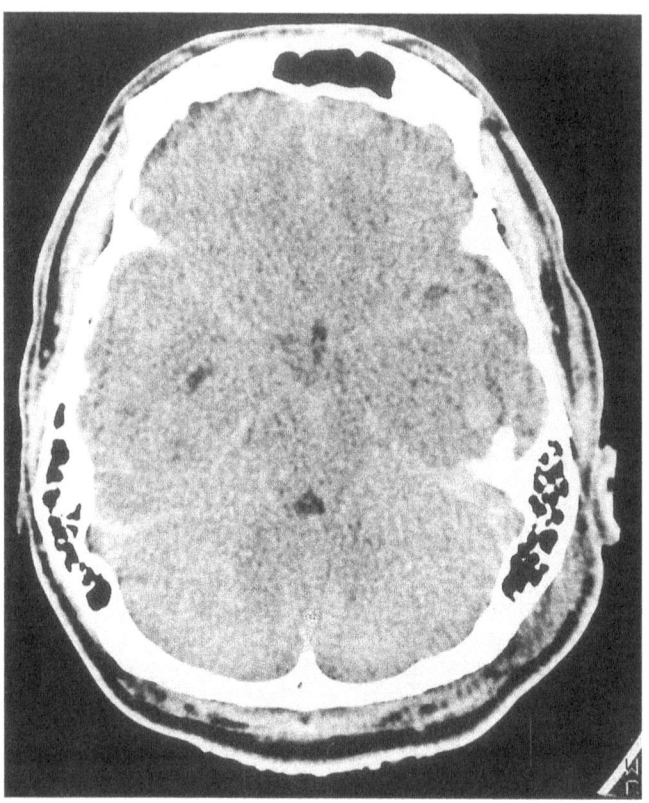

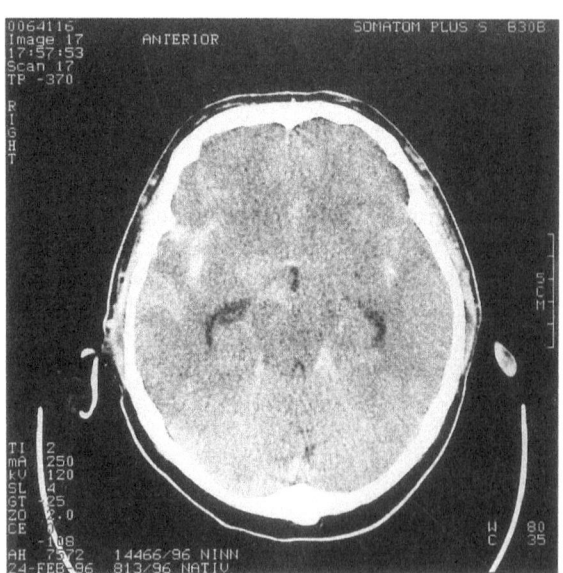

Fig. 2. Computed tomography scan of a high degree subarachnoid hemorrhage (IV/V degree, Hunt-Hess score)

18, 23]. A fiberoptically transmitted light signal is diverted and given back by the diaphragm. A displacement of the diaphragm leads to changes in the returned light signal which enables the microprocessor controlled Camino monitor to calculate and display the related pressure.

$P(ti)O_2$-Monitoring

The LICOX $P(ti)O_2$ probe is a probe of the clark type; oxygen ions can diffundate through a semipermeable diaphragm and generate an electric current. The current intensity is registered and transformed into the tissue oxygen partial pressure by the LICOX device. The first step of the implantation is the calibration of the LICOX probe. During nearly 30 min of the calibration we continue with the surgical part of the implantation. We insert a disposable steel screw into a drill hole, and through the screw we implant a Camino ICP probe, setting zero equal to atmospheric pressure [24].

Fig. 3.
Camino intercranial pressure
(ICP) probe

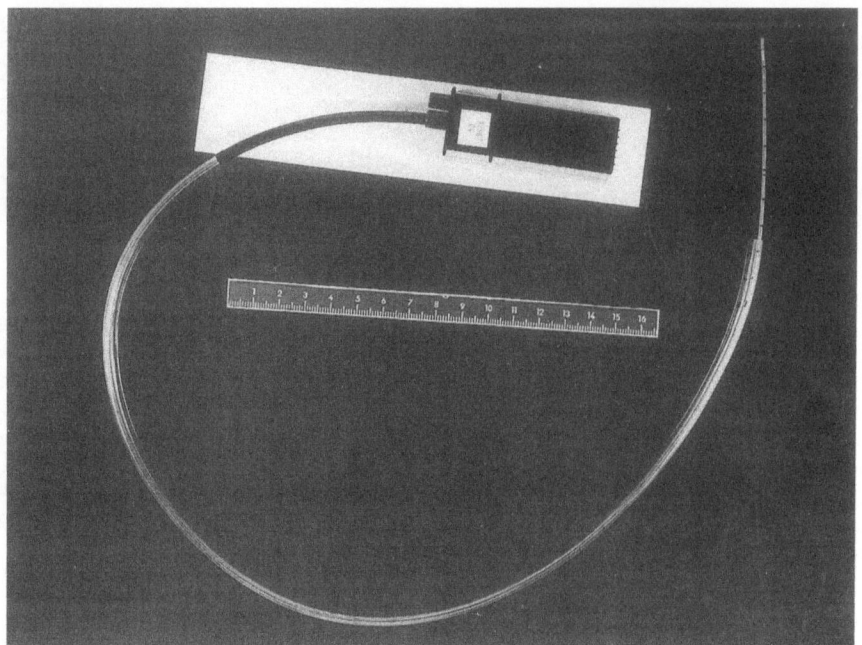

The implantation of the LICOX P(ti)O$_2$ probe is similar to the Camino implantation; a burr hole is placed at Kocher's point. A second screw (steel or titanium) has to be placed where the LICOX probe is implanted. The measuring tips from both probes are placed into the white matter near the border to the grey matter.

The normal values of P(ti)O$_2$ in the human brain were measured by Meixensberger [28] in the normal and pathological cortex.

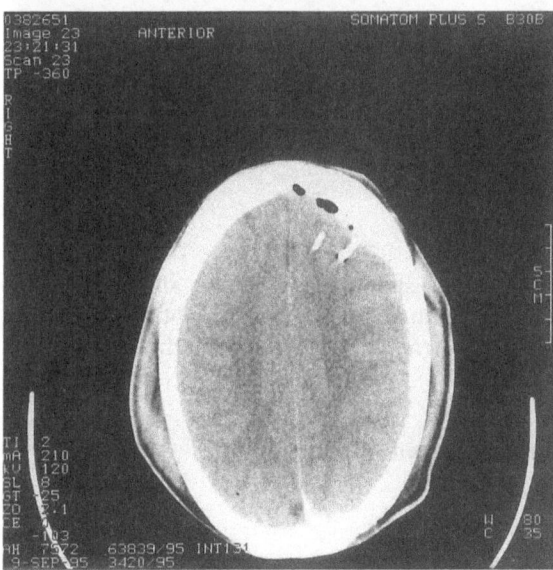

Fig. 4. Computed tomography scan during measurement. Severe closed brain injury, traumatic SAH, inserted Licox and Camino probes into the frontal lobe

The value of P(ti)O$_2$ during intracranial hypertension is between 20 and 30 mmHg. In normal human cortex the values of oxygen partial pressure are from 30–40 mmHg. Values less than 10–15 mmHg are the critical borderline for hypoxia and can be treated either by hemodynamic methods (catecholamine therapy) or increasing FiO$_2$.

Paratrend Monitoring

The Paratrend-7 (P7) probe monitors to parameter pH, PCO$_2$, PO$_2$, and temperature in the white matter. The ph sensor is a single miniature fiberoptic filament. It works on the principle of optical absorption. A section of the filament contains a red dye that is sensitive to ph. The PCO$_2$ sensor operates on the same principle as the ph sensor. It is a single, mirror-tipped filament containing the red dye dissolved in a bicarbonate buffer. As the CO$_2$ diffuses into the sensor through a gas-permeable membrane, the CO$_2$ generates H$^+$ from the buffer, and the dye color changes. The P7 sensor detects PaO$_2$ using a miniature version of the classic Clark electrode, the same type of electrode used in a blood gas analyzer. The P7 sensor contains a miniature thermocouple, consisting of two thin metal wires that read the actual temperature of the blood inside the artery. The P7 monitor corrects all blood gas readings to 37 °C for easy comparison with standard arterial blood gas (ABG) values. With the touch of a button, the P7 will display blood gas values at patient temperature. All of the P7 sensors respond with a slight delay, except the temperature sensor, which responds instantly to any change.

Fig. 5.
Licox probe (Clarke probe)

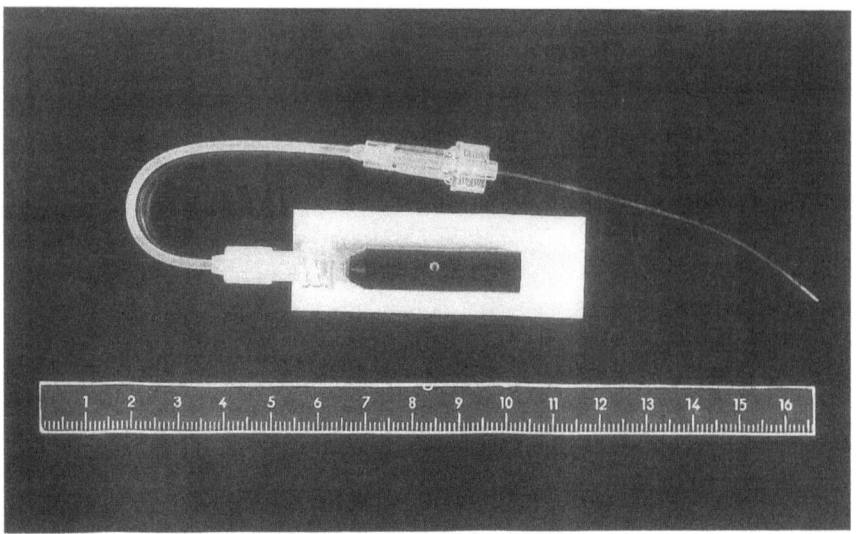

Fig. 6.
Paratrend probe

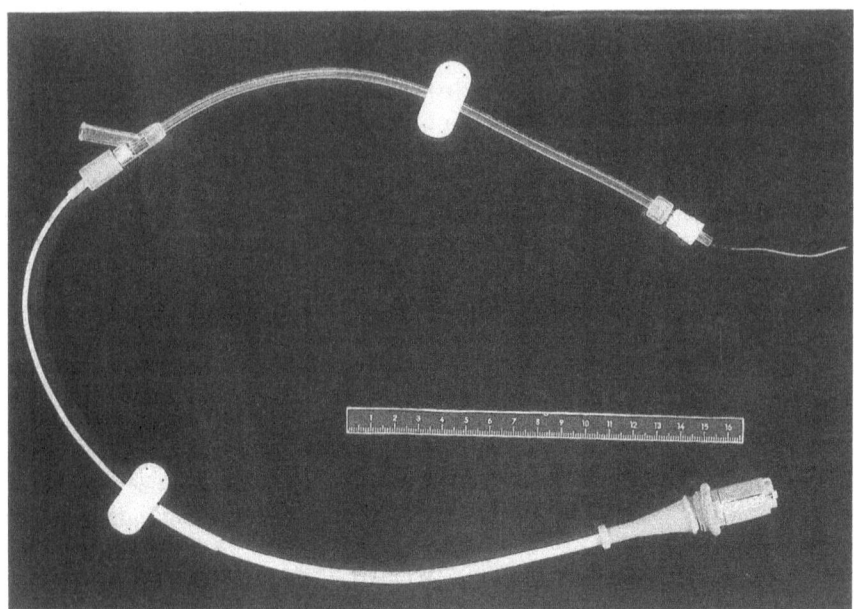

Originally, this sensor was designed for in vivo blood gas analysis. We use this sensor, like the LICOX oximetric probe, in the white matter of the frontal lobe.

CPP Calculation

CPP is a calculated parameter, mathematically defined as the following equation: MABP – ICP = CPP. It is easy to calculate this parameter, either on a PC or within a computerized system, as we use it in the neuromonitor (Zeppelin, Munich, Germany). The

value of CPP should be 60 mmHg and the value of MABP 90 mmHg to guarantee an optimal perfusion during intracranial hypertension periods.

Monitoring and Treatment

Monitoring

All patients are monitored by invasive blood pressure measurement. A central venous catheter (v. subclavia) should be used.

The following circulatory parameters are monitored:

1. ECG (without on-line registration)
2. Systolic/mean/diastolic arterial blood pressure
3. Temperature (via bladder drainage)
4. End-expiratory CO_2
5. ICP
6. CPP
7. $P(ti)O_2$

Parameters 2–7 are served on-line into the neuromonitor system.

8. Continuous EEG (during pharmacological paralysis and barbiturate coma)

In addition, the following laboratory examinations are performed:

9. Blood gas analysis
10. Electrolyte status
11. Serum osmolality
12. Blood analysis

Additional parameters monitored include:

- $P(ti)CO_2$
- brph
- Brain temperature

Treatment

Intracranial hypertension is a medical emergency situation. the treatment of ICP that increases above 20 mmHg has to be consistent. The dosis of drugs, recommendation of hyperventilation levels, and other principles of treatment have to be focused on the following target points:

1. Decreasing ICP
2. Increasing CPP
3. Elevation of MABP to levels >90 mmHg
4. Optimization of $P(ti)O_2$ to levels >25 mmHg

Therapy starts after accurate diagnosis. Clinical status, a Glasgow coma scale assessment, and circulatory status have to be determined immediately and a CT examination given. At the same time treatment of circulatory shock and controlled ventilation must start. In a situation where no correct diagnosis can be made, i.e., when no CT examination is possible, but high ICP levels are suspected, therapy should be begun even without diagnosis.

According to the "Guidelines of severe brain injuries", the following strategies for treating elevated ICP are in practice:

1. Oxygenation and hydration
2. Moderate hyperventilation (PCO_2 30–35 mmHg)

3. Head elevation and correct position
4. Sedation and pharmacological paralysis
5. Osmotic therapy (mannitol, sorbitol)
6. Barbiturates (thiopental, methohexital)

It is the experience of the authors not to use non-osmotic diuretics in treatment of ICP, because disturbances in the homeostasis of water and electrolyte balance have to be avoided. Corticosteroids should also not used in the treatment of raised ICP.

Oxygenation, Hyperventilation and Hydration

Severely brain injured patients with raised ICP are extremely vulnerable to hypoxic and hypodynamic events. All patients are generally intubated and given controlled ventilation with a PO_2 target above 100 mmHg ($F(i)O_2$ <0.4). Moderate hyperventilation starts immediately with PCO_2 levels of 30–35 mmHg. Any hypotension and shock symptoms have to be treated as strong as possible with volume (fluid balance ± 0 ml), blood or blood components, and catecholamines. Therapeutic target is a MABP of 90 mmHg to achieve an optimal CPP of 60 mmHg.

Fig. 7. The neuromonitoring system

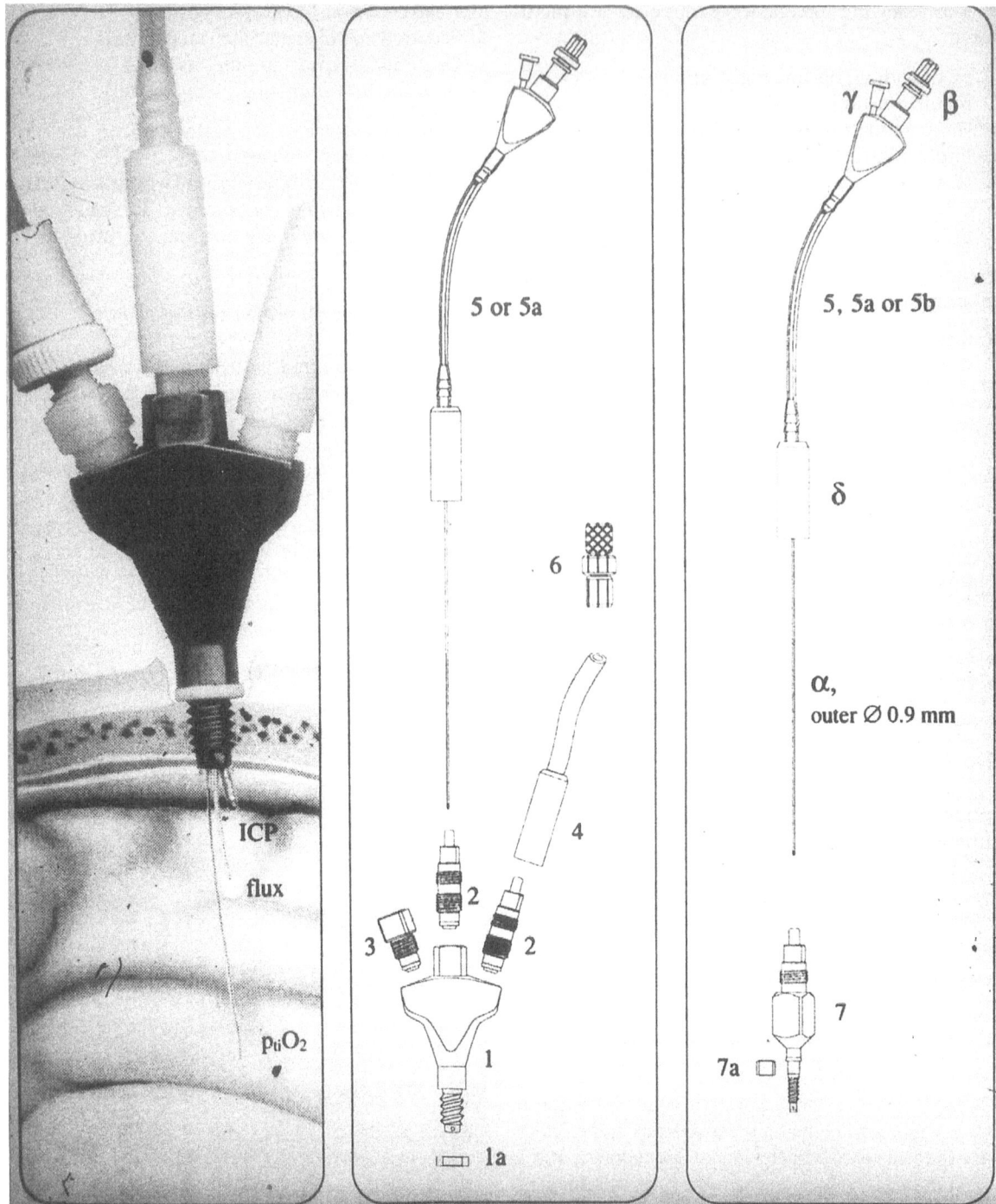

Fig. 8. The Licox and Camino probe inserted into the white matter

Head Elevation and Position

A correct positioning of the head, midline and elevated 30°, is useful and necessary. During head rotation ICP increases up to 10 mmHg and the same effect can be observed with head elevation of less than 30°.

Sedation and Pharmacological Paralysis

To control pain, sympathetic stress, and agitation, patients are sedated, usually at first with benzodiazepines or barbiturates, if ICP control is not possible. Pharmacological paralysis must be used in all situations involving problems with controlled ven-

Fig. 9.
"Normal" reaction in pre-oxygenation procedures. Observed in all recording situations with expected intact autoregulation functions. This typical spike was not visible in situations with perfusion stop or with intractable edema with extremely high ICP levels

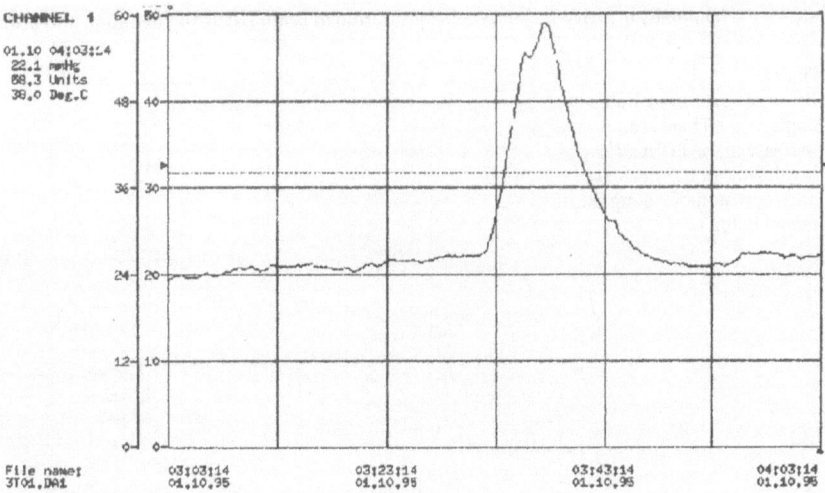

Fig. 10.
Effect of forced and moderate hyperventilation. There is an increase of oxygen partial pressure changing from forced to moderate hyperventilation (PCO$_2$ 30 – 35 mmHg)

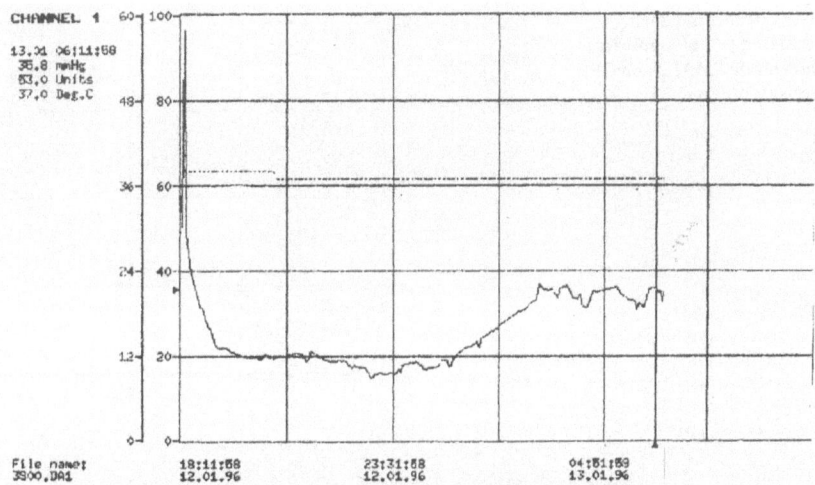

Fig. 11.
Effect of application of sorbitol bolus. Increase of P(ti)O$_2$ is for a relatively short period with a visible return of values to original values

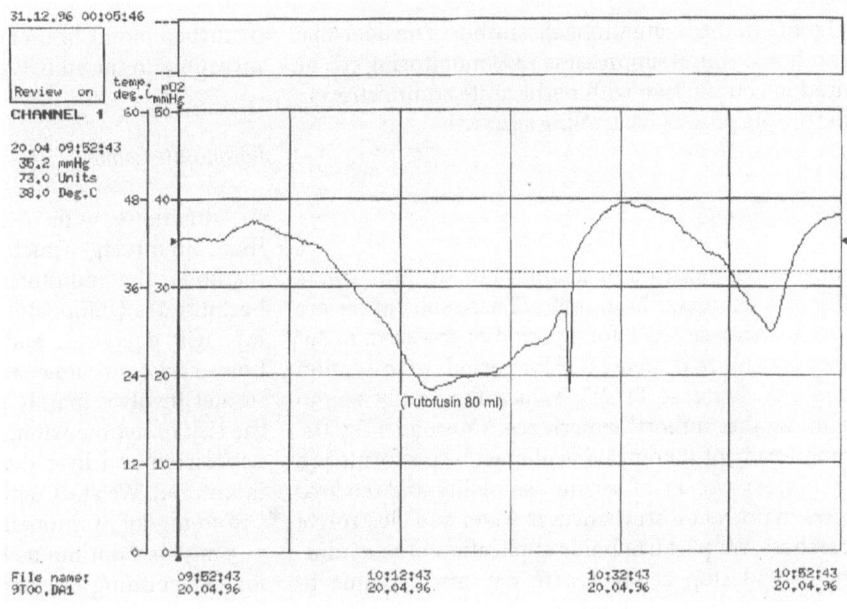

Fig. 12.
Effect of barbiturate bolus application. There is an increase in the P(ti)O$_2$ recording during an ICP decrease and perfusion of 150 mg thiopental bolus

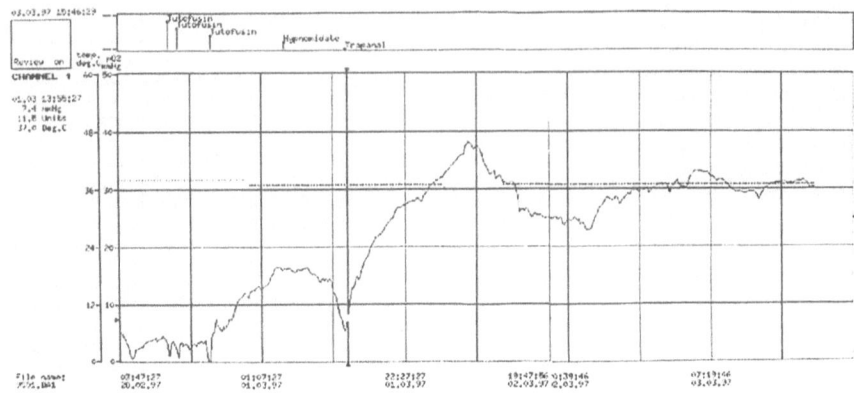

Fig. 13.
P(ti)O$_2$ recording during decreased (< 60/40 mmHg) and elevated MABP and CPP levels (> 90/60 mmHg)

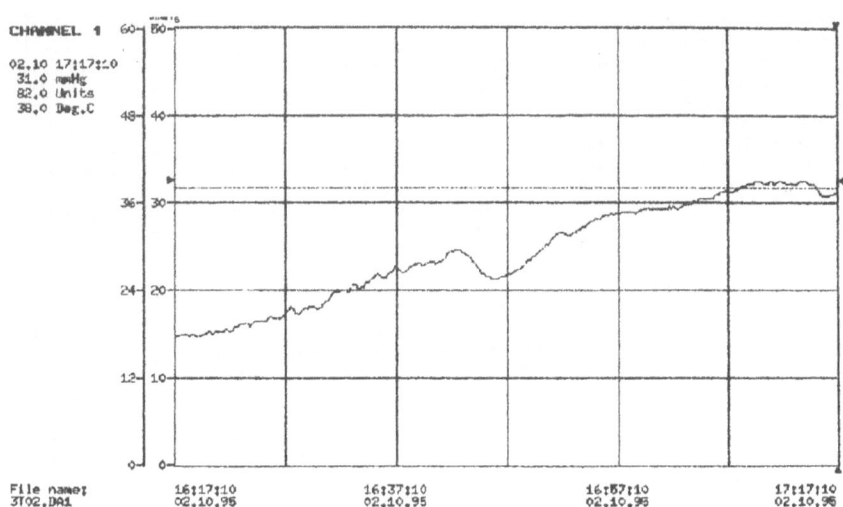

tilation. In these situations the authors consider that continuous burst suppression EEG monitoring can be used in conjunction with barbiturate administration for the purpose of controlling seizures.

Osmotic Therapy

Intravenous infusion of mannitol or sorbitol lowers ICP and decreases brain mass. These substances are able to increase SABP for a period of 30–40 min. But they are able to decrease ICP for periods of 10–15 mm and also increase P(ti)O$_2$ values for nearly 20–30 min, in the authors' experience. Consequently, the monitoring of therapy control must be performed in laboratory checks of serum osmolality and electrolytes to prevent disturbances in water and electrolyte balances. We perform bolus application of osmodiuretics and stop at the fourth day after trauma to prevent rebound phenomena with increasing ICP in

disturbed blood-brain areas. These effects have been measured in raised ICP and decreased P(ti)O$_2$ levels.

Barbiturate Coma

In situations with extremely raised ICP larger than 40 mmHg which are intractable by other methods, the monitoring effects under high-dose barbiturates (thiopental or methohexital) are amazing. ICP decreases and P(ti)CO$_2$ increases during bolus and continuous application. The use of barbiturates involves mainly potential complications, and the risk of hypotension, infarction, infection, cardiac depression and liver dysfunction have to be closely monitored. We start with a bolus-loading dose from 5 to 10 mg/kg of thiopental and a continuous drip of 3–5 mg/kg. Continuous EEG monitoring is necessary for determining the optimal therapy level of burst suppression EEG.

Fig. 14.
P(ti)O$_2$ recording during cerebral perfusion stop – before clinical herniation signs are visible

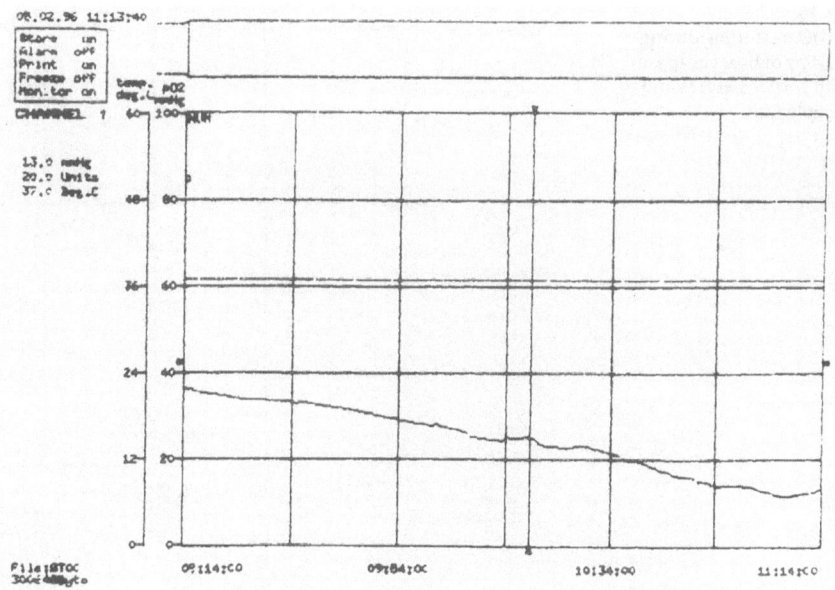

Fig. 15.
P(ti)O$_2$ recording of cerebral hypoxia in case of intractable ICP and cerebral perfusion stop

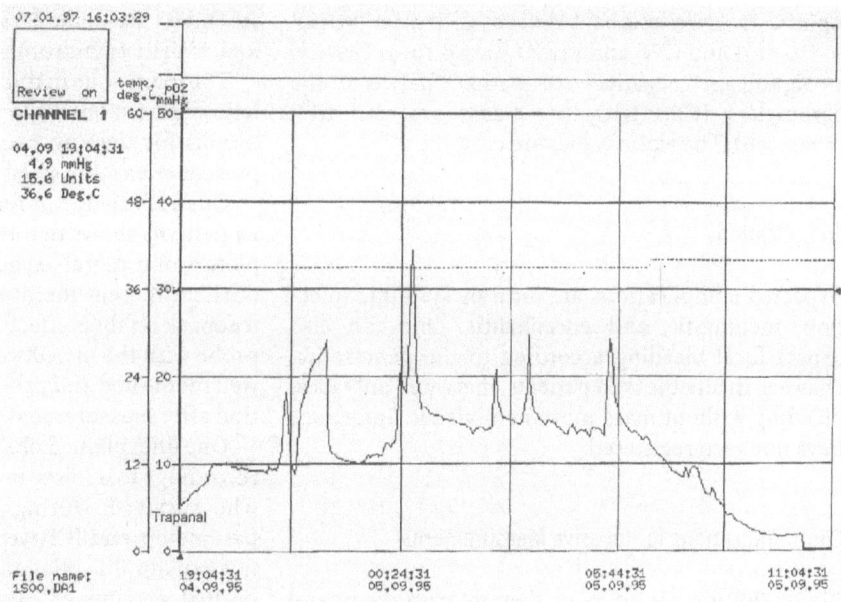

Results

Effects on Increasing and Decreasing P(ti)O$_2$ Levels

These levels can be illustrated by the original data recording diagram from the LICOX computer. In all illustrated situations we expected an intact or approximately intact autoregulatory function of brain vessels. Two patients who presented "fixed" and unmodulatable oxygen recordings died immediately after trauma.

In all patients data recording of oxygen partial pressure declines to lower, but stable levels (5–8 mmHg) after a 3 to 4-day recording period. In this time all other measured data (CPP, MABP, ICP) during therapy were at compensated levels. This might be a system-intrinsic effect, because the working of the probe tip is adapted to a diffusion process. It may be that blood or cell damage in the implantation region causes in correct oxygen diffusion. Precise reasons cannot be given at this time.

Fig. 16.
P(ti)O$_2$ recording during
therapy of bronchospasm
with corticosteroids and
theophylline

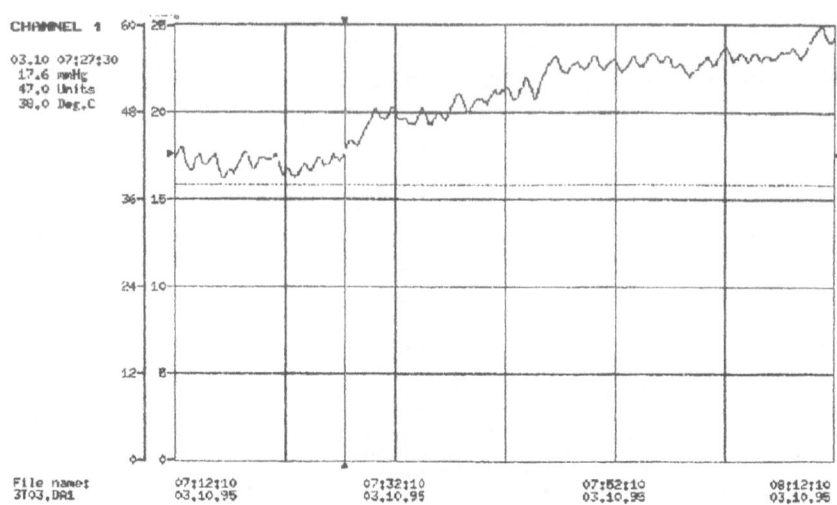

Preliminary Statistic Management

We found during secondary data management significant correlations between CPP and MABP ($r = 0.710$) and CPP and P(ti)O$_2$ ($r = 0.190$). There is a significant negative correlation between the parameters ICP/P(ti)O$_2$ ($r = 0.228$) and ICP/CPP ($r = 0.596$). The alpha values are 0.05.

Complications

Expected complications are local or systemic infection, meningitis, and encephalitis. One can also expect local bleeding according to the penetration channel. In all observed patients, there was only local bleeding without mass movement visible. Infections have not been registered.

Contraindications for Invasive Measurements

Blood clotting disorders or therapy with dicumarol are contraindications for invasive monitoring systems. Low-dose heparin therapy is no contraindication. All observed patients received low-dose heparin during their ICU period by continuous drip of 420 IU/h.

Discussion

The pathophysiological development which occur after severe brain injury causes difficulties and medical problems caused various intracerebral changes. The brain trauma itself can become critical. In 15 % of the cases brain trauma is combined with other traumatic lesions, such as thorax trauma. The critical care of these patients is a medical challenge. The goal is either to control raised ICP and to prevent secondary ischemic lesions of the brain and neurological defect syndromes or not.

The main issue in multimodality neuromonitoring with different observed parameters gives better conditions for therapy, because a lot of information is present at an early stage of development.

Our experience in long time neuromonitoring in 24 patients shows that P(ti)O$_2$ monitoring provides a picture of cerebral oxygenation and microcirculation during different therapeutical procedures and gives feedback on their effects. It is difficult to calibrate the probe with the LICOX computer, but it is, in this way, well monitored and provides reliable data (Recalibration after measurement gives a drift of ± 1.5 mmHg).

One unexplained observation is the shift in P(ti)O$_2$ recordings to a lower level after 3 days in all patients who survived. During this time all hemodynamic parameters and ICP were at good levels so they cannot explain this phenomenon. One explanation may be that detritus or minimal bleeding at the probe point diminishes the diffusion process after a few days.

Tissue oxygen measurement provides an indirect monitor of the autoregulatory functions of brain vessels, given by the "hyperoxygenation curve" during oxygenation procedures with FiO$_2$ of 1.0. In initial monitoring of patients who died during observation we recorded quite different developments of tissue oxygen reaction. The initial values are in hypoxic state, i. e., less than 15 mmHg. There is no reaction to hyperoxygenation or hyperventilation maneuvers. We expect an already disturbed autoregulatory function in these situations. The only effect we observed was a dramatical decrease of P(ti)O$_2$ during cerebral perfusion stop, before clinical signs of herniation, e.g., wide pupils, occurred. These observations have been confirmed with transcranial Doppler sonography and perfusion scintigram. We will have to discuss in future whether tissue oxygenation monitoring is a diagnostic tool in the diagnosis of brain death.

A very impressive discovery is the observation of a dramatic increase of $P(ti)O_2$ in situations with circulatory treatment. The parameters of middle arterial blood pressure (MABP) and cerebral perfusion pressure (CPP) are a subject of interest in this context. Theories about the danger of arterial hypotension in brain injuries [29–31] are confirmed by these data recordings. So-called "hemodynamic treatment" in situations of raised ICP has become a new tool. Circulatory stabilization is realized by catecholamine therapy and fluid balance. The central parameters are MABP and CPP, which have to be calculated by software. Volume therapy is regulated in terms of central venous pressure (CVP). Problems will occur in patients suffering from heart insufficiency. The catecholamine regime has to be observed by a Swan-Ganz catheter.

The effects of strong and moderate hyperventilation as one therapeutical tool to decrease raised ICP is also impressive. The modulation of hyperventilation can now be performed without the risk of cerebral hypoxia, using tissue PO_2 monitoring. We have seen the critical borderline when PCO_2 drops below 28 mmHg: here, a dramatic decrease in $P(ti)O_2$ of 10 mmHg and more immediately ensures.

We confirm the effects of barbiturates and osmodiuretics during therapy in increasing $(P(ti)O_2$ and decreasing ICP levels. It is of interest that the raised level of tissue oxygen after bolus application of e.g. sorbitol is restricted to a short period of 20–30 min (Fig. 11). A similar effect is to be described after bolus injection of e.g., thiopental or methohexital (Fig. 12). After the initial raising effect on tissue PO_2, values return to their initial position. Another interesting observation is the same but shorter effect that etomidate has in bolus or continuous-drip application.

The significance of the patient's head position has to be noticed. In situations with head rotation to one side or in a flat position without 30° elevation an increase in ICP can be observed as well as a decrease in tissue oxygenation after a period of more than 20 min. Venous backing-up causes brain swelling and intracranial hypertension occurs.

The patients who died (3/20 with SBI, 6/8 with SAH IV°/V°) suffered from severe intractable brain edema. In the surviving group there were no patients with apallic syndrome. Eleven patients with severe brain injury survived without a defect syndrome and three patients survived with a neurological defect syndrome. In the SAH group six patients died and one patient survived in a state of rehabilitation. This high mortality is easily explained by the low initial Hunt and Hess score. In future we will focus on low-grade SAH in order to control hypoxic danger following vasospasm.

The previous statistical calculations give clear information. We found significant correlations between CPP and MABP ($r = 0.710$) and CPP/$P(ti)O_2$ ($r = 0.190$). There is a significant negative correlation between the parameters ICP/$P(ti)O_2$ ($r = 0.228$) and ICP/CPP ($r = 0.596$) The alpha values have been 0.05.

The question whether it is possible to improve the outcome, i.e. morbidity and mortality, of brain-injured patients undergoing multimodal neuro-monitoring methods cannot be answered at this time. A follow-up study will give a report on results in morbidity and quality of life 6 months after trauma in the patients who survived. All brain injured patients who will undergo this study have been treated in a special neuroreabilitation center after leaving our ICU.

Conclusions

1. The first and eminent target of monitoring $P(ti)O_2$ [$P(ti)CO_2$, brph, br temp] in correlation with ICP and the hemodynamic parameters MABP and CPP, however, is to gain experience with the pathophysiological events during brain trauma and the development of intracranial hypertension and to use this experience in order to prevent secondary tissue damage.

2. The second target is to provide support, if provable effects are detected in therapeutic measure that aim to lower ICP and to raise $P(ti)O_2$. Here we can provide a wealth of experience in reference to expected pathophysiological changes during brain injury. By comparing them to other situations, the physician is able to make out dangerous situations during continuous neuromonitoring. There are possibilities to react to these developments in therapy before the patient suffers from a bad outcome caused by a neurological defect syndrome.

References

1. Barin ES et al. (1987) Physical characteristics and clinical evaluation of a new disposable fiberoptic transducer-tipped catheter system. Anaesth Intens Care 15 : 323–329
2. Becker DP, Miller JD et al. (1977) The outcome form severe head injury with early diagnosis and intensive management. J Neurosurg 47 : 491–502

3. Bouma GJ, Muizelaar JP (1995) Cerebral blood flow in severe clinical head injury. New Horiz 3 (3): 384–394
4. Bullock R (1995) Mannitol and other diuretics in severe neurotrauma. New Horiz 3 (3): 448–452
5. Chambers IR et al. (1993) An evaluation of the Camino system in clinical practice. Neurosurgery 33 (5): 866–868
6. Chambers IR et al. (1990) A clinical evaluation of the Camino subdural screw and ventricular monitoring kits. Neurosurgery 26 (3): 421–423
7. Chesnut RM (1995) Secondary brain insults after head injury. Clinical perspectives. New Horiz 3 (3): 366–375
8. Chesnut RM (1995) Medical management of severe head injury: present and future. New Horiz 3 (3): 581–593
9. Chesnut RM (1996) The guidelines in severe brain injuries. J Neurotrauma (Dec Suppl)
10. Cold GE (1989) Does acute hyperventilation provoke cerebral oligaemia in comatose patients after acute head injury? Acta Neurochir 96: 100–106
11. Crutchfield JS et al. (1990) Evaluation of a fiberoptic intracranial pressure monitor. J Neurosurg 72: 482–487
12. Cruz J, Miner ME (1990) Continuous monitoring of cerebral oxygenation in acute brain injury injection of mannitol during hyperventilation. J Neurosurg 73: 725–730
13. Ghajar J (1995) Intracranial pressure monitoring techniques. New Horiz (3): 395–399
14. Gambardella G et al. (1992) Monitoring of brain tissue pressure with a fiberoptic device. Neurosurgery 31 (5): 918–922
15. Genneralli TA, Spielman GM et al. (1982) Influence of the type of intracranial lesion on outcome from severe head injury. J Neurosurg 56: 26–32
16. Gotoh F, Tanaka K (1988) Regulation of cerebral blood flow. In: Handbook of clinical Neurology. Elsevier, Amsterdam, pp 47–77
17. Graham DI, Hume Adams J (1971) Ischaemic brain damage in fatal head injuries. Lancet I: 265–266
18. Hollingsworth-Fridlund P et al. (1988) Evaluation of new technologies in critical care: use of fiber-optic pressure transducer for intracranial pressure measurements: a preliminary report. Heart Lung 17 (2): 111–120
19. Jennett B, Bond M (1975) Assessment of outcome after severe brain damage. Lancet 1: 480–484
20. Kuhn TJ, Bauer BL (1996) Multimodal neuromonitoring based on CPP and P(ti)O$_2$. Intensive Care Med [Suppl]: 687–691
21. Lang EW, Chesnut RM (1995) Intracranial pressure and cerebral perfusion pressure in severe head injury. New Horiz 3 (3): 400–414
22. Langfitt TW, Obrist WD (1977) Correlation of cerebral blood flow with outcome in head injured patients. Ann Surg 186 (4): 411–414
23. Luerssen TG (1990) Fiberoptic intraparenchymal pressure monitoring for the measurement of intracranial pressure in children. Concepts Pediatr Neurosurg (Basel) 10: 204–213
24. Maas AIR, Fleckenstein W, de Jong DA, van Santbrink H (1993) Monitoring cerebral oxygenation: experimental studies and preliminary clinical results of continuous monitoring of cerebrospinal fluid and brain tissue oxygen tension. Acta Neurochir Suppl (Wien) 59: 50–57

25. Marion DW et al. (1995) Hyperventilation therapy for severe traumatic brain injury. New Horiz 3 (3): 439–447
26. Marshall LF et al. (1979) The outcome with aggressive treatment in severe head injuries. I. The significance of intracranial pressure monitoring. J Neurosurg 50: 20–25
27. Marshall LF et al. (1991) A new classification of head injury based on computerized tomography. J Neurosurg 75: 14–20
28. Meixensberger J, Dings J (1993) Studies of tissue PO$_2$ in normal and pathological human brain cortex. Acta Neurochir Suppl (Wien) 59: 58–63
29. Miller JD (1992) Evaluation and treatment of head injury in adults. Neurosurg Q 2 (1): 28–43
30. Miller JD (1986) Minor, moderate and severe head injury. Neurosurg Rev 9: 135–139
31. Miller, JD, Becker DP et al. (1977) Significance of intracranial hypertension in severe head injury. J Neurosurg 47: 503–516
32. Muizelaar JP (1984) Effect of mannitol on ICP and CBF and correlation with pressure autoregulation in severely head-injured patients. J Neurosurg 61: 700–706
33. Muizelaar JP, Marmarou A et al. (1993) Improving the outcome of severe head injury with the oxygen radical scavenger polyethylene glycol-conjugated superoxide dismutase: a phase II trial. J Neurosurg 78: 375–382
34. Piper IR et al. (1990) Systems analysis of cerebrovascular pressure transmission: an observational study in head-injured patients. J Neurosurg 73: 871–880
35. Prielipp RC, Coursin DB (1995) Sedative and neuromuscular blocking drug use in critically ill patients with head injuries. New Horiz 3 (3): 456–468
36. Robertson CS et al. (1989) Cerebral arteriovenous oxygen difference as an estimate of cerebral blood flow in comatose patients. J Neurosurg 70: 222–230
37. Sheinberg M et al. (1992) Continuous monitoring of jugular venous oxygen saturation in head-injured patients. J Neurosurg 76: 212–217
38. Sutton LN et al. (1990) Cerebral venous oxygen content as a measure of brain energy metabolism with increased intracranial pressure and hyperventilation. J Neurosurg 73: 927–932
39. Teasdale G, Jennett B (1974) Assessment of coma and impaired consciousness. Lancet 2: 81–83
40. Tonnesen AS (1995) Hemodynamic management of brain-injured patients. New Horiz 3 (3): 499–505
41. Unterberg A, Kiening K et al. (1993) Long term observations of intracranial pressure after severe head injury. The phenomenon of secondary rise of intracranial pressure. Neurosurgery 32 (1): 17–24
42. Wahl M, Schilling Unterburg A, Baethmann A (1993) Mediators of vascular and parenchymal mechanisms in secondary brain damage. Acta Neurochir Suppl (Wien) 57: 64–72
43. Wilberger JE, Cantella D (1995) High-dose barbiturates for intracranial pressure control. New Horiz 3 (3): 469–473

Part 2
Neuroendoscopic Treatment of Hydrocephalus

Endoscopic Third Ventriculostomy in Treatment of Obstructive Hydrocephalus Caused by Primary Aqueductal Stenosis

D. Hellwig, A. Heinemann, and T. Riegel

Summary

Primary aqueductal stenosis is a rare cause of obstructive hydrocephalus. Etiology of this lesion remains unclear till today. Possible factors causing primary aqueductal stenosis such as genetic disorders, infectious disease or vitamin shortage during pregnancy are described. A total of 22 patients with primary aqueductal stenosis are presented, 8 of whom are treated by CSF shunt placement and 14 with an endoscopic third ventriculostomy. The efficacy of third ventriculostomy compared to conventional CSF diversion techniques in treatment of chronic obstructive hydrocephalus is discussed.

Introduction

The Sylvian aqueduct is an anatomical notch of CSF pathways. It has an average length of 1.1 cm and a diameter of 0.15–0.8 cm [27]. The aqueduct can be divided into three segments:

- The upper third below the epithalamic commissure has a triangle-shaped lumen.
- The middle segment near to the upper colliculi is ampullate.
- The caudal segment is U-shaped.

For a long time stenosis or occlusion of the aqueduct had been a diagnostic and therapeutic problem. The diagnosis "aqueductal stenosis" was often made by the pathologist. Prenatal diagnosis of aqueductal stenosis carries a grave prognosis, with an overall mortality of 40% and normal development in only 10% of cases [28].

In 1936 Stookey and Scarff [42] stated that up to present there is no uniform clinical syndrome, which is characteristic for this disease, and a suitable operative method is missing. Up to now the etiology of benign aqueductal stenosis in adolescence is an unsolved problem and an adequate standardized therapeutic regimen does not exist [7]. There are various theories of origin:

1. In X-chromosomal transmitted aqueductal stenosis, occurrence is partly associated with thumb anomalies. The lumen of the aqueduct is narrowed, slit-shaped, and covered with ependyma [3, 33, 43, 48].
2. Neonatal or infantile meningoencephalitis could be due for aqueductal stenosis. In experimental studies mumps virus or *Toxoplasma gondii* infections effect a stenosis of the aqueduct of Silvius. The prenatal inflammation leads to a destruction of ependyma cells and consequently to aqueductal stenosis [21].
3. In animal experiments vitamin A, vitamin B_{12}, and folic acid shortage during pregnancy can lead to aqueductal stenosis [34, 47].

There are many problems in diagnosis and therapy of primary aqueductal stenosis. Uneventful pregnancy, delivery, and postnatal development are differential diagnostic presuppositions. No hints of meningitis, brain injury, intracerebral bleedings, or neoplasia are present. The membraneous occlusion of the aqueduct is a rare cause of hydrocephalus [48]. In most cases first manifestation of PAS is during the second decade of life [7].

Common clinical symptoms of primary aqueductal stenosis (PAS) due to obstructive hydrocephalus are headache, gait disturbance, vertigo, ataxia, and cognitive dysfunction. Papilledema

Table 1.
X-chromosomal transmitted aqueductal stenosis

Only male patients affected; women transmitters
Familiarly frequency partly associated with thumb anomalies, mental retardation, spasticity
Lumen of the aqueduct is narrowed, slit-shaped, and covered with ependyma
No ependyma channels as seen in septated aqueduct

Table 2.
Symptomatology of hydrocephalus due to primary aqueductal stenosis

Headache
Vertigo
Ataxia
Tremor
Gait disturbances
Papilledema (50%)
Cognitive dysfunctions
Psychomotoric changes
Dysendocrine symptoms
Seizures (10%)

(50%), seizures (10%), tremor, dysendocrine symptoms, and psychomotoric changes occur frequently (Table 2).

Radiological diagnosis of PAS can be established by sagittal MRI combined with ECG-gated CSF flow sequences and intraoperative digital dynamic subtraction ventriculography or ventriculoscopy. Indirect methods are cranial computed tomography (CCT) or skull X-ray, which show the signs of chronic obstructive hydrocephalus.

Conventional treatment of chronic occlusive-hydrocephalus due to PAS is the application of a CSF shunt. The main problem with this therapy is the acute intervention in a balanced system, which has been established over a long period. Many complications, such as overdrainage, slit ventricle syndrome [8, 19] or subdural fluid collections, are common after shunting procedures.

In obstructive hydrocephalus the resorptive capacity for CSF is intact. Consequently, so-called "inner shunting" methods can be applied. Jason Mixter was the first one to perform a third ventriculostomy in 1923, he did so by perforation the floor of the third ventricle to the prepontine cistern [30]. In following years many attempts had been made to establish third ventriculostomy as the standard treatment of obstructive hydrocephalus [6, 11, 16, 17, 20, 38, 40, 45]. However, most of them were bound to fail due to unsuitable instruments.

With the availability of highly sophisticated neuroendoscopic instruments and technologies, third ventriculostomy is now witnessing renaissance in treatment of obstructive hydrocephalus, especially in cases of PAS.

Patients and Methods

Patients

In a retrospective study 22 patients suffering from obstructive hydrocephalus caused by PAS were examined. There were 10 female and 12 male patients. The average age was 22 years. Diagnosis of aqueductal stenosis was established by CCT or MRI examination with sagittal planes and shows a widespread morphological variation of aqueductal stenosis. The patients had been operated on between January 1988 and May 1997. In 8 patients (up until March 1993) a shunting procedure had been performed. A total of 14 patients (after March 1993) were treated by endoscopic third ventriculostomy.

Fig. 1.
Flexible endoscope fixed in the Marburg neuroendoscopy fixation and guiding device. In the background, stenoscope for digital dynamic subtraction ventriculography

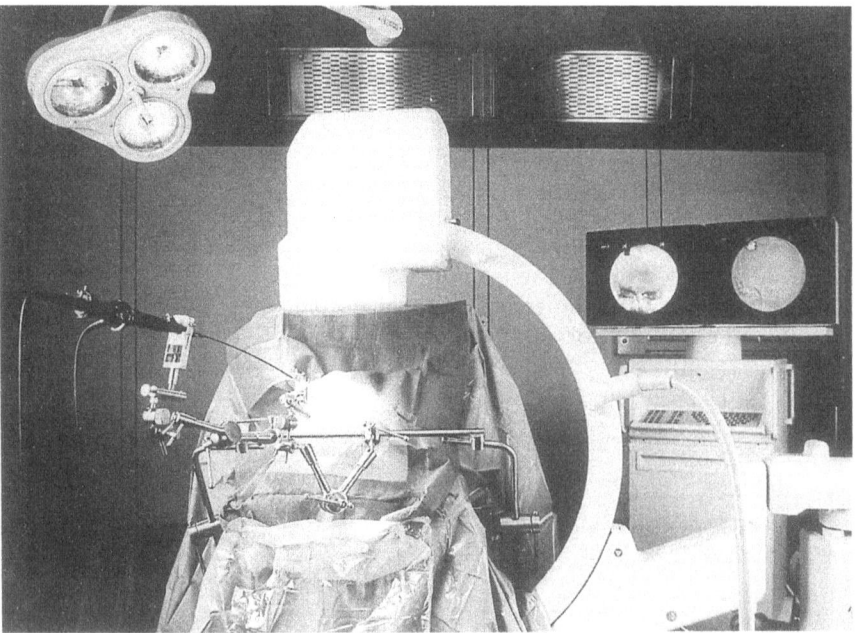

Operative Technique

In combination with the Marburg endoscopy fixation and guiding device, a flexible endoscope is introduced through a frontal burr hole (Fig. 1). The foramen of Monro is identified by the surrounding choroid plexus, and the typical Y-shape of the septal and thalamostriate vein. After reaching the third ventricle, the mamillary bodies serve as a landmark for the positioning of the third ventriculostomy. The localization for third ventriculostomy is between the dorsum sellae and the mamillary bodies and away from the bifurcation of the basilar artery (Fig. 2). There are different techniques for perforating the floor of the third ventricle [5, 22, 38]. The easiest way is to use a rigid endoscope for blunt opening of the thin floor. This technique has the danger of vascular injury in the prepontine cistern. Other methods are vaporization with bare laser fibers in a contact more or coagulation with ultrathin monopolar or bipolar probes [15, 29]. Once the floor of the third ventricle is perforated, one can recognize the flow of CSF. During this procedure, touching of the basilar artery or one of its branches must be avoided. Widening the artificial porus is also necessary either with a Fogarty catheter or a double balloon catheter (Fig. 3a, b). Intraoperatively, patency of the ventriculostomy is proven by digital subtraction ventriculography (Fig. 4). The postoperative CSF flow dynamics are studied by ECG-gated dynamic MRI examination (Fig. 5).

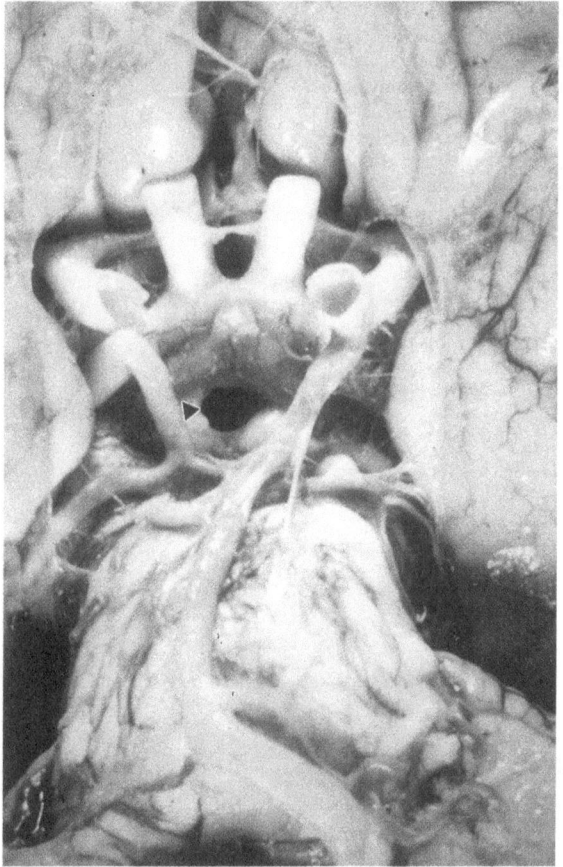

Fig. 2. Localization of third ventriculostomy in front of the mamillary bodies *(arrow)* (view from caudal)

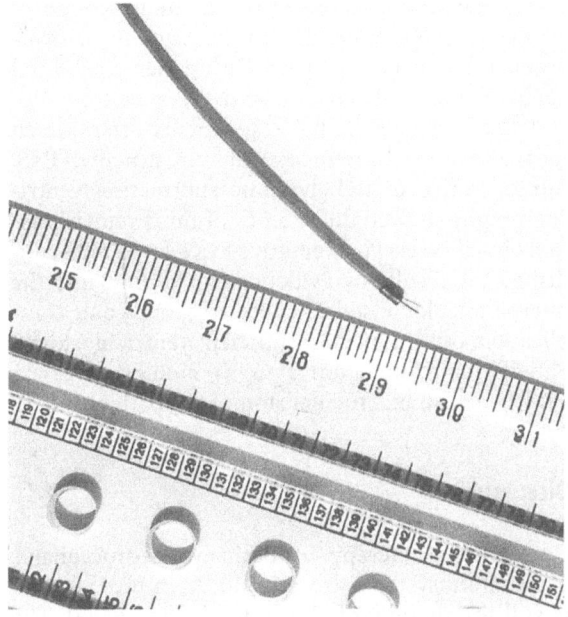

a

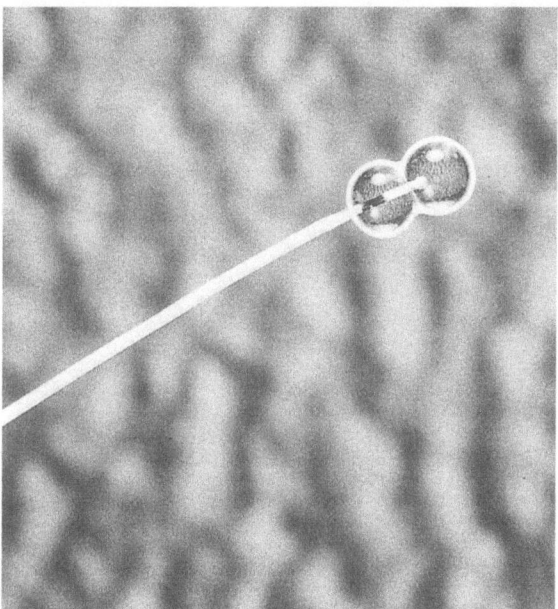

b

Fig. 3. a Flexible microbipolar electrocoagulation probe (Erbe Co.). **b** Double balloon catheter (Cordis Co.) for widening the ventriculostoma

Fig. 4.
Intraoperative digital
dynamic subtraction ven-
triculography demonstrates
the aqueductal stenosis and
the CSF flow into the interpeduncular cistern after third
ventriculostomy

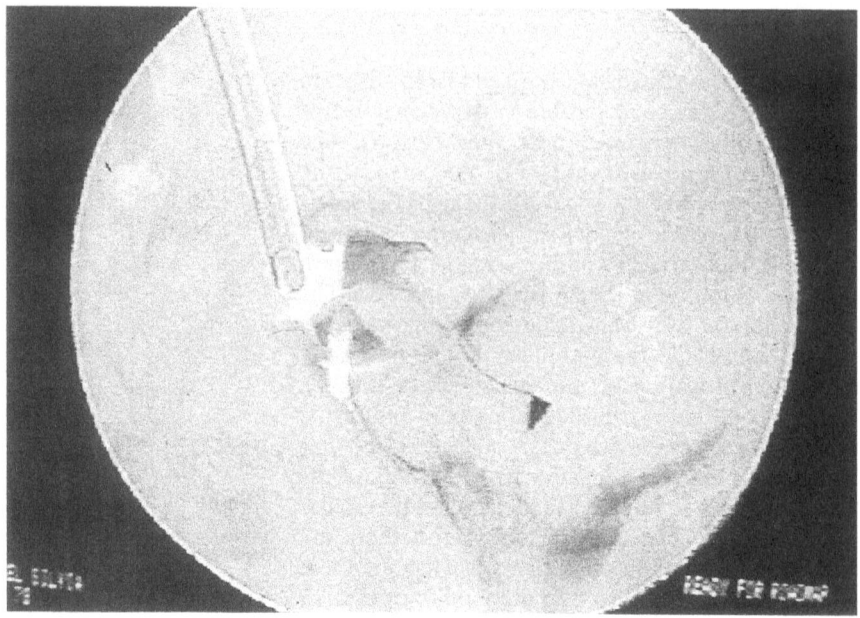

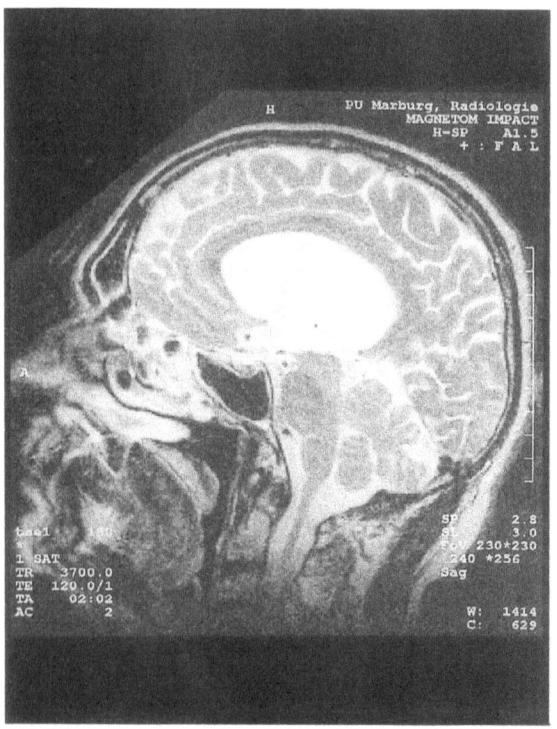

Fig. 5. Postoperative ECG-gated MRI shows patency of third
ventriculostomy

Results

When admitted to hospital all patients showed signs
of raised intracranial pressure (ICP) caused by
obstructive hydrocephalus. MRI investigations revealed different types of aqueductal stenosis (Fig.
6 a–d) in typical localization.

After application of CSF shunts (8 patients until
March 1993), reoperation rate was high. A total of
17 revision operations due to dysfunction or overdrainage had to be performed. In 4 patients, subdural
hematomas or hydromas occurred (Fig. 7 a, b).

Third ventriculostomy (14 patients after March
1993) was much more successful in treatment of PAS.
Intraoperative digital dynamic subtraction ventriculography showed the patency of third ventriculostomy in all cases. Postoperative ECG-gated MRI confirmed the results by evidence of CSF flow into the
interpendicular cistern. There was only one complication, caused by an insufficient ventriculostoma.
Patient follow-up from 1 to 43 months gives no
evidence of an insufficient stoma in 13 patients.

Discussion

Conventional therapy of occlusive hydrocephalus
using shunt systems was introduced in 1952 [31] and
is still the therapy of choice today [35, 36, 44]. The
advantage of the available shunts is the rapid reduction of clinical symptoms related to raised ICP.
Peri- and postoperative morbidity and letality rates

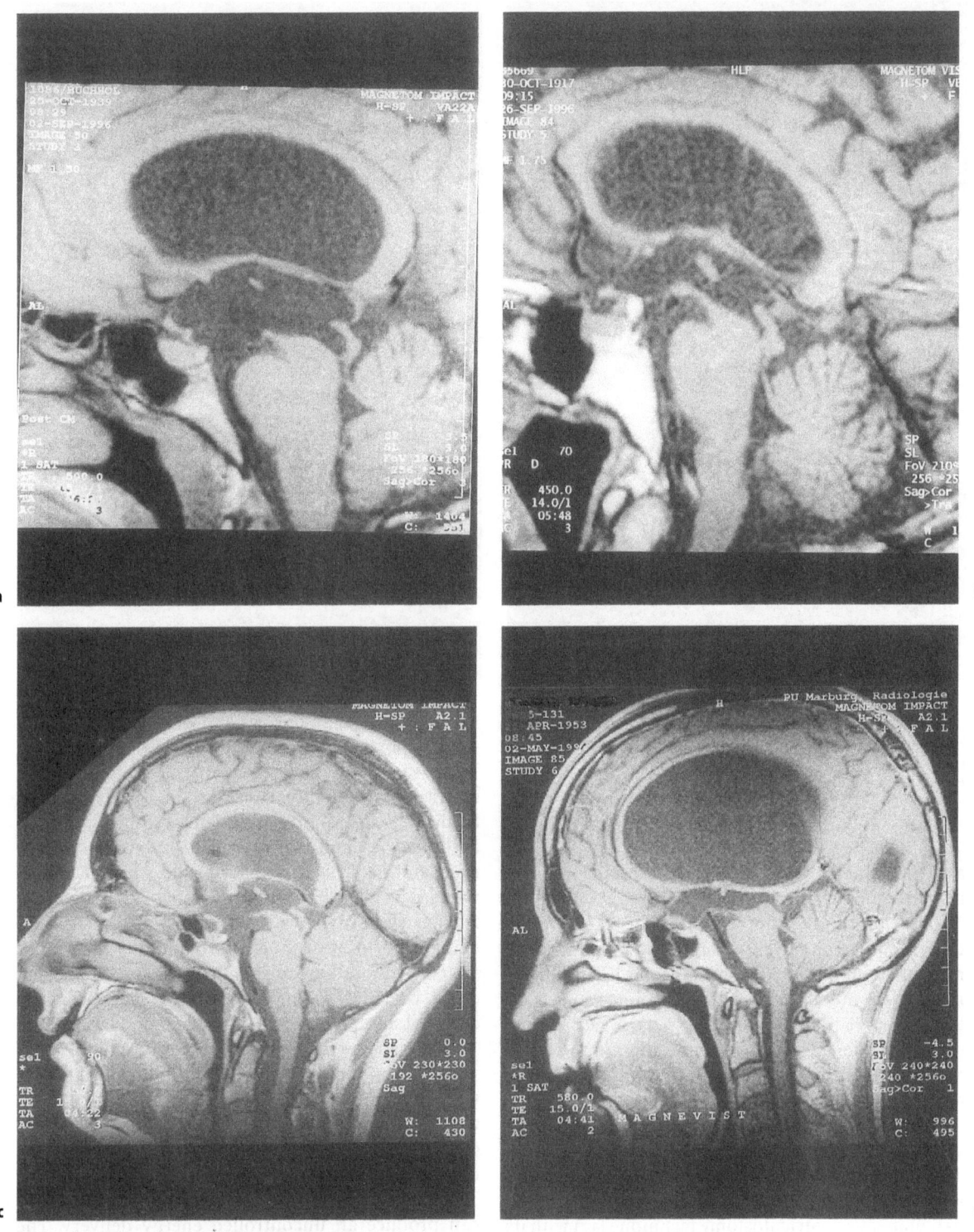

Fig. 6a–d. Different types of aqueductal stenosis

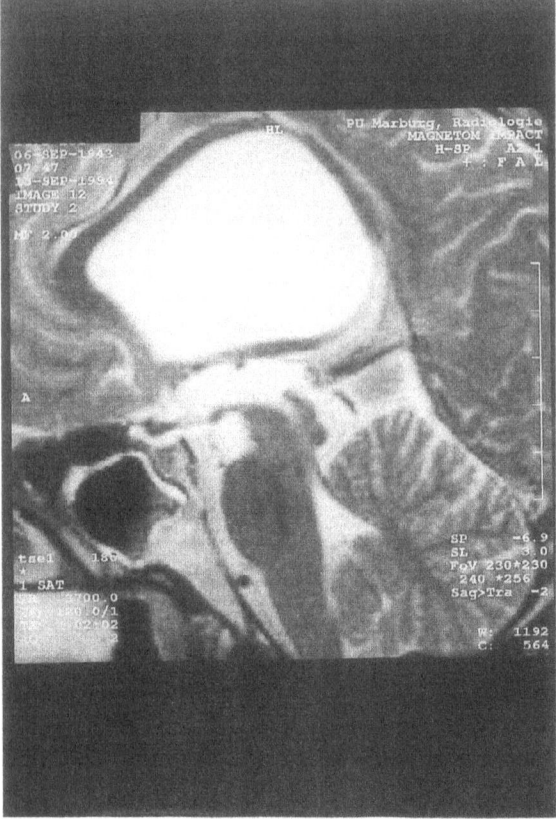

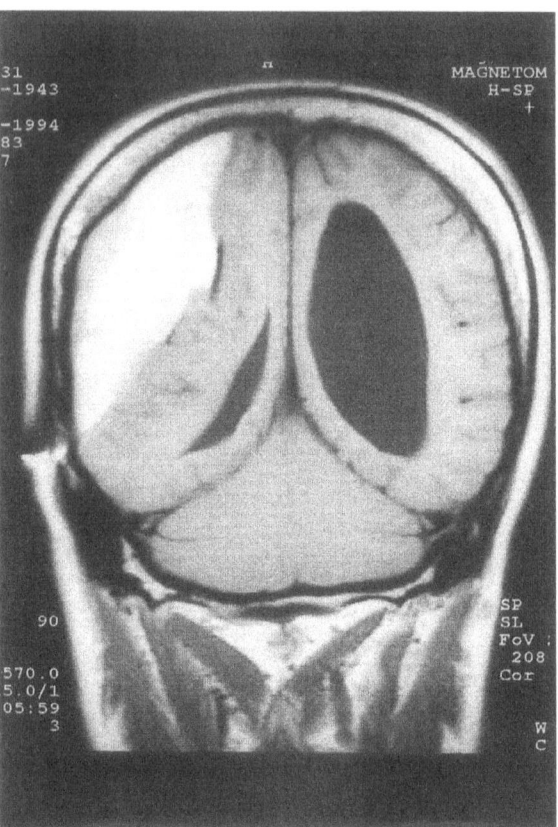

a   b

Fig. 7. a Occlusive hydrocephalus caused by primary aqueductal stenosis. **b** After application of a VP shunt, a large space occupying subdural hematoma occurred

are low, but complication rates should not be neglected in the long term follow-up [25].

Infection of the shunt system is the most important complication of CSF diversion system. Infection rates of about 8% were reported [10, 37]. The letality rate of patients suffering from shunt infections is up to 30%–40% [46]. Dysfunction [9, 13], overdrainage as well as the slit ventricle syndrome are the other significant complications after shunting of chronic obstructive hydrocephalus [8, 19].

To avoid these complications, endoscopic third ventriculostomy has seen a renaissance recent years. This is due to the availability of suitable endoscopes and refined supplementary instruments. Today the percutaneous ventriculostomy technique [17] with its high complication rate or the stereotactic method [20, 24] without endoscopes should be considered obsolete. As a large cavity, the chronic obstructive hydrocephalus provides optimal conditions for neuroendoscopic interventions.

The type of endoscope used for ventriculostomy is a point of controversy. Caemaert, Cohen and Jones [4,

5, 22] reported good results with rigid endoscopes. The advantage of the rigid endoscope is obviously the bright picture quality. However, the disadvantage is the lack of steerability and maneuverability [14, 15]. It should be emphasized that the combination of a rigid guiding system and a flexible steerable endoscope provides the option of penetrating the ventricular system with a stiff cannula and afterwards being able to steer freely through the ventricular system without harming the cortex, something that could occur when moving a rigid endoscope in different directions.

The instrumental opening of the floor of the third ventricle can be performed by different techniques with the help of special microinstruments [4, 5, 16, 29]. None of these methods is really superior to the others. We perform third ventriculostomy with the help of microbipolar probes and balloon catheters. The use of Laser probes for vaporization of the floor can produce an uncontrolled energy delivery. The blunt destruction of the floor with the tip of the endoscope should be avoided and is obsolete.

Efficacy of third ventriculostomy is demonstrated routinely with postoperative ECG-gated dynamic MRI [18, 23, 26, 32]. The other option is to confirm patency of third ventriculostomy intraoperatively, using digital dynamic ventriculography [1, 2].

The success of third ventriculostomy in treatment of obstructive hydrocephalus caused by PAS can finally be judged by the decline of clinical symptoms. CCT or MRI control examinations give no clear evidence because, in ca. 60% of the patients with improved neurology after third ventriculostomy, the size of the ventricular system is unchanged [38]. Radiographic restoration of the ventricular system after third ventriculostomy may steady over a period of 2 years [32].

Conclusion

Etiology of primary aqueductal stenosis is still unknown. Many hypotheses about its pathogenesis have to be discussed. In this study we found a widespread morphological variation of PAS, which leads us to assume that this pathology could be affected by noxa during different steps of embryogenesis. Furthermore, the therapeutic results prompt us to recommend third ventriculostomy as the operative method of choice in treatment of primary aqueductal stenosis.

References

1. Bauer BL, Hellwig D (1994) Minimally invasive neurosurgery – a survey. Acta Neurochir Suppl 61:1–12
2. Bauer BL, Hellwig D (1995) Intracerebral and intraspinal endoscopy. In: HH Schmidek, WH Sweet (eds) Operative neurosurgical techniques, 3rd edn. Saunders, Philadelphia, pp 695–713
3. Castro-Gago M, Alonso A, Eiris-Punal J (1996) Autosomal recessive hydrocephalus with aqueductal stenosis. Childs Nerv Syst 12 (4):188–191
4. Camaert J, Abdullah J (1994) A multipurpose cerebral endoscope and reflections on technique and instrumentation in endoscopic neurosurgery. Acta Neurochir Suppl (Wien) 61:49–53
5. Cohen A (1993) Endoscopic laser third ventriculostomy. N Engl J Med 328:552
6. Dandy WE (1922) Cerebral ventriculoscopy. Bull Johns Hopkins Hosp 33:189–190
7. Dauch WA, Wranze-Bielefeld E, Lütcke E, et al. (1992) Indiopathischer Hydrozephalus in der Adoleszenz. Monatsschr Kinderheilkd 140:357–362
8. Epstein F, Lapras C, Wisoff JH (1988) "Slit ventricle syndrome." Etiology and treatment. Pediatr Neurosci 14:5–10
9. Forrest DM, Cooper DGW (1968) Complications of ventriculo-atrial shunt infections. A review of 455 cases. J Neurosurg 29:506–512
10. George R, Leibrock L, Epstein M (1979) Long-term analysis of cerebrospinal fluid shunt infections. J Neurosurg 51:804–811
11. Griffith HB (1987) Endoscopic intracranial surgery. In: Advances of neurosurgery, vol 14 Springer, Berlin Heidelberg New York, pp 2–24
12. Haverkamp F, Krämer A, Fahnenstich H et al. (1996) X-chromosomal recessive hydrocephalus internus: a separate disease picture? 2 further case reports and review of literature. Klin Pediatr 208 (3):93–96
13. Hayden PW, Shurtleff DB, Stunts TJ (1983) A longitudinal study of shunt function in 360 patients with hydrocephalus. Dev Med Child Neurol 25:334–337
14. Hellwig D, Bauer BL, List-Hellwig E, Mennel HD (1991) Stereotactic procedures on processes of the cranial midline. Acta Neurochir Suppl (Wien) 53:23–32
15. Hellwig D, Bauer BL (1992) Minimally invasive neurosurgery by means of ultrathin endoscopes. Acta Neurochir Suppl (Wien) 54:63–68
16. Hirsch JF, Hirsch E, Sainte-Rose C et al. (1986) Stenosis of the aqueduct of Sylvius. J Neurosurg Sci 30:29–39
17. Hoffmann HJ, Harwood-Nash D, Gilday DL (1980) Percutaneous third ventriculostomy in the management of noncommunicating hydrocephalus. Neurosurgery 7:313–321
18. Hofmann E, Becker T, Meixensberger J, Jackel M et al. (1995) Disturbances of cerebrospinal fluid (CSF) circulation - neuropsychiatric symptoms and neuroradiological contribution. J Neural Transm Gen Sect 99 (1–3):79–88
19. Hyde-Rowan MD, Rekate HL, Nulsen FE (1982) Reexpansion of previously collapsed ventricles: the slit ventricle syndrome. J Neurosurg 56:536–539
20. Jack CR, Kelly PJ (1989) Stereotactic third ventriculostomy: assessment of patency with MR imagery. AJNR Am J Neuroradiol 10 (3):515–522
21. Johnson RT, Johnson KP (1969) Hydrocephalus as a sequela of experimental myxovirus infections. Exp Mol Path 10:68–80
22. Jones RFC, Kwok B, Stening W et al. (1994) Neuroendoscopic third ventriculostomy. A practical alternative to extracranial shunts in non-communicating hydrocephalus. Acta Neurochir Suppl (Wien) 61:79–83
23. Kadowaki C, Hara M, Numoto M et al. (1995) Cine magnetic imaging of aqueductal stenosis. Childs Nerv Syst 11 (2):107–111
24. Kelly PJ (1991) Stereotactic third ventriculostomy in patients with nontumoral adolescent/adult onset of aqueductal stenosis and symptomatic hydrocephalus. J Neurosurg 75:865–873
25. Keucher TR, Mealey J (1979) Long-term results after ventriculo-aterial and ventriculoperitoneal shunting for infantile hydrocephalus. J Neurosurg 50:186–197
26. Kunz U, Goldmann A, Bader C et al. (1994) Endoscopic fenestration of the 3rd ventricular floor in aqueductal stenosis. Minim Invasive Neurosurg 37:42–47
27. Lang J (1991) Topographic anatomy of performed intracranial spaces. Acta Neurochir Suppl (Wien) 54:1–10
28. Levitsky DB, Mack LA, Nyberg DA (1995) Fetal aqueductal stenosis diagnosed sonographically: how grave is the prognosis? AJR 164 (3):721–730
29. Manwaring K (1922) Endoscopic ventricular fenestration. In: Manwaring K, Crone KR (eds) Neuroendoscopy 1. Mary Ann Liebert, New York, pp 79–79
30. Mixter WJ (1923) Ventriculoscopy and puncture of the third ventricle. Boston Med Surg J 188:277–278
31. Nulsen FE, Spitz EB (1952) Treatment of hydrocephalus by direct shunt from ventricle to jugular vein. Surg Forum 2:399–403
32. Oka K, Go Y, Utsunomiya H et al. (1995) The radiographic restoration of the ventricular system after third ventriculostomy. Minim Invasive Neurosurg 38 (4):1568–162
33. Okamoto N, Wada Y, Kawabata H (1996) A novel mutation in LICAM gene in a Japanese patient with X-linked hydrocephalus. Jpn J Hum Gent 41:431–437
34. Overholser MD, Whitley JR, O'Dell BL et al. (1954) The ventricular system in hydrocephalic rat brains produced by a deficiency of vitamin B_{12} or folic acid in the maternal diet. Anat Rec 120:917–933
35. Pudenz RH (1980) The surgical treatment of hydrocephalus – a historical review. Surg Neurol 15:15–26

36. Raimondi AJ, Matsumoto S (1967) A simplified technique for performing the ventriculoperitoneal shunt (technical note). J Neurosurg 26 : 357 – 360

37. Renier D, Lacombe J, Pierre Kahn A et al. (1984) Factors causing acute shunt infection. Computer analysis of 1174 operations. J Neurosurg 61 : 1072 – 1078

38. Sainte-Rose C (1992) Third ventriculostomy. In: Manwaring KH, Crone KR (eds) Mary Ann Liebert, New York, pp 47 – 62

39. Sayers NP, Kosnik EJ (1976) Third ventriculostomy as the rational treatment of obstructive hydrocephalus. Childs Brain 2 : 24 – 30

40. Scarff JE (1935) Third ventriculostomy as a rational treatment of obstructive hydrocephalus. J Pediatr 6 : 870 – 871

41. Seow WK, Needleman HL, Smith LE (1995) Enamel hypoplasia, bilateral cataracts, and aqueductal stenosis: a new syndrome? Am J Med Genet 58 (4) : 371 – 373

42. Stookey B, Scarff J (1936) Occlusion of the aqueduct of Sylvius by neoplastic and non-neoplastic processes with a rational surgical treatment for relief of the resultant obstructive hydrocephalus. Bull Neurol Inst NY 5 : 348 – 377

43. Takechi T, Tohyama J, Kurashige T et al. (1996) A deletion of five nucleotides in the LICAM gene in a Japanese family with X-linked hydrocephalus. Hum Genet 97 (3) : 353 – 356

44. Torkildson A (1960) A follow-up study 14 to 20 years after ventriculo-cisternostomy. Acta Psch Neurol Scand 35 : 113 – 121

45. Vries JK (1978) An endoscopic technique for third ventriculostomy. Surg Neurol 9 : 165 – 168

46. Walters BC, Hoffmann HJ, Hendrick EB et al. (1984) Cerebrospinal fluid shunt infection. Influences on initial management and subsequent outcome. J Neurosurg 60 : 1014 – 1021

47. Woodard J, Newberne PM (1967) The pathogenesis of hydrocephalus in newborn rats deficient in vitamin B_{12}. J Embryol Exp Morph 17 : 117 – 187

48. Yamasaki F, Kodama Y, Hotta T et al. (1996) Adult onset aqueductal stenosis caused by membraneous occlusion in the aqueduct: a case report. No Shinkei Geka 24 (8) : 745 – 748

Third Ventriculostomy in the Treatment of Hydrocephalus: Experience with More than 120 Cases

C. Teo

Summary

There are many controversial issues surrounding endoscopic third ventriculostomy (ETV) in the treatment of non-communicating hydrocephalus. To help clarify some of these issues, 121 consecutive patients who underwent ETV were reviewed.

Introduction

Although the concept of creating a hole in the floor of the third ventricle to bypass an obstruction at the level of the fourth ventricle has been acknowledged for decades, the technique of endoscopic third ventriculostomy (ETV) has resurrected this approach. However, with this new technique come doubts and unanswered questions regarding patient selection, surgical technique, post-operative assessment and long-term success.

This retrospective study aims to clarify some of these issues: specifically, the question of patient selection on both clinical and radiological grounds, the advantages and disadvantages of several different techniques and the clinical and radiological assessment of outcome.

Material and Methods

The charts of 133 patients that had been considered candidates for ETV were reviewed. All patients had been treated by the author. Patients were excluded if they had inadequate follow-up, e.g. less than 6 months, or poor documentation of selection criteria and inadequate assessment of outcome. Patients were considered candidates for surgery on the basis of CT scans, MRI or ventriculograms suggesting a non-communicating hydrocephalus with the obstruction distal to the third ventricle. On these grounds several patients were included who may have been expected to have communicating hydrocephalus on the basis of their primary disease. For example, infants with hydrocephalus secondary to intraventricular haemorrhage or neonatal meningitis who had a CT or MRI that suggested a blockage at the level of the aqueduct were still considered candidates. A list of the patients and their primary diagnoses can be seen in Table 1. In the last 4 Years, CT scan alone has been accepted as a preoperative study without compromising results. The features that are favourable include (a) enlargement of the lateral and third ventricles, (b) normal-sized or small fourth ventricle, (c) characteristic "ovoid" shape to the third ventricle, and (d) minimal or barely discernible subarachnoid space. Of the 133 patients who met all the criteria, four had inadequate follow-up and eight underwent ventriculoscopy but did not have a ventriculostomy. Most of the failed attempts were due to abnormal anatomy. Two patients had clinically asymptomatic intraventricular haemorrhages that precluded safe completion of the third ventriculostomy.

Table 1. Aetiology of hydrocephalus

Primary diagnosis	No. of patients
Spina bifida	55
Aqueduct stenosis	38
Tumour	17
IVH	5
Post-meningitic	4
NPH	2

NPH, normal pressure hydrocephalus;
IVH, intraventricular haemorrhage of prematurity.

Surgical Technique

The patients were placed supine on the operating table with their head in a horseshoe headrest. Unless there was a pre-existing left frontal shunt, a standard right frontal burr hole was made just anterior to the coronal suture and 3 cm from the midline. Once the dura and pia were opened, a brain needle was used to enter the ventricle, and then the scope, in a sheath, was passed down the same tract. Anatomical structures that were identified in the lateral ventricle before passage of the scope into the third ventricle were the thalamostriate and septal veins and the choroid plexus. The author makes the ventriculostomy just posterior to the infundibular recess rather than just anterior to the mammillary bodies. By using this technique, one hopes to avoid the basilar bifurcation complex lying directly below the floor of the third ventricle. The scope itself was used to create the hole. If the patient had a frontal shunt, then it was removed through the same incision. Any doubt about shunt function was clarified before ETV.

Results

Of the 121 patients, 55 were male and 66 were female. Ages of the patients ranged from 1 week to 72 years with a mean of 16.5 years and a median of 12.3 years. All patients were followed post-operatively for at least 6 months, the longest being 6 years. Mean follow-up was 22 months (median follow-up 19 months). All patients were assessed post-operatively with clinical examination and repeat imaging. Thirteen patients had neuropsychological evaluation before and after ETV. Patients were then divided into two groups: success or failure. For the operation to be considered a success, the following criteria were required: resolution of clinical symptoms and signs and decrease in ventricular size ($n = 82$), or resolution of clinical symptoms and signs, improvement on psychometric testing and decrease or no change in ventricular size ($n = 6$). Using these criteria, 88 patients (73 %) had a successful outcome from ETV. On the basis of "intent to treat by ETV", 88 patients had a successful outcome out of 129 attempts (68 %). Of the 33 failures, most ($n = 30$) had declared themselves within 6 weeks of surgery. Two patients with malignant brain tumours and secondary hydrocephalus were initially managed with ETV and remained asymptomatic. One child developed symptomatic, progressive hydrocephalus 12 weeks after ETV and 6 weeks after cranial irradiation for leptomeningeal primitive neuroectodermal tumour. Another child had a successful ETV vor hydrocephalus secondary to a fourth ventricular ependymoma, with complete resolution of pressure symptoms. Although the child remained asymptomatic, when a follow-up CT scan showed no improvement in the ventricular size, another neurosurgeon placed a shunt in the child. This patient has had multiple shunt revisions and remains shunt dependent. Another patient with spina bifida had recurrence of her pre-ETV symptoms one year later. Her ventricular size never changed after ETV and never increased with clinical decompensation. Furthermore, there was no deterioration in her neuropsychological profile. Eventually the patient was admitted for

Table 2. Results by diagnosis

Diagnosis	Patients	
	(%)	(No.)
Spina bifida	70 %	(38/55)
Aqueduct stenosis	84 %	(32/38)
Tumour	88 %	(15/17)
IVH	20 %	(1/5)
Post-meningitic	50 %	(2/4)
NPH	0 %	(0/2)

IVH, intraventricular haemorrhage of prematurity;
NPH, normal pressure hydrocephalus.

intracranial pressure monitoring which demonstrated Lundberg "B" waves without hypertension. Due to persistent headaches, she was reluctantly re-shunted with complete resolution of her symptoms.

Further analysis was performed in an attempt to identify prognostic factors for success. Patients were categorized according to primary diagnosis, age and previous insertion of a shunt. While there was no statistical difference in results by diagnosis (Table 2), there appears to be a trend that suggests ETV will be less successful in conditions where one would expect to have impairment of CSF absorption, e.g. IVH and post-meningitic hydrocephalus. Age certainly plays a role in outcome from ETV. Of the 32 patients <2 years of age, 16 (50 %) had successful outcomes, whereas those > 2 years of age had an 81 % success rate (72/89; $p < 0.02$). In support of this trend, if patients were divided into those <6 months of age ($n = 15$) and compared to those >6 months of age ($n = 106$), a stronger difference was demonstrated (26 % success vs 81 % success; $p < 0.005$). Finally, patients were divided into those that had been previously shunted ($n = 82$) and those who had never been shunted ($n = 39$). Those who had been previously shunted, returning with clinical and radiological malfunction, had many more successful ventriculostomies (72/82 or 88 %) than those never shunted (16/39 or 41 %). These data were then analyzed using a stepwise linear regression analysis (Statistica data analysis program). The two parameters that showed significant difference in outcome were age >2 years ($p < 0.02$) and a history of previous shunting ($p < 0.02$).

Complications

There were no deaths in this series and an overall complication rate of 6 %. Most complications were related to transient or permanent hypothalamic dysfunction (Table 3). Patient GG had ETV for hydrocephalus associated with neurofibromatosis, which

Table 3.
Complications of hydro-
cephalus

Patient	Diagnosis	Complications	Outcome
GG	NF	Thirst	Complete resolution
CC	MM	Hyperphagia	Complete resolution
ST	MM	D.I.	Complete resolution
RM	AS	Amenorrhea	Complete resolution
SR	AS	Cardiac arrest	Good
PD	AS	CSF leak	Failed ventriculostomy
CA	MM	Amenorrhea	Permanent

NF, neurofibromatosis; MM, myelomeningocele; AS, aqueduct stenosis.

was presumed to be secondary to a tectal tumour. However, he returned to hospital with a serum sodium of 197 mmol/l with concentrated urine and no thirst. It transpired that he was mentally retarded and living alone and had not drunk water for 2 weeks. One month later he required a shunt for progressive hydrocephalus. His thirst mechanism returned to normal after 18 months. Patients RM and CA had normal menstrual cycles before ETV. RM has returned to normal after 18 months and CA is still amenorrheic with a normal prolactin level 8 months after ETV. Patient SR became severely bradycardic when pressure was applied to the third ventricular floor. The second attempt to push through the floor resulted in 6 s of cardiac arrest with reversion to sinus rhythm when the scope was retracted. Eventually a hole was made in the floor and the patient remains shunt independent.

Discussion

Any new technique is open to scrutiny and ETV is certainly no exception. The concept of bypassing an intraventricular obstruction by placing a hole in the floor of the third ventricle was proposed at the start of this century [1, 9, 12]. Why then has it taken so long to gain acceptance with the neurosurgical community? Despite recent publications supporting the role of ETV in the management of hydrocephalus [3, 4, 6–8, 13, 14, 17] there are still many unanswered questions and neurosurgeons are justified in their trepidation.

Of particular concern is patient selection. More invasive and more expensive tests do not appear to improve the success rate. In this study, seven patients had ventriculography, five had CSF isotope studies and five had CSF infusion studies in an attempt to refine selection criteria. Of these 17 patients, who all demonstrated non-communicating hydrocephalus and apparently adequate CSF absorptive capacity, the success rate was no better than for those who had CT or MRI alone (10/17 or 59% compared to 78/104 or 75%). CSF infusion studies do not allow enough time for the absorptive mechanisms to start functioning.

A patient can sometimes take weeks to become asymptomatic after ETV. The primary diagnosis is as good as any other factor as a prognosticator of success. If a patient has previously demonstrated an ability to absorb CSF and has a condition now that is most likely causing obstruction, then ETV should be effective. The exceptions to this rule are those cases that have dual pathology, e.g. the child with neurofibromatosis, a tectal glioma and triventricular hydrocephalus. Unfortunately, from the natural history of this disease we know that sometimes the hydrocephalus is communicating in type. ETV may be ineffective (see patient GG), the child with a posterior fossa tumour causing secondary non-communicating hydrocephalus. Both of the patients in this series who had tumours, ETV, radiotherapy and then required shunts my have developed an absorptive problem from radiotherapy or from spread of tumour to the leptomeninges. More advanced age and history of previous shunting have been found consistently to positively affect outcome. Once again, the reason for this is unknown, but one can speculate that the older a child, the more mature are the arachnoid villi. It may be related to generation of an adequate "driving" pressure of CSF across the superior sagittal sinus that cannot be achieved in an infant with open fontanelle and sutures.

Surgical technique may play a role in success. It certainly plays a role in the incidence of complications. The hole needs to be at least 5 mm in diameter and as far from the basilar bifurcation as possible. Following these rules may increase the incidence of hypothalamic complications, which are not insignificant in this series. However, the absence of any vascular complications in this series is noteworthy [16]. Furthermore, all but one of the hypothalamic complications were temporary.

Long-term follow-up is mandatory. This series highlights the importance of maintaining a close vigil on these patients, especially those whose ventricles remain large. Jones and colleagues have reported very late failures, as long as 5 years after ETV [8, 14]. In this series one patient failed 1 year after ETV without radiological evidence of deterioration. This raises the next important issue of ventricular size. There are some neurosurgeons who are resistant to ETV because historically the procedure does not necessarily return the ventricles to normal size. In this series, of the 72 patients who had been previously shunted and who had successful ETVs, only 16 had return of their ventricular size of baseline. Four patients had resolution of signs and symptoms and return of psychometric tests to normal after ETV but no decrease in the size of their ventricles. Of the 16 patients who had never been shunted, all had decrease in their ventricular size, but none had slit or small ventricles. The conclusion from this observation is that successful ETV does not always result in return of ventricular size to baseline or normal. Whether this equates with poorer cognitive outcome is beyond the scope of this paper and would require shunting those individuals whose ventricles remained large and documenting neuropsychological improvement.

Conclusions

The overall success rate of ETV using radiological characteristics of non-communicating hydrocephalus as selection of suitable patients is 73%. It is a safe procedure with very few permanent side effects and mostly hypothalamic temporary complications. Assessment of success is difficult, controversial and most dependent on long-term follow-up. When ETV fails it usually does so within 6 weeks of surgery due to a co-existent absorptive problem. However, late failures were seen in this series and have been reported by others. Success rates may be improved by selecting patients over the age of 6 months who have had previous shunts and whose primary diagnosis is consistent with a non-communicating hydrocephalus.

References

1. Dandy WE (1922) An operative procedure for hydrocephalus. Johns Hopkins Hosp Bull 33 : 189 – 190
2. Griffith HB (1975) Technique of fontanelle and persutural ventriculoscopy and endoscopic ventricular surgery in infants. Childs Brain 1 : 359 – 363
3. Hirsch JF (1982) Percutaneous ventriculocisternostomies in noncommunicating hydrocephalus. Monogr Neural Sci 8 : 170 – 178
4. Hoffman HJ, Hardwood-Nash D, Gilday DL (1980) Percutaneous third ventriculostomy in the management of noncommunicating hydrocephalus. Neurosurgery 7 : 313 – 321
5. Jack CR Jr, Kelly PJ (1989) Stereotactic third ventriculostomy: assessment of patency with MR imagery. AJNR Am J Neurodadiol 10 : 515 – 522
6. Jones RFC, Stening WAS, Bryden M (1990) Endoscopic third ventriculostomy. Neurosurgery 26 : 86 – 92
7. Jones R, Teo C, Kwok B, Stening W (1992) Neuroendoscopic third ventriculostomy. In: Manwaring KH, Crone KR (eds) Neuroendoscopy. Liebert, New York
8. Jones RFC, Kwok BCT, Stening W, Vonau M (1994) The current status of endoscopic third ventriculostomy in the management of noncommunicating hydrocephalus. Minim Invasive Neurosurg 37 : 28 – 369
9. Mixter WJ (1923) Ventriculoscopy and puncture of the floor of the third ventricle. Boston Med Surg J 188 : 277 – 278
10. Perlan BB (1968) Percutaneous third ventriculostomy in the treatment of a hydrocephalic infant with aqueductal stenosis. Int. Surg 49 : 443 – 448
11. Sayers MP, Kosnik EJ (1976) Percutaneous third ventriculostomy: experience and technique. Childs Brain 2 : 24 – 30
12. Stookey B, Scarff J (1936) Occlusion of the aqueduct of Sylvius by neoplastic and nonneoplastic processes with a rational surgical treatment for relief of the resultant hydrocephalus. Bull Neurol Inst NY 5 : 348 – 377
13. Teo C (1996) Neuroendoscopy. In: Andrews RJ (ed) Intraoperative neuroprotection. Williams and Wilkins, Baltimore, pp 423 – 443
14. Teo C, Jones R (1996) Management of hydrocephalus by endoscopic third ventriculostomy in patients with myelomeningocele. Pediatr Neurosurg 25 : 57 – 63
15. Teo C, Jones R, Stening W, Kwok B (1991) Neuroendoscopic third ventriculostomy. In: Matsumoto S, Tamaki N (eds) Hypdrocephalus: pathogenesis and management. Springer, Berlin Heidelberg New York, pp 672 – 683
16. Teo C, Rahman S, Boop FA, Cherny WB (1996) Complications of endoscopic neurosurgery. Childs Nerv Syst 12 : 248 – 253
17. Vries J (1978) An endoscopic technique for third ventriculostomy. Surg Neurol 9 : 165 – 168

The Role of Third Ventriculostomy in Previously Shunted Hydrocephalus

U. Kehler, J. Gliemroth, U. Knopp, and H. Arnold

Summary

The problem of third ventriculostomy is that exact preoperative prediction of CSF outflow resistance is not possible [1]. This is of even more importance in patients with previously shunted noncommunicating hydrocephalus who present with shunt malfunction.

We describe the course of two patients who received a third ventriculostomy after malfunction/infection of a shunt. Both outcomes were satisfying, proving that CSF absorption had remained intact even years after shunting. However, one patient needed some time for adaptation: the third ventriculostomy with simultaneous shunt explantation was followed by 2 days of apparently increased ICP before the CSF outflow had completely adapted. In such cases, preoperative CSF absorption studies would have shown an elevated CSF outflow resistance. Preoperative dynamic CSF studies are not reliable in predicting whether CSF resorption will normalize under normal CSF load of the subarachnoid space.

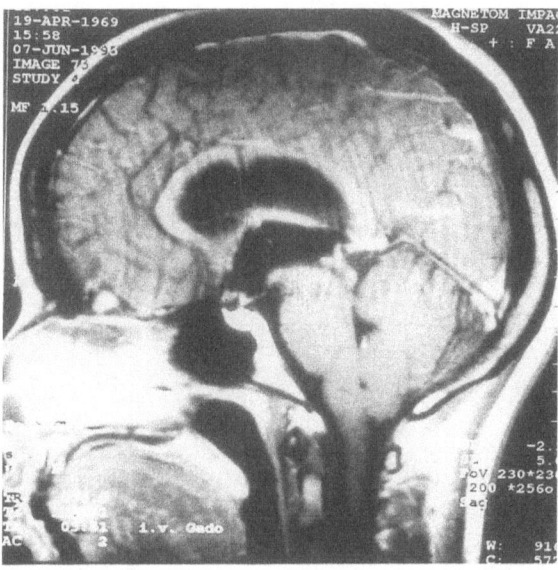

Fig. 1. Medial sagittal MRI scan showing an aqueductal stenosis and a downward bulged floor of the third ventricle (case 1)

Introduction

Long-term shunting for noncommunicating hydrocephalus may lead to atrophy or involution of the resorption sites of CSF. Therefore, the success of third ventriculostomy in previously shunted patients with noncommunicating hydrocephalus is questionable. However, there are many reports of successful third ventriculostomies in previously shunted patients [2–5, 7]. The postoperative course shows some particularities in our patients that were not mentioned in the literature and may have serious impact on the management of patients (i.e., interpretation of CSF resorption studies and postoperative management).

We report on two patients in whom we performed a third ventriculostomy and removed the malfunctioning/infected shunt that was implanted years before.

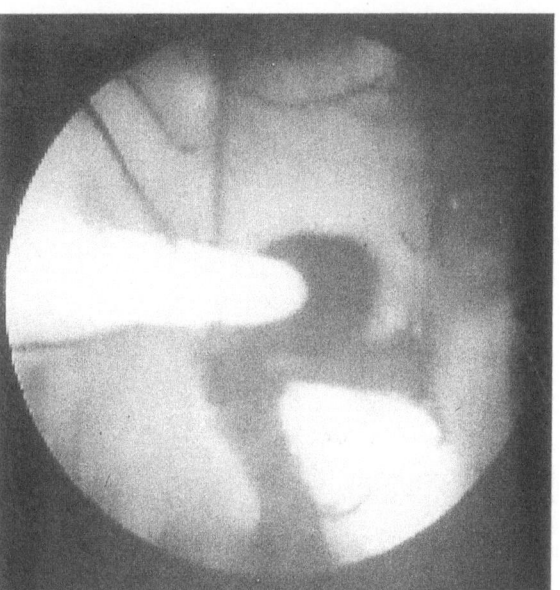

Fig. 2. View of the foramen Monro and the ventricular catheter (case 1)

Case 1

A 24-year-old woman was shunted for symptomatic hydrocephalus caused by aqueductal stenosis. Six months later, she developed intermittent shunt dysfunction complaining about headache, sleepiness, and vomiting. One year and 8 months after the shunt implantation, she was admitted to our hospital with severe headache, vomiting, and vertical gaze paresis. MRI showed a triventricular hydrocephalus with a downward bulged floor of the third ventricle (Fig. 1). A third ventriculostomy was performed (Figs. 2–4), and the shunt was removed simultaneously.

In the first 2 days after surgery, the patient complained of very severe headache, sleepiness and bradycardia requiring atropine. A new shunt implantation was already being discussed when the patient suddenly improved markedly, and she is without any complaints in the present 1.5-year follow-up.

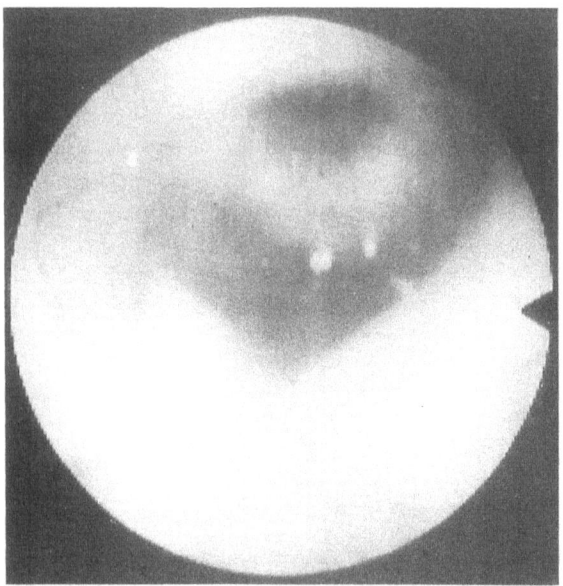
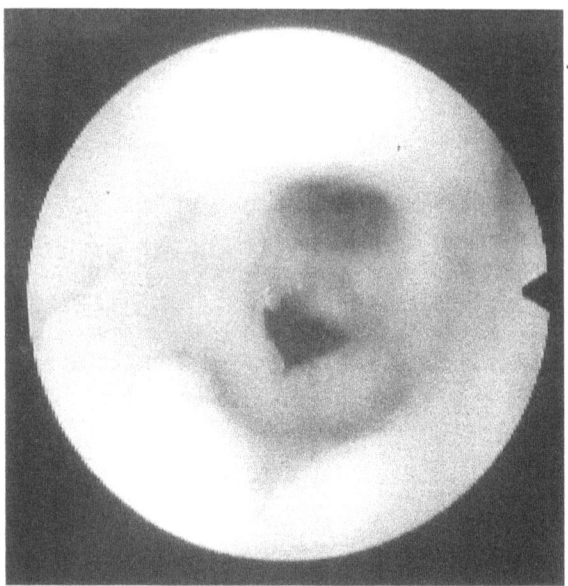

Fig. 3 a, b. Floor of the third ventricle **a** before and **b** after fenestration (case 1)

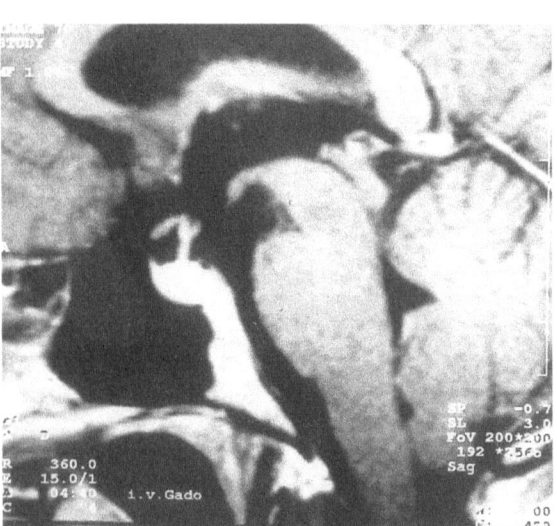

Fig. 4. MRI scan postoperatively: the floor of the 3rd ventricle with the chiasma and pituitary stalk once again in normal position (ct. Fig. 1), the fenestration between the pituitary stalk and the mamillary bodies is visible (case 1)

Probably, an adaptation to the new CSF resorption modalities took place in the first 2 days. The resorption sites of the CSF were again opened. The clinical course shows without any doubt that CSF resorption may remain intact or may regain its former resorption capacity even after long-term shunting.

Case 2

A 3-year-old boy with a Down-syndrome, a Dandy-Walker malformation, and a noncommunicating hydrocephalus was shunted. At the age of 6 years, the patient showed a shunt malfunction and the complete shunt was changed. Three months after the revision, a shunt infection occurred which could not be treated conservatively. MRI showed the Dandy-Walker malformation, an immense hydrocephalus with the downward bulged floor of the third ventricle, and a suspected aqueductal stenosis (Fig. 5). The shunt was removed, a third ventriculostomy was performed, and the aqueduct was inspected but did not show any

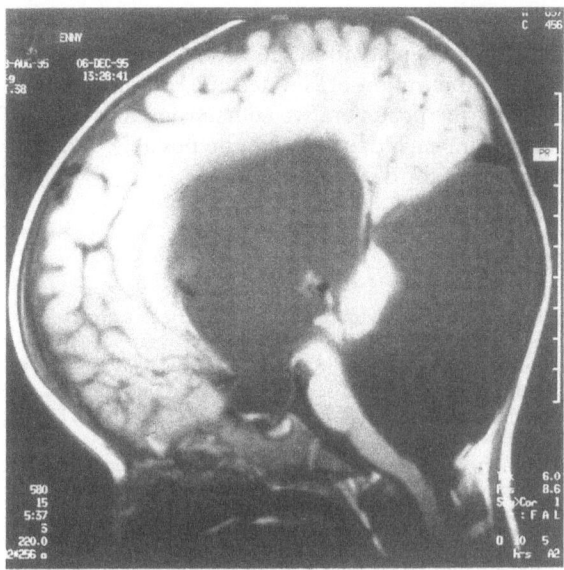

Fig. 5. Medial sagittal MRI scan showing a downward bulged floor of the third ventricle and a suspected aqueductal membrane (case 2)

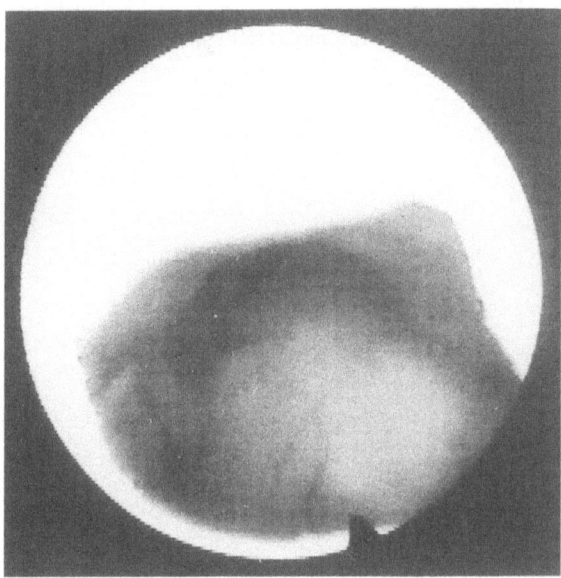

Fig. 6. Endoscopic view through the aqueduct showing no obstruction (case 2)

obstruction (Fig. 6). Additionally, an external ventricular drainage was introduced.

The infection could be treated effectively, but a CSF leak appeared along the external catheter that could not be stopped by additional sutures. Therefore, the drainage was removed and the patient observed closely. No signs of elevated ICP appeared, and the patient could be discharged. The follow-up time of presently 10 months was uneventful. After a 3-year period of shunting including a shunt infection, the patient is now shunt-independent after the third ventriculostomy.

Discussion

The clinical courses of the two patients prove that the CSF absorption capacity may remain intact or may regain its former function even years after shunting. Therefore, third ventriculostomy can be effective in noncommunicating hydrocephalus even years after shunting, and in the authors' opinion, it should be considered as an alternative to shunt revision.

Especially the first case showed an important observation: clinical signs of elevated ICP (headache, bradycardia, sleepiness) were observed in the first 2 days after surgery (third ventriculostomy and shunt explantation) before the patient improved markedly. This "transition time" is interpreted as an adaptation to the new CSF absorption modalities, where the absorption sites have to "reopen" once more. An elevated ICP for a certain time may be necessary for such reopening.

The adaptation could explain why a patient needs some time to recover after shunt explantation and third ventriculostomy. However, many questions arise for postsurgical management: How much time does the adaptation take? Does the adaptation time vary from one patient to another? How long can we tolerate an adaptation time with elevated ICP without any harm to the patient? Until now, no one can answer these questions, but they underline the necessity of close observation of patients in the postoperative course. A new shunt should be implanted if there is any doubt about further toleration.

A temporary external ventricular drainage could be helpful in the early postoperative follow-up: ICP could be slowly raised by elevating the drainage system to obtain a "smooth" adaptation without risking harmful ICPs or emergency shunt reimplantations. However, it is not clear what amount of pressure is necessary to "reopen" the absorption sites. This means we do not know how far we have to raise CSF pressure by elevating the overflow of the external ventricular drainage before we can conclude that there will not be any reopening.

Conversely, if there is no sufficient flow through the ventriculocisternostomy (because CSF leaves via the external drainage), it might reclose again [5].

The preoperative testing of CSF dynamics might be useless if there is an adaptation to the new CSF absorption modalities, because in the preadaptation situation, the resorption sites do not seem to be sufficiently open. In contrast, CSF outflow resistance has to be pathologically elevated. Consequently, an elevated CSF outflow resistance in such cases is not a contraindication to third ventriculostomy.

Furthermore, a presurgical "volume loading" of the CSF space seems to be nonsensical to demonstrate a possible adaptation (of CSF resorption) because the required time and pressure are unknown, and the test is even more complex than the third ventriculostomy itself.

In the authors' indication for third ventriculostomy and simultaneous shunt explantation at present is independent of CSF dynamic studies and should be done in cases of noncommunicating hydrocephalus with clinical signs of shunt malfunction. The MRI should show a downward bulged floor of the third ventricle, which seems to be an important predictive sign for a successful procedure [1, 3].

If the indication after clinical and MRI findings remains doubtful, third ventriculostomy is also acceptable because in some cases it may spare the patient the life-long threat of shunt complications. It must be emphasized that the procedure has to be performed by well-trained hands in order to be safe [2, 3, 5–7].

In the literature, there is only very little information about the postoperative course of patients who underwent a third ventriculostomy with simultaneous shunt explantation. However, this information (signs and duration of elevated ICP) is important to interpret the postoperative course (possible adaptation). From such data, it might be possible to define the limits of third ventriculostomy.

References

1. Cohen AR (1995) Endoscopic third ventriculostomy – guest editor commentary. Techn Neurosurg 214–215
2. Jones RFC, Kwok BCT, Stenung WA, Vonau M (1994) The current status of endoscopic third ventriculostomy in the management of non-communicating hydrocephalus. Minim Invasive Neurosurg 1:28–36
3. Kehler U, Gliemroth J, Arnold H (1996) Endoskopische Behandlung des Verschlußhydrocephalus. Focus MUL (Lübeck) 13:82–87
4. Kehler U, Gliemroth J, Wagner R, Arnold H (1996) 3rd Ventriculostomy versus ventriculoperitoneal shunt. An analysis of results, complications, and costs. Zentralb Neurochir Suppl: 84
5. Kelly PJ (1991) Stereotactic third ventriculostomy in patients with nontumoral adolescent/adult onset aqueductal stenosis and symptomatic hydrocephalus. J Neurosurg 75:865–873
6. Kunz U, Goldmann A, Bader C, Waldbaur H, Oldenkott P (1994) Endoscopic fenestration of the 3rd ventricular floor in aqueductal stenosis. Minim Invasive Neurosurg 37:42–47
7. Sainte-Rose C, Chumas P (1995) Endoscopic third ventriculostomy. Techn Neurosurg 1:176–184

Ultrasound-Guided Endoscopic Fenestration of the Third Ventricle in Obstructive Hydrocephalus

A. Rieger, N. G. Rainov, L. Sanchin, G. Schöpp, and W. Burkert

Summary

Endoscopic fenestration of the third ventricular floor became a standard procedure for treatment of hydrocephalus caused by aqueductal stenosis or obstruction of CSF pathways. Endoscopic ventriculostomy restores physiological CSF circulation, and implantation of external shunt valve systems can be avoided. We employed this technique for treatment of 20 patients with obstructive hydrocephalus caused by aqueductal stenosis (17 cases) or suprasellar arachnoid cysts (three cases). Diagnosis and choice of this particular type of surgery were based on standard and movement-sensitive MRI scans. Intraoperative ultrasound (US) guidance was used to identify anatomic structures such as the foramen of Monro and to position the rigid endoscope. US-guided endoscopy retains all advantages of stereotactic endoscopy and, in addition, appears to be faster and simpler than stereotaxy, allowing for real-time control of endoscope position. In our hands, US-guided endoscopic fenestration of the third ventricle yielded very good short-term and long-term results. In this series, no complications were noted immediately after surgery except for two cases with a slight psychosyndrome. Long-term follow-ups demonstrated a cure in 12 cases and a considerable improvement of clinical signs and symptoms in five cases. In an additional three cases, initial improvement was followed by recurrence of clinical and neuroradiologic symptoms. These were treated by implantation of a standard external valve system.

In conclusion, if careful preoperative selection according to specific criteria is carried out, US-guided fenestration of the third ventricular floor is a curative procedure with minimal complications and a high success rate.

Introduction

Shunt valve implantation in hydrocephalus patients carries a rather high risk of early postsurgical complications and late valve failures [7, 19]. Endoscopic fenestration of the third ventricular floor represents a minimal invasive technique applicable especially in obstructive hydrocephalus with normal CSF resorption. This technique is believed to be relatively safe to the patient, and only a few complications have been described [10]. Prerequisites for success with endoscopic fenestration are preoperative selection of suitable cases and a ventriculostomy large enough to ensure permanent restoration of CSF circulation.

Walter Dandy was the first to describe the use of an endoscopic device for cerebral ventriculoscopy [1]. Mixter reported single patients with obstructive hydrocephalus treated by endoscopic perforation of the ventricular floor, and proved restored circulation by intraventricular dye injection and subsequent lumbar CSF sampling [18]. The development of sophisticated endoscopic equipment in the past two decades radically changed the original procedure and enabled neurosurgeons to perform it safely in large groups of patients. Thus, risks and gains of endoscopic ventriculostomy were evaluated in a broad range of diseases, and the technique was proposed as a standard treatment in obstructive hydrocephalus and aqueductal stenosis [5, 9, 11, 20, 22]. The standard endoscopic ventriculostomy employs a ventricular puncture to introduce the device with endoscopic orientation inside the CSF spaces [5, 9, 22]. If there is a ventricular wall or choroid plexus bleeding or if the CSF is of high protein content, direct visualization of anatomical structures inside the ventricles is rather difficult [17].

The present study is aimed at evaluating a modified ventriculostomy procedure, the US-guided endoscopic fenestration of the third ventricular floor, with respect to its results and possible limitations in hydrocephalic patients with aqueductal stenosis or space-occupying arachnoid cysts.

Patients and Methods

Patients

We treated 17 patients with hydrocephalus due to stenosis or occlusion of the aqueduct and three pa-

Table 1. Descriptive data of the patient population ($n = 20$)

Sex: male/female	8/12
Age (years) (mean, range)	37 (0.5–62)
Genesis of hydrocephalus	
Pineal tumor	3
Space-occupying arachnoid cyst	3
Genuine aqueductal stenosis	14
Primary third ventriculostomy	15
Shunting prior to ventriculostomy	2
Ventriculocystocisternostomy	3

tients with suprasellar space-occupying arachnoid cysts. Some of these patients had previously been treated with external shunt valves (Table 1). In two children and one adult, there was a tumor (pinealoma) that caused aqueductal stenosis. In the remaining 14 cases, no obvious cause of stenosis could be identified. Six patients were subjected to surgery because of acute elevation of intracranial pressure (ICP). Two of them were children with shunt failures. In 14 cases, there were considerable neurological signs and deficits such as headaches, ataxia, and memory disturbances.

Ventriculostomies were performed in all 17 cases with stenosis or occlusion of the aqueduct. In the patients with arachnoid cysts, a ventriculocystocisternostomy [2] was carried out.

Diagnostics and Selection

Three ventricular hydrocephalus was diagnosed by CT and/or MRI prior to surgery. Tumor diagnosis was performed by contrast-enhanced MRI only. Movement-sensitive MRI oriented in the sagittal plane parallel to the aqueduct was carried out in all selected patients (Fig. 1). A 1.0-T MR scanner (Magnetom Impact, Siemens, Erlangen, Germany) with head coil was employed using the PSIF 2D sequence and the GMR sequence in read-out and slice-select directions within TR only.

Cases of post-meningoencephalitic hydrocephalus with aqueductal stenosis were excluded from the present series because of assumed disturbances in normal CSF resorption. A prerequisite for endoscopic surgery was the proof of a dilated third ventricle (>1 cm in the coronal plane) which facilitates manipulations inside the ventricle and diminishes the risk of damaging the hypothalamus [15].

Surgical Technique

The head of the anesthetized patient was positioned in a Mayfield head holder. A right frontal burr-hole was placed 15 mm lateral to the bregma, and a small bone flap (3 × 2 cm) was elevated using the Elan craniotome (Aesculap, Tuttlingen, Germany). After opening the dura, an ultrasound device (Combison 311, Kretztechnik, Trefenbach, Austria) with a 7.5-MHz surgical transducer was used to investigate the intracranial anatomy and trace direction of ventricular puncture. A Cushing cannula was forwarded under US guidance until the foramen of Monro was reached, and the needle track was bluntly widened. CSF samples were taken for paraclinical investigation, and the cannula was replaced with a 4-channel

Fig. 1.
Flow-sensitive MRI scan in a child with a pineal tumor prior to endoscopic third ventriculostomy. No CSF flow exists in the aqueduct and the small fourth ventricle

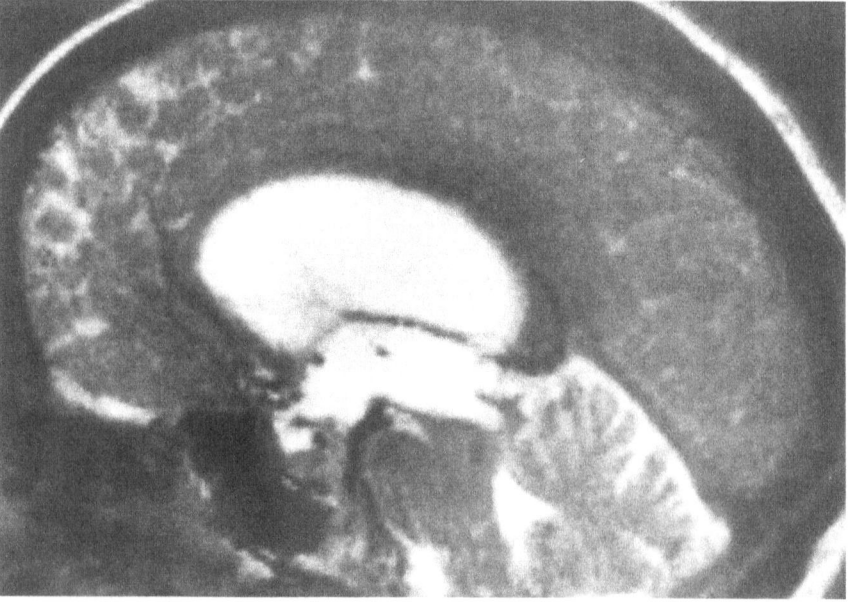

rigid endoscope (Aesculap) with an outer diameter of 6.2 mm that was then introduced under US guidance into the right lateral ventricle. The foramen of Monro was identified by direct visualization, and the tip of the device was forwarded into the third ventricle. Anatomical landmarks such as the mamillary bodies and the infundibular recess allowed correct placement of the monopolar electrode. After US control of correct endoscope placement, an area of approximately 4 mm² was punctured several times with the tip of the coagulation electrode. Bridging membrane tissue between the single holes was removed by flexible microforceps. Care was taken to avoid gross displacement of the floor of the third ventricle. The arachnoid layer was then fenestrated in the same manner. The basilar artery and the clival surface were inspected endoscopically.

In cases with an arachnoid cyst, the lateral ventricle was accessed from the respective side where the cyst protruded more strongly into the ventricle. The cyst wall, which was rather thick in all three cases, was coagulated and opened with flexible microscissors. The endoscope was then forwarded under US guidance into the cyst cavity (Fig. 2). The tip of the device was positioned just proximal to the clivus, where the cyst wall appeared to be thinner than in the intraventricular parts. The same perforation technique was used as in standard third ventriculostomy. After gaining access to the basal cisterns, the skull base and the adjacent structures could be overlooked from the intradural endoscopic point of view. Thus, the course

of the pituitary stalk, the abducent and oculomotor nerves, and the posterior communicating artery could be followed visually. This procedure of a "two-floor" cyst fenestration is termed ventriculocystocisternostomy after Deco et al. [2]. CSF bypass after fenestration was confirmed in some cases by cisternal ventriculography with an X-ray contrast agent. The endoscope was then slowly withdrawn, and the remaining canal in the brain was occluded with oxymethylcellulose (Sorbacel; Hartmann, Heidenheim, Germany) and sealed with fibrin glue (Tissucol, Immuno, Vienna, Austria). CSF flow discontinued immediately after gluing. Dura was tightly sutured and the bone flap put back and secured in place.

Outcome Assessment

The success of third ventriculostomy is not immediately clinically evident as is the case for external shunting. However, long-term results can be evaluated using following criteria: reversal of clinical signs of elevated ICP [3], structural evidence of improved cortical mantle as seen on CT or MRI scans [9], reversal of memory disturbances [10], and ventriculostomy patency as assessed by movement-sensitive MRI [8,13].

All patients were followed for at least 3 months after surgery, and evaluated according to the above criteria. MRI was carried out 3, 4, 8, and 12 weeks after surgery.

Results

Twelve patients were permanently cured, another five reported significant improvement. In one patient with aqueductal stenosis, there was a deterioration after initial improvement of symptoms. A conventional shunt procedure had to be carried out because of progressing enlargement of ventricle I – III. Two further patients had a transient organic psychosyndrome, but recovered without permanent deficits. The remaining patients stayed free of surgery-related complications.

Postoperative CT and/or MRI scans demonstrated some decrease in the size of the lateral ventricles and slight gyral expansion. Movement-sensitive MRI confirmed permanent CSF flow through the ventriculostomy into the interpeduncular cistern and down the clivus (Fig. 3). No spontaneous closures of the ventriculostomy were observed. The size of the arachnoid cysts was shown to decrease immediately after ventriculocystocisternostomy, as shown by intraoperative US (Fig. 4).

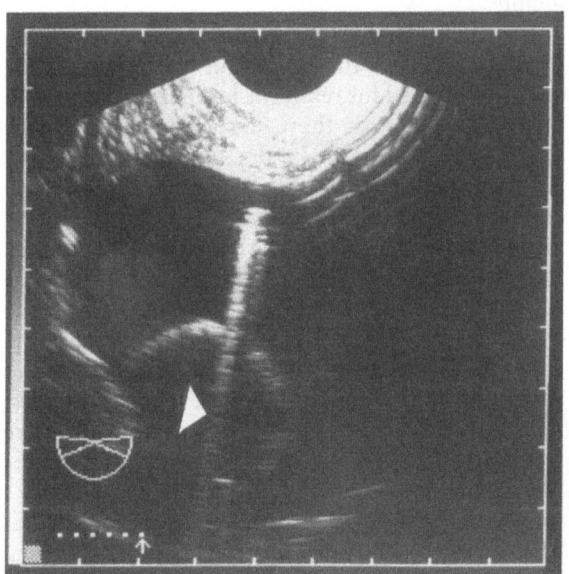

Fig. 2. Intraoperative US image of the endoscopic device with the tip located in a suprasellar arachnoid cyst (*white arrow*)

Fig. 3.
Flow-sensitive MRI scan
8 weeks after ventriculostomy
Note the signal void in the
third ventricle and the basal
cisterns indicating intense
CSF flow (*white arrows*)

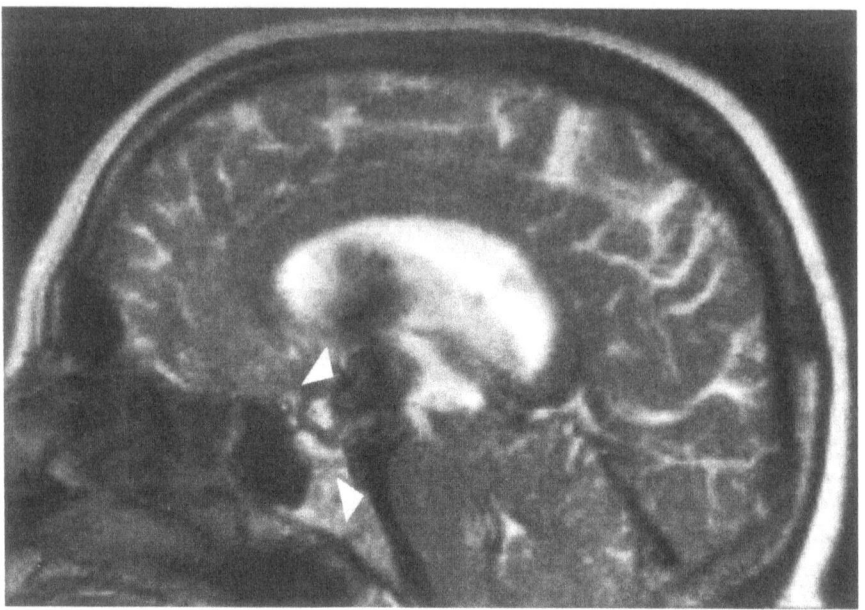

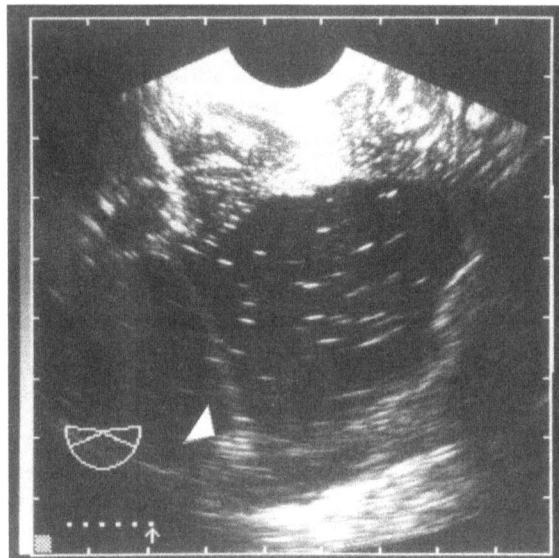

Fig. 4. US image of an arachnoid cyst immediately after endoscopic cyst-wall fenestration. Note the extreme volume reduction of the cyst (*white arrow*)

Discussion

This study was aimed at evaluating a modified endoscopic technique for third ventriculostomy and at reducing surgery-related complications. We were able to achieve a clinical cure in 14 hydrocephalic patients and a lasting improvement in another five cases. No permanent morbidity resulted from endoscopic surgery. Ultrasonic guidance of the endoscope allowed fast and simple real-time intraventricular navigation even under conditions of reduced direct visibility.

Endoscopic ventriculostomies have some surgery-related complications. Sayers and Kosnik [21] reported an early postoperative death caused by bleeding and hypothalamic damage. In their series of 46 patients, three had a temporary diabetes insipidus and two a transient mesencephalic syndrome with oculomotor and abducent nerve dysfunction. Jones et al. [9] experienced severe intraoperative bleeding in one of 24 patients, and this patient developed a postoperative hemiparesis. In three other cases, ventriculostomy was impossible because of strong vegetative reaction (bradycardia) to perforation attempts.

Spatial localization of the endoscope is being considered crucial to the success of surgery. The first device to be used for real-time localization relative to bone structures was the C-arm fluoroscope. It was applied to endoscopy in the same setting as in pituitary surgery [10, 15]. Stereotactic guidance of the endoscope has been widely used with or without intraoperative CT scans [6, 12] and believed to minimize surgery-related complications [11, 12]. MRI has also been utilized in combination with stereotactic endoscopy for third ventriculostomy [4] and appears to be superior to all previous modalities in respect to spatial resolution and anatomical detail. The most recent developments in the ventriculostomy field are frameless real-time stereotaxy [16, 17] and 3 D neuro-navigation [23].

The endoscopic view of the ventricle is a somewhat distorted close-view anatomy [14, 17], although visual orientation in the enlarged lateral ventricle and third ventricle is not difficult under normal circumstances. Localization of the foramen of Monro is greatly facilitated by the choroid plexus and the concomitant veins. Large cysts or tumors, however, may displace anatomical landmarks and may impose some problems on close-view endoscopic orientation [10,17].

Manwaring [14] first used intraoperative US guidance in pediatric endoscopy and placed the US array over the great fontanelle. In adults and older children, a small craniotomy is necessary if US is to be used [16, 17]. In our approach, the small burr-hole-type craniotomy is easy to perform and allows direct visualization and avoidance of cortical vessels as well as minimization of trauma to the cortical surface. US navigation aids in choosing the proper trajectory of the endoscopic device towards the foramen of Monro. Kelly [11] emphasized the importance of the endoscope entry angle which has to be parallel to the foramen-premamillary triangle line. If this is not the case, the fornix may be injured during endoscopic movements and manipulations. In our hands, US guidance was sufficient to properly position the endoscopic device and avoid injuries to adjacent structures. The most precise endoscopic technique, stereotactic endoscopy, was shown to be safe enough to avoid any neurological morbidity in a series of 38 patients [11, 12]. Our series yielded comparable results, except for two cases with temporary psychosyndrome of unknown genesis. A possible cause for these disturbances may be rapid changes in intraventricular pressure caused by saline flushing of the endoscope. Furthermore, the method of ventricular floor perforation appears to be important in respect to postoperative neurological morbidity [4,10,13,17,23].

In ventriculocystocisternostomy, US guidance and navigation are of importance for cases of distorted ventricular anatomy only. In our three cyst patients, however, ventricular anatomy was easy to recognize and follow, and no bleeding occurred to obscure endoscopic view.

In conclusion, US-guided endoscopic ventriculostomy and ventriculocystocisternostomy are faster and easier to perform than stereotactic or CT/MRI-guided techniques and offer a rather inexpensive alternative to 3 D neuronavigation. One of the few disadvantages of US as a guiding method seems to be the poor resolution of deep-brain structures.

References

1. Dandy WE (1922) Cerebral ventriculoscopy. Johns Hopkins Hosp Bull 33 X: 189
2. Deco P, Brugieres P, Le Guerinel C, Djindjian M, Keravel Y, Nguyen JP (1996) Percutaneous endoscopic treatment of suprasellar arachnoid cysts: ventriculocystostomy or ventriculocystocisternostomy? J Neurosurg 84: 696–701
3. Garcia-Bengochea F (1987) Persistent memory loss following section of anterior fornix in humans. Surg Neurol 27: 361–364
4. Goodman RR (1993) Magnetic resonance imaging-directed stereotactic endoscopic third ventriculostomy. Neurosurgery 6:1043–1047
5. Hirsch JF (1982) Percutaneous ventriculocisternostomies in non-communicating hydrocephalus. Monogr Neurol Sci 8:170–178
6. Hoffman HJ, Harwood-Nash D, Gilday DL (1980) Percutaneous third ventriculostomy in the management of non-communicating hydrocephalus. Neurosurgery 7: 313–321
7. Illingworth RD, Logue VS (1971) The ventriculocaval shunt in the treatment of adult hydrocephalus. Results and complications in 101 Patients. J Neurosurg 35: 681–685
8. Jack CR, Kelly P (1989) Stereotactic third ventriculostomy: assessment of patency with MR imaging. Radiology 10: 512–522
9. Jones RFC, Stening WA, Brydon M (1990) Endoscopic third ventriculostomy. Neurosurgery 26: 86–92
10. Jones RFC, Kwok T, Stening WA, Vonau A (1994) The status of third ventriculostomy in management of non-communicating hydrocephalus. Minim Invasive Neurosurg 37: 28–36
11. Kelly PJ (1991) Stereotactic third ventriculostomy in patients with nontumoral adolescent/adult onset aqueductal stenosis and symptomatic hydrocephalus. J Neurosurg 75: 865–873
12. Kelly PJ, Goerss S, Kall BA, Kispert DB (1986) Computer tomography based stereotactic third ventriculostomy: technical note. Neurosurgery 18: 791–794
13. Kunz U, Goldmann A, Bader C, Walbaur H, Oldenkott P (1994) Endoscopic fenestration of the 3rd ventricle floor in aqueductal stenosis. Minim Invasive Neurosurg 37: 42–47
14. Manwaring KH (1991) Endoscopic-guided placement of ventriculoperitoneal shunt. In: Manwaring KH, Crone KR (eds) Neuroendoscopy, vol 1. Liebert, New York, pp 29–40
15. Manwaring KH (1991) Endoscopic ventricular fenestration. In: Manwaring KH, Crone KR (eds) Neuroendoscopy, vol 1. Liebert, New York, pp 79–88
16. Manwaring KH, Manwaring ML, Moss SD (1994) Magnetic field-guided endoscopic dissection through a burr hole may avoid more invasive craniotomies – a preliminary report. Acta Neurochir (Wien) 61:18–19
17. Manwaring KH, Hamilton AJ (1996) Neurosurgical endoscopy. In: Tindall GT, Cooper PR, Barrow DL (eds) The practice of neurosurgery, vol 1. Williams and Wilkins, Baltimore, pp 233–242
18. Mixter WJ (1923) Ventriculoscopy and puncture of the third ventricle. Boston Med Surg J 188: 277–278
19. Renier D, Lacombe J, Pierre-Kahn A (1984) Factors causing acute shunt infection: computer analysis of 1174 operations. J Neurosurg 61: 1072–1078
20. Sainte-Rose C (1991) Third ventriculostomy. In: Manwaring KH, Crone KR (eds) Neuroendoscopy. Liebert, New York, pp 47–52
21. Sayers MP, Kosnik EJ (1976) Percutaneous third ventriculostomy: experience and technique. Childs Brain 2: 24–30
22. Vries JK (1978) An endoscopic technique for third ventriculostomy. Surg Neurol 9: 165–168
23. Walker ML, Carey L, Brochmeyer DL (1995) The neuronavigational 1.2-mm neuroview neuroendoscope. Neurosurgery 3: 617–618

Endoscopic Stenting in Aqueductal Stenosis

M. J. Fritsch and K. H. Manwaring

Introduction

Extracranial shunting is the most common procedure for treatment of hydrocephalus. Alternative procedures such as endoscopic septostomy and endoscopic third ventriculostomy not only provide a more natural route for CSF flow but also provide a possibility of reduction in the number of inserted shunts, and reduction in the number of surgeries for shunt revision [1]. Furthermore, low pressure syndromes secondary to overdrainage can be avoided. Shunt independence in an increasing number of patients with hydrocephalus was achieved by septum pellucidum septostomy and third ventriculostomy and reported by several authors [2–4, 7, 9].

Endoscopic septostomy and third ventriculostomy have the risk of more significant complications than ventriculoperitoneal shunts [8]. However, endoscopic fenestration when used in appropriate selected patients and when successful is usually a one-time procedure unlike a shunt which may require multiple revisions and bears increased risk of infection.

In patients with membranous occlusion of the aqueduct, a combined third ventriculostomy and aqueductal plasty has been described [5]. In this technique, a percutaneous transluminal angioplasty balloon catheter was inserted into the aqueduct of Sylvius, piercing through the membranous tissue, and followed by balloon inflation. This procedure resulted in CSF flowing through the aqueduct. In Oka's report [5], an additional third ventriculostomy was performed. After the successful combined approach of third ventriculostomy and aqueductal plasty, clinical symptoms of hydrocephalus disappeared, the ventricular system was reduced in size, and the basal cisterns reopened.

All treated patients showed patency to CSF flow at the aqueduct and floor of the third ventricle on postoperative MRI imaging studies. Only one patient out of 11 subsequently required a lumboperitoneal shunt.

In patients with recurrent membranous occlusion or in cases with more extensive obstruction or stenosis over a longer segment of the aqueduct, aqueductal plasty alone might not be sufficient to provide adequate long-term CSF flow. In these patients, aqueductal stenting may be considered as one future option of treatment.

Indications

Endoscopic aqueductal stenting is indicated in patients with obstructive hydrocephalus secondary to occlusion of the aqueduct of Sylvius of congenital or acquired nature. The hydrocephalus could be caused either by membranous occlusion of the aqueduct, primary aqueductal stenosis of unknown etiology, or more extensive obstruction typically secondary to tumor compression. The typical location of tumor leading to compression of the aqueduct is in the tectal plate, the dorsal thalamus, or the pineal region (Figs. 1–6). Further on, isolated fourth ventricle hydrocephalus may be treated by stenting.

Preoperative CT and MRI studies should reveal a hydrocephalic supratentorial ventricular system, a dilated third ventricle and the confirmation of obstruction at the aqueduct level. MRI should confirm the absence of a CSF-flow void signal in the aqueduct or a cine MRI study should show no flow. the third ventricle should be dilated to a transverse diameter of about 4–5 mm to allow safe manipulation and passage of the endoscope.

The risks of aqueductal stenting include infection, vascular injury, and injury to the eloquent brainstem structures adjacent to the third and fourth ventricle and the aqueduct.

Endoscopic Anatomy

The third ventricle communicates with the lateral ventricles through the foramen of Monro and posteriorly with the fourth ventricle through the aqueduct. The anterior floor of the third ventricle is formed by the optic recess, the optic chiasm, the infundibulum and the infundibular recess, and the mamillary bodies. The posterior floor is formed by the aqueduct of Sylvius, the posterior comissure, and the pineal recess. The cerebral aqueduct in an adult is about

Fig. 1.
Sagittal gadolinium-enhanced
MRI of a 14-year-old boy with
obstruction of the aqueduct
and moderate hydrocephalus
secondary to a tumor in the
tectal region

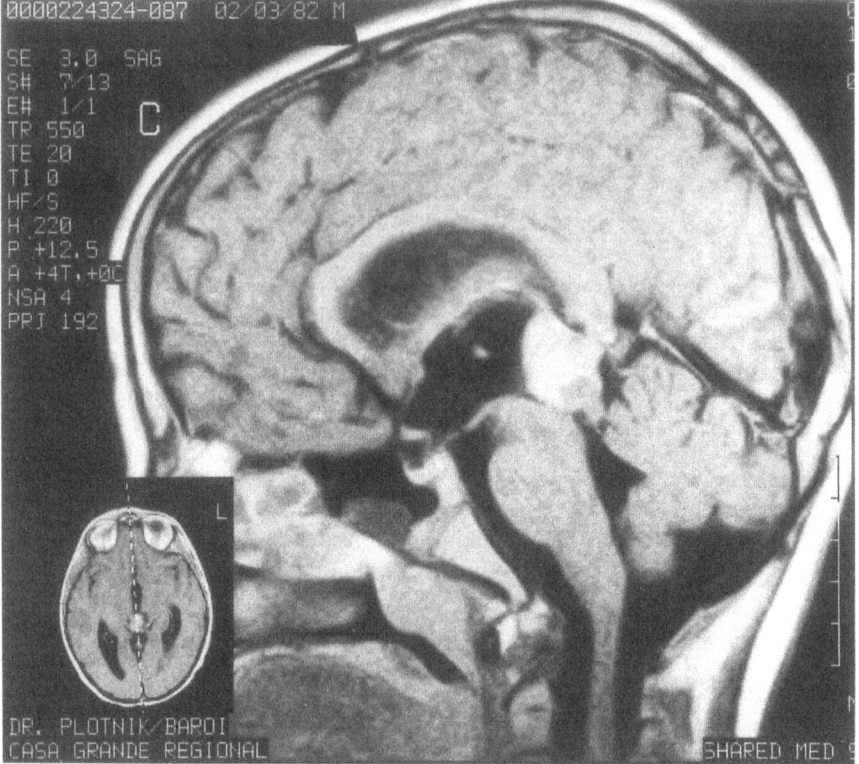

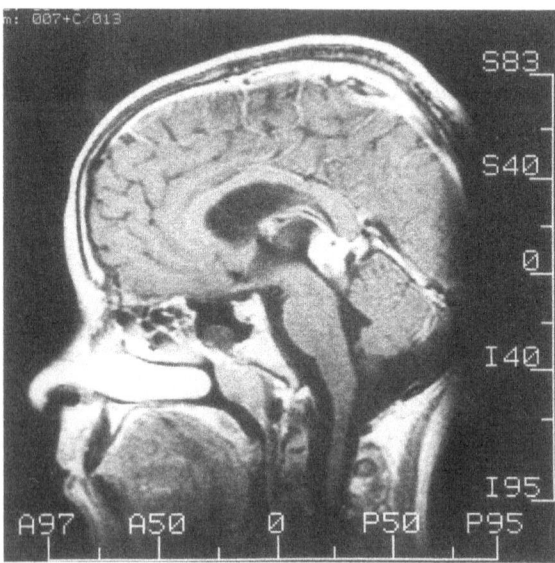

Fig. 2. The same patient as in Fig. 1 underwent right frontal burr-
hole approach endoscopic biopsy and aqueductal stenting. The
tumor was diagnosed as germinoma. MRI shows the stent within
the aqueduct

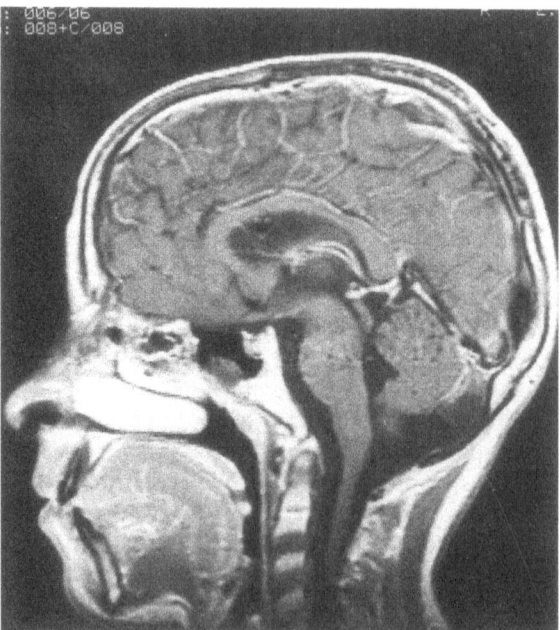

Fig. 3. Sagittal gadolinium-enhanced MRI of the same patient as
in Fig. 1 after radiation therapy

Fig. 4.
CT series on the same patient as in Fig. 1, 1 day after endoscopic aqueductal stenting

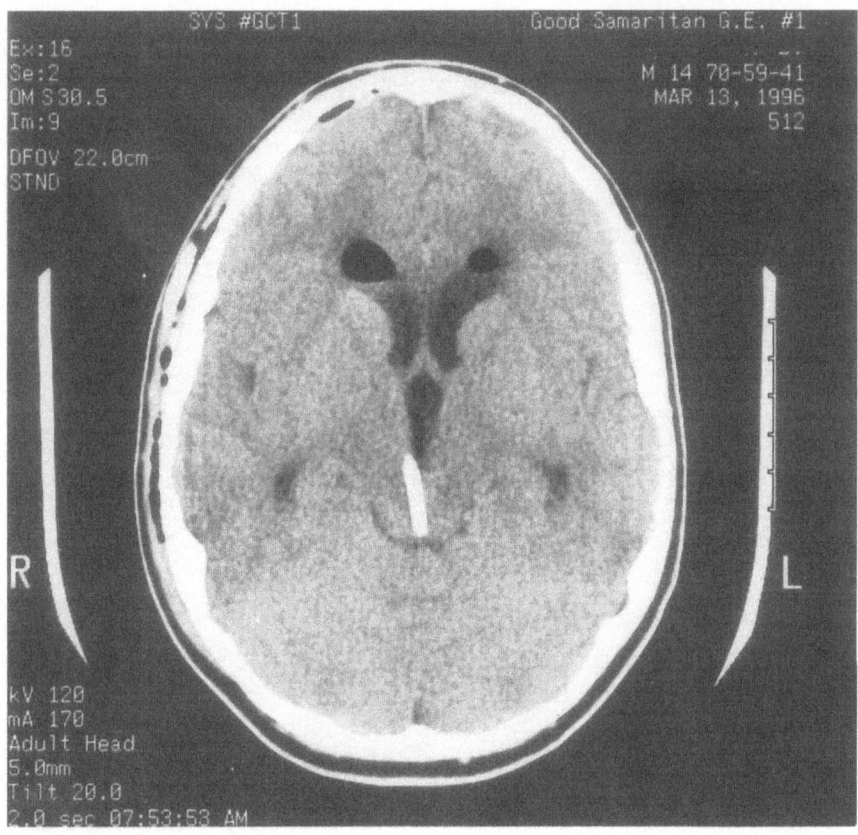

Fig. 5.
CT series on the same patient as in Fig. 1

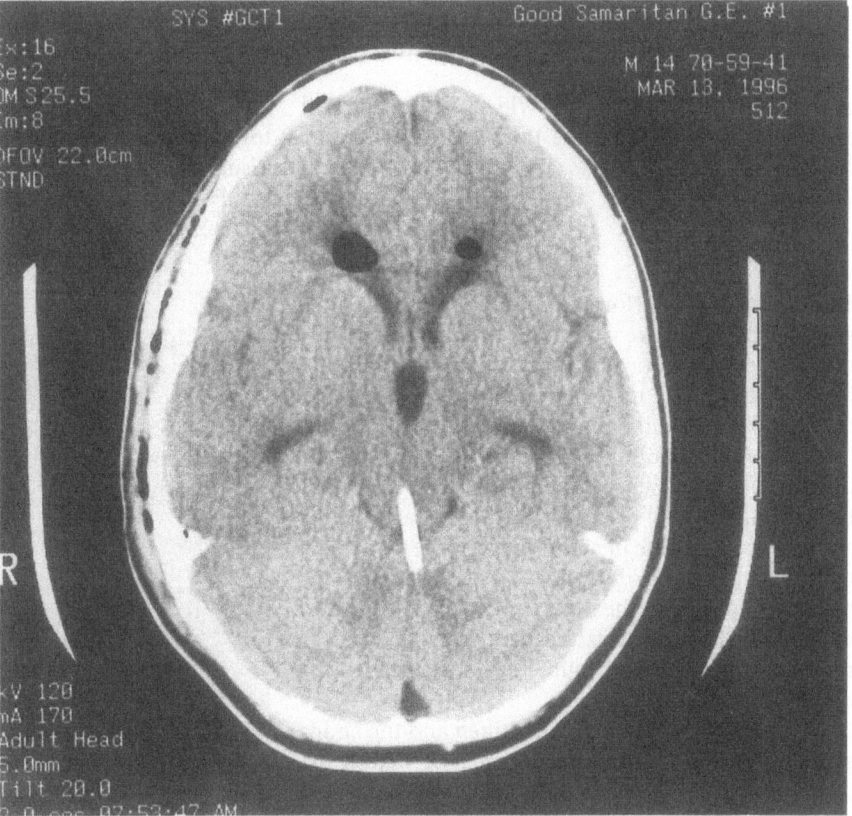

Fig. 6.
CT series on the same patient
as in Fig. 1

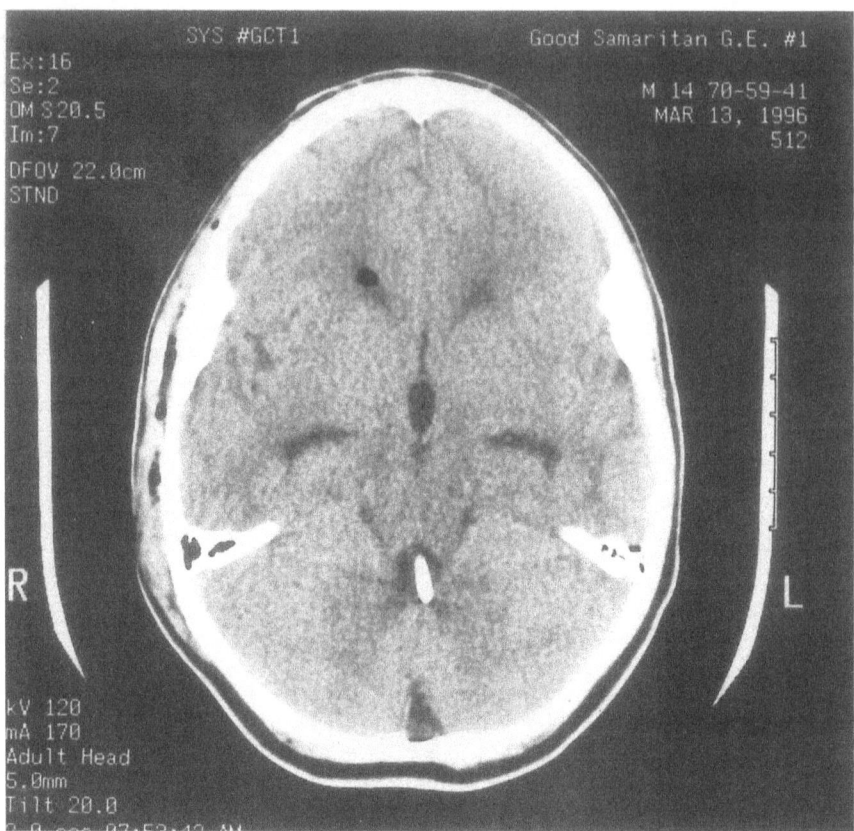

15 mm long and about 1 – 2 mm wide. A 4-cm stent will effectively communicate the third and fourth ventricle.

Operative Technique

The procedure is performed under general anesthesia, and preoperative antibiotic prophylaxis is administered. The patient is in a supine position, the head is neutral and rested upon a padded horseshoe frame. A right frontal burr-hole is place about 3 cm lateral from the midline and about 3 – 5 cm in front of the coronal suture, just behind the hairline. This more frontally than usual placed burr-hole allows a better straight access to the posterior part of the third ventricle when using a rigid endoscope. The dura is incised. A 12-peelaway sheath with stylus is passed into the anterior horn of the lateral ventricle. The stylus is removed, CSF is released, and the endoscope is introduced into the ventricle.

The foramen of Monro is identified and the endoscope is passed into the third ventricle. Within the third ventricle, the posterior anatomy is identified: the orifice of the aqueduct, the posterior comissure,

and the pineal recess. The endoscope is moved slowly towards the aqueduct for further inspection. After visual confirmation of occlusion, a soft-tipped 2- or 3-F Fogarty balloon catheter is passed through the instrument channel of the endoscope and the

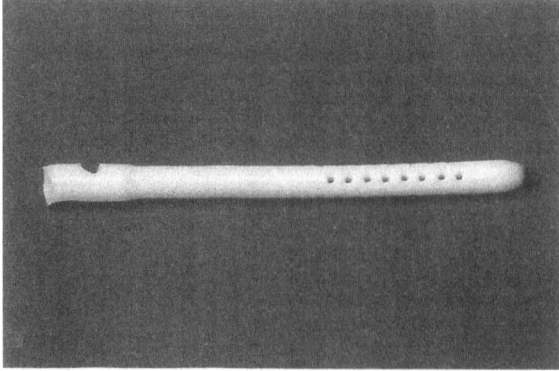

Fig. 7. The 4-cm-long proximal ventricular catheter of modified design. The proximal tip is rounded to prevent tissue damage during the process of advancement. An additional side hole was created with a rongeur on the distal-thickened reinforcing sleave

aqueduct is probed with the deflated balloon. If the deflated balloon passes readily, balloon dilation is commenced carefully until patency of the aqueduct is demonstrated by CSF flow. The endoscope is removed and a 4-cm long proximal ventricular catheter (stent) of modified design is carefully advanced through the peelaway sheath. It is tamped gently into the aqueduct by the reinserted endoscope.

This procedure leaves a straight stent with a widened sheath at the proximal end of the catheter. The widened part of the catheter holds the stent in place. No further fixation is needed. The distal catheter tip is uncut to prevent tissue damage or dislocation and disorientation during the process of the advancement of the stent into the direction of the fourth ventricle (Figs. 7–12).

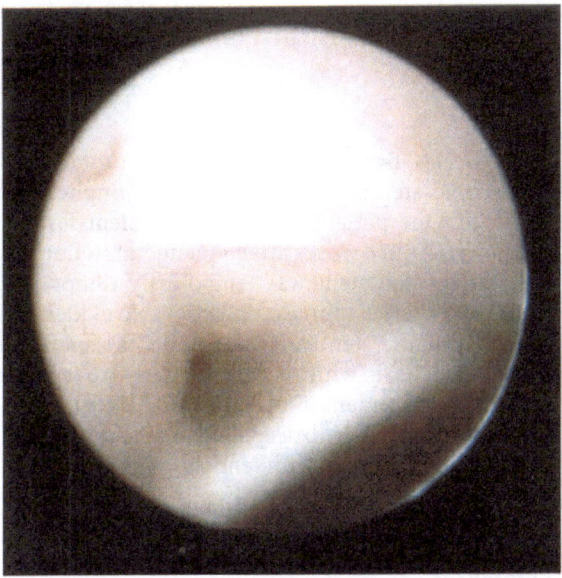

Fig. 8. Posterior floor of the third ventricle (orifice of the aqueduct and posterior comissure)

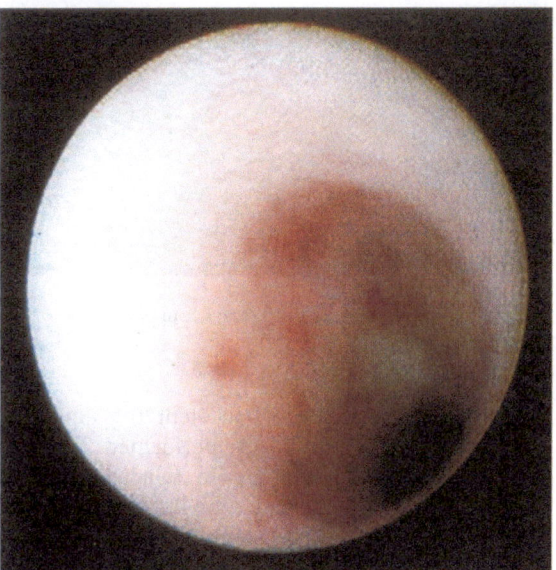

Fig. 10. Patency of the aqueduct of the third ventricle is seen after careful balloon dilation

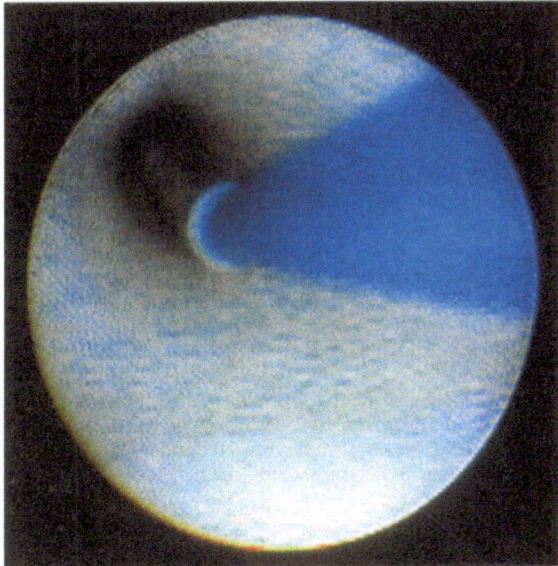

Fig. 9. The aqueduct of the third ventricle is probed with a soft-tipped deflated Fogarty balloon

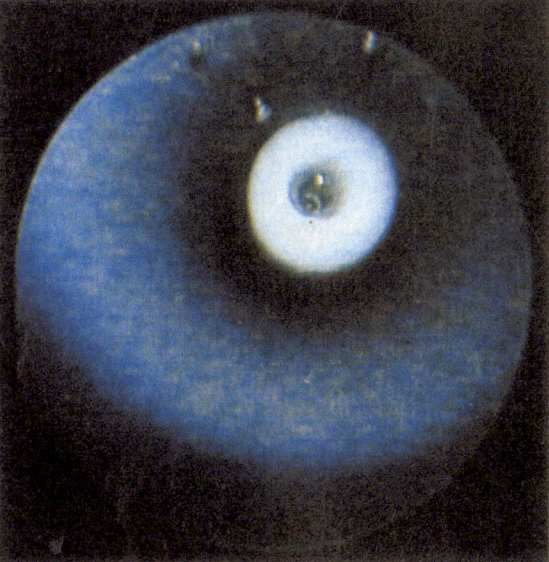

Fig. 11. The stent is advance through the peelaway sheath and tamped gently into the aqueduct of the third ventricle by the endoscope

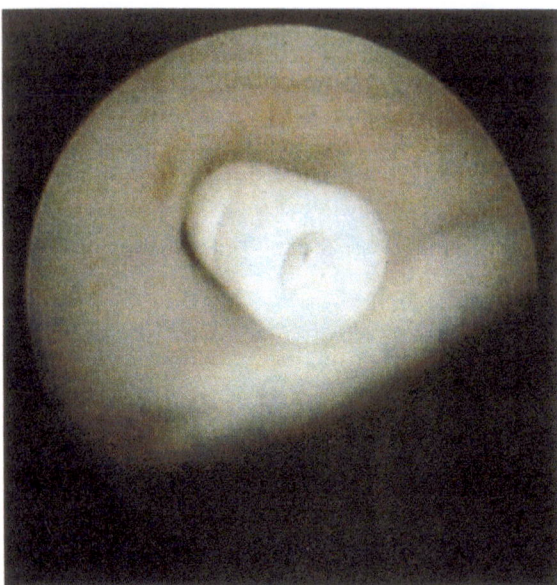

Fig. 12. Aqueductal stent of third ventricle in place

If it appears impossible to stent the aqueduct, a third ventriculostomy can be performed. A lumbar peritoneal shunt is required in selected cases to control the hydrocephalus by the lumbar route post aqueductal plasty, and stenting is required if communicating hydrocephalus is also present, such as in some inflammatory etiologies of aqueduct stenosis. A temporary external ventricular drain (EVD) is positioned through the peelaway sheath at the end of some stenting procedures to allow quantification of effective hydrocephalus control by the stent.

Postoperative Management

The patient is transferred to the ICU. If an EVD is employed, it is set at 10–15 cm of water and raised subsequently over the following 1 to 3 days. The adequate function of the aqueductal stent is demonstrated by decreasing CSF output and a neurologically stable patient without clinical findings of increased intracranial pressure. To prove patency of the stent, a water-soluble contrast medium (iohexol) can be injected into the supratentorial ventricular system and a CT scan can be obtained 1 to 2 hours later to demonstrate contrast within the fourth ventricle.

All patients should undergo MRI to show stent position, diminished ventricle caliber, and rule out injuries in the adjacent brain-stem parenchyma. Long-standing postoperative ventriculomegaly without clinical findings for increased intracranial pressure is typically described for patients after third ventriculostomy. Restoration of the size of the third ventricle, followed by restoration of the size of the lateral ventricles, may be a process of years [6]. Similar long-standing ventriculomegaly may be seen in patients after aqueductal stenting.

Results

In our series, five patients underwent a total of six endoscopic stenting procedures. No complications occurred. One patient developed transient fever of unknown origin 6 weeks after aqueductal stenting. In this patient, the stent was temporarily removed to rule out a possible source of infection. No evidence of infection of the stent or within the CSF space was found. Restenosis with recurrence of hydrocephalus required replacement of the aqueductal stent. Symptoms and signs of hydrocephalus disappeared in all five patients treated with endoscopics aqueductal stenting.

References

1. Hoffman HJ, Harwood Nash D, Gilday DL (1980) Percutaneous third ventriculostomy in the management of noncummunicating hydrocephalus. Neurosurgery 7:313–321
2. Jones RFC, Stening WA, Brydon M (1990) Endoscopic third ventriculostomy. Neurosurgery 26:86–92
3. Kelly PJ (1991) Stereotactic third ventriculostomy in patients with nontumoral adolescent/adult onset aqueductal stenosis and symptomatic hydrocephalus. J Neurosurg 75:865–873
4. Manwaring KH (1992) Endoscopic ventricular fenestration. In: Manwaring KH, Crone KR (eds) Neuroendoscopy, vol 1. Liebert, New York, pp 79–89
5. Oka K, Yamamoto M, Ikeda K, Tomonaga M (1993) Flexible endoneurosurgical therapy for aqueductal stenosis. Neurosurgery 33:236–243
6. Oka K, Go Y, Kin Y, Utsunomiya H, Tomonaga M (1995) The radiographic restoration of the ventricular system after third ventriculostomy. Minim Invasive Neurosurg 38:158–162
7. Sainte-Rose C (1992) Third ventriculostomy. In: Manwaring KH, Crone KR (eds) Neuroendoscopy, vol 1. Liebert, New York, pp 47–62
8. Teo C, Rahman S, Boop FA, Cherny B (1996) Complications of endoscopic neurosurgery. Childs Nerv Syst 12:248–253
9. Vries JK (1978) An endoscopic technique for third ventriculostomy. Surg Neurol 9:165–168

Endoscopic Treatment of Posthaemorrhagic Hydrocephalus in Premature Newborns

V. Urban, I. Tijsma, C. Busch and R. Schönmayr

Summary

Implantation of a ventriculoaterial or ventriculo-peritonal shunt system is the general procedure for treating posthaemorrhagic hydrocephalus in newborns. There is still a question as to whether this form of hydrocephalus is merely a CSF circulation blockage or is caused by dysfunctional resorption. With modern CT and dynamic MRI it is possible to obtain more exact evidence pertaining to changes of the ventricular system and basal cisterns.

In our clinic, between May and July 1995, five premature newborns with posthaemorrhagic hydrocephalus were treated by endoscopic ventriculocisternotomy. Our goal was to restore CSF circulation. The ages of the newborns ranged from the 30th week of pregnancy to 3 months of age. In one case an additional endoscopic fenestration was performed on a large prepontine cyst. In another case, it was possible to partially remove blood clotting found in the third ventricle endoscopically. In all five cases it was possible to restore CSF circulation. However, defective growths and partial bleeding were also discovered during surgery outside the ventricular cavities. Therefore it was necessary to insert ventriculoperitonal shunts in four of five children within 3 weeks – 6 months after the initial procedure. The case with the prepontine cyst also needed to be endoscopically reopened. Our results indicates that newborn posthaemorrhagic hydrocephalus seems to be caused by malresorption.

Introduction

The standard procedure for the treatment of post-haemorrhagic hydrocephalus in premature newborns is to allow first open drainage, with later insertion of either a ventriculoatrial or ventriculope-ritonal shunt [5].

Our goal is to prevent ventricular shunting by performing ventriculocisternostomy. If hydrocephalus is caused by CSF circulation disturbance, cisternostomy would be sufficient therapy.

Methods and Material

The ventriculocisternostomy was performed with a custom-made neonatoscope developed at our clinic (Zeppelin). The rigid-lens endoscope measures 3 mm in its outer diameter. It contains a 1-mm working canal. Other instruments used were scissors, grusping forceps, and a balloon catheter. During the endoscopic operation, the floor of the third ventricle was aimed. Fenestration of the membrane directly in front of the corpora mamillaria led to the prepontine basal cistern.

Ventriculostomies were performed on a total of five premature newborns with posthaemorrhagic hydrocephalus. Their ages ranged from the 30th week of pregnancy to 3 months of life. All operations were performed after sonographic and dynamic MRI diagnoses indicated necessary open ventricular drainage. No additional open-drainage surgery was performed postoperatively. Ventricular size and head circumference were measured daily. In the case of the prepontine cyst that was displacing the brainstem with its space-occupying lesion, additional fenestration was required. In another case blood clots could be endoscopically removed from the third ventricle prior to ventriculostomy (Fig. 1, 2).

Results

Unfortunately, four of five children needed postoperative shunting. During surgery serious lesions and blood clotting were discovered in the basal cisterns. In one case the lamina terminalis was additionally opened, where suprachiasmic adhesions and clotting were also found. Shunting was performed within 3 weeks – 6 months after the ventriculostomies. In one case requiring additional fenestration the cyst began to grow again, which resulted in a repeated brain stem displacement. Therefore renewed endoscopic fenestration preceded the shunting. In two cases hygromas formed near the endoscopic approach site within days after the otherwise problem-free ventriculostomies.

Fig. 1

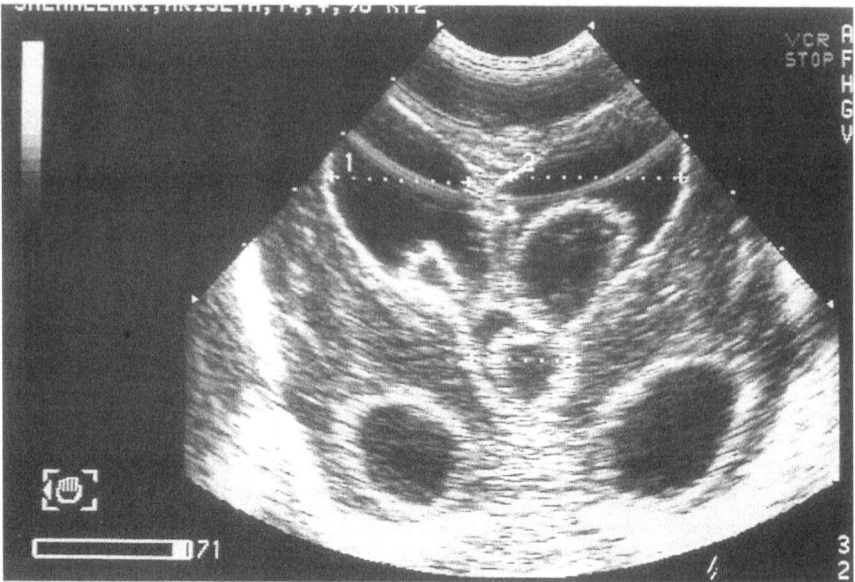

Fig. 2

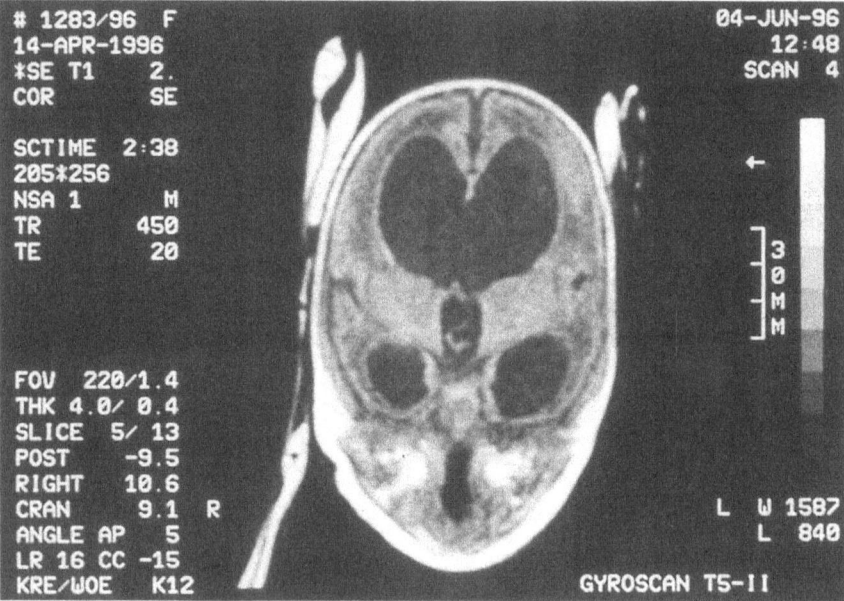

Discussion

Based upon our findings to date, endoscopic ventriculostomy alone is not sufficient as therapy for posthaemorrhagic hydrocephalus. Whether additional coagulation of the choroid plexus could improve results remains questionable and is a matter of controversy in the literature [1–4]. Ventriculostomy was performed safely in each case. We suspect that posthaemorrhagic hydrocephalus is caused by CSF malresorption.

References

1. Griffith et al. (1990) The treatment of childhood hydrocephalus by choroid plexus coagulation and artificial cerebrospinal fluid perfusion. J Neurosurg 4:95–100
2. Milhorat TH (1996) Choroid plexus and cerebrospinal fluid production. Science 166:1514–1516
3. Milhorat TH et al. (1976) Normal rate of cerebrospinal fluid formation five years after bilateral choroid plexectomy. J Neurosurg 44:735–739
4. Pople et al. (1993) Control of hydrocephalus by endoscopic choroid plexus coagulation – long-term results and complications. Eur J Pediatr Surg 3 [Suppl]:17–18
5. Pudenz RH (1981) The surgical treatment of hydrocephalus – an historical review. Surg Neurol 15 (1):15–26
6. Walker et al. (1995) Neuroendoscopic third ventriculostomy: a nursing perspective. J Neurosci Nurs 27 (2):78–82

Part 3
MIEN for Space-Occupying Lesions

Endoscopy of the Ventricular System: Indications, Operative Procedure, and Technical Aspects

F. Oppel, H. J. Hoff, and H. W. Pannek

Summary

This paper describes our technique of interventional endoscopy of the ventricular system through a 10-mm burr hole. This minimally invasive technique requires special tools developed in our clinic for comfortable and safe procedures.

Following MR image reconstruction, a 10-mm burr hole is placed frontally 3 cm parasagittally, and a working sleave is carefully inserted. With a 30° angled endoscope, intraventricular lesions can be identified and, if necessary, removed through the sleave.

This technique has been used for treatment of 14 patients: six with colloid cysts, six with hydrocephalus due to blocking of the foramen of Monro, and two on whom tumor biopsies were performed with partial or total tumor removal. Thirteen patients had either minimal or significant improvement. Due to bleeding (one patient), morbidity was 6%.

Interventional endoscopy with its minimal surgical trauma and short intra-operative time span represents an effective alternative treatment for intraventricular lesions to be completely resected.

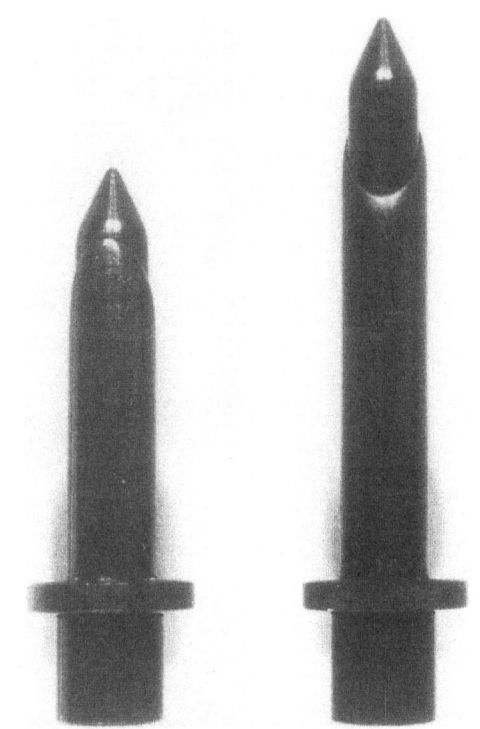

Fig. 1. Working sleeve of our group with outer diameter of 10 mm, inner diameter 9 mm, overall length 80/50 mm (top)

Introduction

To minimize surgical trauma, endoscopic techniques have been used as long ago as 1922 by Dandy in the treatment of intraventricular lesions [2–5, 7, 10]. The first therapeutic approach to the basal cisterns under endoscopic control was performed in 1978 [9–15].

Traditionally, wide-exposure craniotomies have been used to treat the majority of intra-ventricular lesions. Current technology allows for safer, less traumatic surgery through a 10-mm burr hole with a shortened operation time span, but the gold standard are the results of wide exposure by craniotomy.

To avoid intraventricular lesions, we developed a working sleeve (Fig. 1) used in endoscopic systems with special adapted tools (Fig. 2).

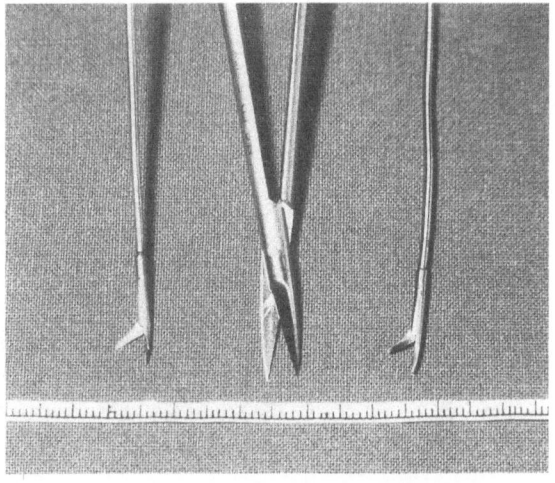

Fig. 2. Our specially adapted tool on the right and left in comparison with a standard microscissors (middle)

Fig. 3.
42-year-old male with a
colloid cyst

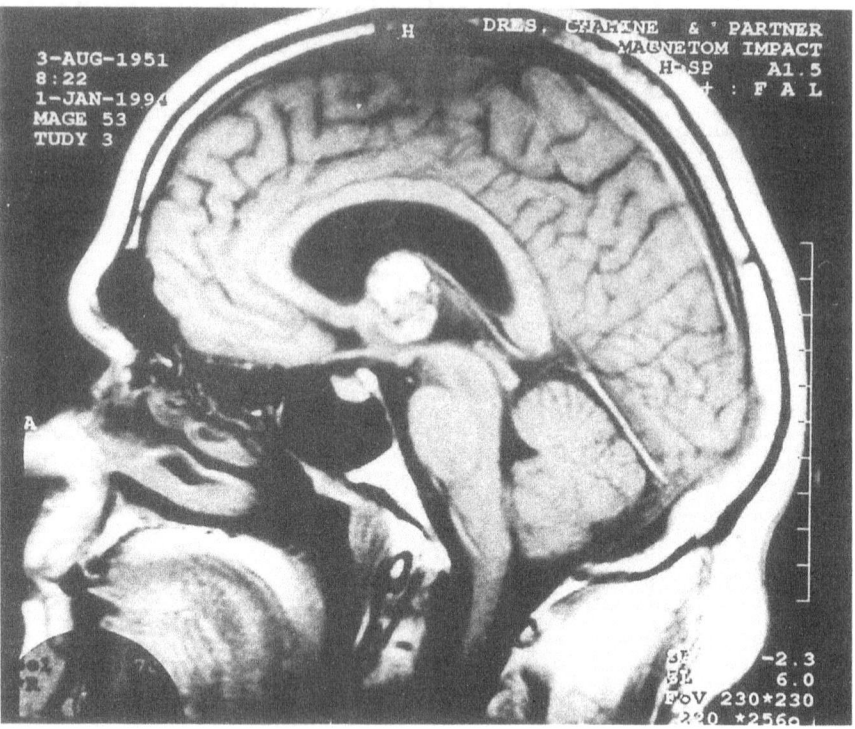

Fig. 4.
Same patient as in Fig. 3.
2 months after procedure

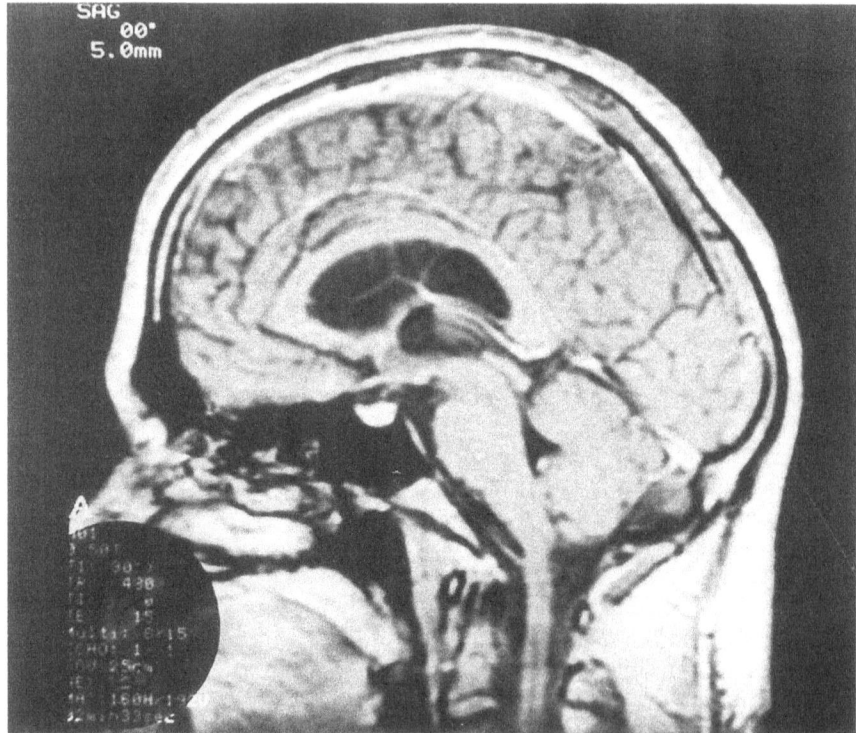

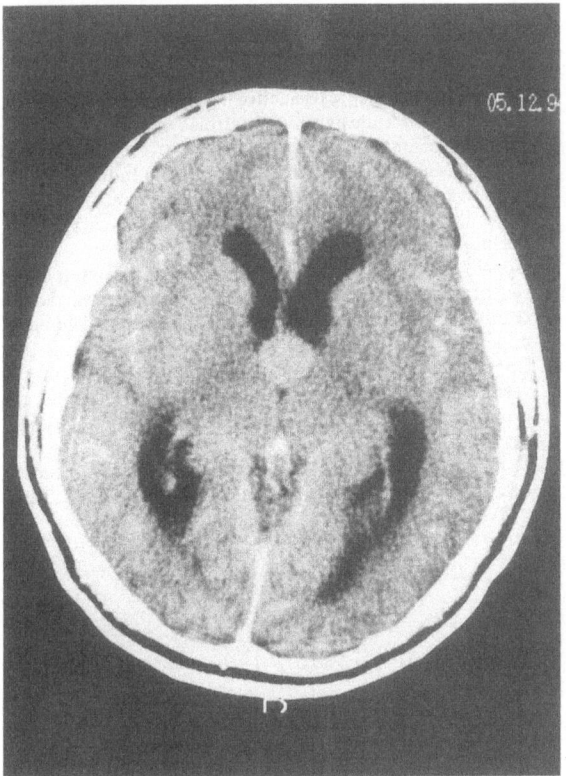

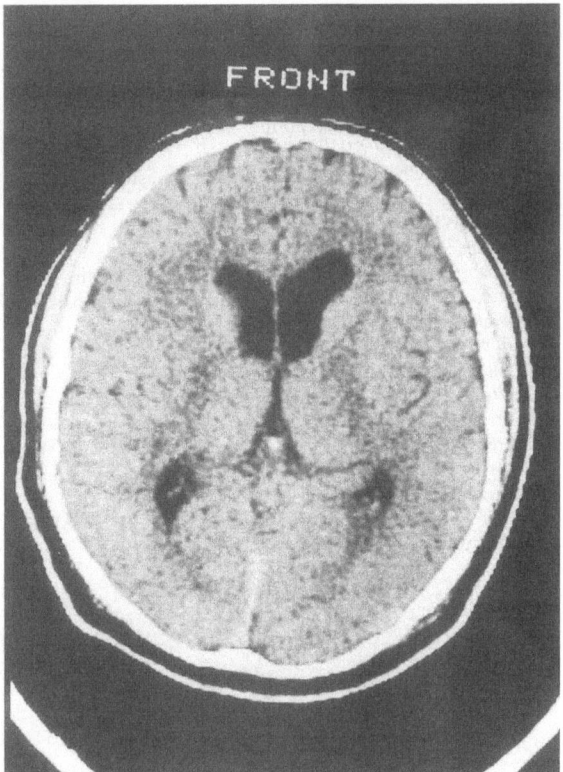

Fig. 5. 24-year-old male with a colloid cyst

Fig. 6. The same patient as in Fig. 5. 13 days after procedure at the day of demission

Methods and Materials

Through a 10-mm right or left frontal burr-hole, a working sleeve with stylet which was developed in our clinic is inserted. The outer sleeve diameter is 10 mm, the inner diameter 9 mm, and the overall length 80 mm.

Placement can be controlled stereotactically or with ultrasound guidance. A video unit connected to a rigid endoscope with a 30° angled optical system is used. Microscopic instruments and bipolar coagulation inserted through the sleeve can be controlled in the visual field of the scope. This provides enough space and flexibility to separate tumor tissue or membranes and remove intraventricular tumors through the tube.

Results

From July 1993 to May 1996, 14 patients (seven male, seven female) between 3 and 71 years of age were operated on intraventricularly using an 30° angled optical endoscope. The indications were colloid cysts in 6 cases, arachnoidal cysts in six cases, oc-clusion of the foramen of Monro in six cases, and intra- and periventricular tumors in two cases (Fig. 3, 4).

All patients were successfully treated; colloid cysts were totally removed (Fig. 5, 6). One patient received a shunt due to a persistent hydrocephalus. Thirteen of 14 patients had good results, one patient experienced bleeding which was successfully managed by immediate craniotomy. There was no mortality.

Discussion

Bleeding is a major problem in endoscopic surgery. With well-adapted tools, this problem can be controlled without wide exposure, but one should be prepared to perform a craniotomy if necessary. Minimal surgical trauma and a short operation time span are the main advantages for patients with colloid cysts and foramen of Monro membranes [1]. Endoscopic surgery presents a method for patients to remain shuntfree. The total removal of lesions normally performed in wide-exposure craniotomy can also be performed with our method using a working sleeve and a 10-mm burr-hole [6, 8].

References

1. Auer LM, Holzer P, Ascher PW, Heppner F (1988) Endoscopic neurosurgery. Acta Neurochir 90 : 1 – 14
2. Bauer BL, Hellwig D (1992) Minimally invasive neurosurgery I. Acta Neurochir (Wien) 54
3. Bauer BL, Hellwig D (1994) Minimally invasive endoscopic neurosurgery – a survey. Acta Neurochir 61 : 1 – 12
4. Dandy WE (1922) Cerebral ventriculoscopy. Bull Johns Hopkins Hosp 33 : 189
5. Fukushima T, Ishijima B, Hirakawa K, et al. (1973) Ventriculo-fiberscope: a new technique for endoscopic diagnosis and operation. J Neurosurg 38 : 251 – 256
6. Hall WA, Lunsford LD (1987) Changing concepts in the treatment of colloid cysts. J Neurosurg 66 : 186 – 191
7. Mixter WJ (1923) Ventriculoscopy and puncture of the third ventricle. Boston Med Surg J 188 : 277
8. Powell MP, et al. (1983) Isodense colloid cysts of the third ventricle: a diagnostic and therapeutic problem resolved by ventriculoscopy. Neurosurgery 13 : 234 – 237
9. Oppel F, Zeytountchian C, Mulch G, Kunft H (1978) Adv Neurosurg 5 : 269 – 275
10. Oppel F, Mulch G (1979) Selective trigeminal root section via an endoscopic transpyramidal retrolabyrinthine approach. Acta Neurochir Suppl (Wien) 28 : 565 – 571
11. Oppel F, Mulch G (1979) Selective trigeminal root section via an endoscopic transpyramidal retrolabyrinthine approach. Acta Neurochir Suppl (Wien) 28 : 565 – 571
12. Oppel F, Mulch G, Brock M, Zühlke D (1981) Indications and operative technique for endoscopy of the cerebellopontine angle. In: Samii M, Janetta PJ (eds) Cranial nerves. Springer, Berlin Heidelberg New York, 429 – 437
13. Oppel F, Mulch G, Brock M (1981) Endoscopic section of the sensory trigeminal root, the glossopharyngeal nerve and the cranial part of the vagus for intractable facial pain caused by upper jaw carcinoma. Surg Neurol 16 : 92 – 95
14. Oppel F, Handrock M (1984) Endoscopic section of the vestibular nerve by transpyramidal retrolabyrinthine approach in Mènière's disease. Adv Otorhinolaryngol 34 : 234 – 241
15. Temple F, Grant FC (1923) Ventriculoscope and intraventricular photography in internal hydrocephalus. JAMA 80 : 461

Endoscopic Management of Intracranial Arachnoid Cysts

H. W. S. Schroeder, and M. R. Gaab

Summary

Eleven patients with congenital arachnoid cysts were treated endoscopically. Seven of the cysts were located in the middle cranial fossa, two in the posterior cranial fossa, and two in the suprasellar prepontine region. The patients presented with headache, seizures, vomiting, nausea, dizziness, paresthesia, aphasia, balance problems, and precocious puberty. Using the Gaab universal neuroendoscopic system, we performed cysto-cisternostomies, ventriculocystostomies, and ventriculocystocisternostomies. Frameless infrared-based computerized neuronavigation has proved helpful for orientation. In the middle fossa cysts, a catheter was implanted between cyst and basal cisterns. In two cases, significant bleeding occurred and the procedure had to be stopped and instead continued microsurgically. The follow-up period ranged from 1 to 45 months. There was no mortality. One case of meningitis, one of transient oculomotor palsy, and a chronic subdural hematoma occurred after microsurgical cystocisternostomy, but no permanent morbidity. The symptoms resolved in nine patients and improved in one. Precocious puberty continued in one case. In ten cases, the cysts decreased in size. In one patient who presented with increasing headache 7 months after endoscopy, endoscopic examination revealed a sufficient cystocisternostomy. The results indicate that the endoscopic management of arachnoid cysts is a safe and effective treatment option and should seriously be considered as initial therapy. In case the endoscopic procedure fails, microsurgical fenestration or shunting can subsequently be performed without additional risk.

Introduction

Arachnoid cysts are developmental intraarachnoid collections of cerebrospinal fluid [7, 20, 28]. Due to the increased use of magnetic resonance (MR) imaging and computerized tomography (CT), arachnoid cysts are recognized with increasing frequency. These cysts may become symptomatic by expanding and compressing the surrounding brain tissue [17]. Many operative procedures have been recommended for the treatment of arachnoid cysts including microsurgical cyst excision [11, 21, 25, 30] or fenestration [2], stereotactic aspiration [18, 26], cyst fenestration with arachnoid plasty [33], cystocisternostomies [4], ventriculocystostomies [27], cystosubdural shunting [37], and cystoperitoneal shunting [1, 7, 17]. However, major complications and high failure rates have been reported. Hence, the effectiveness of a neuroendoscopic approach is investigated. We report our experience of 11 patients with arachnoid cysts treated endoscopically.

Material and Methods

From April 1993 to January 1997, 11 consecutive patients with arachnoid cysts underwent endoscopic treatment at our institution. All patients were followed prospectively. There were seven males and four females. Three patients were younger than 15 years at the time of surgery. The patients presented with headache, seizures, vomiting, nausea, dizziness, paresthesia, aphasia, balance problems, and precocious puberty. The neurological examination on admission was unremarkable in all cases except one, in which an unsafe Unterberger's and Romberg's test and blind gait deviation were observed. CT and MR imaging demonstrated seven middle fossa cysts, two posterior fossa cysts, and two suprasellar prepontine cysts with mass effect on the neighboring brain tissue.

Endoscopic Procedures

Cystocisternostomies, ventriculocystostomies, and ventriculocystocisternostomies were performed with the aid of the neuroendoscopic system developed by the senior author [14] and manufactured by Karl Storz (Tuttlingen, Germany). For the treatment of arachnoid cysts we prefer rigid rod-lens scopes because of their excellent optical quality. Details con-

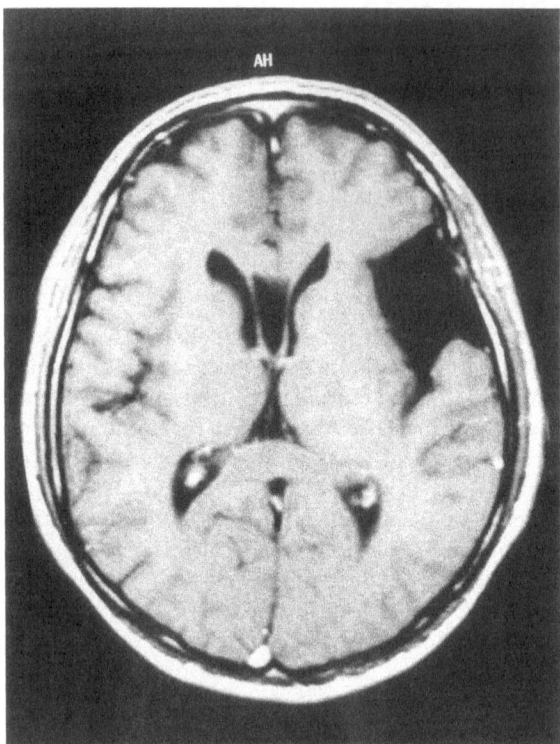

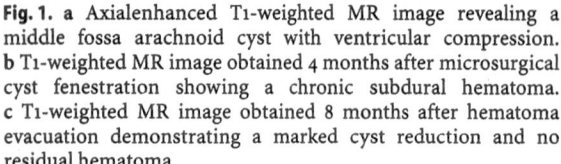

a

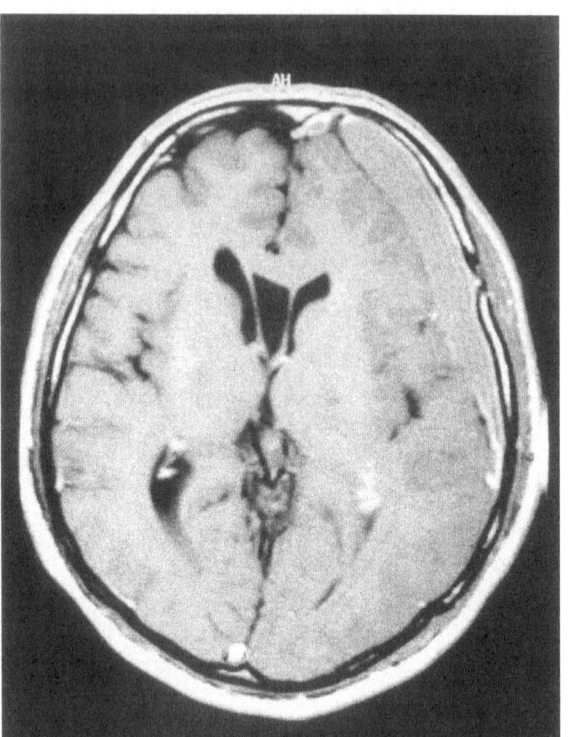

b

Fig. 1. a Axialenhanced T1-weighted MR image revealing a middle fossa arachnoid cyst with ventricular compression. **b** T1-weighted MR image obtained 4 months after microsurgical cyst fenestration showing a chronic subdural hematoma. **c** T1-weighted MR image obtained 8 months after hematoma evacuation demonstrating a marked cyst reduction and no residual hematoma

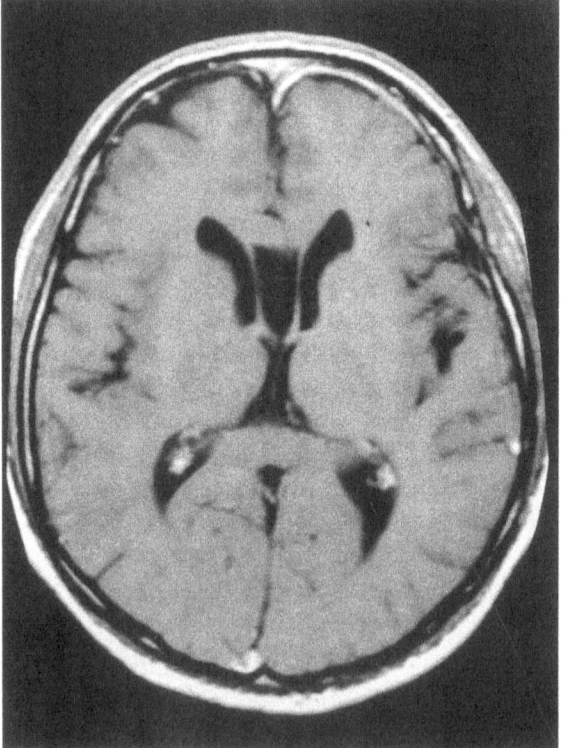

c

cerning the operative technique were reported previously [32]. Briefly, all procedures were done under general anesthesia. The operating field was prepared to allow immediate open microsurgical intervention in case of complications. In general, the endoscopic sheath was inserted free-hand according to the best trajectory seen on MR images. In two middle fossa cysts, we used infrared-based computerized neuronavigation(Surgical Microscope Navigator or SMN) manufactured by Carl Zeiss (Oberkochen, Germany). Neuronavigation was especially helpful in a case with minor bleeding which blurred vision. The suprasellar prepontine cysts were approached transcortically via the right lateral ventricle. In the middle and posterior fossa cysts, the burr hole was placed just over the cyst. The operating sheath was fixed with two LEYLA retractors. Once the trocar was removed, the diagnostic scope was inserted and the cysts inspected. After orientating, the diagnostic scope was replaced by the operative scope. In the posterior fossa cysts, a cystocisternostomy was performed by producing a wide opening between cyst and cisterns with scissors and grasping forceps. In the middle fossa cysts, a fimbrial ventricular catheter (Cordis, Miami, Fla., USA) was placed between cyst and adjacent basal cisterns to prevent occlusion of the cystocisternos-

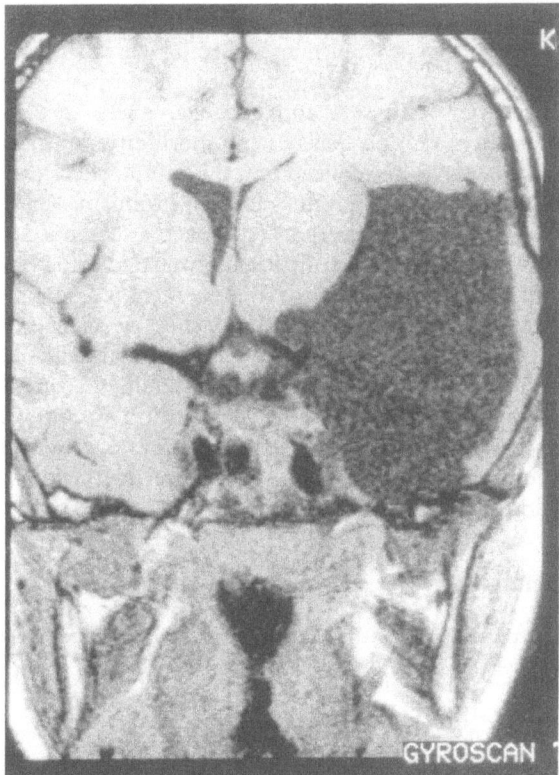

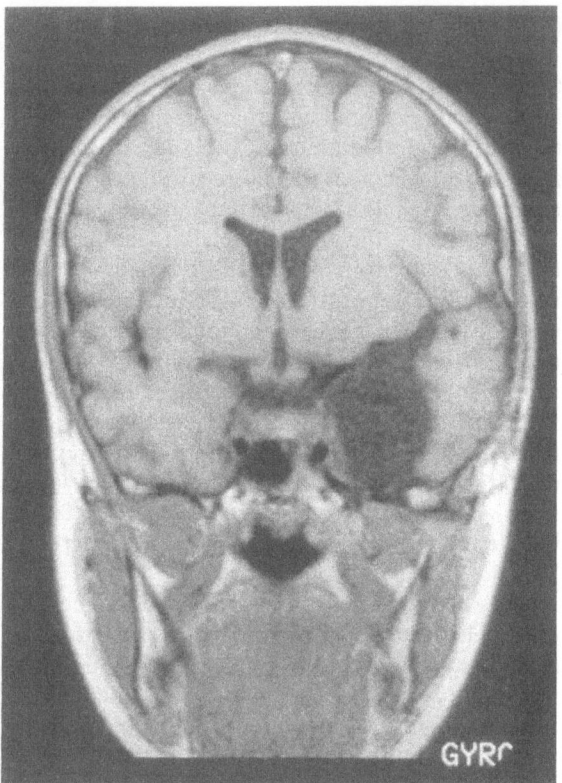

a

b

Fig. 2. a Coronal T1-weighted MR image showing a large middle fossa cyst with midline shift and ventricular compression. **b** T1-weighted MR image obtained 12 months after endoscopy revealing a marked decrease in size of the cyst with expansion of the ventricular system and nearly no midline shift

tomy by subsequent scarring. In one suprasellar-prepontine cyst, a ventriculocystostomy was performed in the bulged floor of the third ventricle in front of the mamillary bodies and a fimbrial catheter was inserted between cyst and ventricle. In the other suprasellar prepontine cyst, the cyst membrane that protruded through the enlarged foramen of Monro was widely resected and a second opening was created at the floor of the cyst (ventriculocystocisternostomy). The mean operation time was 70 min with a range of 35 to 120 min.

All procedures were performed under continuous irrigation with Ringer's solution. Minor bleedings stopped spontaneously. Larger vessels were coagulated with a bipolar coagulation probe. In two middle fossa cysts, the endoscopic procedure had to be abandoned because a significant bleeding blurred the vision and prevented orientation and safe operation. Therefore, the procedures were continued microsurgically.

Results

There was no mortality and no permanent morbidity. After inserting a catheter between carotid artery and oculomotor nerve in a middle fossa cyst, we observed a transient oculomotor palsy which resolved completely within 3 weeks. A case of meningitis occurred but responded quickly to antibiotics. In one middle fossa cyst in which the procedure had to be continued microsurgically, a chronic subdural hematoma developed after 4 months (Fig. 1). The hematoma was evacuated via a burrhole. Postoperative MR imaging revealed a marked decrease in size of the arachnoid cyst and no subdural hematoma.

The follow-up periods ranged from 1 to 45 months. The cysts diminished in size in ten patients (Fig. 2). To date, the cyst-related symptoms have resolved in nine patients and improved in one. In one patient who presented with increasing headache 7 months after endoscopy, endoscopic examination revealed a sufficient cystocisternostomy. The headache is of lower intensity and different kind. Precocious puberty continued in spite of a marked decrease in the size of the cyst. This patient was referred to an endocrinologist for endocrine treatment. The two patients who presented with seizures had no seizures after surgery and required no antiepileptic medication.

Discussion

Only a few reports on the endoscopic treatment of intracranial arachnoid cysts have been published in the neurosurgical literature [3, 5, 6, 9, 10, 12, 13, 15, 22, 24, 27, 36, 39]. Mostly, no details regarding endoscopic technique, complications, and outcome were given.

In our series, the endoscopic approach to arachnoid cysts has proven to be an effective and safe treatment option. Ten cysts decreased in size; one cyst remained unchanged. The symptoms resolved in nine patients and improved in one. Precocious puberty continued, although the cyst decreased markedly in size. Therefore, we consider this endoscopic ventriculocystostomy to be successful. Complete reversal of precocious puberty after decompression of suprasellar arachnoid cysts has been reported [8, 34, 35], but often the endocrine dysfunction never regressed [17, 27].

In two middle fossa cysts, the endoscopic procedure had to be abandoned due to significant bleeding that obscured vision. In spite of forced irrigation, a clear image could not be maintained. Hence, a microsurgical cyst fenestration (cystocisternostomy) was performed. In one of these cases, a subdural hematoma occurred after 4 months. After hematoma evacuation via a burrhole, the patient has been completely free of complaints. The postoperative MR image revealed a marked cyst reduction. In the other "microsurgical case", the headache which initially disappeared recurred after 7 months. As the cyst seemed to enlarge slightly, we performed an endoscopic revision, but we found the catheter connecting cyst and basal cisterns to be in place and patent. Since the headache is less intense than that experienced preoperatively and other factors are probably responsible for the more recent headache, no further interventions such as shunting were undertaken. In the strictly endoscopic procedures, we observed a case of meningitis and one of transient oculomotor palsy, but no permanent morbidity.

In the suprasellar prepontine cysts, we clearly observed a slit-valve-like arachnoid structure which opened and closed synchronously with arterial pulsation [31]. As other authors [6, 29], we suspect this slit-valve mechanism to be responsible for cyst formation and enlargement. In contrast, we did not find a valve-like structure in any of the middle or posterior fossa cysts. For the treatment of suprasellar arachnoid cysts, creation of a communication between ventricle and cyst (ventriculocystostomy) as well as between cyst and basal cisterns (cystocisternostomy) with destruction of the slit-valve mechanism is crucial [10].

In general, arachnoid cysts are treated by microsurgical fenestration/resection [2, 16, 19, 25, 30, 35, 38] or by cystoperitoneal shunting [1, 7, 17, 23]. However, controversy continues regarding which surgical treatment is best. The endoscopic procedures used in our series were safe and effective. There was no mortality and no permanent morbidity. Surgical trauma could be reduced to a minimum. Hence, we find endoscopic treatment to be a promising alternative to the microsurgical approach and shunting. However, the number of patients evaluated is too low and the follow-up periods are too short to compare our results with standard approaches. Nevertheless, we recommend endoscopic treatment of arachnoid cysts as therapy of first choice. Should the endoscopic procedure fail, established treatment options such as microsurgical fenestration or cystoperitoneal shunting can subsequently be performed without additional risk to the patient.

References

1. Arai H, Sato K, Wachi A, Okuda O, Takeda N (1996) Arachnoid cysts of the middle cranial fossa: experience with 77 patients who were treated with cystoperitoneal shunting. Neurosurgery 39 : 1108 – 1113
2. Artico M, Cervoni L, Salvati M, Fiorenza F, Caruso R (1995) Supratentorial arachnoid cysts: clinical and therapeutic remarks on 46 cases. Acta Neurochir (Wien) 132 : 75 – 78
3. Auer LM, Holzer P, Ascher PW, Heppner F (1988) Endoscopic neurosurgery. Acta Neurochir (Wien) 90 : 1 – 14
4. Barth A, Seiler RW (1994) Surgical treatment of suprasellar arachnoid cyst [letter]. Eur Neurol 34 : 51 – 52
5. Caemaert J, Abdullah J, Calliauw L, Carton D, Dhooge C, van Coster R (1992) Endoscopic treatment of suprasellar arachnoid cysts. Acta Neurochir (Wien) 119 : 68 – 73
6. Caemaert J, Abdullah J, Calliauw L (1994) Endoscopic diagnosis and treatment of para- and intra-ventricular cystic lesions. Acta Neurochir Suppl (Wien) 61 : 69 – 75
7. Ciricillo SF, Cogen PH, Harsh GR, Edwards MSB (1991) Intracranial arachnoid cysts in children. J Neurosurg 74 : 230 – 235
8. Clark SJ, Van Dop C, Conte FA, Grumbach MM, Berger MS, Edwards MS (1988) Reversible true precocious puberty secondary to a congenital arachnoid cyst [letter]. Am J Dis Child 142 : 255 – 256
9. Cohen AR (1993) Endoscopic ventricular surgery. Pediatr Neurosurg 19 : 127 – 134
10. Decq P, Brugières P, Le Guerinel C, Djindjian M, Kéravel Y, Nguyen JP (1996) Percutaneous endoscopic treatment of suprasellar arachnoid cysts: ventriculocystostomy or ventriculocystocisternostomy? Technical note. J Neurosurg 84 : 696 – 701
11. Dei-Anang K, Voth D (1989) Cerebral arachnoid cyst: a lesion of the child's brain. Neurosurg Rev 12 : 59 – 62
12. Dhooge C, Govaert P, Martens F, Caemaert J (1992) Transventricular endoscopic investigation and treatment of suprasellar arachnoid cysts. Neuropediatrics 23 : 245 – 247
13. Eiras Ajuria J, Alberdi Vinas J (1991) Traitement endoscopique des lésions intracranienne. A propos de 8 cas. Neurochirurgie 37 : 278 – 283
14. Gaab MR (1994) A universal neuroendoscope: development, clinical experience, and perspectives. Childs Nerv Syst 10 : 481
15. Grunert P, Perneczky A, Resch K (1994) Endoscopic procedures through the foramen interventriculare of Monro under stereotactical conditions. Minim Invasive Neurosurg 37 : 2 – 8

16. Hanieh A, Simpson DA, North JB (1988) Arachnoid cysts: a critical review of 41 cases. Childs Nerv Syst 4:92–96

17. Harsh GR IV, Edwards MSB, Wilson CB (1986) Intracranial arachnoid cysts in children. J Neurosurg 64:835–842

18. Iacono RP, Labadie EL, Johnstone SJ, Bendt TK (1990) Symptomatic arachnoid cyst at the clivus drained stereotactically trough the vertex. Neurosurgery 27:130–133

19. Jallo GI, Woo HH, Meshki C, Epstein FJ, Wisoff JH (1997) Arachnoid cysts of the cerebellopontine angle: diagnosis and surgery. Neurosurgery 40:31–38

20. Krawchenko J, Collins GH (1979) Pathology of an arachnoid cyst. Case report. J Neurosurg 50:224–228

21. Kurokawa Y, Sohma T, Tsuchita H, Kitami K, Suzuki S, Ishikawa A (1990) A case of intraventricular arachnoid cyst. How should it be treated? Childs Nerv Syst 6:365–367

22. Lange M, Oeckler R, Beck OJ (1990) Surgical treatment of patients with midline arachnoid cysts. Neurosurg Rev 13:35–39

23. Martinez Lage JF, Poza M, Sola J, Puche A (1992) Congenital arachnoid cyst of the lateral ventricles in children. Childs Nerv Syst 8:203–206

24. Merienne L, Leriche B, Roux FX, Devaux B (1992) Utilisation du laser Nd-YAG en endoscopie intracranienne. Expérience préliminaire en stéréotaxie. Neurochirurgie 38:245–247

25. Oberbauer RW, Haase J, Pucher R (1992) Arachnoid cysts in children: a European co-operative study. Childs Nerv Syst 8:281–286

26. Pell MF, Thomas DG (1991) The management of infratentorial arachnoid cyst by CT-directed stereotactic aspiration. Br J Neurosurg 5:399–403

27. Pierre-Kahn A, Capelle L, Brauner R, Sainte-Rose C, Renier D, Rappaport R, Hirsch J-F (1990) Presentation and management of suprasellar arachnoid cysts. Review of 20 cases. J Neurosurg 73:355–359

28. Robinson RG (1971) Congenital cysts of the brain: arachnoid malformations. Prog Neurol Surg 4:133–174

29. Santamarta D, Aguas J, Ferrer E (1995) The natural history of arachnoid cysts: endoscopic and cine-mode MRI evidence of a slit-valve mechanism. Minim Invasive Neurosurg 38:133–137

30. Sato H, Sato N, Katayama S, Tamaki N, Matsumoto S (1991) Effective shunt-independent treatment for primary middle fossa arachnoid cyst. Childs Nerv Syst 7:375–381

31. Schroeder HWS, Gaab MR (1997) Endoscopic observation of a slit-valve mechanism in a suprasellar prepontine arachnoid cyst: case report. Neurosurgery 40:198–200

32. Schroeder HWS, Gaab MR, Niendorf W-R (1996) Neuroendoscopic approach to arachnoid cysts. J Neurosurg 85:293–298

33. Shigemori M, Okura A, Takahashi Y, Tokutomi T (1996) New surgical treatment of middle fossa arachnoid cyst. Surg Neurol 45:189–192

34. Sweasey TA, Venes JL, Hood TW, Randall JB (1989) Stereotactic decompression of a prepontine arachnoid cyst with resolution of precocious puberty. Pediatr Neurosci 15:44–47

35. Turgut M, Ozcan OE (1992) Suprasellar arachnoid cyst as a cause of precocious puberty and bobble-head doll phenomenon [letter]. Eur J Pediatr 151:76

36. Walker ML, Petronio J, Carey CM (1994) Ventriculoscopy. In: Cheek WR, Marlin AE, McLone DG et al (eds) Pediatric neurosurgery, 3rd edn. Saunders, Philadelphia, pp 572–581

37. Wester K (1996) Arachnoid cysts in adults: experience with internal shunts to the subdural compartment. Surg Neurol 45:15–24

38. Yamakawa H, Ohkuma A, Hattori T, Niikawa S, Kobayashi H (1991) Primary intracranial arachnoid cyst in the elderly: a survey on 39 cases. Acta Neurochir (Wien) 113:42–47

39. Zamorano L, Chavantes C, Moure F (1994) Endoscopic stereotactic interventions in the treatment of brain lesions. Acta Neurochir Suppl (Wien) 61:92–97

The Use of Endoscopes During Surgery in the Suprasellar Region

J. A. Grotenhuis

Summary

During microneurosurgical procedures, especially at the skull base, lesions or parts of them may be hidden behind unretractable structures. Even the microinstruments themselves obscure the view at the involved structures, mainly because the operating microscope allows only a coaxial visual control.

Careful and judicious use of endoscopes extends the range of view and will lessen the need for extensive dissection and retraction, thus reducing the inevitable surgical trauma. For example, the use of endoscopes during surgery in the suprasellar region with its congeries of important neural and vascular structures will be described.

Introduction

Neurosurgeons are frequently confronted with lesions in the suprasellar region. Among the most common lesions encountered are pituitary adenomas, meningiomas, craniopharyngiomas, and aneurysms of the anterior circle of Willis. All of these lesions lead to displacement of numerous structures within the suprasellar region, such as the optic nerves and chiasm, the internal carotid arteries and their branches, the anterior cerebral arteries, the anterior communicating artery, the upper part of the basilar artery, the origin of both posterior cerebral and superior cerebellar arteries, the oculomotor nerves, the pituitary stalk, and the adjacent hypothalamus.

Depending on the exact site and size of a lesion, several surgical approaches can be used, but they all have their blind corners. The variety of viewing angles gained by using endoscopes, either through the same or a different approach, provides the neurosurgeon with an overview of otherwise hidden parts of the region invisible under the operating microscope [1, 2, 3].

Materials and Methods

Besides the usual setup for microneurosurgical operations, more technical equipment is necessary for endoscope-assisted operation. The most important part is the endoscope itself. Generally, one has to decide between an endoscope based on fiberoptics or one based on a glass-rod lens. The latter gives the best optical image, but the smallest diameter currently available is 4 mm, although reduction towards 3 mm will be soon a reality. Fiberscopes with an acceptable, although unequivocally less bright and clear image, have diameters as small as 1 mm. Incorporation of more fibers increases, image quality, but the fiberscope itself will become stiffer and thicker, hence losing its sole advantage over a lenscope.

Presently, we use the Neuroview Endoscope (Aesculap, Tuttlingen, Germany), a 4-mm lenscope, with 0° and 30° lenses and a 90° angled eyepiece, and the Murphy scope (Clarus, Minneapolis, Minn. USA), a fiberscope available in different shapes and lengths with a diameter of 1.5 mm. A light-weighted CCD color camera, a Xenon light source, and a high-resolution video monitor are further essential parts of the equipment.

We incidently started to use endoscopes during microneurosurgical procedures early in 1992, but only since the middle of 1994 is it considered as a standard tool in all open cranial procedures. Up to December 1996, it has been used to examine 297 patients (excluding the intraventricular endoscopic procedures).

As an example, we will describe two patients in detail to illustrate the use of an endoscope during an open procedure in the suprasellar region.

Case 1

A 47-year-old woman was admitted to the hospital with an acute subarachnoid hemorrhage (SAH). The CT scan (Fig. 1) revealed the SAH, but also a round lesion at the left anterior clinoid process, which was suspected to be a giant aneurysm. Angiography did not show any abnormality. MRI and MRA examina-

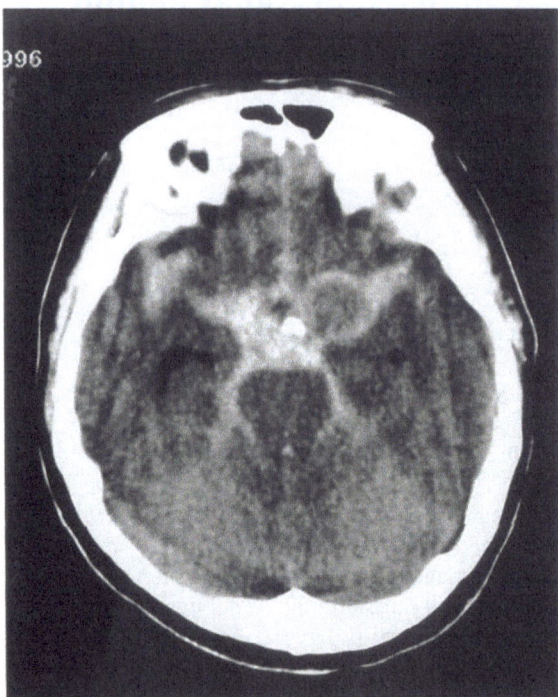

Fig. 1. CT of subarachnoid hemorrhage, more pronounced at the right side, as well as of the round lesion at the left clinoid region, which was initially thought to be a giant aneurysm

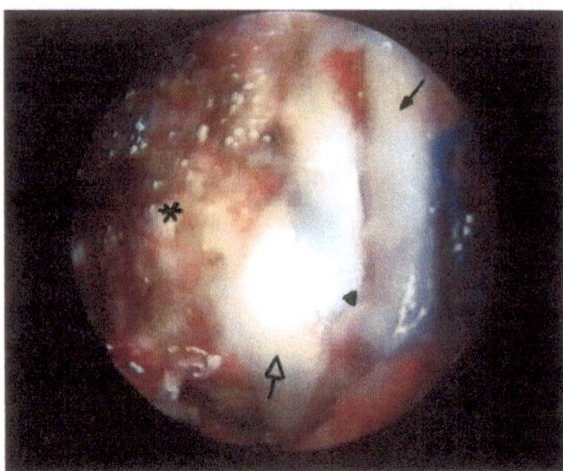

Fig. 3. Endoscopic picture of a meningioma with its attachment to the oculomotor nerve. *Asterisk*, meningioma; *small arrow*, tip of suction tube; *double arrow*, left internal carotid artery with the offspring of the posterior communicating artery; *large arrow*, left oculomotor nerve

tion also failed to show any vascular malformation. The round lesion at the left side showed the typical aspect of a meningioma (Fig. 2). Repeat angiography failed to demonstrate an aneurysm.

The patient was operated on through a small left-sided peritonal approach. The meningioma, which was firmly attached to the dura at the cavernous sinus, was dissected under the operating microscope. After some debulking of the tumor, the dense attachment to the oculomotor nerve was seen. Under endoscopic control the tumor could be dissected from the nerve (Fig. 3). After total removal of the meningioma, the anterior circle of Willis was inspected endoscopically. We found a thick arteria

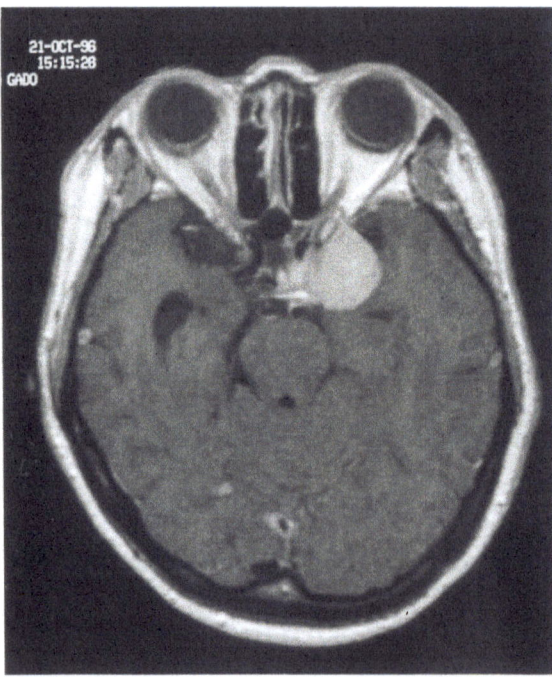

Fig. 2. MRI of the typical aspect of a meningioma at the left medial anterior tentorium

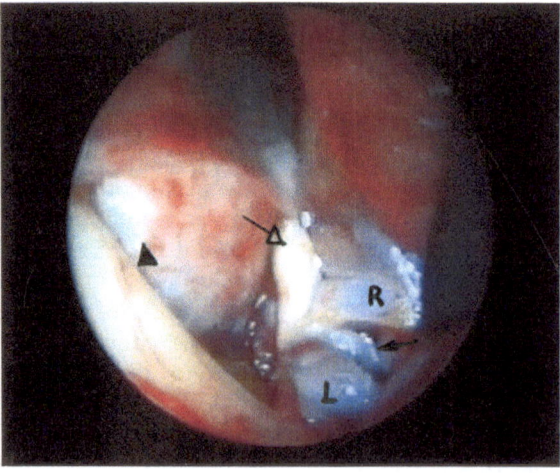

Fig. 4. Endoscopic picture of the anterior communicating artery complex *L* left A-2 segment of anterior cerebral artery; *R* right A-2 segment of the anterior cerebral artery; *large arrow*, A. mediana corporis callosi; *small arrow*, right optic nerve

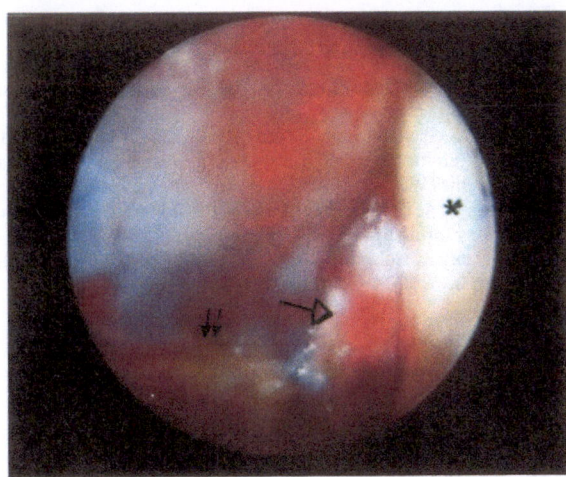

Fig. 5. Endoscopic picture after removal of meningioma. *Asterisk* dural attachment of the meningioma; *small arrow*, left internal carotid artery; *large arrow*, left oculomotor nerve

mediana corporis callosi at the anterior communicating artery and the typical yellow-colored adhesions that follow a subarachnoid hemorrhage, but no aneurysm in this region (Fig. 4). The right carotid artery showed a small cherry-red spot at its medial wall that could be considered to be a so-called blister aneurysm (Fig. 5). In this part, there were dense adhesions around the sellar diaphragm and the vessel itself. The offspring of the ophthalmic artery at the right side could be seen clearly with the endoscope, and there was no aneurysm. Finally, we decided to wrap the small blister aneurysm at the carotid artery wall.

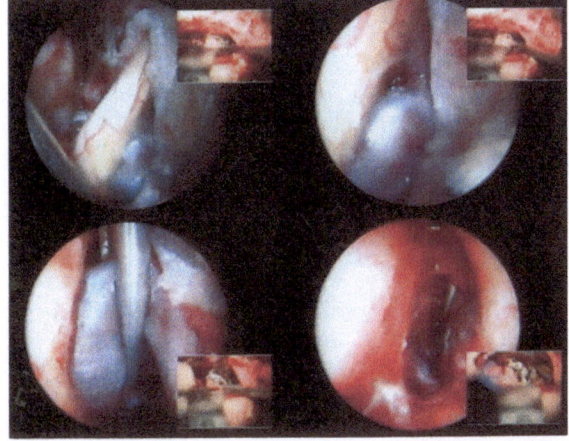

Fig. 6a–d. a Endoscopic picture of right optic nerve with opening of the optic canal. **b** Close-up picture of right optic nerve, ophthalmic artery and aneurysm, showing the small space between those structures where the clip blade has to be placed **c** Position of clip on aneurysm. The ipsilateral part of the clip can be seen perfectly. **d** Contralateral side of the clip. The endoscope was placed between both optic nerves in front of the (prefixed) chiasm, showing the exact position of the clip blade

Case 2

A 42-year-old woman was admitted to the hospital because of several attacks of acute headache and slowly decreasing vision of her right eye. CT scan, MRI examination, and angiography revealed a large aneurysm at the offspring of the right ophthalmic artery.

The patient was operated through a small right-sided supra-orbital craniotomy using an eyebrow incision. After initial dissection of the arachnoid membranes around the optic nerves, chiasm, and carotid artery and unroofing of the right optic canal (Fig. 6a), endoscopic inspection clearly showed the available space between the optic nerve, aneurysm neck, and ophthalmic artery for passing the clip blade (Fig. 6b). After placement of the clip, its correct position could be verified not only by viewing coaxial to the microscopic view along one clip blade (Fig. 6c) but also by viewing between both optic nerves (Fig. 6d) despite the fact that this patient had a prefixed chaism which would have been impossible to overlook by turning the operating microscope or even by enlarging the craniotomy. Besides the excellent endoscopic visualization, it should be noted that this verification could be achieved without retraction of brain, nerves or arteries.

Discussion

The concept of minimally invasive neurosurgery has evolved from developments in neurosurgical instrumentation, computer technology, revival of stereotactic techniques towards frameless navigation techniques and robotics, and neuroimaging allowing preoperative visualization of the individual patient's anatomy and leading to an individually planned surgical approach [3].

Within this concept of minimally invasive neurosurgery, there was a revival of interest in the use of endoscopes. The optical technology provided by modern endoscopes fit perfectly into the concept of minimizing the inevitable surgical trauma without loss of therapeutic efficacy or even with increasing it.

The usual microneurosurgical dissection is thwarted by the fact that it provides only a coaxial view and hence a coaxial control of the tip of the microinstruments. However, these microinstruments themselves can also obscure the view of important structures, which makes more retraction of neural and vascular structures necessary.

The ability of endoscopes to provide a wide variety of viewing angles allows visualization of extended regions of the brain surface and its subarachnoid spaces, complementary to those seen under the operating microscope.

However, endoscopes not only provide us with the possibility to "peek around corners", but they show the complete anatomy in a completely different visual dimension [4], which should be used in addition to the more familiar microscopic view for neurosurgical dissection. This will alleviate the inevitable operative trauma and will enhance the final outcome for the patient.

References

1. Grotenhuis JA (1996) Endoscope-assisted craniotomy. Techn Neurosurg 1 : 201 – 212
2. Matula C, Tschabitscher M, Day JD, Reinprecht A, Koos WT (1995) Endoscopically assisted microneurosurgery. Acta Neurochir (Wien) 134 : 190 – 195
3. Perneczky A, Tschabitscher M, Resch KDM (1993) Endoscopic anatomy for neurosurgery. Thieme New York
4. Perneczky A, Cohen A, George B, Kanno T (1994) Editorial. Minim Invasive Neurosurg 37 : 1

Endoscopic Management of Intracranial Arachnoid Cysts

N. J. Hopf, K. D. M. Resch, K. Ringel, and A. Perneczky

Summary

Large intracranial arachnoid cysts (ACs) are found to enlarge in the course of life. Inadequate communication between the cyst and the subarachnoid space is thought to be responsible for neurological symptomatology. An increasing number of patients present with space-occupying ACs but non-localizing or no symptoms at all. With standard neurosurgical techniques, i.e. microsurgical fenestration or shunting, clinical outcome and complication rate were unsatisfactory.

We evaluated 24 patients (seven female, 17 male) with intracranial ACs treated endoscopically between June 1993 and December 1996. Patient age ranged from 4 months to 65 years (mean 29 years), and follow-up time was 3 months to 2 years (mean 9 months). ACs were located in the anterior cranial fossa (two cases), in the temporal region (nine cases), suprasellar (six cases), and in the posterior cranial fossa (seven cases). Indications for surgery were hydrocephalus (eight cases), focal neurological deficit (three cases), obvious space-occupying character but non-localizing symptoms (11 cases), or no symptoms at all (two cases).

Surgical strategy was to create broad communication between the cyst and the subarachnoid space. Different techniques were used, i.e. endoscopic surgery (ES) in ten cases, endoscopic controlled microsurgery (ECM) in five cases, and endoscopic assisted microsurgery (EAM) in nine cases. In all 24 patients sufficient fenestration or partial resection of the cyst could be achieved. Favourable outcome was seen in 17 patients (71%), including the two asymptomatic patients, and six patients (25%) had unchanged symptomatology. One patient (4%), who is additionally operated on an astrocytoma of the quadrigeminal plate, deteriorated. Complications included infection in three patients (12%), bleeding into the cyst in one patient (4%), and subdural haematoma or hygroma in four patients (16%).

Our conclusions are: (a) different endoscopic techniques, i.e. ES, ECM and EAM do provide sufficient treatment of selected arachnoid cysts, (b) endoscopy makes looking and working "around the corner" possible and therefore enables inspection, fenestration or resection of cysts in different locations during the same minimally invasive approach, (c) patients with hydrocephalus or focal neurological deficits show the best results with improvement in 82%, and (d) the most frequent complications are infection and subdural hematoma.

Introduction

Intracranial arachnoid cysts (ACs) are characterized as cysts rising between the two sheets of the arachnoid membrane. Histologically, the membrane consists of fibrous filaments and arachnoid cells. Many theories have been proposed for the development and growth of ACs, including agenesis of brain structures [12, 14, 17], active fluid secretion [9] and arachnoiditis [7]. The pulsatile pump and the slit-valve mechanism [18] presently seem to be the most probable explanation. The pulsatile pup mechanism is responsible for CSF movements. A slit-like opening of the cyst may act as a functional one-way valve capable of building up a pressure gradient between the cyst and the subarachnoid space [19]. Imaging studies (cisternography, MRI) as well as intraoperative findings have confirmed these theories [6, 18]. However, inadequate communication seems to be responsible for either acute deterioration or a slowly progressive symptomatology long after first recognition of ACs.

The natural history of AC is unpredictable. The majority of patients with incidentally diagnosed asymptomatic ACs do not develop symptoms throughout life. Others become symptomatic with or without radiological evidence of enlargement of the cyst. There are also reports of spontaneous disappearance [3]. A study including 86 patients demonstrated that the majority of small ACs remains radiologically unchanged, whereas large ACs tend to grow [2].

Treatment strategies are based on two major considerations: inadequate communication between the cyst and the subarachnoid space or increased intracranial pressure despite sufficient communication. Surgical techniques include fenestration and resection of the cyst in order to create sufficient communication or shunting capable of dealing with both inadequate communication and increased intracranial pressure. Recent progress in minimally invasive techniques, i.e. keyhole surgery and neuronavigation, as well as the introduction of neuroendoscopy, has led to more frequent recommendations of fenestration or resection of cysts and makes shunting unnecessary. This is described especially for suprasellar cysts [1, 4, 11, 15, 18].

In this retrospective study, 24 patients with endoscopically treated ACs were evaluated. Endoscopic surgery (ES), endoscopic-assisted microsurgery (EAM), and endoscopic-controlled microsurgery (ECM) are differentiated and defined. Indications for endoscopy are summarized according to specific anatomical and pathophysiological features of cysts.

Clinical Materials and Methods

Patient Population and Preoperative Symptoms

Between June 1993 and December 1996, 24 patients (seven female, 17 male) had endoscopic procedures for ACs in the Department of Neurosurgery, University of Mainz, Germany (Tables 1, 2). Cysts were located in the anterior cranial fossa in two (8%) patients (one interhemispheric, one convexity), in the middle cranial fossa in 15 (62%) patients (nine sylvian fissure, six supra-

Table 1. Arachnoid cysts of the anterior and posterior cranial fossa

Pt	Localization	Indication	Proc.	OC clin	OC rad	Complic.	Comments
RAm21	Retrocereb.	HC, FND	ES	+	+	/	/
OMf40	Retrocereb.	FND	ES	+	+	/	/
FFf25	Supracereb.	HC, FND	EAM	=	=	/	Stp shunt, VO
RAm19	Retrocereb.	FND	EAM	+	=	Bleeding	VO
ARm30	Left CPA	SOC	EAM	+	+	/	/
HHm52	Supracereb.	FND	ECM	+	+	/	/
AMm23	Pineal reg.	SOC	EAM	+	+	/	/
KPm1	Frontal interh	HC	ES	+	+	Hygroma	SP shunt
WNm27	Frontal left	SOC	EAM	=	+	/	/

Proc., procedure; OC, outcome, complic, complications; HC, hydrocephalus; FND, focal neurological deficit; SOC, space-occupying character; ES, endoscopic surgers; ECM, endoscopic controlled microsurgery; EAM, endoscopic assisted microsurgery; +, improved or free; =, unchanged; −, deteriorated; Stp, status post; VO, ventriculostomy; SP, subduro-peritoneal.

Table 2. Arachnoid cyst of the middle cranial fossa

Pt	Localization	Indication	Proc.	OC clin	OC rad	Complic.	Comments
SPm2	Suprasellar	HC	ES	+	+	/	/
BAf15	Suprasellar	HC	EAM	+	+	/	Stp shunt
HAf54	Suprasellar	SOC	EAM	=	+	/	/
BTf2	Suprasellar	HC	ES	+	+	/	/
KJf3	Suprasellar	HC	EAM	+	+	/	/
TSm11	Suprasellar	HC	ECM	−	−	Sup.inf.	Astrocytoma
DCm30	Sylvian, left	SOC	ES	+	+	/	/
SCf34	Sylvian, left	SOC	ES	+	+	/	Preop. asym.
TEm64	Sylvian, left	PNLS	ECM	=	=	Inf., SDH	Operat. SDH
WWm20	Sylvian, left	SOC	ECM	+	+	/	Preop. asym.
SRm65	Sylvian, right	SOC	ES	+	=	/	/
SMm36	Sylvian, left	PNLS	ES	=	=	SDH	Operat. SDH
SKm46	Sylvian, left	PNLS	ES	=	=	Inf., SDH	Operat. SDH
MSm33	Sylvian, right	SOC	EAM	+	=	/	/
CMm43	Sylvian, right	SOC	ECM	+	+	/	/

Proc., procedure; OC, outcome; complic., complications; HC, hydrocephalus; PNLS, progressive non-localizing symptoms; SOC, space-occupying character; ES, endoscopic surgery; ECM, endoscopic controlled microsurgery; EAM, endoscopic assisted microsurgery; +, improved or free; =, unchanged; −, deteriorated; Sup., superficial wound; Inf., infection; SDH; subdural hematoma; Stp, status post; Operat., operation.

Table 3.
Definition of endoscopic
procedures

Procedure	Abbr.	Approach	Description
Endoscopic surgery	ES	Burr hole	Manipulation through endoscope
Endoscopic-controlled microsurgery	ECM	Burr hole, keyhole, all	Microsurgery under endoscopic control (= videosurgery)
Endoscopic-assisted microsurgery	EAM	Keyhole, all	Alternating use of microsurgery and endoscopy
Endoscopic inspection	EI	Burr hole, keyhole, all	No manipulation

sellar), and in the posterior cranial fossa in seven (30%) patients (three retrocerebellar, two supracerebellar, one cerebello-pontine angle, one pineal region). Patients ranged in age from 4 months to 65 years (mean 29 years). The follow-up period ranged from 3 months to 2 years (mean 9 months).

Symptoms were most frequently headaches (75%), followed by hearing disturbances (30%), gait disturbances (16%) and visual disturbances (16%). Other symptoms were seizures, nausea, dizziness and focal neurological deficits such as cranial nerve palsy. Two patients were asymptomatic (Tables 1, 2).

Indication for surgical treatment was hydrocephalus (HC) in eight cases, including two patients previously shunted presenting with shunt dysfunction, in three cases focal neurological deficit (FND) compatible with the cyst location and obvious space-occupying character (SOC) of the cyst in 13 cases, including three cases with progressive non-localizing symptoms (PNLS) and two cases with no symptoms at all (Tables 1, 2).

Preoperative radiological assessment (US, CT, MRI) demonstrated hydrocephalus in eight cases, three of them with unilateral widening of the lateral

ventricle. An additional two patients had a pre-existing shunt system. Patients were preoperatively investigated with CT cisternography or MRI flow sequence studies to demonstrate inadequate communication between the cyst and the subarachnoid space.

Follow-up evaluation was performed at discharge as well as 3 months and 1 year after surgery. Clinical outcome was differentiated as improved, unchanged and deteriorated on the basis of neurological examination and subjective statements of patients. CT scans of the head, performed in all patients before discharge, were indicated to screen for complications. All patients were asked to have CT cisternography or MR flow sequence studies 3 months after surgery, but results were available only from 11 patients.

Surgical Technique

Surgery was performed as exclusive endoscopic surgery (ES) through burr-hole trepanations, as endoscopic-assisted microsurgery (EAM) or as endoscopic-controlled microsurgery (ECM) through

Fig. 1.
Multifunctional fixation
device connected to a standard Leila retractor for use
with different endoscopes;
9.5-F Wolf cystoscope

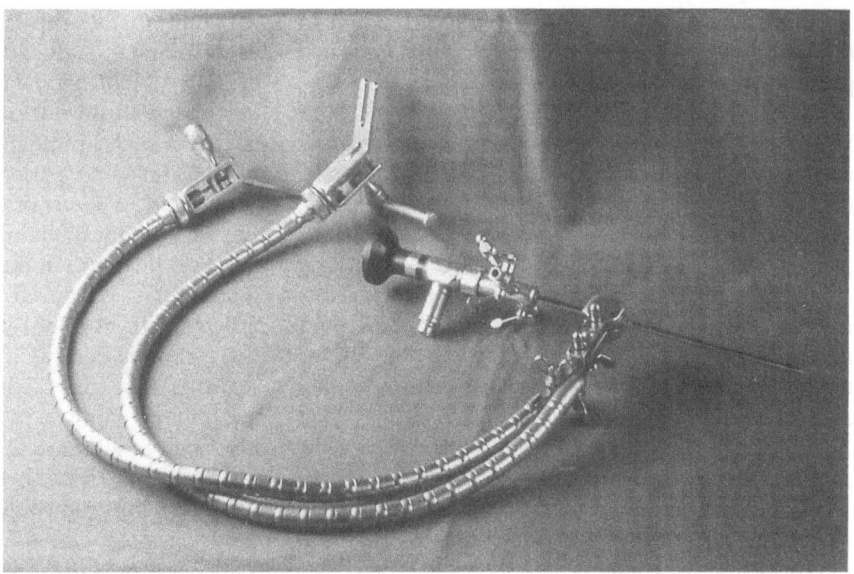

Fig. 2.
Endoscopic picture of a perforation made by bipolar coagulation and enlarged by ballooning with a Fogarty catheter

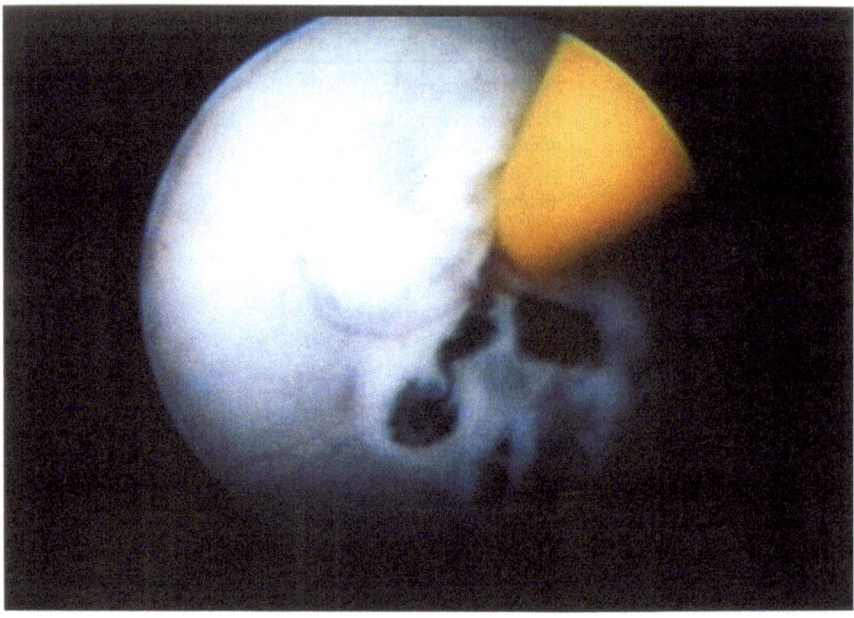

keyhole approaches. Procedures were considered to be ES when all manipulations were performed through the endoscope itself. Procedures were considered to be EAM when microsurgery and endoscopic surgery were performed independently during the same operation. They were defined as ECM when microsurgical technique and instruments were used under endoscopic control (also referred to as videosurgery). Cases where endoscopes were used exclusively for inspection are not included in this retrospective study (Table 3).

All procedures were performed under general anesthesia, on cysts in the anterior and middle cranial fossa in supine position with or without fixation of the head and on ACs in the posterior cranial fossa in a prone position with fixation of the head.

Rigid venticuloscopes with an outer diameter of 18 French (F) with 7-F optics, 5-F working channel and two 3-F channels for suction and irrigation (Aesculap, Tuttlingen, Germany; Wolf, Knittlingen, Germany), and a 13-F rigid cystocope (Wolf, Knittlingen Germany) with 8-F optics and two 4-F working channels were used. A fixation device (Aesculap, Tuttlingen, Germany) was used for the ventriculoscopes and another made by one of the authors [N. J. H.] for the cystoscope. The latter device was connected to a standard Leila retractor (Fig. 1).

Fenestration of cysts was achieved in cases of ES by monopolar and bipolar coagulation or by blunt perforation with grasping forceps or Fogaty catheters (Fig. 2) and in cases of EAM or ECM with standard microsurgical instruments.

Results

Endoscopic procedures were performed in ten cases as ES (42%), in nine cases as EAM (37%), and in five cases as ECM (21%). In all cases, sufficient fenestration, i. e. two or more locations with a minimal diameter of 1 cm, could be achieved (Fig. 3 a, b).

Fifteen patients improved (63%). Three of these patients showed improvement after a prolonged period of unchanged symptomatology. The two asymptomatic patients stayed asymptomatic (8%). Six patients had unchanged symptomatology (25%) throughout follow-up, and one patient (TSm11) who was additionally operated on an astrocytoma of the quadrigeminal plate after the cyst operation deteriorated (Tables 1, 2). Death or permanent neurological deficit did not occur, but transient nerve-III palsy was seen in one patient (TEm64). Best results were achieved in patients with posterior fossa or suprasellar ACs, or those presenting with hydrocephalus or focal neurological deficits compatible with cyst location. In the latter group, nine of 11 patients (82%) improved clinically. In the group of patients with non-localizing symptoms, only 5 of 11 (45%) showed clinical improvement (Tables 4, 5).

Out of six patients with untreated hydrocephalus, one became shunt-dependent (TSm11) and another developed a hygroma that required subduro-peritoneal shunting (KPm1). In one patient with treated hydrocephalus and multiple shunt revisions due to recurrent ventriculitis (FFf25), an endoscopic third ventriculostomy had to be performed before removal

Fig. 3.
Endoscopic picture of an
arachnoid cyst before **a** and
after **b** endoscopic perforation; 18-F Aesculap ventriculoscope

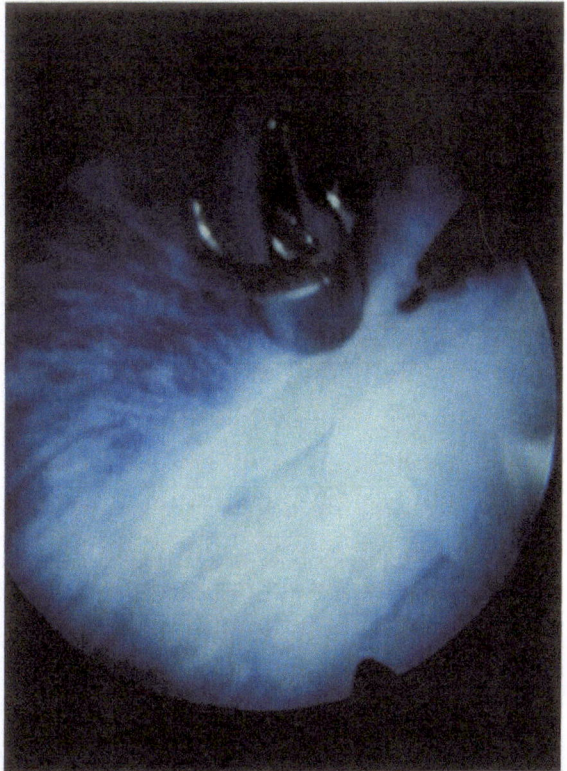

a

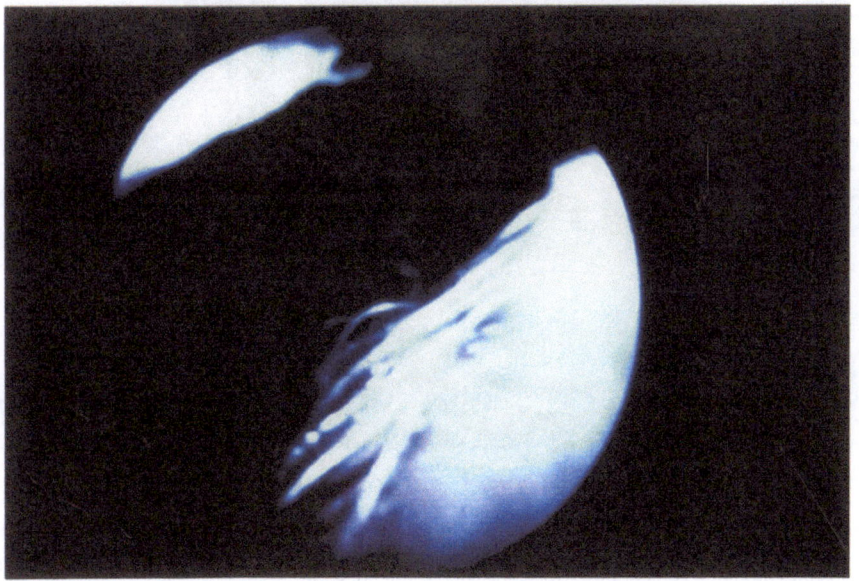

b

of the shunt. In the other patient with shunt-treated hydrocephalus (BAf15), the shunt could be removed, and the patient has since stayed shuntfree.

Radiological outcome showed improvement, i.e. reduction of size and space-occupying character and/or broad communication of the former cyst with the subarachnoid space in 16 patients (67%). Communication was demonstrated by MRI flow sequence studies or cisternography (Fig. 4a, b). In seven patients (29%), no radiological change was seen. Four of these patients also failed to improve clinically. One patient developed progressive hydrocephalus after a second operation for an astrocytoma of the quadrigeminal plate and became shunt-dependent (TSm11).

Simultaneous complications in two patients were meningitis and subdural haematoma, in a third pa-

Table 4.
Correlation of cyst location and clinical outcome

Localization	Improved/free	Unchanged	Deteriorated	n
Posterior fossa	6	1	0	7
Sylvian	6	3	0	9
Suprasellar	4	1	1	6
Frontal	1	1	0	2
Total	17	6	1	24

Table 5.
Correlation of indication for surgery and clinical outcome

Indication	Improved/free	Unchanged	Deteriorated	n
HC	6	1	1	8
FND	3	0	0	3
PNLS	0	3	0	3
SOC	8	2	0	10
Total	17	6	1	24

HC, hydrocephalus; FND, focal neurological deficit; PNLS, progressive non-localizing symptoms; SOC, space-occupying character.

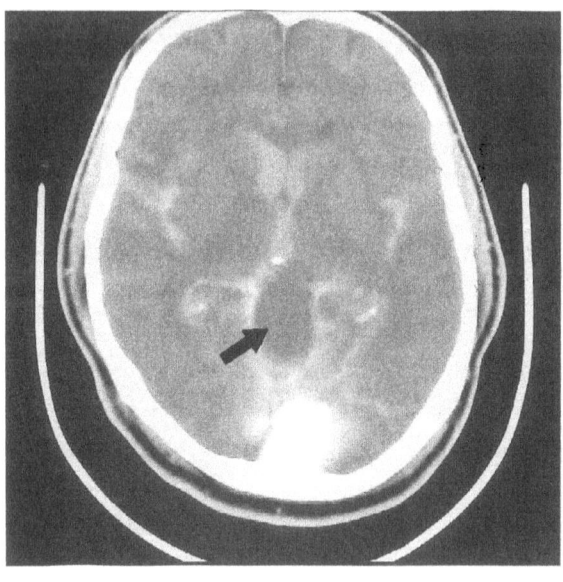 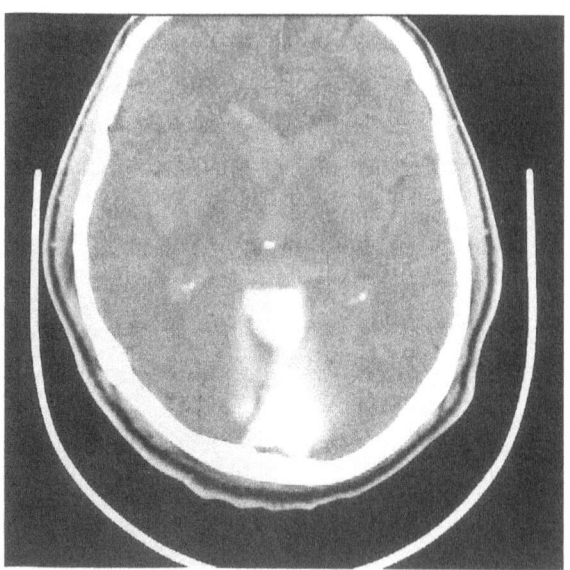

a   b

Fig. 4a, b. CT cisternography of a patient (HHm52) with retro- and supracerebellar arachnoid cysts. a Before endoscopic surgery showing communication of the retrocerebellar cyst but not with the supracerebellar cyst (*arrow*). b After endoscopic-controlled fenestration a broad communication of both cysts is demonstrated

tient bleeding into the cyst, in a fourth subdural haematoma, in a fifth bilateral subdural hygroma, and in a sixth superficial wound infection. Both patients with meningitis had previously been discharged in good health (TEm64, SKm46). On readmission, subdural haematoma was present at the site of burr-hole trepanation. After drainage of the haematoma and treatment of the meningitis, both patients recovered without additional neurological deficits but retained unchanged clinical symptomatology compared to the preoperative status. Bleeding occurred in a patient with a retrocerebellar AC (RAm19). The patient developed hydrocephalus and reclosure of the fenestration site. After treatment of hydrocephalus by endoscopic third ventriculostomy and a second endoscopic fenestration of the cyst, the patient recovered completely and finally improved compared to the preoperative status. Subdural haematoma was found in a 36-year-old patient

(SMm36) who initially became well. Three months later, he again developed headaches. After drainage of the haematoma, he did not significantly improve compared to his preoperative status. Bilateral subdural hygroma occurred in a 4-month-old boy with an interhemispheric AC (KPm1). He recovered completely after bilateral subduro-peritoneal shunting. Superficial wound infection occurred in a patient with a suprasellar AC and an astrocytoma of the quadrigeminal plate (TSm11).

Discussion

Indication

Commonly accepted indications for surgical treatment of intracranial ACs are focal neurological deficits compatible with cyst location and hydrocephalus with or without additional symptomatology [5, 8, 10, 13, 16]. There is still an ongoing discussion on patients with space-occupying ACs presenting with non-localising or minor symptomatology and those who are completely free of symptoms. These patients are at risk of being underestimated since the natural history and pathophysiology of ACs is still not fully understood. In addition, prevention of neurological symptomatology seems to be the future concept for more and more diseases due to the tremendous improvements in neuroradiological diagnostics and development of minimally invasive surgery, i.e. keyhole surgery and neuroendoscopy. This led us to investigate the outcome of patients with non-localising symptoms or, in two cases, no symptoms at all except obvious space-occupying character of the cyst. Treatment of such a lesion may be of utmost importance for patients with high-risk professions who are not allowed to continue their jobs despite good clinical condition, i.e. pilots, surgeons, truck drivers etc. In addition, electrophysiological studies and proper psychological testing may reveal subclinical neurological deficits in some of these patients.

Treatment Strategies and Diagnostics

Surgical strategies directed towards ACs have the goal (1) to establish a free CSF passage by extirpation or partial resection of cysts and (2) to create communication and a consecutive pressure balance by cyst fenestration or shunting. Surgical strategies dealing with hydrocephalus are external or internal shunting. Thus, CSF dynamics, including communication of the cyst with the subarachnoid space, and intracranial pressure are of diagnostic interest. Furthermore,

extension of the cyst, relation to and reaction of the surrounding structures, as well as thickness and vascularity of the cyst wall need to be evaluated preoperatively. MRI in combination with a flow sequence study has proved sufficient in demonstrating anatomical details but is questionable in analysis of CSF dynamics. Communication of the cyst may be falsely demonstrated in the case of thin membranes that are not detectable by MRI, and flow void signals may falsely be present, if a mobile membrane carries on CSF motion. Therefore, intrathecal iodine injection, i.e. cisternography, by ventricular or lumbar route is necessary to prove and quantify cyst communication. Comparison of MR flow sequence studies and cisternographies in selected patients of our series demonstrated such a discrepancy between MRI and cisternography. Our observations indicate that filling of the cyst even 6 hours after application still demonstrates inadequate communication, since intraoperative findings fail to visualize communication in these patients. ICP may be measured preoperatively or intraoperatively by standard techniques. In our series, elevated ICP was found intraoperatively only in those cases where MRI or CT studies typically demonstrated active hydrocephalus. Therefore, routine preoperative ICP measurements are not recommended. MRI including a flow sequence study should be the diagnostic study of first choice. When cysts create hydrocephalus due to direct obstruction of the CSF pathways, communication of the cyst with the subarachnoid space is less important. Cisternography is not mandatory in these patients, as long as surgery is directed towards the cyst itself. In all other cases, cisternography is crucial.

Surgical Technique

Different surgical techniques for the treatment of ACs have been described. Presently recommended surgical treatment is shunting or craniotomy with fenestration or resection of the cyst. Simultaneous shunting of cyst and ventricles or ventricular shunting with additional cyst fenestration is still associated with all complications of shunt procedures. A revision rate of 30% is commonly accepted [5]. Craniotomy and resection of the cyst may be difficult and is associated with major surgical complications, but simple fenestration of ACs poses the risk of reclosure of the fenestration site or elevated intracranial pressure [5, 9, 15, 16]. Therefore, we evaluated the use of endoscopic techniques in the management of ACs. Different endoscopic techniques were found to be useful: endoscopic surgery (ES), endoscopic-controlled microsurgery (ECM), and endoscopic-assisted microsurgery (EAM). ES, where all manipulations are performed through the endoscope it-self, has the advantage that only a burr-hole trepanation is required. In the case of an operation site deep within the brain parenchyma or close to vulnerable structures, the endoscope itself provides a reliable protection. Once the endoscope is brought into the right position and fixed thoroughly (Fig. 1), instruments may be changed ad libitum without danger to any of the structures passed. A disadvantage is the limited range of motion and size of instruments fitting through the working channels. This is es-pecially important for the control of bleeding. ECM, where microsurgical manipulations are performed under endoscopic control (i. e. videosurgery) requires only small keyhole approaches but may be performed during any microsurgical procedure. Advantages are the use of standard microsurgical instruments and the wide range of motion. A disadvantage is the risk of damage to the structures passed, which may be controlled by an additional microscopic picture. EAM, where endoscopic surgery and microsurgery is used alternatively during the same procedure, is per-formed to combine the advantages and to reduce the risks of both techniques (Table 3).

Outcome

In this retrospective study, we used all the techniques described, i. e. ES, ECM and EAM. The surgical goal was a fenestration of the cyst in order to create a pres-sure balance between the cyst and the subarachnoid space as well as to restore free CSF passage. Techni-cally sufficient fenestration or resection could be achieved in all 24 cases. Fenestration was performed with different techniques. Because of the elasticity and toughness of the membranes, blunt perforation is often not possible. Coagulation with a monopolar or bipolar device as well as other standard microsurgi-cal instruments were frequently used. Enlargement of the perforation is most quickly and safely done by ballooning with a Fogarty catheter. Endoscopic operations led to a favourable outcome in 17 patients (71%) (15 improved, 2 stayed asymptomatic) after a mean follow-up period of 9 months. Reclosure of the fenestration site was seen only in one patient with a retrocerebellar AC, following an intracystic bleeding from a dural sinus (RAm19). A second endoscopic operation led to successful fenestration and clinical improvement. Best results were achieved in patients with hydrocephalus or focal neurological deficits compatible with cyst location. In this group, nine of 11 patients (82%) improved clinically. In the group of patients with non-localizing symptoms, only five of 11 (45%) showed clinical improvement (Table 5). We also found a difference in the outcome related to cyst location. Patients with ACs of the posterior fossa showed improvement in six of seven cases (86%), whereas those with cysts of the sylvian region clinically improved only in five of nine cases (55%) (Table 4).

The relatively high overall success rate of 71% compared with microsurgical series [5, 10] was also seen in other endoscopic series [4, 11, 15] and may be due to the following aspects. Firstly, endoscopic tech-niques allow inspection and manipulation in regions that are difficult or impossible to reach with micro-surgical techniques. For example, the prepontine cistern down to the foramen magnum region may be inspected endoscopically but not microsurgically during a fenestration of a suprasellar cyst, ensuring a free CSF outflow to the spinal canal. Secondly, the use of ECM or EAM presents a sophisticated micorsurgi-cal technique for difficult manipulations that may be impossible through the endoscope alone. Thirdly, mean follow-up time is relatively short at 9 months. Due to the retrospective character of this study, careful interpretation is requested.

Conclusions

The term "minimally invasive neurosurgery" implies a concept directed at causing as little trauma as pos-sible to the patient [1]. Reduction in size of the trepanation, time of surgery, and hospital stay are some aspects of realizing this goal. Endoscopic tech-niques provide these requirements compared with our own microsurgical results in ACs in the past 4 years (n = 14). Fifteen patients (63%) had only burr-hole trepanation in the endoscopic group. The mean time of surgery was 1:59 h, i. e. 1:29 h for ES (ten cases) and 2:32 h for EAM (nine cases), but 2:40 h for microsurgical operations. The mean hospital stay was 8.2 days, including 0.5 days in the intensive care unit for the patients treated endoscopically, versus 15.2 days, including 3.7 days in the intensive care unit, for microsurgically treated patients.

Endoscopic techniques also have some disadvan-tages. Infection may be a problem, since sterilization of endoscopes and the associated equipment is still not completely solved. The infection rate was 8% for meningitis (two cases) and 4% for superficial wound infection (one case) in the present study. Single-shot antibiotic prophylaxis may reduce this complication. In addition, proper handling of irrigation is crucial. The most frequent complications were subdural haematoma (three cases) and hygroma (one case). However, subdural effusions seem to be more closely related to the specific aspects of ACs than to endo-scopic technique, since this is also a known problem in microsurgically treated patients [10].

In the present study, endoscopic techniques were found to be useful in the treatment of ACs. Hydrocephalus and focal neurological deficits compatible with cyst location are associated with the best results, i. e. favorable outcome in 82% of the cases. Diagnostic studies must be adapted to surgical strategy, i.e. individual anatomy, CSF pathways, and communication must be evaluated by MRI and cisternography. Endoscopic techniques should be chosen accordingly, i. e. ES, ECM or EAM. Risk of infection (8%) and subdural haematoma or hygroma (16%) have to be taken into account when patients are chosen for surgery. Further knowledge of CSF dynamics and pathophysiology of ACs may lead to adaptation of indication and surgical strategy in the treatment of ACs.

References

1. Bauer BL, Hellwig D (1995) Minimal invasive endoskopische Neurochirurgie (MIEN). D Ärtzeblatt 92 : 2062–2077
2. Becker T, Wagner M, Hofmann E, Warmuth-Metz M, Nadjmi M (1991) Do arachnoid cysts grow? Neuroradiology 33 : 341–345
3. Beltramello A, Mazza C (1985) Spontaneous disappearance of a large middle fossa arachnoid cyst. Surg Neurol 24 : 181–183
4. Caemaert J, Abdullah J, Calliauw L, Carton D, Dhooge C, van Coster R (1992) Endoscopic treatment of suprasellar arachnoid cysts. Acta Neurochir 119 : 68–73
5. Ciricillo SF, Cogen PH, Harsh GR, Edwards MSB (1991) Intracranial arachnoid cysts in children. J Neurosurg 74 : 230–235
6. Du Boulay GH, O'Connell J, Currie J, Bostick T, Verity P (1972) Further investigations on pulsatile movements in the cerebrospinal fluid pathways. Acta Radiol 13 : 496–523
7. Fox JL, Al-Mefty O (1980) Suprasellar arachnoid cysts: an extension of the membrane of Liliequist. Neurosurgery 7 : 615–618
8. Galassi E, Tognetti F, Frank F, Fagioli L, Nasi MT, Gaist G (1985) Infratentorial arachnoid cysts. J Neurosurg 63 : 210–217
9. Go KG (1995) The diagnosis and treatment of intracranial arachnoid cysts. Neurosurg Q 5 : 187–204
10. Go KG, Houthoff HJ, Hartsuiker J, Blaauw EH, Havinga P (1986) Fluid secretion in arachnoid cysts as clue to cerebrospinal fluid absorption at the arachnoid granulation. J Neurosurg 65 : 642–648
11. Jones RFC, Warnock TH, Nayanar V, Gupta JM (1989) Suprasellar arachnoid cysts: management by cyst wall resection. Neurosurgery 25 : 554–561
12. Kato M, Nakada Y, Ariga N, Kokuba Y, Makino H (1980) Prognosis of four cases of primary middle fossa arachnoid cyst in children. Childs Brain 7 : 195–204
13. Manwaring KH (1992) Endoscopic ventricular fenestration. In: Manwaring KH, Crone KR (eds) Neuroendoscopy, vol 1. Liebert, New York, pp 79–89
14. Markakis E, Heyer R, Stoeppler L (1979) Die Aplasie der perisylvischen Region. Neurochirurgia (Stuttgart) 22 : 211–220
15. Pierre-Kahn A, Capelle L, Brauner R, Sainte-Rose C, Renier D, Rappaport R, Hirsch JF (1990) Presentation and management of suprasellar arachnoid cysts. J Neurosurg 73 : 355–359
16. Rappaport ZH (1993) Suprasellar arachnoid cysts: options in operative management. Acta Neurochir (Wien) 122 : 71–75
17. Robinson RG (1964) The temporal lobe agenesis syndrome, Brain 88 : 87–106
18. Santamarta D, Aguas J, Ferrer E (1995) The natural history of arachnoid cyst: endoscopic and cine-mode MRI evidence of a slit-valve mechanism. Minim Invasive Neurosurg 38 : 133–137
19. Williams B, Guthkelch AN (1974) Why do central arachnoid pouches expand? J Neurol Neurosurg Psychiatry 37 : 1085–1092

Endoscopic Pituitary Surgery: Present and Future

G. S. Rodziewicz, R. T. Kelley, R. M. Kellman, and M. V. Smith

Summary

An approach to endoscopic pituitary surgery is outlined, with comments on otolaryngology (ENT) and neurosurgical operative technique. Problems encountered during this type of surgery are illustrated, and future developments of this technique are discussed. During transsphenoidal pituitary surgery, the use of endoscopes provides additional safety and operative ability.

Introduction

Endoscopic transnasal pituitary surgery combines transseptal transsphenoidal approaches to the pituitary [4, 7] with techniques adapted from endoscopic nasal sinus surgery [12]. Many surgeons have reported initial results in combining these approaches [1–3, 5, 6, 8, 9, 10, 11, 13]. In this article, the authors will present one approach to endoscopic transnasal pituitary surgery, discuss problems encountered during endoscopic transnasal pituitary surgery, and speculate on its future.

Technical Discussion

The techniques discussed here are more fully covered in [10]. Briefly, this approach to endoscopic transnasal pituitary surgery uses instruments introduced through a submucosal tunnel via one nostril and an endoscope introduced through the other nostril for visualization. The patient is placed in the reclining position with the head neutral and rotated to the patient's right. Both otolaryngologist and neurosurgeon stand on the patient's right, about level with the patient's chest (Fig. 1). A lumbar drain can be placed preoperatively, and the right lower quadrant of the abdomen may be prepared for a fat graft. An X-ray unit for lateral fluoroscopy is positioned rostral to the anesthesia machine on the side of the table opposite the surgeons. The operating microscope is balanced, but left undraped.

Fig. 1.
Operating room setup. Anesthesiology machine must be placed caudally enough to allow lateral fluoroscopy unit use during the operation

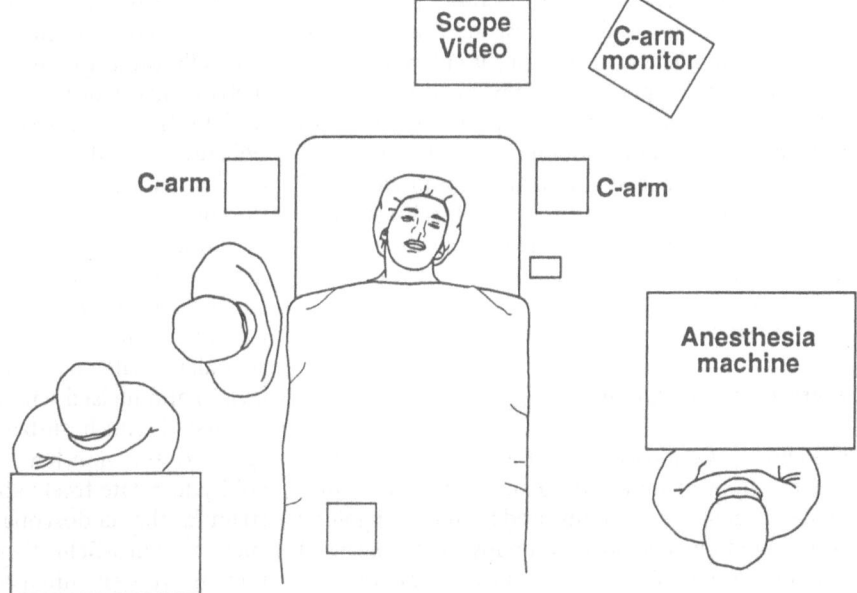

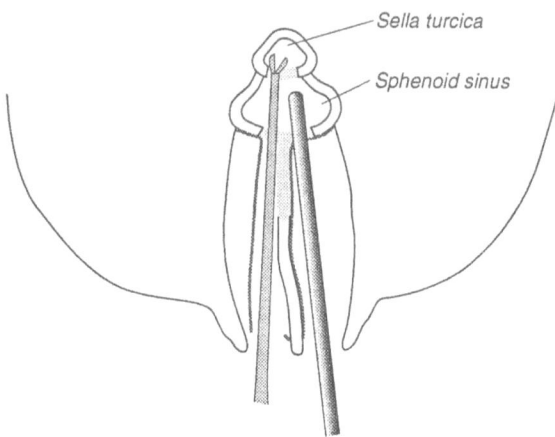

Fig. 2. Transnasal approach. Removal of the bony septum and wide bilateral sphenoidotomies allow both endoscope and instruments adequate room in the sphenoid sinus

ENT Technique

The nose is packed with 4% cocaine on cottonoid pledgets; then the membranous and bony septum bilaterally, as well as the caudal end of the cartilaginous septum, are infiltrated with 1% lidocaine with 1:100 000 epinephrine. The eyes are lubricated, the patient's face is prepared with betadine paint, and the patient is draped. A hemi-transfixion incision is made at the caudal end of the quadrilateral cartilage on the right and a mucoperichondrial flap then elevated on the left side (Fig. 2). The bony-cartilaginous junction is identified and subsequently separated vertically using a Cottle elevator. Bilateral mucoperiosteal flaps are then raised form the perpendicular plate of the ethmoid. Dissection is continued posteriorly along both sides of the bony septum to the sphenoid rostrum, and the bony septum is resected. The spenoid sinus is entered with an osteotome whose placement may be verified with lateral fluoroscopy. Kerrison rongeurs are used to perform a wide sphenoidotomy and the sphenoid sinus mucosa then is removed. At this point, the nasal speculum is replaced with a small self-retaining speculum which is opened only enough to maintain access to the sphenoid sinus for instruments or endoscopes.

Neurosurgery Technique

The choice of endoscope for this procedure is often dictated by availability. In general, we have found rigid endoscopes commonly used for otolaryngology or urology to be adequate for endoscopic transnasal pituitary surgery. These rigid endoscopes are typically 17 cm long, 2.7–4 mm in diameter with viewing

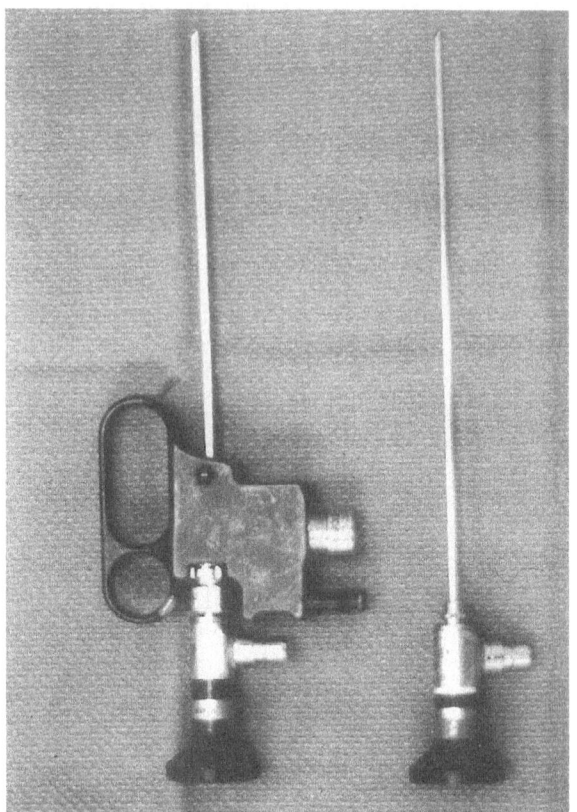

Fig. 3. Standard rigid endoscope on *right* and suction-irrigation sleeve (Wolf) placed over endoscope on the *left*. Such devices save time and frustration during endoscopic surgery

angles from 0° to 30° (Wolf or Storz). The endoscope is threaded through a 14-F peelaway catheter, then introduced via the left nostril through an 8-mm incision in the left posterior nasal mucosa and into the sphenoid sinus. Inspection of the sphenoid sinus and sella turcica identifies bony landmarks, and an instrument can be introduced into the sphenoid sinus to delineate the superior and inferior extent of the sella turcica by fluoroscopy. An auto-irrigating sleeve over the endoscope (e. g., Wolf: Fig. 3) is useful to clear the endoscope lens of blood.

Standard pituitary instruments are introduced through the submucosal tunnel in the right nostril. A small chisel and Kerrison rongeurs are used to open the anterior face of the sella turcica; then the dura is coagulated with an insulated suction coagulator (Valleylab) and incised with a scalpel. The tumor can be evacuated with both bayonet and non-bayonet ring curettes; the non-bayonet ring curettes have the ability to rotate freely about their long axis without striking the endoscope. Scissor-grip instruments such as Shea-Belushi scissors and micro-biopsy forceps are well-suited to this operation. A Frazier-type suction is used to evacuate blood pooling in the

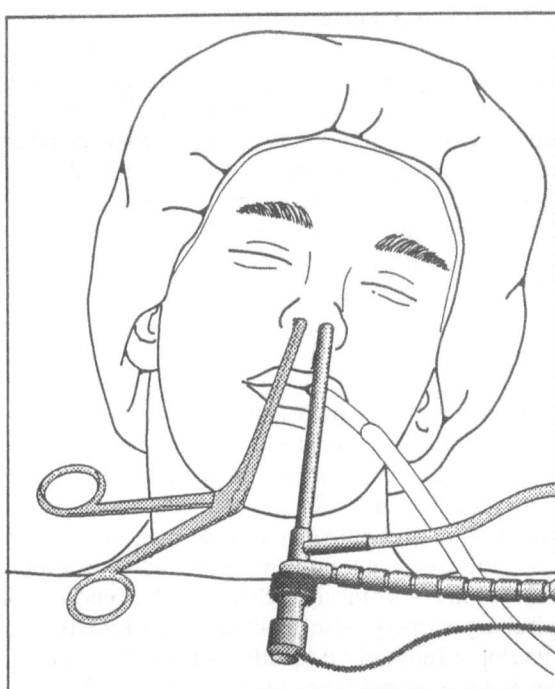

Fig. 4. Bilateral nasal approach. Horizontal angle between instruments and endoscope decrease interference with instrument use. If necessary, endoscope and instruments can be switched from one side to the other

sphenoid sinus, while brisk venous bleeding is best controlled by packing with cottonoids or Gelfoam. After the intrasellar tumor is debulked, the endoscope can be passed directly into the sella turcica to inspect the area of the cavernous sinuses laterally and the diaphragma sellae superiorly. After tumor debulking, we prefer to pack the sella turcica with fat.

Although endoscopic transnasal pituitary surgery can be performed through one nostril, operation through both nostrils allows a horizontal angle of 15°–20° between the axis of the endoscope and the axis of operating instruments (Fig. 4). This decreases the tendency to entangle instruments or hands in the endoscope-camera apparatus. We frequently use rigid fixation of the endoscope; however, it is often advantageous to operate inside the sella turcica with an instrument in one hand and the endoscope freely mobile in the other hand to provide optimal visualization. It may also be advantageous to introduce both endoscope and instruments through the same nostril during part or all of these operations.

Careful preoperative planning and discussion is imperative for planned endoscopic procedures. Both teams must understand each other's requirements and goals. Operative contingencies, particularly indications and methods for converting the endoscopic procedure to the sublabial microscopic transseptal transsphenoidal approach, must be understood. At

best, each team feels that it has learned much more than it has taught.

Problems Section

A variety of problems may be encountered during endoscopic transnasal pituitary surgery. Some problems are unique to this approach, while others are common to all transsphenoidal operations. Problems discussed include:

- Unfamiliarity with endoscopy
- Getting lost
- Lens obstruction
- Venous bleeding
- Inadequate exposure
- Cerebrospinal fluid leak
- Need to convert to sublabial transseptal approach

The first problem a developing pituitary endoscopist encountes is unfamiliarity with endoscopy. It is very helpful to obtain training in endoscopic neurosurgery at a standard course: this will provide basic familiarity with the principles of endoscopy as well as practical experience with the tools. Second, it is useful to have the surgeon's (or the operating room's) endoscope system on the field during standard sublabial microscopic pituitary surgery. This allows the surgeon to gain experience with the endoscope during his standard approach with his standard exposure and landmarks available. Borrowing endoscopic equipment from other operating services also allows one to obtain the experience of operating room staff already familiar with this equipment.

During either endoscopic-assisted or endoscopic pituitary surgery, the surgeon may lose his landmarks and have questions about the location of the endoscope or instruments. At this point, it is imperative to stop, withdraw, and reestablish one's orientation within the operative field before continuing the inspection or resection. The surgical principle of always knowing where your are and seeing what you are doing is doubly important in these small spaces bordered by important structures, particularly when using a two-dimensional representation of a three-dimensional space.

The most common problem encountered during endoscopic transnasal pituitary surgery is obstruction of the lens by blood. The tip of the endoscope is almost always the most dependent portion of the system, and blood oozing from the mucosa will invariably obscure the endoscope lens at frequent intervals during surgery. This problem can best be eliminated by using a lens-cleaning sheath with irrigation and suction ports. In this way, the endoscope can be cleaned in place rather than repeatedly withdrawn and reintroduced only to provoke more bleeding from the nasal mucosa.

Venous bleeding from the sella or parasellar area is encountered often with pituitary surgery and can be controlled by suction and tamponade with cottonoids or gelfoam. The authors have not yet had the challenge of dealing with carotid artery bleeding in this location, but would plan to pack the bleeding vessel should such an event occur.

During endoscopic transnasal pituitary surgery, the surgeon occasionally notices an annoying tendency of his endoscope to impede the excursion of his instruments, or a tendency of pituitary tumor fragments to obscure his endoscope lens. At this point, he should recognize that his anterior sphenoid exposure is probably too limited and perform a really generous bilateral sphenoidotomy and removal of the posterior bony septum. This additional bone opening will ensure that the endoscope lens can provide visualization of the sella turcica without impeding the movement of curettes or rongeurs, as well as keeping the tip of the endoscope lens free of tumor fragments during removal.

Arachnoid tears causing cerebrospinal fluid leaks are not uncommon in pituitary surgery, particularly when operating on macroadenomas or performing reoperations. The ability of the endoscope to view inside the sella turcica allows the surgeon to diagnose small cerebrospinal fluid leaks in time to repair them during the original surgery (Fig. 5).

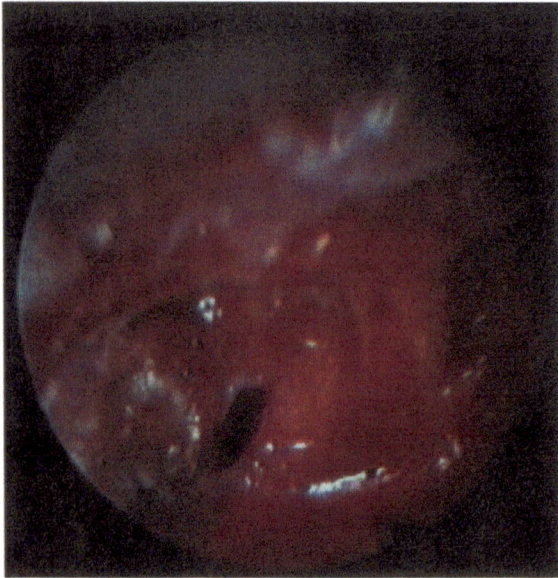

Fig. 5. View into sella turcica. This view with a 25° endoscope shows the diaphragma sellae after reoperation for an ACTH-producing tumor. A tear in the diaphragma sellae is apparent near the *bottom left* of the field. Anterior dural border of the sella turcica is seen in the *upper left corner* of the field

The surgeon should always be prepared to convert from endoscopic transnasal pituitary surgery to standard sublabial transseptal transsphenoidal surgery. If the surgeon at any point during the procedure thinks that the microscopic approach would gain him more advantage, it is easy to perform sublabial incisions, dissect the inferior nasal tunnels, and proceed under microscopic visualization. To date, we have not found this necessary.

The Future

Developments of endoscopic transnasal pituitary surgery include integration of endoscopes and instruments with various image-guided systems, both freehand and robotic. In the authors' opinion, registration and guidance of operating instruments with infrared light-emitting diodes adds a level of safety and technical ability to the endoscopic pituitary surgery. Also of interest are initiatives to develop a fully robotic platform for endoscopes, allowing precise guidance and placement of the endoscope with both the surgeon's hands free for operating.

Most current image-guided surgical systems reconstructing data from preoperative imaging studies do not update tumor coordinates during operative debulking. Thus, changes in tumor location and changes in the location of important structures adjacent to (and moving with) the tumor are not available through these systems. Developments in intraoperative ultrasound may allow the surgeon to visualize both residual tumor and important structures adjacent to the tumor at frequent intervals during surgical debulking, greatly increasing operative safety.

Neurosurgical endoscopists are currently placing great demands on their instrument makers. Many adaptations of pre-existing instruments work well, but the development of instruments specifically for this type of precise, delicate surgery is proving a formidable challenge to our engineering colleagues. As with all neurosurgical endoscopy, instrument development represents a significant constraint to the development of endoscopic transnasal pituitary surgery.

Conclusion

Whether the surgeon decides to use the endoscope as an adjunct to microscopic sublabial transseptal pituitary surgery or as the primary operating tool, familiarity with the equipment and techniques of neurosurgical endoscopy benefits the pituitary surgeon. The authors' experience with this technique suggest that, properly used, endoscopes can increase the safety and decrease the postoperative facial discomfort of transsphenoidal pituitary surgery. Minimally invasive pituitary surgery is rapidly developing its own set of tools for visualizing and operating on lesions in and around the sella turcica. We must continue to think, borrow, and adapt both our equipment and ourselves to the possibilities unfolding before us.

References

1. Carrau RL, Jho HD, Ko Y (1996) Transnasal-transsphenoidal endoscopic surgery of the pituitary gland. Laryngoscope 106:914–918
2. Dhooge C, Govaert P, Martens F, Caemaert J (1992) Transventricular endoscopic investigation and treatment of suprasellar arachnoid cysts. Neuropediatrics 23:245–247
3. Gamea A, Fathi M, el-Guindy A (1994) The use of the rigid endoscope in transsphenoidal pituitary surgery. J Laryngol Otol 108:19–22
4. Hardy J (1971) Transsphenoidal hypophysectomy. J Neurosurg 34:582–594
5. Jankowski R (1995) Endoscopic pituitary surgery. In: Stankiewicz JA (ed) Advanced endoscopic sinus surgery. Mosby, New York, pp 95–102
6. Jankowski R, Auque J, Simon C, Marchal JC, Hepner H, Wayoff M (1992) Endoscopic pituitary tumor surgery. Laryngoscope 102:198–202
7. Laws ER, Kern EB (1976) Complications of transsphenoidal surgery. Clin Neurosurg 23:401–416
8. Mennel HD, Hellwig D, Bauer BL (1994) Results and reliability of stereotactic and endoscopic biopsies in brain tumors. Zentralbl Neurochir 55:79–90
9. Papay FA, Benninger MS, Levine HL, Lavertu P (1989) Transnasal transseptal endoscopic repair of sphenoidal cerebral spinal fluid fistula. Otolaryngol Head Neck Surg 101:595–597
10. Rodziewicz GS, Kelley RT, Kellman RM, Smith MW (1996) Transnasal endoscopic surgery of the pituitary gland. Neurosurgery 39:189–193
11. Shikani AH, Kelly JH (1993) Endoscopic debulking of a pituitary tumor. Am J Otolaryngol 14:254–256
12. Stankiewicz JA (1995) Advanced endoscopic sinus surgery. Mosby, New York
13. Wurster CF, Smith DE (1994) The endoscopic approach to the pituitary gland [letter]. Arch Otolaryngol Head Neck Surg 120:674

Endoscopic Approach to the Mesencephalon Compared to Standard Microsurgical Procedures

M. Scholz, U. Wildförster, M. Hardenack, K. Schmieder, and A. Harders

Summary

The midbrain is a highly vulnerable part of the brain and can be reached in standard microneurosurgery only with difficulties. Neuroendoscopy offers easier approaches to this area. Four case illustrations shall give some impression of this successful method. The first three patients underwent endoscopic interventions that combine third ventriculostomy and operative procedure in the region of the aqueduct. The last case of a 71-year-old woman was an endoscopic-assisted microsurgical operation of an arachnoidal cyst of the cisterna quadrigemina. As a conclusion the endoscope is a useful tool for operations in the mesencephalic region without or with only a minimum damage to cerebral structures.

Introduction

Cerebral structures around the aqueduct belong to the mesencephalon. This part of the brain is highly vulnerable and can be reached in standard microsurgery only with difficulties. Anatomically, mesencephalon can be subdivided in Crura cerebri, tectal plate and tecmentum. Substantia grisea centralis around the aqueduct, nucleus ruber and substantia nigra are other structures we have to keep in mind if we want to intervene in this delicate area [8]. Vascularization of the brain stem occurs via paramedian branches of the basilar artery [3]. Another important point to remember is that both the oculomotor and trochlear nerve have their origin in this part of the brain.

Second in line to the classic microsurgical approaches described in detail by Yarsagil [13], Seeger [11] and others, the interfissural transcallosal approach to the floor of the third ventricle and aqueduct entrance is an elegant established procedure. However, requires osteoplastic trepanation and bears the risk of fornix damage, especially in the hands of unexperienced neurosurgeons. Along the dorsal route, the supracerebellar infratentorial approach is favorable for lesions in the pineal region and tectal plate [7]. Parinaud Syndrome can be observed when traversing the tectal plate.

With the endoscope, especially mesencephalic structures can be reached easier and quicker with a simple burrehole. Endoscopic approaches to date can be differentiated into a) trans-third ventricular, b) trans-aqueductal, and c) dorsal supracerebellar approach.

Materials and Methods

Between December 1995 and September 1996, four patients were operated endoscopically on lesions in the region of the mesencephalon at the Department of Neurosurgery, Ruhr University of Bochum. These cases are presented to underline limitations and operative possibilities of endoscopic procedures in this region.

Case 1

A 62-year-old man presented with a history of headache, progressive gait disturbances and incontinence. Magnetic resonance imaging (MRI) revealed a hypointense lesion dorsal to the aqueduct and development of a hydrocephalus caused by aqueduct stenosis (Fig. 1). We planned endoscopic intervention with combined third ventriculostomy and endoscopic biopsy.

A right frontal burrehole treparation was positioned 2 cm anteriorly than in common third ventriculostomy. The endoscope was passed through the foramen of Monro to the floor of the third ventricle. After ventriculostomy had been performed, the rigid Wolf endoscope is moved to the aqueduct, which was masked with a blood clot. After rinsing and sucking, pathologic mesencephalic surface structure were observed and partly coagulated. Biopsy was performed from the tumor tissue could be easily detected (Fig. 2). Neuropathologic evaluation revealed grade I astrocytoma. The outcome of the patient was excellent. Symptoms decreased and he remained shunt-free. Biopsy was performed from the tumor tissue which could be ensily detected.

Fig. 1.
Magnetic resonance imaging
(MRI) **a** sagittal- **b** coronalsection, T1-weighted images.
Hypointense lesion dorsal of
the aqueduct with aqueduct
stenosis and hydrocephalus

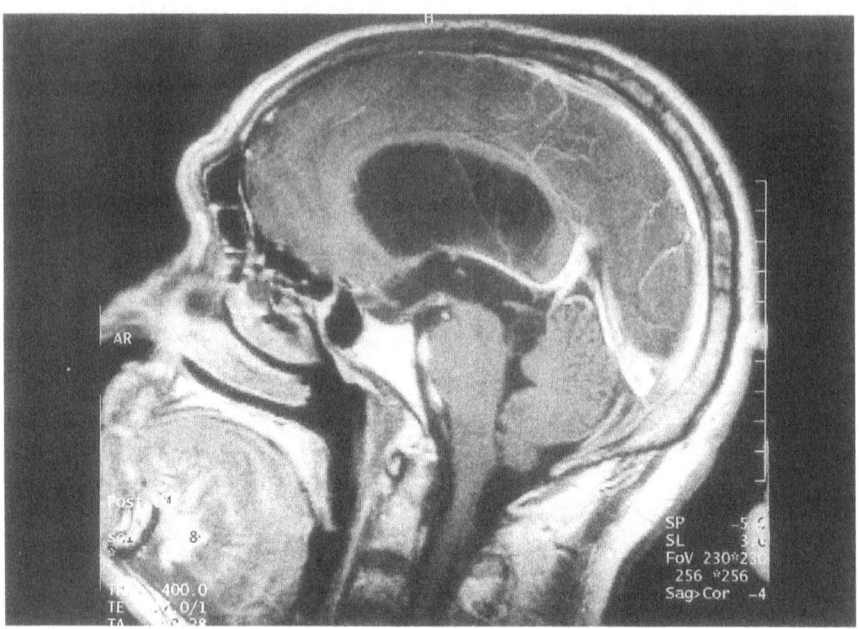

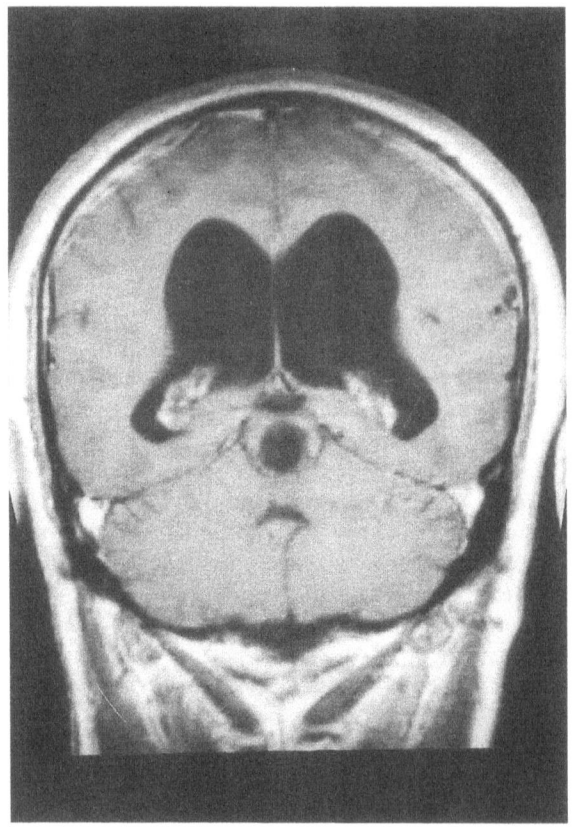

Case 2

A 30-year-old man was hospitalized with recurrent
episodes of headache. Neurologic examination was
normal. MRI confirmed the diagnosis of a cystic
process in front of the aqueduct causing intermittent
aqueductal stenosis (Fig. 3). First, we performed
tumor biopsy. Histologic diagnosis was grade II
astrocytoma. The floor of the third ventricle was
then perforated. The patient's postoperative course

Fig. 2.
Pathologic surface structure
with tumor tissue in the
region of the aqueduct

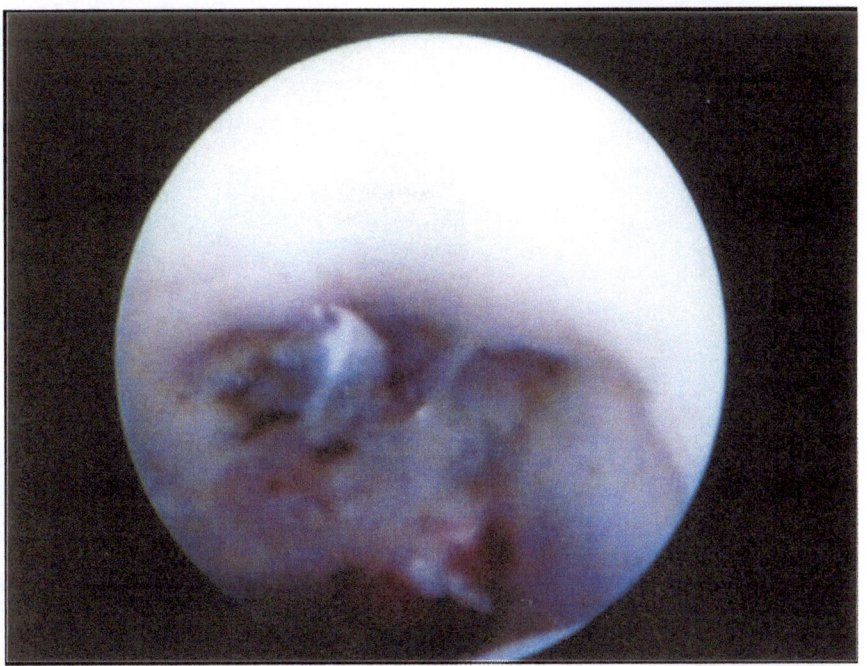

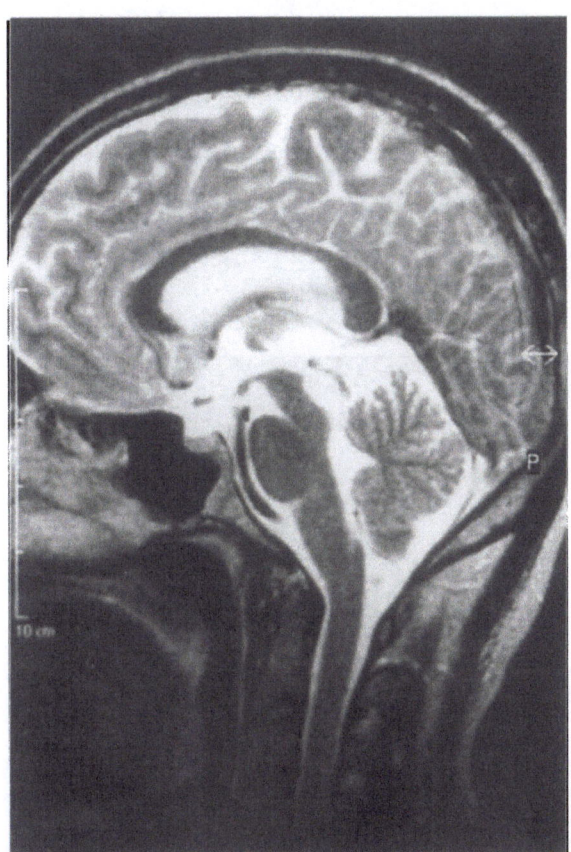

Fig. 3. MRI sagittal-section, T2-weighted images. Cystic lesion in front of the aqueduct

was uneventful. He remained symptom-free in a follow-up examination.

Case 3

In this case, hydrocephalus of a 9-year-old boy was treated with ventriculostomy and perforation of a superficial aqueductal membrane with a Fogarty catheter. Superficial white spots on the ventricular wall represent residual changes after meningitis caused by shunt infection. The balloon was inflated very carefully to a quarter of maximum size at most (Fig. 4).

Case 4

A 71-year-old woman suffered diplopia and gait disturbances. MRI demonstrated an arachnoidal cyst of the cisterna quadrigemina (Fig. 5).

Operative intervention was performed via a small 2 × 2-cm osteoplastic trepanation. The rigid endoscope was introduced in the supracerebellar infratentorial angle. After endoscopic perforation of the cystic membrane, the quadrigeminal plate could be visualized. The operation was then continued microsurgically. Fenestration to the third ventricle was performed with the bipolar forceps.

Fig. 4.
a Aqueductal membrane
(*arrow*) which is **b** perforated
by a Fogarty balloon.
c Situs after perforation

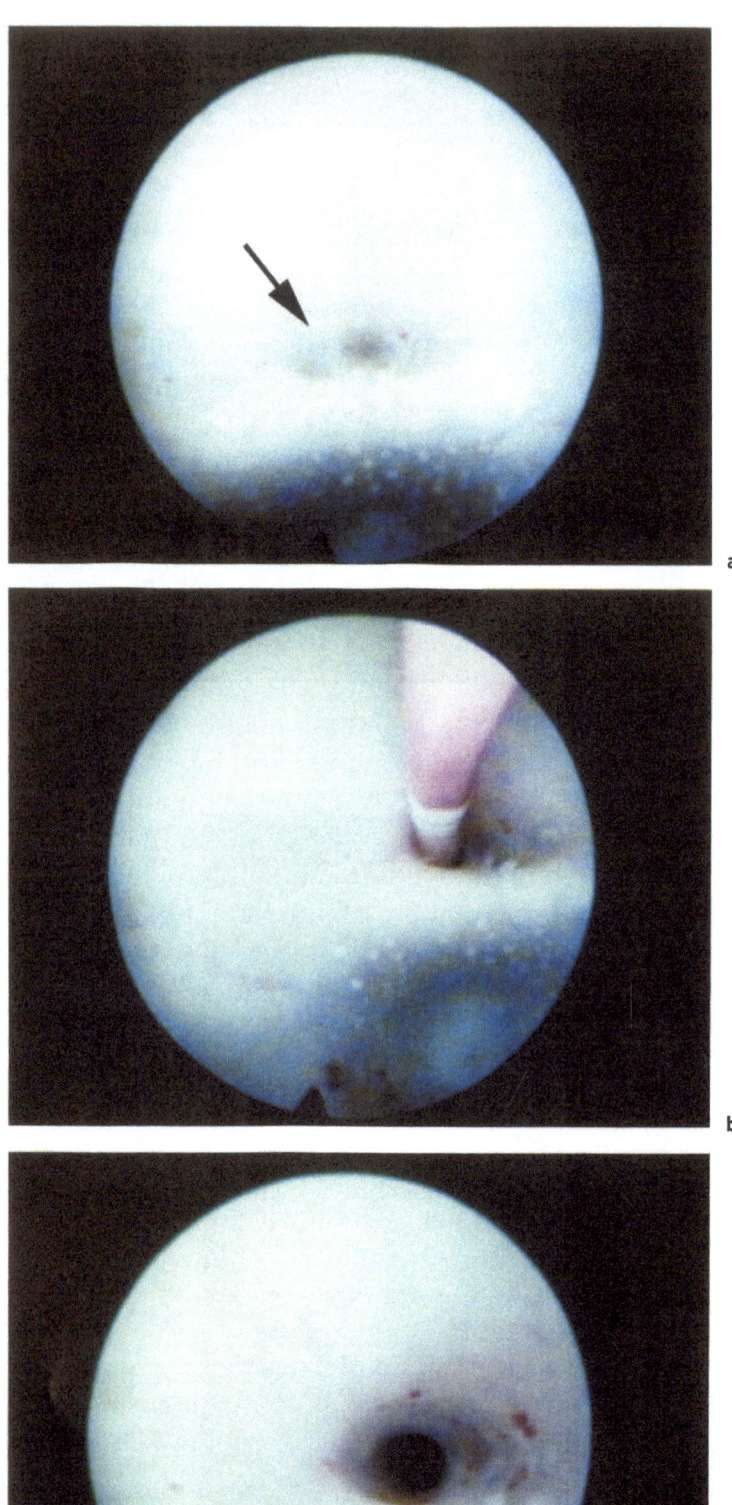

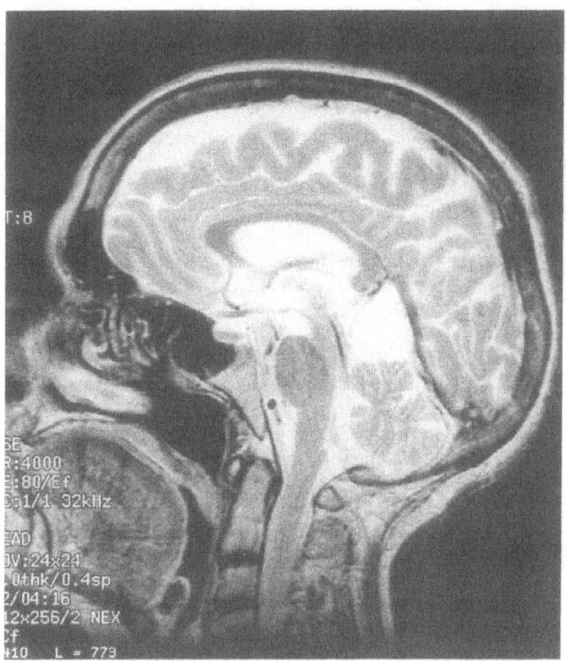
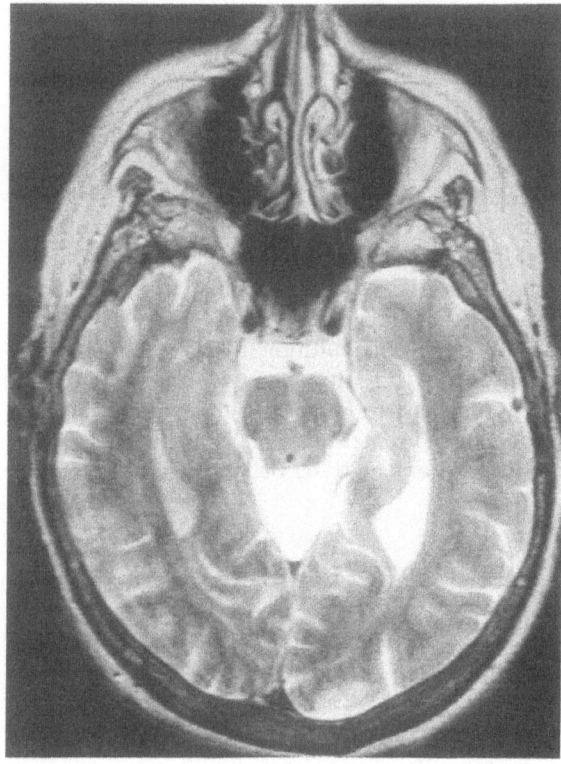

Fig. 5. MRI **a** sagittal and **b** coronalsection, T2-weighted images. Arachnoidal cyst of the cisterna quadrigemina

Results

Operation time was short in all cases. In the three combined procedures (case 1–3) the average time was 50 min. The endoscopic-assisted microsurgical operation took 2.5 h. All interventions could be performed without operationrelated postoperative neurologic deficit. In the last mentioned patient, extrapyramidal symptoms of Morbus Parkinson increased postoperatively and could be managed after intense physiotherapy and change of medication. Complications such as infection or bleeding were not observed.

Discussion

With the endoscope, mesencephalic structures can be reached easy and quick.

Passage through the aqueduct to the fourth ventricle was demonstrated in the cadaver model by Perneczky and Tschabitscher [9]. For this procedure, thin flexible endoscopes are needed. In comparison, the transaqueductal approach is possible in micro-surgery only with difficulties. In classic stereotactic procedures [4] vascular structures cannot be visualized and coagulated directly as in endoscopic interventions. To fill this gap and reduce the risk of vascular damage, mirovascular Doppler monitoring [12] and endoscopy [5] have been introduced in stereotaxis.

The combination of ventriculostomy and biopsy in the region of the aqueduct remains controversial as to whether biopsy or ventriculostomy has to be performed first in the same operation. It may be argued that symptoms are caused by hydrocephalus, so treatment of hydrocephalus by ventriculostomy (the easier step) has to be performed first. Conversely, the cause of symptoms may be suspected to be a tumor in the region of the aqueduct, so tumor biopsy has to be done first and preferably accompanied by reopening of the aqueduct. More experience with this special kind of combined operation has to be made in order to find the ideal recommendation for such procedures.

In every type endoscopic procedure, it is important to calculate the size of the foramen of Monro prior to intervention because the possibility of endoscopic movement is limited in this area. In our opinion, the region of the aqueduct is the most dorsal area that can be operated on with the rigid endoscope. For lesions such as pineal cysts, we prefer flexible endoscopes. Further technical developments are needed to perform safe endoscopic procedures in the midbrian region. In the past years, navigation systems using preoperatively acquired digital images (CT or MR data) have been developed for open microneurosurgery [1, 6]. The combination of an endoscope with such a system would be of interest especially in pathologic anatomy and cases lacking natural landmarks. The control of the endoscopic tip in difficult situations such as bleeding or complex tumor resection seems possible with the novel techniques of image-guided surgery such as open MRI [2]. Intelligent multifunctional instruments with the possibility of intraoperative sonography [10], sensation transfer and fine tissue separation have to be developed to enter this "area of no touch" as senior neurosurgeons have described the mesencephalon.

References

1. Barnett G, Kormos D, Steiner C, Weisenberger J (1993) Use of a frameless, armless stereotactic wand for brain tumor localization with two-dimensional and three-dimensional neuroimaging. Neurosurgery 33 : 674 – 678
2. Dumoulin CL, Souza SP, Darrow RD (1993) Real-time position monitoring of invasive devices using magnetic resonance. Magn Reson Med 29 : 411 – 415
3. Duvernoy HM (1978) Human brainstem vessels. Springer, Berlin Heidelberg New York
4. Grunert P, Ungersböck K, Bohl J, Kitz K, Hopf N (1994) Results of 200 intracranial stereotactic biopsies. Neurosurg Rev 17 : 59 – 66
5. Hellwig D, Eggers F, Bauers BL (1991) Endoscopic stereotaxis. Preliminary results. Stereotact Funct Neurosurg 54, 55 : 418
6. Laborde G, Gilsbach JM, Harders A, Klimek L, Mösges R, Krybus W (1992) Computer assisted localizer for planning of surgery at intra-operative orientation. Acta Neurochir (Wien) 119 : 166 – 170
7. Lapras F, Patet JD (1987) Controversies, techniques and strategies for pineal tumor surgery. Posterior approaches. In: Apuzzo MLJ (ed) Surgery of the third ventricle. Williams and Wilkins, Baltimore, pp 649 – 662
8. Mayr R (1985) Makroskopische Anatomie des unteren Hirnstammes: Mittelhirn, Brücke und verlängertes Mark. In: Benninghoff A et al. (eds) Makroskopische Anatomie des Menschen, vol 3. Urban und Schwarzenberg, Munich, pp 117 – 132
9. Pernezky A, Tschabitscher M, Resch KDM (1993) Approach through the aqueduct. In: Endoscopic anatomy for neurosurgery. Thieme, Stuttgart, pp 239 – 242
10. Resch KDM, Reisch R, Hertel F, Perneczky A (1996) Endo-Neuro-Sonographie: eine neue Bildgebung in der Neurochirurgie. Endoskopie Heute 2 : 152 – 158
11. Seeger W (1980) Supratentorial operations in ventricular system. In: Microsurgery of the brain. Anatomical and technical principles, vol 1. Springer, Berlin Heidelberg New York, pp 142 – 180
12. Ungersböck K, Grunert P, Bohl J et al. (1992) CT-guided stereotactic biopsy aided by Doppler ultrasonic vascular monitoring. Acta Neurochir (Wien) 115 : 152 – 155
13. Yasargil MG (1996) Intraventricular tumors. In: Yarsagil MG (ed) Microneurosurgery, vol 4 b. Thieme, Stuttgart, pp 313 – 322

The Removal of Intracerebral Lateral Hematomas by an Endoscopic Method in Combination with Microsurgical Techniques and Video System

A-G. Danchin, G.V. Zervigun, A.V. Afonasiev, and A.A. Danchin

Summary

The research data presented here is based on analysis of the medical histories of 62 patients with intracerebral hematomas that were operated on by various surgical methods. In 60 patients the hemorrhages occurred spontaneously; lateral intracerebral traumatic hematomas were diagnosed in two cases. Fifty-one patients were operated on by conventional methods of open or resectional craniotomy. Eleven patients with lateral intracerebral hematomas were operated on using a bipolar endoscopic technique. Hematomas were removed completely in all patients operated on by traditional methods and in those undergoing biportal endomicrosurgical interventions The comparative analysis of treatment outcome deals in part with traumatization by various methods of surgical interventions for intracerebral lateral hematomas. Biportal endomicrosurgical approachs for the removal of intracerebral lateral hematomas can considerably minimize surgical trauma (the length of soft tissue exposure, the trephinement defect area of the skull, and the loss of blood are much less than in traditional craniotomies) and allow radical evacuation of the intracerebral lateral hematomas.

Introduction

Among the methods of operative interventions for removing spontaneous and traumatic intracerebral hematomas the overwhelming majority of cases still employ such classic procedures as open and resectional craniotomies with subsequent evacuation of hematomas using various tools and equipment. The development of applied endomicroneurosurgery has enabled us to minimize operative trauma while maintaining the possibility of radical removal of intracerebral hematomas. In this context the methods of endoscopic removal of intracerebral hematomas using a one-portal approach have found wide application in clinical practice [1, 2].

We present a new technique for biportal endoscopic evaluation of intracerebral spontaneous and traumatic hematomas, increasing the range of possibilities for manipulations inside a pathological region by various microsurgical tools under visual monitor control.

Methods and Materials

We analyzed 62 cases of intracerebral hematomas operated on by various surgical methods. The volume of hematomas ranged from 60 to 160 ml. In 60 patients the hemorrhages had occured spontaneously; however, patients who developed intracerebral hematomas as a result of vascular malformation's rupture were not included in these observations. Lateral intracerebral traumatic hematomas were diagnosed in two patients. Fifty-one patients underwent operations by conventional methods such as open and resectional craniotomies. The skull defects of these operations were usually quite large or corresponded to the projective sizes of intracerebral hematomas were operated on with the help of a biportal endoscopic technique using a video system; in 9 cases the hematomas were caused by spontaneous hemorrhages and in two by trauma.

We used a Zimmer one-chip camera, Wolf and Karl Storz rigid endoscopes with tubus implanter (diameter 4 mm) torsional arrangement of objectives at 0°, 30°, and 70°, a working tube 2.7 mm in diameter, and a length of 180 mm, Aesculap microsurgical tools, and Sony video system. The exact landmarks of projections of the intracerebral lateral hematoma borders onto the skin of the head and stereotactical calculations carried out under computed tomographic (CT) guidance. Linear skin aponeurotical sections (10–15 mm long) were carried out, and the approach was performed through two 6- to 10-mm burrholes deviating 10 mm from the projection of hematoma borders. After crosswise section of dura the rigid endoscope with tubus implanter was entered through one of the burrholes to the calculated depth. The liquid part of hematoma and the small blood clots were extracted partially by continuous irrigation and aspiration. The aspirator was used for evacuation of the main part of the hematoma, which presented as large and fixed residual blood clot under visual monitor control. Various microsurgical manipulations associated with hemostasis were carried out, by the same approach.

The operation was ended with the establishment of a double lumen catheter for drainage in the hematoma cavity after removal. The drainage was removed 1–2 days after the intervention.

Results and Discussion

We do not present a detailed analysis of our treatment results with intracerebral hematomas operated on by traditional methods because these results coincide with those of other researchers. We are interested merely in analyzing some results of treatment by biportal endomicrosurgical interventions concerning intracerebral lateral hematomas and comparing them to treatment results of patients operated on by traditional techniques.

The hematomas that were removed via biportal endoscopic approaches were localized in lateral regions of the hemispheres: in seven cases several millimeters underneath the cortex and in four cases in communication with subarachnoidal space. In all 11 patients the intracerebral lateral hematomas were removed completely.

As an example, we present a brief extract from the medical history of a 63-year-old patient admitted to our Department of Neurosurgery and Neurology (Main Military Clinical Hospital, Defense Ministry of Ukraine) on 11 April 1996 in a critical condition due to a spontaneous intracerebral lateral hematoma of the right parieto-occipital region.

The patient was referred from the Clinic of Cardiology where he had undergone treatment for idiopathic hypertension. Three hours prior to hospitalization the patient suddenly lost consciousness. On arrival at the clinic the patient was comatose and presented a left sided hemiplegia. CT showed the oval focus of hemorrhage with clear borders in the right parieto-occipital area communicating with the subarachnoidal space, with a sizes of 45 × 59 × 35 mm and with small perifocal brain edema (Fig. 1). Two hours after admission the patient was operated on: two burrhole trepanations were performed (diameter up to 10 mm) directly in the region of the intracerebral lateral hematoma projection. The endoscope was guided into the hematoma cavity, and the hematoma was evacuated partially by irrigation-aspiration. The aspirator was installed and with its help the main part of hematoma was removed under visual control (Fig. 2). Hemostasis was carried out by mono- and

bipolar coagulation (Fig. 3). The cavity of the removed hematoma was drained by a double lumen catheter for 1 day with continuous irrigation using saline solution. CT control showed the radical removal of the hematoma (Fig. 4).

The intraoperative blood loss of patients operated on with biportal endoscopy was not more than 10 ml. With classic approaches the minimal blood loss is 50 ml. The mean total length of skinaponeurotical sections using endomicrosurgical intervention was

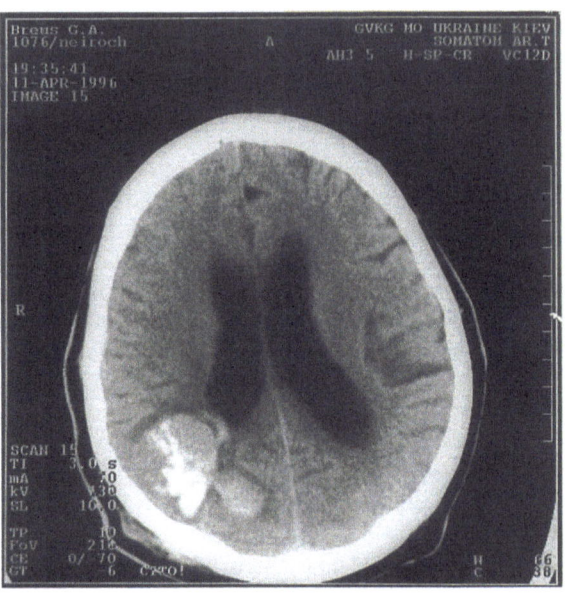

Fig. 1. CCT of an intracerebral lateral hematoma of the right parieto-occipitalis area (sizes in 45, 59, and 35 mm)

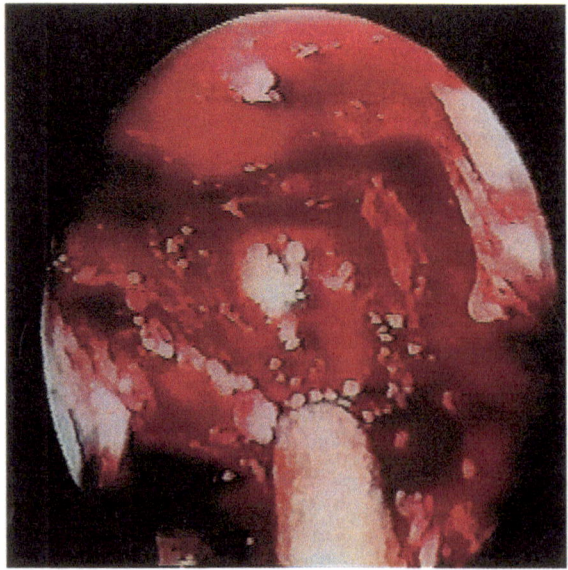

Fig. 2. The removal of an intracerebral hematoma by an aspirator under endoscopic control

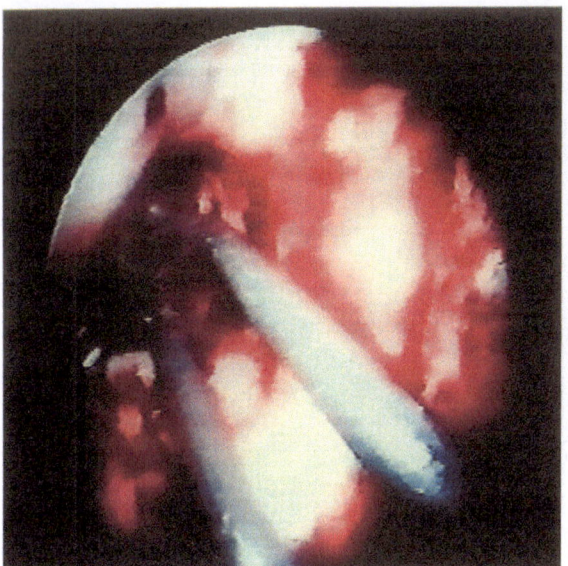

Fig. 3. Hemostasis carried out by monopolar diathermy

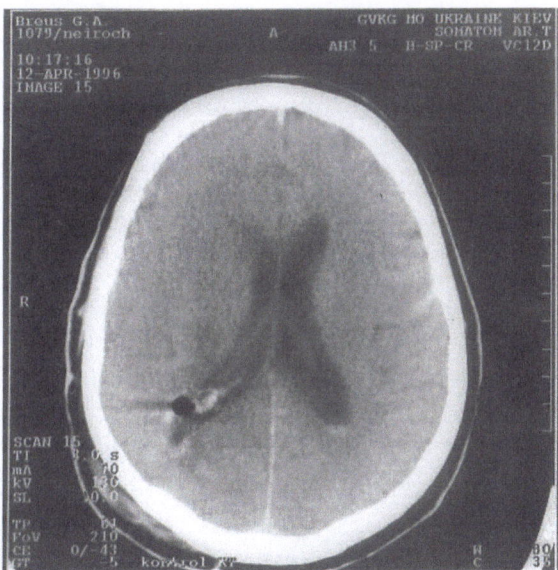

Fig. 4. CT control shows radical removal of the intracerebral hematoma on the second day after surgery

36 mm, the mean area of burrholes was less than 2 cm². In classical operations these parameters thus entailed a minimum of 140 mm and 20 cm², respectively.

The crucial step for successful completion of the biportal endoscopic operation is localizing the burrholes chosen for the evacuation. The burrholes were placed in regions where the rigid endoscope could be introduced most expedient, such as at important functional anatomical areas of the brain and positions of practical convenience for surgical manipulations by the neurosurgeon. (Thus the optimum site was based on the projection of intracerebral lateral hematoma.)

It is important to stress that all manipulations in the hematoma cavity were carried out under monitor control in such way that the endoscope serves not only as a tool for partial removal of hematoma, a source of light in the operative field, and transmitter of video signal but also as a microscope allowing a 10- to 15-fold increase in the size of the operative field. Furthermore, methods of hemostasis in the cavity of the evacuated hematoma during biportal endomicrosurgical interventions proved successful in all 11 cases with the use of mono- or bipolar coagulation or compression of small bleeding vessels by a balloon catheter. However, in the postoperative period hemorrhage recurred in one patient; it was again removed endoscopically. In one patient meningocephalitis developed and had to be treated by antibiotics. In two patients operated on by conventional methods the postoperative period was complicated by development of epidural hematomas which were repeatedly removed. In no case did an epidural hematoma occur during the postoperative period after biportal endoscopic microsurgical operation.

Conclusions

In the treatment of intracerebral lateral hematomas preference is given to biportal endoscopic microsurgical removal as the safest and least invasive operative procedure. The use of portal endoscopic techniques of intracerebral lateral hematoma evacuation can minimize surgical trauma.

References

1. Auer LM (1985) Endoscopic evacuation of intracerebral haemorrhages. High-tech surgical treatment – a new approach to the problem? Acta Neurochir (Wien) 74 : 124–128
2. Auer LM, Ascher PW, Heppner F, Ladurner G, Ebone G, Lechner H, Tolly E (1985) Does acute endoscopic evacuation improve the outcome of patients with spontaneous intracerebral hemorrhage? Eur Neurol 24 : 254–261
3. Auer LM, Holzer P, Ascher PW, Heppner F (1988) Endoscopic neurosurgery. Acta Neurochir (Wien) 90 : 1–14
4. Karahan VB (1994) Endoscopic operation at head and brain injury. Neurotraumatology. The directory. In: Moscow Institute of Neurosurgery (eds) IPC "Vasar-Ferro". Moscow, pp 219–220
5. Karahan VB (1995) Operative endoscopy in surgery. Endoscop Surg (Moscow) 1 : 24–32
6. Lantuch AV, Potapov AA (1994) Intracerebral hematomas. Neurotraumatology. The directory. In: Moscow Institute of Neurosurgery (eds) IPC "Vasar-Ferro". Moscow, pp 35–37

Endoscopic Removal of Subdural Hematomas and Hydromas by Microsurgery and Video System

A.G. Danchin, A.V. Afonasiev, and A.A. Danchin

Summary

Observations in 80 patients with traumatic sub-dural hematomas operated on by various methods were analyzed. Traditional surgical interventions such as open craniotomies were performed in 68 patients. Twelve hematomas were removed by our newly developed biportal and multiportal endoscopic techniques, involving rigid endoscopes, video systems, and micro-surgical tools. In all of our interventions the subdural hematomas were completely re-moved. Biportal or multiportal endomicrosurgical evacuation of subdural hematomas is the method of choice, offering greater opportunities of radical hematoma while surgical traumatization is significantly decreased.

Introduction

Substantial experience has recently been amassed in neuroendoscopic interventions. The hope of re-searchers and general practitioners in developing surgical methods to increase safety and to reduce tis-sue trauma has been justified. These problems can be addressed by introduction of endoscopic operations that provide a greater ability and freedom of action, i.e., to control an operating field under significantly increased safety outside direct visibility. Neuro-endoscopy is systematically applied in principle only for the treatment of noncommunicating hydroce-phalus. Work devoted to the endoscopic removal of intracranial hematomas remains rare [1, 2].

Traditional methods of removing subdural hema-tomas are the single-portal surgical approaches based on open craniotomies and resectional tre-phinations of the skull, and endoscopic surgical inter-ventions. However, results of microneurosurgical subdural endoscopic procedures using biportal techniques have been published [1].

The goals at our Department of Neurosurgery and Neurology (Main Military Clinical Hospital, Defense Ministry of Ukraine) were: the development of novel techniques of biportal and multiportal endoscopic operative interventions in combination with micro-surgical techniques in cases of subdural hematomas and comparative analysis of the results of various methods of surgical treatment. An intensive theoretical discussion preceded practical clinical application of these operations.

Methods and Materials

Analysis of the follow-up of 80 patients with trau-matic subdural hematomas operated on with various surgical methods was carried out. Depending on the organization of hematomas, the acute ones were found in 26 patients, subacute in 31, and chronic in 23. Traditional surgical interventions such as resectional craniotomies were carried out in 68 patients; in 12 cases hematoma were removed by our newly developed bi- or multiportal endoscopic methods. The volumes of hematomas ranged from 50 to 170 ml. Computed tomograph (CT) was used for hematoma diagnosis and stereotactical calculations.

We used a complete set of a one-chip camera. Zimmer source of light, Wolf and Karl Storz rigid endoscopes (0°, 30°, and 70°) a working channel 2.7 mm in diameter, and a length of 180 mm. A tubus of 4 mm external diameter with two fixed valves was fitted to the working canal of the endoscope at the beginning of the operation. This allows constant irrigation with saline solution and aspiration of the hematoma contents. Aesculap microsurgical tools were used for surgical manipulations. The Sony video system helped to control all steps of the operation. Postoperative CT investigations were performed for all patients.

Our technique of endoscopic removal of subdural hematomas with microsurgery and video control dif-fers in principle from that described by other authors who have used simple portal approaches and flexible endoscopes [1, 2, 4]. Our methods of bi- or multi-portal endoscopic removal of subdural hematomas use a rigid endoscope. This is based on the spherical structure of the skull and geometric form of the hematoma while a single- or double-prominent lens occupies the space created between dura and arachnoid. This space in the absence of brain edema is generally large enough and remains during all operative interventions. It is possible to insert the rigid endoscope through one burrhole and various microsurgical tools for evacuation of the hematoma through another under visual monitor control. On the basis of CCT examination the localization and extension of the hematoma borders are defined, and stereotactical calculations determine the exact direction of both the endoscope and the aspirator and microtools. The projection of the hematoma borders on the skin of the skull is plotted in the CT research department.

Two, three or four burrholes (diameter 6–10 mm) are used, depending on the hematoma size. The size of the linear skinaponeurotical sections does not exceed 15–20 mm for each section. Gently sloping throughs are made on the edges of the burrholes by microchisel and allow the rigid endoscope and microsurgical tools to be inserted parallel to the dura into the subdural space without traumatizing the cortex. The rigid endoscope with the two fixed valves of the tubus implanter is entered into the burrholes, and the liquid part of the hematoma and some small

blood clots are evacuated by irrigation and aspiration. Under visual monitor control the aspirator is inserted to remove large remaining blood clots fixed at the dura in the same way via these burrholes other microtools, scalpels, dissectors, spatulas, scissors, curettes, spoons, coagulation probes, clips, etc. are placed and used under endoscopic control for manipulations such as coagulation of bleeding vessels. Hemostasis can also be carried out with the help of an inflated balloon catheter that is left in situ for some time in the region of bleeding vessels. After radical removal of the subdural hematoma a double lumen drainage is left in situ for 1–2 days.

The brief extract from the medical history of an 87-year-old patient who was operated on using our technique offers an example: On 9th November 1996 the patient was admitted to our Department of Neurosurgery and Neurology in a serious condition. The soporous patient suffered from a left-sided hemiparesis, showed a meningeal syndrome, and had a body temperature of 40 °C. Four days before hospitalization he had suffered a head injury. The subdural hematoma was diagnosed by CCT. Its parameters were: extention over the whole right cerebral

Fig. 1.
CCT image of a subdural hematoma of the right hemisphere

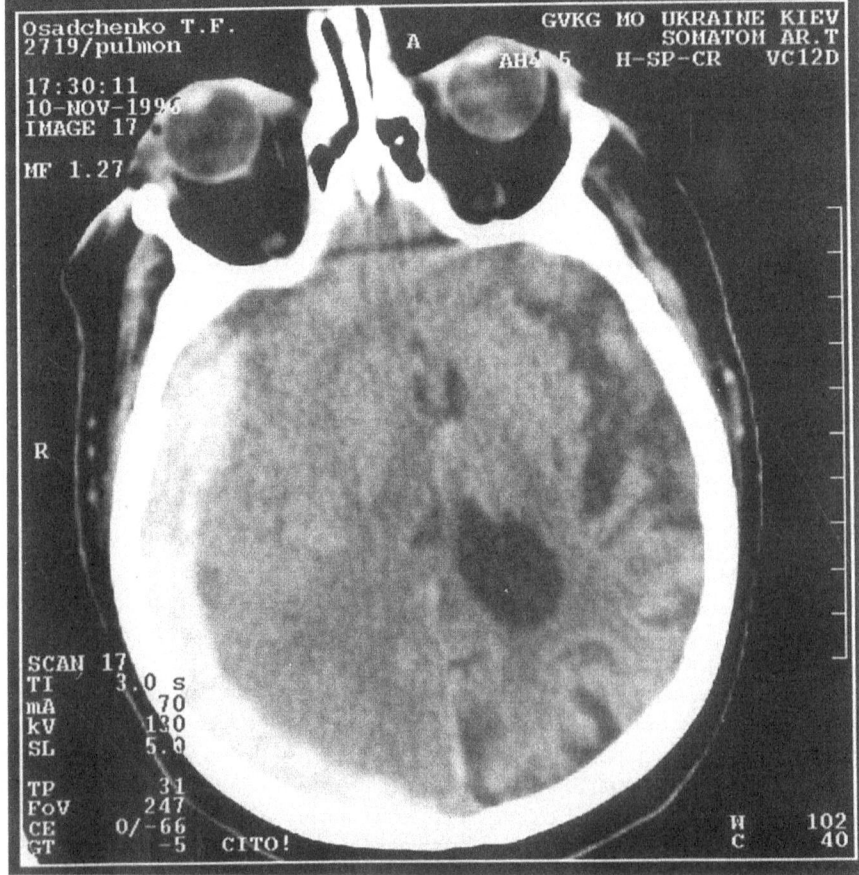

hemisphere, maximum length 24 cm, maximum width 16 cm, thickness 18 mm, volume 170 ml, density 55–65 H. A significant dislocation of midline structures up to 20 mm and a deformation of the right lateral ventricle were obvious (Fig. 1).

The endoscopic intervention according to the technique described above was performed immediately (Fig. 2). Burrholes (diameter 8–10 mm) were made through four skinaponeurotical sections,

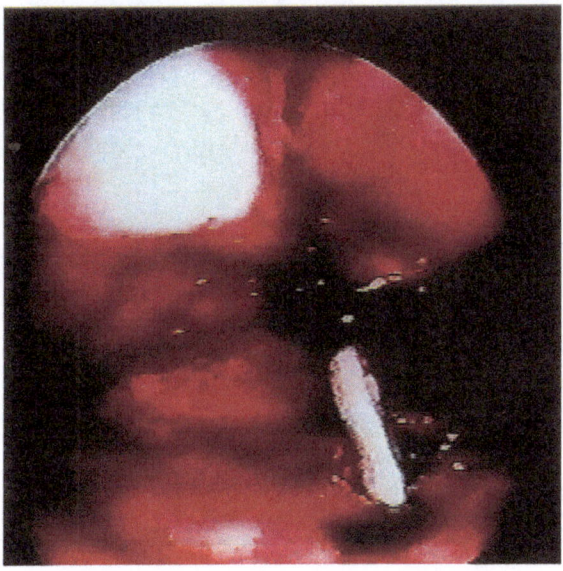

Fig. 2. Insertion of aspirator into a subdural space under visual monitor control through one of the burrholes, while the endoscope in inserted into a second burrhole

each 15 mm in length. The dura was open by X-figurative section. The liquid contents of hematoma and a few clots were removed partially in aspiration-irrigation technique under endoscopic control. The most difficult part of the operation was removal of the remaining solid hematoma clots, which as a rule were fixed to the dura. The aspirator was put into the hematoma cavity under visual monitor control and the remaining part of the hematoma was evacuated step by step (Fig. 3). Hemostasis was carried out using bipolar and monopolar diathermy (Fig. 4). The surface of the cortex in the region of the removed hematoma showed small contusions with widened cortical vessels and signs of subarachnoidal hemorrhage. The double drainage was inserted through a burrhole under endoscopic control. Postoperative CCT control confirmed the radical removal of the hematoma and an improvement of the midline shifting on the second day after operation (Fig. 5).

Results and Discussion

Among brain-compressing traumatic hemorrhages, subdural hematomas occur most frequently, with an incidence of about 40 %. Reducing raised intracranial pressure by urgent surgical intervention is a basic element in pathogenetic treatment. The efficacy of operations is in direct proportion to their early and radical performance. Radicality is defined as the completeness of hematoma removal. This can be achieved by traditional craniotomies or by bi- or multiportal endomicrosurgical operations.

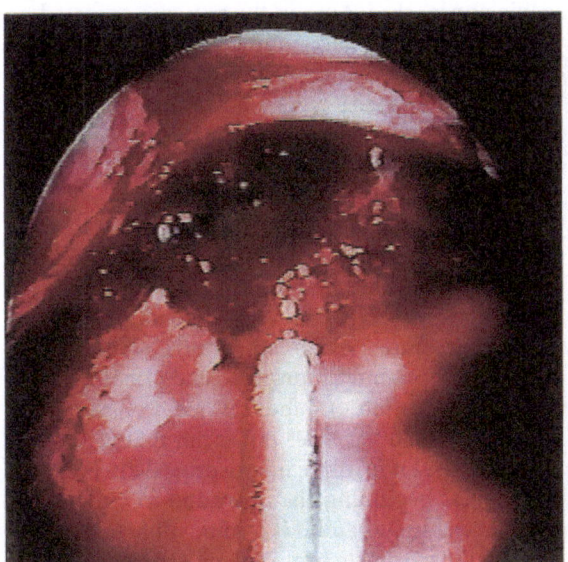

Fig. 3. Removal of remaining parts of hematoma and blood clots using an aspirator under endoscopic control

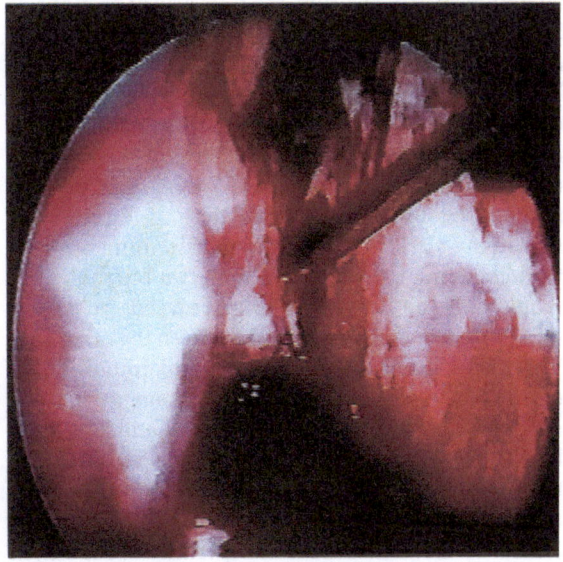

Fig. 4. Hemostasis carried out by monopolar diathermy

Fig. 5.
CCT control on the 2nd post-operative day confirms radical removal of the hematoma and improvement of the midline shifting

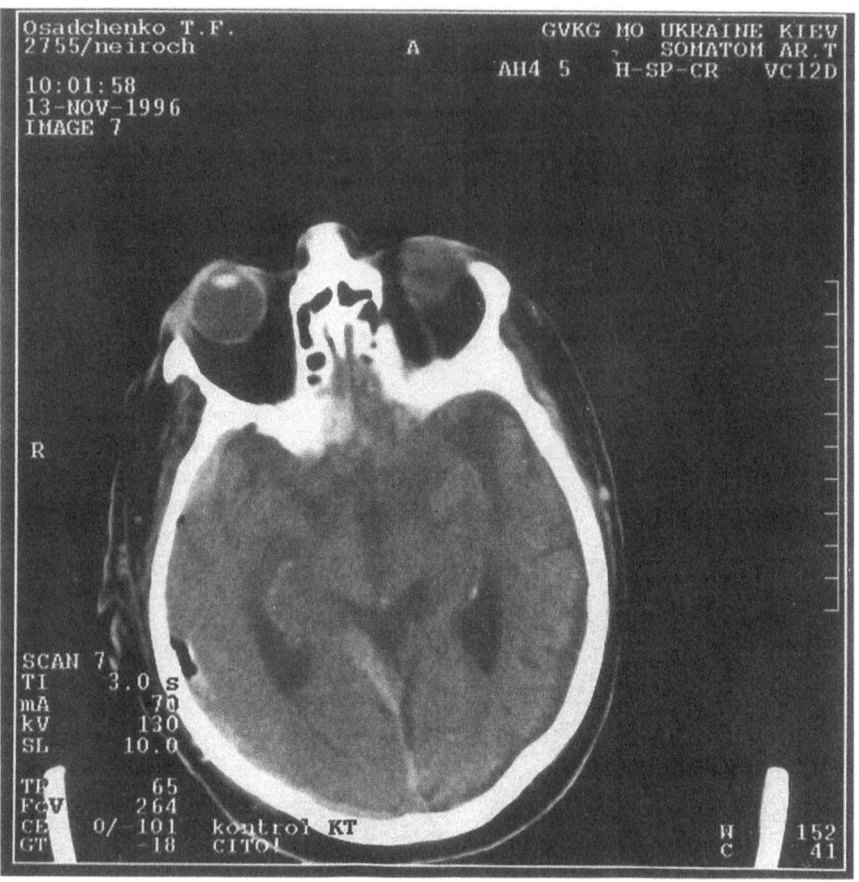

In all of our presented cases the subdural hematomas were evacuated completely. However, while the length of the skinaponeurotical section in traditional operations is 150–280 mm, the total size of sections of bi- or multiportal endomicrosurgical interventions ranges from 30 to 80 mm. As a rule the sizes of burr-holes of traditional surgical approaches are the same or larger than these of the subdural hematomas. The minimum size is 16 cm². The total area of skull defects using endoscopic techniques corresponds to the number of burrholes, which depend directly on hematoma size. If the longitudinal size dimension of the hematoma is less than 70 mm, two burrholes are required. If the hematoma length exceeds 70 mm, a third burrhole is placed exactly between the other two. This makes it possible to work with microtools in the cavitiy of the subdural hematoma. In one patient the subdural hematoma was hemispherical, and four burrholes were necessary for total hematoma removal. The total area of trepanation defects using endoscopic microsurgical techniques ranges from 0.5 to 3 cm². Minimum blood loss with the operative approach, in conventional variants, is estimated a 50 ml, while with endoscopic interventions the maximum is 10 ml.

By the three above criteria traumatization using operative approaches with traditional craniotomies is much greater than in bipolar or multiportal endoscopic approaches. Among the patients operated on with conventional techniques three developed an epidural hematoma in the postoperative period. These hematomas were evacuated repeatedly. Mortality in this group reached 26%. Among patients operated on by biportal or multiportal endoscopic technique the mortality was 16.6%.

However, we did not carry out detailed analysis of mortality in the two groups. First of all, mortality did not depend on the approach but on the duration and the degree of midline shifting and raised intercranial pressure. Extended brain edema is an absolute contraindication for endoscopic microsurgical evacuation of subdural hematomas. Nevertheless, brain edema does not mean that it is necessary to refuse endoscopic microsurgical intervention. We suggest beginning the operation in endomicrosurgical technique, and if it is impossible to enter the subdural space due to brain edema, then changing, to conventional craniotomy to remove the subdural hematoma.

Conclusions

The biportal and multiportal endoscopic removal of subdural hematomas in combination with microsurgical technique and video system has a number of essential advantages in comparison to conventional craniotomies.

The main advantages are: (a) reduction in traumatization (significant reduction in size of soft tissue sections of the head, reduced blood loss, considerable decrease in volume of operative damage to the skull, reduced area of opened dura). (b) decrease in wound closing, and (c) rare occurrence of such life-threatening complications as epidural hematoma.

References

1. Hellwig D, Kuhn TJ, Bauer BL (1996) Endoscopic treatment of septated chronic subdural hematomas. Surg Neurol 45:272–277
2. Karahan VB (1992) Endoscopic removal of epidural hematomas of postcranial fossa (message on 2 cases). Questions Neurosurg (Moscow) 2–3:34–35
3. Karahan VB (1994) Endoscopic operation at head and brain injury. Neurotraumatology. The directory. In: Moscow Institute of Neurosurgery (eds) IPC "Vasar-Ferro". Moscow, pp 219–220
4. Karahan VB (1995) Operative endoscopy in surgery. Endoscop Surg (Moscow) 1:24–32
5. Lihterman LB, Hitrin LH (1994) Subdural hematomas. Neurotraumatology. The directory. In: Moscow Institute of Neurosurgery (eds) IPC "Vasar-Ferro". Moscow, pp 172–176
6. Romodanov AP (1989) Modern aspects of diagnostics and complex treatment of closed head and brain injury in an acute period. Questions Neurosurg (Moscow) 5:35–38

**Part 4
Linear Accelerator Radiosurgery for
Minimally Invasive Treatment
of Arteriovenous Malformations**

Linear Accelerator Radiosurgery for Arteriovenous Malformations: The Relationship of Size, Dose, Time, and Planning Factors onto Outcome

R. Engenhart-Cabillic and J. Debus

Introduction

The management of arteriovenous malformations (AVMs) has been modified lately due to the availability of new techniques such as endovascular embolization, radiosurgery, or the combination of microsurgery, embolization, and radiosurgery [2, 4, 7, 9, 12, 16, 17, 30]. Besides preservation or improvement of neurological function, the ultimate goal in the treatment of AVMs is complete elimination of risk of hemorrhage. It is believed that since Steiner et al. [29] reported the first patient with an AVM to be treated by Gamma knife radiosurgery, more than 15 000 AVM patients worldwide have been radiosurgically treated. The majority of large radiosurgery series report that AVM obliteration rates exceed about 80% within a latency period of 2–3 years, at which point the risk of subsequent hemorrhage is eliminated. In this paper we report our clinical experience of 145 patients with AVMs treated between 1983 and 1993 using the nonconverging arc technique and give an overview of the literature on linear accelerator radiosurgery treatment results for arteriovenous malformations.

Material and Methods

Radiosurgical Treatment

Although the roots of stereotactic radiosurgery date back to the 1950s, utilization of linear accelerator (LINAC), radiosurgery originated in 1980 [1, 4, 15, 22]. LINAC-based techniques are quite varied, using a wide area of stereotactic headframes, X-ray energies, and territory collimators. In general, LINAC radiosurgery techniques fall into three broad categories yet are comparable to the gamma knife. The basic principle of linear accelerator radiosurgery is beam convergence (the nonconverging are technique, static beam technique, and dynamic technique) [15, 18, 22]. Mechanical adjustment of both gantry and LINAC table is absolutely necessary to achieve the required stability of the isocenter, i.e., the point where all converging collimated beams cross.

Mechanical overall accuracy must be achieved, which does not permit deviations of the measured target point of more than 1 mm from the calculated target point. Considering the pixel and voxel size of the CT and MR imaging planning techniques used as well as the inaccuracy of X-ray systems, which limits the precision of target delineation and target point calculation to ±1.0–1.5 mm under optimal conditions, mechanical inaccuracy of more than 1 mm should not be exceeded.

The goal of each of these radiosurgery techniques is a distinct gradient of irradiation dose at the treatment field edges which markedly reduces the dose of irradiation to the surrounding normal structures. Depending on the size of the collimator, the dose falloff beyond the peripheral isodose line is rapid. Average distances of about 2 mm are reported for the 90%–50% isodose lines and about 4 mm for the 90%–20% isodose lines [18, 22]. As a result, only a relatively small volume receives a significant irradiation dose. However, this is a function of the target volume treated. In comparison, the dose gradients achievable with LINAC radiosurgery systems do not differ significantly from those of gamma knife systems.

The Heidelberg radiosurgical system is based on nine noncoplanar arc irradiations focused onto the stereotactic target [15]. This technique results in a spherical dose distribution. However, in order to conform the prescription isodose to irregular target volumes with convergent beam technique, multiple isocenters are required. Exact matching of target and treatment volume reduces overdosage to organs at risk and underdosage in the target volume and eliminates the need to use multiple treatment volumes with dose overlapping and radiobiological disadvantages of significant inhomogeneity. To overcome this problem, the multiple static field technique was established in 1993 consisting of 14 noncoplanar, irregularly shaped fields [6, 25]. The 14 fields are distributed on five noncoplanar plans (coach positions: −15°, −52°, 90°, 52°, 15°). A schematic representation of a typical set of converging arcs vs. 14 irregular fields is shown in Fig. 1. For field shaping, a small and extremely precise micromultileaf colli-

Fig. 1 a, b.
Treatment techniques used
for radiosurgery in Heidelberg. **a** Multiple arc technique
using nine noncoplanar arcs
of 140°; **b** Multifield technique
using 14 noncoplanar
irregulary shaped fixed fields;
ap, anteroposterior;
VERT, vertical; *LAT*, lateral

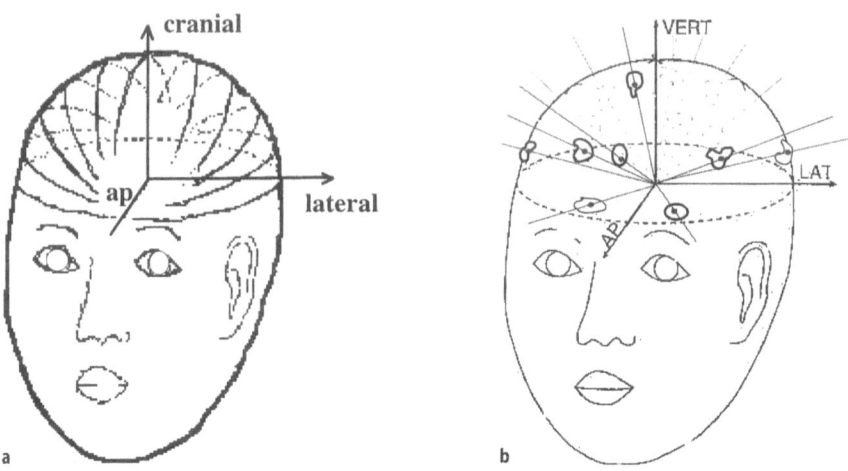

mator with tungsten leaves 1 mm in size has been developed by Pastyr et al. [20] (transmission depth: 8 cm; maximum field size: 6 cm × 6 cm; width of the leaves at the isocenter: 1.5 mm). The high resolution of the leaves makes it possible to define arbitrary irregular fields. For dose calculation, a convolution kernel algorithm was developed allowing the calculation of the primary and the scatter-dose component separately [25]. The resolution of the primary and the scatter kernel is 1 mm. The entire 3 D dose calculation at a Vax Station 3200 requires approximately 15 CPU s. The comparison of this technique with an optimized multiple isocenter treatment technique revealed that dose concentration and dose gradient outside an irregular target volume are equivalent.

Nidus Definition and Treatment Planning

Target volume definition represents one of the crucial steps in radiosurgical treatment of cerebral AVMs. To allow the application of high doses in radiosurgery and minimize side effects of irradiation, precise localization and definition of the target volume (e. g., size and volume of the nidus, origin and hemodynamic relevance of feeding arteries, and venous drainage pattern) is necessary. Currently it is well established that the entire nidus must be enclosed in the target volume, i.e., neither feeding arteries nor draining veins need to be irradiated.

Biplanar intra-arterial angiography is the gold standard for planning treatment of AVMs; however, it provides only two-dimensional information which is critical especially for irregular shaped lesions. Therefore it is difficult or impossible to conform the target volume to the 3-D configuration of the nidus. Computer tomography (CT) and magnetic resonance imaging (MRI) with MR angiography (MRA) have proven to be helpful supplements to conventional X-

ray angiography for more precise target localization [10, 19, 21]. Since MRI is a substantial part of our imaging protocol, the patients are immobilized using a ceramic MR-compatible stereotactic head frame.

Apart from angiography, our planning protocol consists of MRI, MRA, and as of 1994, a 3-D time of flight (TOF) MRA, i. e., a flow measurement based on a phase contrast sequence. The 3-D TOF technique allows assessment of the time-dependent AVM inflow through arterial feeders, precise demarcation of the AVM nidus, and calculation of AVM shunt time by exact and time-dependent visibility of draining veins. Furthermore we use functional MRI in patients with target volumes in eloquent regions. All of the patients were examined on a 1.5-T imager (Magnetom SP 4000, Siemens, Erlangen, Germany) using a specially constructed linear-polarized head coil [10, 26]. All images have to be checked for geometric distortion, corrected if necessary, and prepared for 3-D radiotherapy planning.

The nidus of the AVM is outlined in the different diagnostic imaging modalities for optimized target localization. A separate computer program calculates a fit between individual landmarks on the angiographic and MR images that enables transfer of coordinates between the two imaging modalities. Size and shape of the nidus and stereotactic coordinates were determined by coregistration based on all stereotactically correlated imaging information.

All adult patients had local anesthesia supplemented by intravenous sedation during fixation of the stereotactic coordinate frame and subsequent diagnostic imaging modalities and radiosurgery. Children under 14 years were treated under general anesthesia. Patients rested with the stereotactic frame in place during optimization of dose planning. Before each treatment the exact position of the frame was carefully checked as was mechanical accuracy of all gantry and coach motions.

Table 1. Volumes of 145 AVMs treated with radiosurgery

Patients		AVM volume (cc)
(No.)	(%)	
30	20.7	< 1.0
52	35.9	1–14.2
62	42.8	14.2–110
1	0.69	210

AVM volume range 0.4–110 cc and one case of 210 cc. Median volume (80% isodose) 6.1 cc.
AVM, arteriovenous malformation.

Table 2. Clinical summary of 145 patients treated with LINAC radiosurgery

Initial signs and symptoms	Patients	
	(No.)	(%)
Hemorrhage	75	52
Seizure	44	30
Paresis	19	13
Migraine	8	5

All of our treatments were prescribed to the 80% isodose. Maximum dose ranged from 12.5–36.0 Gy whereas minimum dose to the periphery of the lesion was 10 Gy. The dosage delivered depend on localization, size, and radiosensitivity of critical structures in the vicinity of the nidus, changing with increasing experience. Median treatment volume with an 80% isodose was 6.1 cc and ranged from 0.4 to 110 cc with one exception (210 cc). The volume of the treated lesions is listed in Table 1.

Patient Population

Between 1983 and 1996, 350 patients with cerebral AVMs were treated with stereotactic LINAC radiosurgery in our institution. In 1993 we were able to use the multiple isocenter treatment technique for irregularly shaped AVMs. Thus our experience consists of two distinct periods. All of the 145 patients treated during the first decade were prepared for treatment and treated with the nonconvergent arc technique.

The age of our patients ranged from 7 to 66 years, with a mean age of 34 years. Seventeen patients (8%) were 60 years old or younger. In Table 2 the initial signs and symptoms of these patients are listed. Seventeen patients (12%) had subtotal resection, another 16 patients (11%) underwent embolization to reduce the size of the AVM before radiosurgical treatment, and four patients were treated by the multimodality approach of surgery embolization and radiosurgery. All AVMs were graded according to Spetzler and Martin [28] with respect to size,

proximity to critical structures, and venous drainage pattern.

Follow-up Evaluation

Follow-up included neurological examinations and contrast-enhanced CT or MRI at intervals of 6 months for the first 2 years after radiosurgery. A high-resolution intraarterial angiogram was requested of the AVM patients 2 years after radiosurgery or prior to that if their MRIs suggested obliteration.

Results

Twenty-four months of control angiography are available in 97 of the 120 cases with a follow-up of more than 2 years. Of these 55% demonstrated complete obliteration of their malformation, 22% had partial reduction in nidus size, and 19% showed only slight or no change in size. Prognostic factors for complete obliteration were complete coverage of the nidus with an irradiation dose adequate for obliteration, a prescribed marging dose of > 18 Gy, and the volume of the nidus treated. In the group treated with all three favorable prognostic factors, complete obliteration was achieved in 83% of patients. AVMs have been subdivided into three different ranges of volumes. Total obliteration rate at 24 months follow-up decreased from 71% in AVMs of less than 20 mm in maximum dimension to 42% in large AVMs of more than 40 mm in dimension (Table 3).

Table 3. Angiographic obliteration vs AVM volume over time following radiosurgery

AVM		Complete obliteration rate	
Diameter	Volume	2 years (%)	3 years (%)
< 20 mm	< 1 cm³	71	83
20–40 mm	4.19–33.5 cm³	63	75
> 40 mm	> 33.5 cm³	42	50

Fig. 2.
Clinical outcome of 138
patients with cerebral AVMs
treated with LINAC radio-
surgery

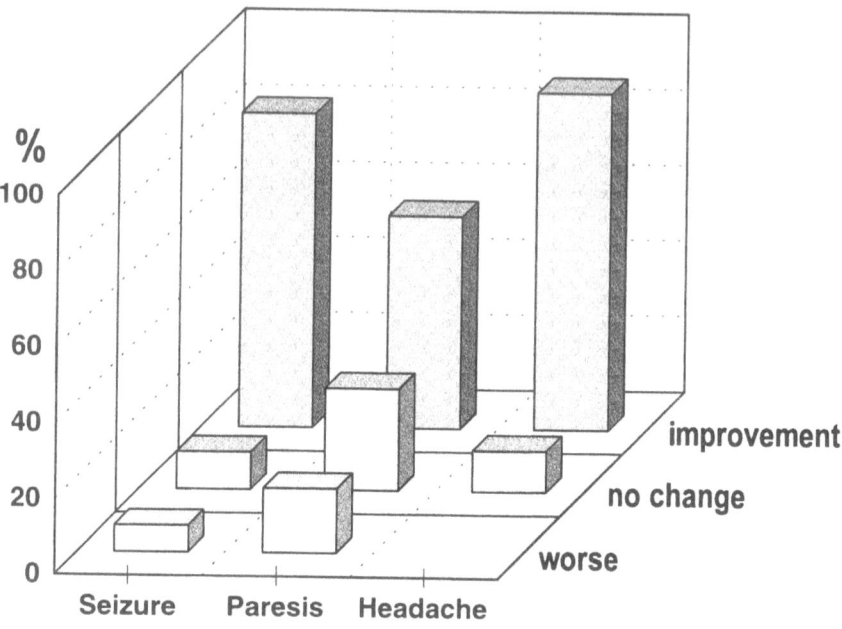

Clinical Outcome

With a mean value of 44 months and a maximum
value of 9 years, clinical follow-up of 138 patients
(94%) was available. Of 145 patients, 92% regained at
least their previous activity level after radiosurgery.
Clinical follow-up monitoring revealed significant
improvement in neurological symptoms in 65 patients (47%) either objectively due to reduction or

elimination of seizures or paresis, or subjectively due
to relief from severe vascular headaches. Neurological outcome as a consequence of treatment results
and post-treatment hemorrhage and treatment-
related side effects are combined in Fig. 2. These
results seemed to be strongly affected by the obliteration process. The majority of patients showing
clinical improvement belonged to the group with
complete obliteration in follow-up angiography as

Fig. 3.
Outcome regarding epilepsy
in patients with pre-existing
seizure disorder. This benefit
relates to obliteration grade.
CO, complete obliteration;
PO, partial obliteration;
NC, no change

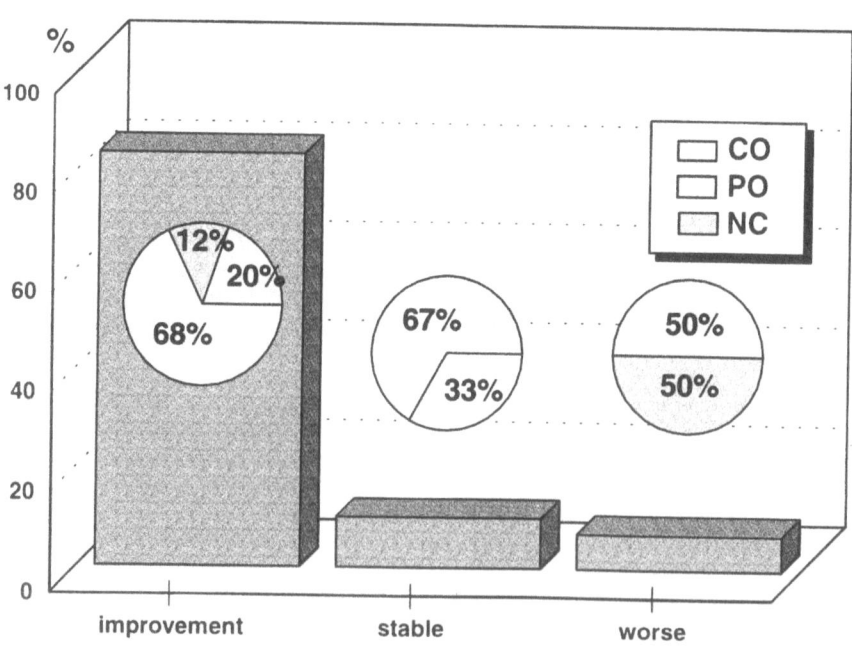

shown in Fig. 3 with respect to seizures. The condition of 46 patients (33%) remained stable or showed only slight improvement in neurological status. Sixteen patients (11.6%) developed new neurological deficits directly attributable to treatment 6–24 months after single high dose irradiation; ten of these patients (7.3%) recovered completely from neurological deterioration involving, e.g., aphasia and weakness, temporary increase in hemiparesis, and dysphasia. Permanent neurological deterioration was found in six patients (4.3%). Neurological deterioration varied inversely with AVM grade according to Spetzler and Martin: in our series, grade I and grade II lesions were irradiated without any incidence of minor or major deficit, nor did new cases of hemorrhage occur during the latency period in this group. Treatment-related deterioration of the condition was observed in five of 20 patients (25%) who had grade V angiomas. Two of four patients with brain necrosis belonged to this group. Development of cerebral necrosis (four patients) correlated closely with treatment volume and the dose delivered. Even when target volumes of less than 33 cc were irradiated, no irreversible side effect was observed with a dose of 20 Gy at the margin of the nidus. Eleven patients had an AVM hemorrhage between 4 months and 7 years after radiosurgery. In six patients bleeding occurred within the first 2 years before complete obliteration and five patients died of hemorrhage, of whom three had previously unruptured AVMs. Three patients experienced a first rebleeding, and one patient experienced a fourth rebleeding.

Discussion

A review of the literature during the past two decades demonstrates progressive evolution in the therapeutic management of AVMs. Optimum treatment of patients with AVMs must stress protection from the risk of hemorrhage and preservation or improvement in neurological function. In microsurgical series, overall results improved with diminished morbidity and mortality as a result of improved neurosurgical

techniques and anesthesia [16, 28]. Endovascular embolization is rarely the only treatment modality. Prevailing opinion is that embolization technique be combined with surgical resection, mainly preoperatively [2, 7].

Single fraction radiosurgery has shown clear efficacy for AVMs irrespective of the technique used. Stereotactic radiosurgery can obliterate AVMs by inducing a sclerosing process causing hyperplasia of angioma vessels that leads to thrombosis. The thickening of vessels is a dynamic process which takes place over a period of several months to 2–3 years. There is a general agreement that partial obliteration does not protect against the risk of hemorrhage. Using the gamma unit, obliteration rates between 59% and 85% are reported. Results based on the LINAC radiosurgical system range from 55% to 81% (Table 4). The results published by the respective institutions are not necessarily directly comparable because of significant variations in patient characteristics and treatment data. Prognostic parameters for estimating the obliteration rate of cerebral AVMs after radiosurgery depend on several factors including complete coverage of the nidus, the dose prescribed, and the volume of the nidus treated.

However, stratification of obliteration rate by AVM volume was performed by investigators who used the gamma unit and LINAC radiosurgery system. The summary of the results shows that there is a clear relationship between AVM volume and dose delivered. Our overall obliteration rate of AVMs at 2 years was 55%, which relates to the large amount of patients with larger volumes. For AVMs of less than 4.2 cc (less than 20 mm in diameter), the obliteration rate was 71% at 2 years and 83% at 3 years. Colombo et al. [5] reported an overall obliteration rate of 80% at 2 years. However, for AVMs larger than 25 mm in mean diameter, the obliteration rate decreases to only 33%. The Pittsburgh group [23] using the gamma knife reported an 81% obliteration rate for lesions less than 2 cm in diameter and a 45% rate for patients treated with lesions between 20 and 44 mm in diameter. Yamamoto and co-workers [31], Betti et al. [1], and Friedman et al. [13] reported a correlation be-

Table 4.
Two-year complete obliteration rates for AVMs with linear accelerator radiosurgery systems

Institution	No. of patients	Dose (Gy)	Obliteration rate (%)	Side effects
University of Florida	158	10–25	81	1.3
Vincenza	79	18.7–40	80	3.3
McGill University	33	50–55		6.0
Harvard	6	15–25	73	
Heidelberg	97/120	10–28	55[a]	4.0
Lyon	20	–	87	10.0

[a] 72% for doses above 18 Gy.

tween smaller nidus volumes and higher 2-year obliteration rates. Apart from location, treatment volume correlates significantly with radiosurgery-induced toxicity. Complication rates, reported for small lesions are low. However, the drawback of radiosurgery in the treatment of AVMs is the long delay between intervention and complete obliteration of pathological vessels. Therefore the combined mortality and morbidity risks from rebleeding during the latency period and irradiation effects have to be considered. Because of the latency period, radiosurgery should be offered to those patients whose lesions are considered difficult to resect. The results for larger AVMs demonstrate higher obliteration rates were likely to be achieved with higher irradiation doses, but at the expense of a higher incidence of irradiation-induced morbidity. For this group of patients a combination of intravascular embolization, microsurgery, and radiosurgery may be appropriate.

References

1. Betti OO, Munari C, Rosler R (1989) Stereotactic radiosurgery with the linear accelerator: treatment of arteriovenous malformations. Neurosurgery 24:311–321
2. Bien S, Voigt K, Caplan R (1996) Interventional neuroradiology in the brain, head, and neck region. In: Brandt (ed) Neurological disorders – course and treatment. Academic, London, pp 333–368
3. Bunge HJC, Chinela AB, Guevara JA et al. (1992) Infratentorial arteriovenous malformations: radiosurgical treatment. In: Lunsford LD (ed) Stereotactic radiosurgery update. Elsevier, New York, pp 169–176
4. Colombo F, Benedetti A, Pozza F et al. (1989) Linear accelerator radiosurgery of cerebral arteriovenous malformations. Neurosurgery 24:833–840
5. Colombo F, Pozza F, Chierego G, Casentini L, De Luca G, Francescon P (1994) Linear accelerator radiosurgery of cerebral arteriovenous malformations: an update. Neurosurgery 34:14–22
6. Debus J, Engenhart-Cabillic R, Rhein B, Schlegel W, Schad L, Pastyr O, Wannenmacher M (1994) Clinical application of conformal radiosurgery using multileaf-collimators. Int J Radiat Oncol Biol Phys 30:265
7. Deruty R, Pelissou-Guyotat I, Amat D, Mottolese C, Bascoulergue Y, Turjman F, Gerard JP (1995) Multidisciplinary treatment of cerebral arteriovenous malformations. Neurol Res 17:169–177
8. Engenhart R, Kimmig BN, Höver KH et al. (1993) Long-term follow-up brain metastases treated by percutaneous stereotactic single high-dose irradiation. Cancer 71:1353–1361
9. Engenhart R, Wowra B, Debus J, Kimmig B, Höver KH, Lorenz W, Wannenmacher M (1994) The role of high-dose, single-fraction irradiation in small and large intracranial arteriovenous malformations. Int J Radiat Oncol Biol Phys 30:521–529
10. Essig M, Engenhart R, Knopp MV et al. (1996) Cerebral arteriovenous malformations: improved nidus demarcation by means of dynamic tagging MR-angiography. Magn Reson Imaging 14:227–233
11. Flickinger JC, Kondziolka D, Lunsford LD (1995) Radiosurgery of benign lesions. Semin Radiat Oncol 5:220–224
12. Friedman WA, Bova FJ (1992) Linear accelerator radiosurgery for arteriovenous malformations. J Neurosurg 77:832–841
13. Friedman WA, Bova FJ, Mendenhall WM (1995) Linear accelerator radiosurgery for arteriovenous malformations: the relationship of size to outcome. J Neurosurg 82:180–189
14. Friedman WA, Blatt DL, Bova FJ, Buatti JM, Mendenhall WM, Kubilis PS (1996) The risk of hemorrhage after radiosurgery for arteriovenous malformations. J Neurosurg 84:912–919
15. Hartmann GH, Schlegel W, Sturm V, Kober B, Pastyr O, Lorenz WJ (1985) Cerebral radiation surgery using moving field irradiation at a linear accelerator facility. Int J Radiat Oncol Biol Phys 11:1185–1192
16. Heros RC, Korosue K, Diebold PN (1990) Surgical excision of cerebral arteriovenous malformations: late results. Neurosurgery 4:570–576
17. Lawton MT, Hamilton MG, Spetzler RF (1995) Multimodal treatment of deep arteriovenous malformations: thalamus, basal ganglia, and brain stem. Neurosurgery 37:29–36
18. Mehta MP, Noyes WR, Mackie TR (1995) Linear accelerator configurations for radiosurgery. Semin Radiat Oncol 5:203–212
19. Morikawa M, Numaguchi Y, Rigamonti D et al. (1996) Radiosurgery for cerebral arteriovenous malformations: assessment of early phase magnetic resonance imaging and significance of gadolinium-DTPA enhancement. Int J Radiat Oncol Biol Phys 34:663–675
20. Pastyr O, Schlegel W, Höver KH, Rhein B, Maier-Borst W (1993) Ein Micro-Multileaf-Kollimator für stereotaktisch geführte Strahlenbehandlungen. In: Müller RG, Erb J (eds) Medizinische Physik. Deutsche Gesellschaft für Medizinische Physik, pp 234–235
21. Phillips MH, Kessler M, Chuang FYS, Frankel KA, Lyman JT, Fabrikant JI, Levy RP (1991) Image correlation of MRI and CT in treatment planning for radiosurgery of intracranial vascular malformations. Int J Radiat Oncol Biol Phys 20:881–889
22. Podgorsak EB, Pike GB, Olivier A et al. (1989) Radiosurgery with high energy photon beams: a comparison among techniques. Int J Radiat Oncol Biol Phys 16:857–865
23. Pollock BE, Lunsford LD, Kondziolka D, Maitz A, Flickinger JC (1994) Patient outcomes after stereotactic radiosurgery for "operable" arteriovenous malformations. Neurosurgery 35:1–8
24. Pollock BE, Lunsford LD, Kondziolka D, Bissonette DJ, Flickinger JC (1996) Stereotactic radiosurgery for postgeniculate visual pathway arteriovenous malformations. J Neurosurg 84:437–441
25. Rhein B, Engenhart R, Debus J et al. (1994) Stereotaktische Hochdosis-Konvergenztherapie bei irregulär geformten Zielvolumina. Beschreibung einer Mehrfeldertechnik mit 11 bis 14 nicht-koplanaren irregulären Stehfeldern. In: Tautz M (ed). Medizinische Physik 94. Deutsche Gesellschaft für Medizinische Physik, pp 224–225
26. Schad LR, Bock M, Baudendistel K et al. (1996) Improved target volume definition in radiosurgery of arteriovenous malformations by stereotactic correlation of MRA, MRI, blood bolus tagging, and functional MRI. Eur Radiol 6:38–45
27. Souhami LA, Olivier EB, Podgorsak MP, Pike GB (1990) Radiosurgery of cerebral arteriovenous malformations with the dynamic stereotactic irradiation. Int J Radiat Oncol Biol Phys 19:775–782
28. Spetzler RF, Martin NA (1986) A proposed grading system for arteriovenous malformations. J Neurosurg 65:476–483
29. Steiner L, Leksell L, Greitz T (1972) Stereotactic radiosurgery for cerebral arteriovenous malformations. N Engl J Med 323:96–101
30. Steiner L, Lindquist C, Adler JR, Torner JC, Alves W, Steiner M (1992) Clinical outcome of radiosurgery for cerebral arteriovenous malformations. J Neurosurg 77:1–8
31. Yamamoto Y, Coffey RJ, Nichols DA, Shaw EG (1995) Interim report on the radiosurgical treatment of cerebral arteriovenous malformations. J Neurosurg 83:832–837

Part 5
Endoscopic Carpal Tunnel Release

Endoscopic Surgery of Carpal Tunnel Syndrome: Technical Report

R. Filippi, R. Schubert, and A. Perneczky

Summary

Experiences with one-portal endoscopic carpal tunnel release are reported. In 1996 we operated 14 patients with carpal tunnel syndrome endoscopically (three males, 11 females; mean age: 55 years). Indications for endoscopic surgery were typical symptoms such as pain and dysesthesia, hypesthesia in the innervation area of the median nerve, no paresis, reduction of conduction velocity of the median nerve in the carpal tunnel and resistance to conservative treatment. Twelve procedures were performed with intravenous regional anesthesia and two were performed with local anesthesia. Three months after surgery 13 patients showed complete recovery without any complications. One patient described persistent dysesthesia in fingers 2 and 3. According to our findings, endoscopic carpal tunnel release is an adequate minimal invasive method for the treatment of carpal tunnel syndrome.

Introduction

Endoscopic carpal tunnel release is a new technique for the treatment of median nerve entrapment in the carpal tunnel. Many studies are published which report advantages and problems associated with endoscopic surgery [1, 10, 11, 19, 22]. It is still a matter of dispute whether this procedure is more suitable than the conventional surgical technique [3, 6, 10, 12, 17, 22]. The necessity of one or two portals proceeding endoscopic carpal tunnel surgery is discussed controversialy [7, 14, 19]. The two-portal endoscopic procedure involves one skin incision at the wrist and a second one in the palmar area distal of the carpal ligament. The incision beyond the carpal tunnel is made to guarantee a complete splitting of the retinaculum. With the one-portal technique the distal end of the retinaculum is identified from the endoscopic picture. No palmar incision is made due to the danger of a subsequent painful scar.

In this study the less-invasive one-portal endoscopic procedure of carpal tunnel release using the Agee instrumentation and the clinical features and outcome of 14 patients treated with this technique are presented. Technical aspects and clinical results are discussed in relation to reports in the literature.

Materials and Methods

In 1996 14 patients with carpal tunnel syndrome were endoscopically operated using the Agee carpal tunnel release system. All of the patients had pain and sensory loss in the territory of the median nerve but no paresis. The conduction velocity of the median nerve was reduced in the area of the carpal tunnel in every case. All patients had undergone unsuccessful trials of conservative therapy before referral for surgery. The indications for endoscopic carpal tunnel release are: (a) typical pain, (b) dysesthesia/hypesthesia, (c) no paresis, (d) reduction of conduction velocity of median nerve and (e) resistance to conservative treatment.

At the start of the procedure the outdrive of the backward-directed knife at the distal end of the endoscope is tested (Fig. 1). The hand is positioned in hyperextension and the anatomical landmarks such

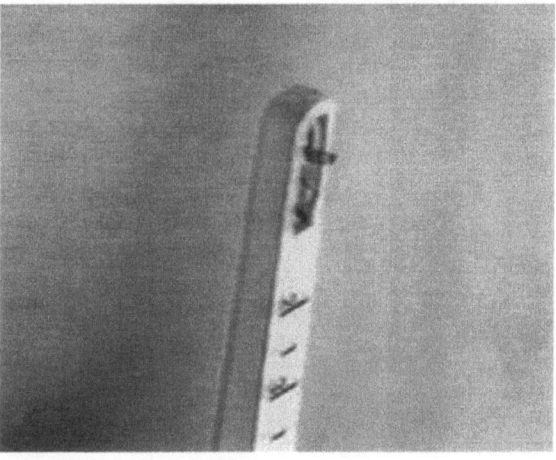

Fig. 1. Outdriven backward directed knife at the tip of the endoscope

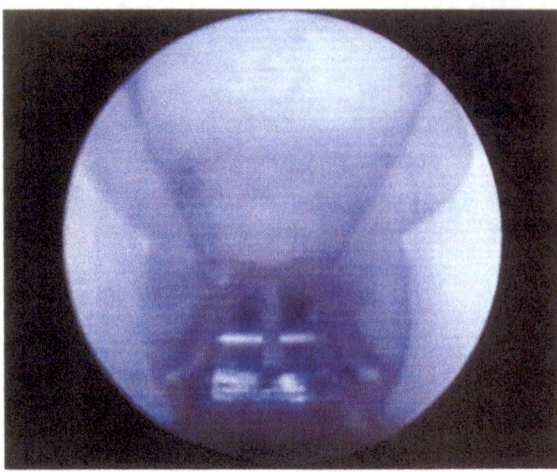

Fig. 2. Endoscopic view of the transverse ligament. Note the distal end of the retinaculum perceptible by prolapsed fat from the palmar area

as the tendon of the long palmar muscle, pisiform bone, the hamulus of the hamate bone, and the line of the fourth finger to the wrist are defined. After making a skin incision of 1–1.5 cm the tendon of the long palmar muscle is dissected, the forearm fascia is opened, and the carpal tunnel is identified using a blunt dissector. The endoscope is introduced into the carpal tunnel above the nerve. The distal end of the transverse ligament is identified by the prolapsed fat of the palm (Fig. 2).

The knife at the top of endoscope is driven out and the carpal retinaculum is split by pulling back the endoscope under constant visual control. Finally, a further drive-in of the endoscope is done to control complete splitting (Fig. 3).

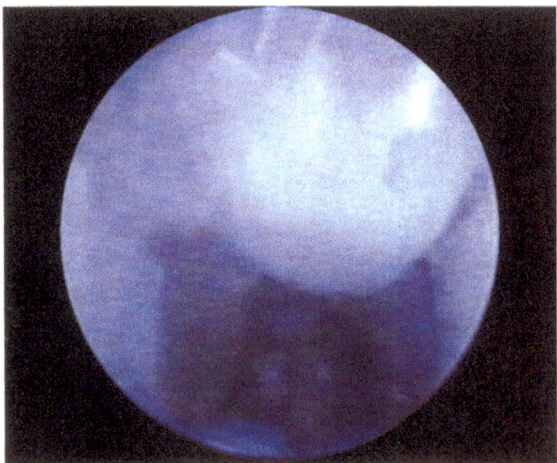

Fig. 3. Endoscopic control of split ligament. The rims of both parts of the retinaculum with palmar fat between them are clearly visible

Twelve procedures were performed with bloodless and intravenous regional anesthesia, two with the patients under local anesthesia. The clinical features and outcome of patients were noted and compared with those reported in literature.

Results

The ages of the patients were between 38 and 74 years (mean age 55 years). The male to female ratio was 3:11. The mean duration of symptoms was 18 months. Nine patients were treated as outpatients and five stayed in hospital for a few days (mean of 2 days) for preoperative examinations in the case of other serious diseases.

Intraoperative handling of the Agee equipment was easy and safe. The endoscopic view was excellent. The distal end of the carpal retinaculum could be identified definitively before splitting the ligament.

The process of cutting the retinaculum was easy because of the very sharp knife. Control of the split-ted ligament was possible by a final drive-in of the endoscope during which the rims of both parts of retinaculum and the palmar fat between them was observed.

Three months after surgery 13 patients showed a complete recovery of the preoperative clinical symptoms. One patient described persistent dysesthesia in fingers 2 and 3. The reason for this dysesthesia was a partial damage of sensible nerves of these fingers which became apparent in a second operation. There were no surgical complications such as rebleeding or wound infection.

Both patients who were operated under local anesthesia felt an uncomfortable narrowness in the carpal tunnel during the procedure. Additionally we had to bear in mind in these cases that every small bleeding worsens the quality of endoscopic view.

Discussion

Both the open and endoscopic carpal tunnel surgical techniques result in relief of pain, paresthesia, and numbness in almost all cases [5, 6, 13, 14, 15]. However, in most of the studies endoscopic surgery was preferred by the patients because of the small skin incision with no painful scar, short operation time, reduced postoperative pain, early return to daily activities, and short hospital stay after surgery [1, 5, 11, 13, 14, 22].

Nevertheless there are possible complications which demand caution during endoscopic procedures. The distal end of the carpal retinaculum must be identified definitively in order to determine whether the ligament was split completely [8, 11]. Injuries to the common digital nerves, median nerve itself, ulnar nerve, and ulnar artery may occur during this procedure [2, 4, 12, 18, 21, 23]. Even the presented study contains one case of injured sensible nerve branches to the second and the third finger. Sympathetic dysfunction is reported only in a small number of cases (0.28 %) [12]. It is very important to consider closely some special aspects of the endoscopic procedure in order to avoid these complications. It is doubtful whether there is enough space for the endoscope inside the carpal tunnel in patients with median-nerve-dependent pareses. Therefore we recommend for the present the exclusion of these cases from endoscopic surgery.

Anatomical landmarks such as the pisiform bone the hamulus of the hamate bone, and the tendon of the long palmar muscle are used to locate the required tissue layers and anatomical structures [4, 9, 16]. Should there be any doubt about anatomical structures in endoscopic view the surgeon should not hesitate to convert to the open technique [21].

It is controversial whether the one-portal or the two-portal technique is the best endoscopic solution [7]. According to other authors [1, 20, 22] we could gain all necessary visual information using only a single portal. Therefore it seems to be unnecessary to risk persistent pain and dysesthesia in the palmar area due to a second skin incision behind the end of retinaculum.

In contrast to others [19] we propose that endoscopic carpal tunnel surgery should be done with bloodless and regional anesthesia because it is more comfortable for the patient and the quality of endoscopic pictures is much better.

Conclusion

According to our results endoscopic carpal tunnel release appears to be an adequate, minimally invasive method for the treatment of carpal tunnel syndrome. Risks of nerve dysfunction are low and can be minimized by selecting patients without paresis and by an exact consideration of the required landmarks and the anatomical structures.

References

1. Agee JM, McCarroll HR Jr, Tortosa RD, Berry DA, Szabo RM, Peimer CA (1992) Endoscopic release of the carpal tunnel: a randomized prospective multicenter study. J Hand Surg [Am] 17 (6): 987–995
2. Arner M, Hagberg L, Rosen B (1994) Sensory disturbances after two-portal endoscopic carpal tunnel release: a preliminary report. J Hand Surg [Am] 19(4): 548–551
3. Bande S, DeSmet L, Fabry G (1994) The results of carpal tunnel release: open versus endoscopic technique. J Hand Surg [Br] 19(1): 14–17
4. Born T, Mahoney J (1995) Cutaneous distribution of the ulnar nerve in the palm: does it cross the incision used in carpal tunnel release? Ann Plast Surg 35(1): 23–25
5. Brock M, Iprenburg M, Janz C (1994) Die endoskopische Behandlung des Karpaltunnelsyndroms. Deutsches Ärzteblatt 91: 2111–2114
6. Brown RA, Gelberman RH, Seiler JG 3rd, Abrahamsson SO, Weiland AJ, Urbaniak JR, Schoenfeld DA, Furcolo D (1993) Carpal tunnel release. A prospective, randomized assessment of open and endoscopic methods. J Bone Joint Surg Am 75(9): 1265–1275
7. Brown MG, Keyser B, Rhotenberg ES (1992) Endoscopic carpal tunnel release. J Hand Surg [Am] 17(6): 1009–1011
8. Cobb TK, Cooney WP (1994) Significance of incomplete release of the distal portion of the flexor retinaculum. Implications for endoscopic carpal tunnel surgery. J Hand Surg [Br] 19(3): 283–285
9. Cobb TK, Knudson GA, Cooney WP (1995) The use of topographical landmarks to improve the outcome of Agee endoscopic carpal tunnel release. Arthroscopy 11(2): 165–172
10. Erdmann MW (1994) Endoscopic carpal tunnel decompression. J Hand Surg [Br] 19(1): 5–13
11. Feinstein PA (1993) Endoscopic carpal tunnel release in a community-based series. J Hand Surg [Am] 18(3): 451–454
12. Friol JP, Chaise F, Gaisne E, Bellemere P (1994) Endoscopic decompression of the median nerve in the carpal tunnel. Aprosos of 1,400 cases. Ann Chir Main Memb Super 13(3): 162–171
13. Futami T (1995) Surgery for bilateral carpal tunnel syndrome. Endoscopic and open release compared in 10 patients. Acta Orthop Scand 66(2): 153–155
14. Hallock GG, Lutz DA (1995) Prospective comparison of minimal incision "open" and two-portal endoscopic carpal tunnel release. Plast Reconstr Surg 96(4): 941–947
15. Kelly CP, Pulisetti D, Jamieson AM (1994) Early experience with endoscopic carpal tunnel release. J Hand Surg [Br] 19(1): 18–21
16. Levy HJ, Soifer TB, Kleinbart FA, Lemak LJ, Bryk E (1993) Endoscopic carpal tunnel release: an anatomic study. Arthroscopy 9(1): 1–4
17. Menon J, Etter C (1993) Endoscopic carpal tunnel release – current status. J Hand Ther 6(2): 139–144
18. Murphy RX Jr, Jennings JF, Wukich DK (1994) Major neurovascular complications of endoscopic carpal tunnel release. J Hand Surg [Am] 19(1): 114–118
19. Okutsu I, Ninomiya S, Takatori Y, Hamanaka I, Genba K, Ugawa Y, Schonholtz GJ, Okumura Y (1993) Results of endoscopic management of carpal tunnel syndrome. Orthop Rev 22(1): 81–87
20. Roth JH, Richard RS, MacLeod MD (1994) Endoscopic carpal tunnel release. Can J Surg 37(3): 189–193
21. Scoggin JF, Whipple TL (1992) A potential complication of endoscopic carpal tunnel release. Arthroscopy 8(3): 363–365
22. Sennwald GR, Benedetti R (1995) The value of one-portal endoscopic carpal tunnel release: a prospective randomized study. Knee Surg Sports Traumatol Arthrosc 3(2): 113–116
23. Viegas SF, Pollard A, Kaminksi K (1992) Carpal arch alteration and related clinical status after endoscopic carpal tunnel release. J Hand Surg [Am] 17(6): 1012–1016

Results of Transillumination-Assisted Carpal Tunnel Release

A. Franzini and G. Broggi

Summary

A novel technique to release the carpal tunnel syndrome is described. The procedure may be considered "minimally invasive" and is assisted by intraoperative transillumination. Advantages compared with endoscopic procedures are suggested to be the absence of tourniquet-induced limb ischemia and the smaller size of the devices inserted within the carpal tunnel. Indications and results in 50 consecutive patients with 2 years follow-up are discussed.

Introduction

Since it was first described by J. R. Learnmonth in 1933 [7] the surgical technique of carpal tunnel release has undergone several modifications and a standard methodology is still lacking [4]. Nevertheless, two main groups of surgical procedures may be distinguished: the first group includes open surgery under loupes or operative microscope [11]; the second group includes endoscopic surgery [1, 3, 8, 9]. Some authors have compared the two procedures in a homogenous series of patients. A slight preference for endoscopy over open surgery was found due to the short postoperative course and the early use of the operated hand [2, 6]. The aim of this report is to analyse the results obtained by an alternative surgical procedure which is similar to endoscopic surgery but is assisted by transillumination and strictly based on anatomic landmarks [5]. The advantages advocated over endoscopic surgery include the abolition of tourniquet-induced limb ischemia and the smaller size of the device inserted within the narrow carpal tunnel.

Methods and Patients

The procedure is performed under local anaesthesia injected subcutaneously at the site of the small wrist incision. The hand is maintained slightly extended and a 1.5-cm longitudinal incision is made at the wrist between the flexor radialis and palmaris longus tendons.

The wrist flexion crease and the area in which the superficial palmar branch of the median nerve may run are avoided by this approach (Fig. 1, upper left). The median nerve and the proximal edge of the transverse carpal ligament are exposed by blunt dissection. A grooved guide is then inserted within the tunnel and by virtue of its convexity it slides over the median nerve pointing to the base of the fourth finger (Fig. 1, upper left). This trajectory originates more laterally than in endoscopic procedures to minimize conflicts with the ulnar nerve and its superficial branches. Note that the small size of the guide (3 mm high and 4 mm wide) allows it to be placed over the nerve while larger endoscopic devices are usually inserted medially to the nerve closer to the ulnar side.

A fiber optic light probe (1.5 – 3 mm in diameter) is then pushed along the groove of the guide and connected to a light source (white light, 60 W, 3100 K). When the probe runs under the ligament the light is absorbed by the tight collagen fibers and just a pale spotlight is visible through the intact palm skin. At the end of the ligament a bright light appears within the palmar skin and the smooth tip notch of the guide may be correctly positioned at the distal edge of the transverse carpal ligament (Fig. 1, upper and lower right). The modified Paine retinaculotome [10] is then pushed within the groove of the guide to cut the ligament until the arrest tip notch is firmly maintained at the predetermined position (Fig. 1, lower left). The shape of the retinaculotome allows subfascial section of the ligament and the guide tip protects the arterial arch from accidental damages (Fig. 2).

Patients subjected to the described procedure fulfilled the following conditions:

- Spontaneous nocturnal pain and numbness the first three fingers
- Reduced nerve conduction velocity at electromyographic recordings
- Absence of major motor involvement of thenar muscles
- Absence of systemic or focal diseases possibly related to carpal tunnel syndrome

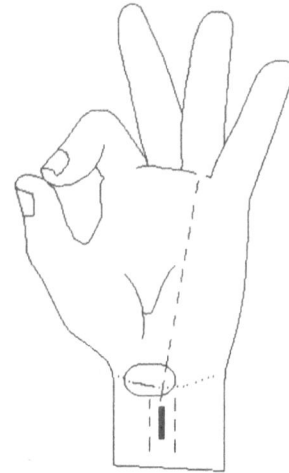

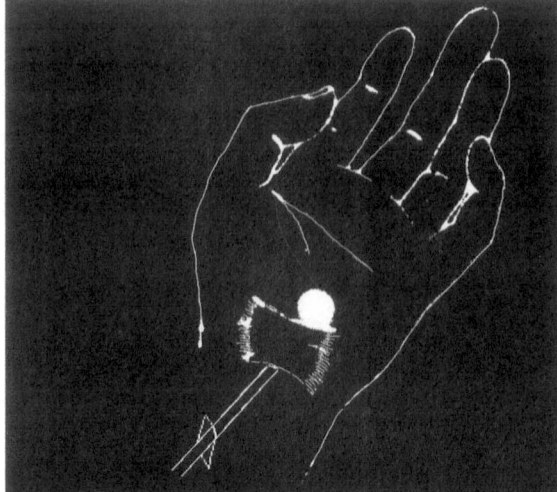

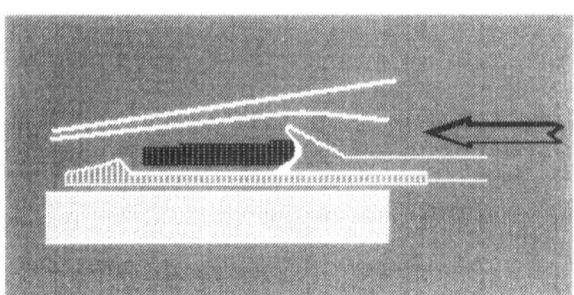

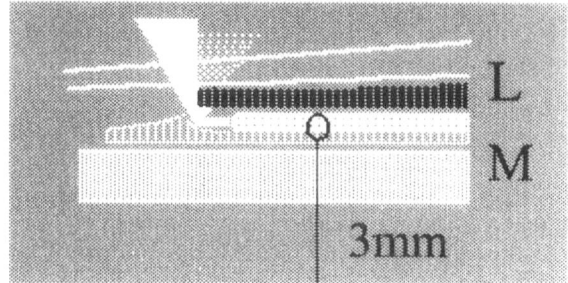

Fig. 1. *Upper left*, skin incision centred between the palmaris brevis and flexor radialis carpi tendons. The area in which the palmar superficial branch of the median nerve may be encountered is spared by the incision at the wrist flexion crease. The trajectory along which the guide is inserted points to the base of the fourth finger. *Upper right*, The fiber optic probe is inserted along the trajectory and a bright spotlight appears at the distal edge of the transverse carpal ligament. *Lower right*, schematic sagittal representation of the guide and light probe inserted between the ligament (L) and the median nerve (M). *Lower left*, the carpalotome is pushed forward along the guide to cut the ligament until the smooth tip notch of the guide reaches the proper position at the distal edge of the ligament

Out of 120 patients operated on since 1994, 50 were included in a retrospective analysis (20 preliminary cases have been withdrawn because of the learning curve phenomenon and 50 other patients have been withdrawn because the follow-up was shorter than 1 year). The age of patients in the final series ranged between 32 and 66 years (mean age 52 years); 80% were females. The duration of follow-up ranged between 1 and 3 years (mean 2 years).

Results and Discussion

There was a complete failure of treatment in four patients in which the preoperative symptomatology recurred soon after surgery (8%). Three other patients complained of persistent tender points at the base of the first finger despite the spontaneous disappearance of symptoms. The procedure was completely successful in 43 patients (86%).

The following points were noted:

- Utilization of the operated hand a few hours after surgery without the need of a splinting device (subfascial section of the ligament and respect of the wrist flexion crease)
- Early return to full activity within 1–3 weeks (mean 3 weeks)
- No dysesthesia within the ulnar side of the hand [12] (the small size of the devices inserted within the tunnel allows a more lateral approach than in endoscopic surgery; Fig. 2)
- No sympathetic dystrophy or neurovegetative symptoms in the operated hand and arm (abolition of tourniquet-induced ischemia?).

In our opinion the transillumination-assisted procedure is a valid surgical technique to treat "cryptogenetic" carpal tunnel syndrome and may be considered as an alternative to open surgery in mild and moderate diseases.

Fig. 2.
The instruments utilized in the transillumination-assisted carpal tunnel release: *A*, the fiber optic light probe; *B*, the grooved guide with the smooth tip notch at its extremity; *C*, the carpalotome

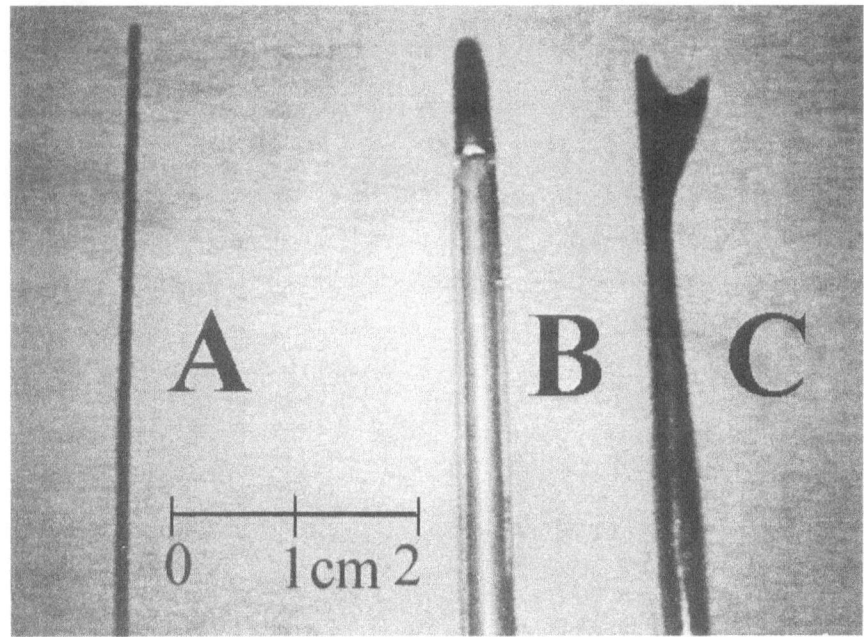

References

1. Agee JM, McCarrol HR, Tortosa RD (1992) Endoscopic release of the carpal tunnel: a randomized prospective multicenter study. J Hand Surg 17 : 987 – 995
2. Brown RA, Gelberman RH, Seiler JC III (1993) Carpal tunnel release: a prospective randomized assessment of open and endoscopic methods. J Bone Joint Surg Am 75 : 1265 – 1274
3. Chow JCY (1990) Endoscopic release of carpal ligament for carpal tunnel syndrome. 22 month clinical result. Arthroscopy 6 : 288 – 297
4. Chabaud B, Flocard F., Dasse Y et al. (1993) Applications chirurgicales des variations anatomiques du nerf mèdian au poignet. Neurochirurgie 39 : 92 – 100
5. Franzini A, Broggi G, Servello D, Dones I, Pluchino MG (1996) Transillumination in minimally invasive surgery for carpal tunnel release. J Neurosurg 85 : 1184 – 1186
6. Futami T (1995) Surgery for bilateral carpal tunnel syndrome. Endoscopic and open release compared in 10 patients. Acta Orthop Scand 66(2) : 153 – 155
7. Learmonth JR (1933) The principle of decompression in the treatment of certain diseases of peripheral nerves. Surg Clin North Am 13 : 905 – 913
8. Menon J (1993) Endoscopic carpal tunnel release: a single portal technique. Contemp Orthop 26 : 109 – 116
9. Okutsu I, Ninomiya S, Takatory Y et al. (1989) Endoscopic management of carpal tunnel syndrome. Arthroscopy 5 : 19 – 24
10. Pagnatelli DM, Barrer SJ (1991) Carpal tunnel syndrome: surgical treatment using the Paine retinaculatome. J Neurosurg 75 : 77 – 81
11. Shapiro S (1995) Micorsurgical carpal tunnel relase. Neurosurgery 37 : 66 – 70
12. Shinya K, Lanzetta M, Conolly WB (1995) Risk and complications in endoscopic carpal tunnel release. J Hand Surg [Br] 20(2) : 222 – 227

Limitations of Endoscopic Release of Carpal Ligament in Carpal Tunnel Syndrome

H. Senyurt and H.-U. Thal

Summary

From 1992 up to June 1996, 292 hands with electro-physiologically proven carpal tunnel syndrome (CTS) were operated on. 250 carpal tunnel syndromes were operated endoscopically by means of the dual portal technique according to Chow. In the case of 42 hands, due to the developed CTS with most severe electrophysiological changes, permanent sensory deficit and starting motor dysfunction up to the point of atrophy with an average distal latency of 7.83 ms we operated using the conventional open technique. Of the 250 hands the operation of which had been started endoscopically, 12 hands had to be continued in the open technique due to anatomic variants and difficulties during the operation. Eight (67%) of the 12 hands had a distal latency of more than 7.5 ms, the carpal tunnel being very narrow and causing difficulties during the operation. Four (33%) out of the 12 hands exhibited anatomic variations due to which it was also impossible to continue operating on endoscopically; therefore we continued using the open technique.

In total 3 (1.2%) nerve lesions occurred in the group in which we had started to operate endoscopically (250). These 3 cases represented anatomic variants; the operation was completed using the open method. In the case of 9 hands an iatrogenic lesion of the superficial arterial palmar arch occurred. The haemorrhages did not change the clinical course. In the case of two hands started endoscopically and completed by open surgery, an open revision had to be carried out 4 months later. In the group where we used the conventional open method of operation, one nerve injury occurred (2.4%). In one patient suffering severe diabetes mellitus there was a wound infection, and two other patients exhibited algodystrophy. In three patients a swelling developed.

In the case of the patients with most severe electrophysiological changes, i.e., with motor deficit to the point of atrophy, endoscopic operation created more problems and increased the risk of complications during the operation. The limit was arbitrarily set clinically and/or electrophysiologically due to a distal latency exceeding 7.5 ms.

Introduction

Since Sir James Paget first described the carpal tunnel syndrome in the year 1854, it has been regarded as the most frequently diagnosed peripheral nerve syndrome. It comprises up to 18% of all peripheral nerve lesions. Fifty per cent of all brachialgiae are due to CTS [3]. According to the latest epidemiological studies, the prevalence is 9.2% among women and 0.6% among men [7].

Therapeutically, the open technique of operation has been the method of choice up to now. For the first time – for the purpose of decompression of the median nerve in the wrist – the flexor retinaculum was severed by Sir James Learmonth in the year 1932. The incision of the flexor retinaculum is intended to decompress the median nerve and to improve the clinical symptoms. After the operation, patients now and then reported a weakness of the hand, inconvenience in the thenar and hypothenar as well as hypersensitivity of the scar. These problems caused a delay before the patients were able to return to work.

As an alternative to the conventional open method Paine developed a new instrument, the retinaculotome, for a less invasive operativ technique. The endoscopic release of the carpal ligament was developed by Chow [5] as a dual portal technique and by Agee [1] as a single opening technique. As less soft tissue is divided, the endoscopic severance of the flexor retinaculum developed by Chow and Agee makes it possible for patients to get back to work earlier, but it includes the potential risk of a lesion of neurovascular structures and of an incomplete incision of the flexor retinaculum [4].

Today, endoscopic incision of the transverse carpal ligament in the case of CTS is an established operation method of operation. Nevertheless there are admonitory and even disapproving opinions warning of possible dangers. In the literature there are reports about some disastrous complications [1, 4, 6, 13, 14, 18, 22]. No report could be found which commented on the limitations of this fine surgery method which, however, requires besides experience a lot of technology. Which patient should be operated on by means of the open technique, which patient should be operated on endoscopically, and where are the limits?

Methods and Material

In the period from 1992 up to June 1996, 292 carpal tunnel syndromes were operated in the Neurosurgical Clinic in Lünen. 250 hands were operated on endoscopically; 42 hands were operated on using the open technique. In the endoscopic group, the endoscopic method had to be discontinued in the case of 12 patients, and the operations were continued by means of the open technique.

The patient group consisted of 87 men and 205 women. Of this patient population 42 had bilateral carpal tunnel incisions. Nineteen patients of the 42 cases with bilateral carpal tunnel incisions had been operated on previously on the other side by means of the open technique. The average age of the patients was 57.7 years (the youngest patient was 19 and the oldest patient was 89 years). The side allocation indicated a higher number on the righthand side (right 172; left 78 both sides 42). The preoperative duration of the clinical symptoms of CTS was 14.7 months on average (minimum 6 weeks, maximum 13 years). In the case of two patients a preceding wrist trauma was present anamnestically. In the case of 25 patients diabetes mellitus was present; 14 patients suffered rheumatoid arthritis; two patients had chronic renal insufficiency; and there were three cases of hypothreosis. In the case of five female patients this occurred post partum; 1 female patient was pregnant.

Preoperative diagnosis included signs and symptoms as well as electro-neurophysiologic examination, which means that all the patients operated on endoscopically showed pathologic values. In the group of patients operated on endoscopically (250), preoperative distal latency was 6.34 ms on average. In the group of patients operated on by means of the open technique (42), preoperative distal latency was 7.83 ms on average. Among the 12 patients in whom we had started to operate on endoscopically and had to complete the operation using the open technique. eight exhibited a distal latency exceeding 7.5 ms.

Those 42 patients with most severe electrophysiological changes (predominantly with a distal latency of more than 7.5 ms of the median nerve), with motoric dysfunction symptoms to the point of atrophy were operated on by means of the open technique from the outset. After the operation, a forearm splint as described by Link was put on for a fortnight.

In the case of 250 patients we started operating endoscopically in view of their anamnesis, the clinical neurological examination (patients exhibiting a trauma in the anamnesis, chronic polyarthritis, an already visible thenar atrophy and a secondary amyloidosis as well as tenosynovitis were excluded from this group), and the electrophysiological measurements. After the operation, a compression bandage was applied for 1 day.

The patients were asked postoperatively after 3 months up to 36 months to name "subjective complaints". In addition, 274 patients (93.8%) underwent a follow-up electrophysiological examination 3 months after the operation. For 18 patients it was not possible to come to the follow-up electrophysiological examination for various reasons. The patients were also asked after the operation when they had been able to take up their former jobs again. Moreover, those patients who had bilateral CTS (19) and who had been operated on by means of the open technique previously where asked about the difference between the two methods of operation.

Results

Using the dual portal technique according to Chow, operations on 250 hands were started endoscopically. In the case of 12 patients, endoscopic surgery had to be stopped due to difficulties during the operation and the open technique was applied to complete the operation. In the case of four patients this was due to anatomic variants; in the case of eight patients the carpal tunnel was very narrow (average distal latency with these eight patients was 7.5 ms), so that it was not possible to push the trocar smoothly through. In this group, three (1.2%) nerve lesions occurred, which were put down to anatomic variants. Two of the nerve lesions were noticed during the operation and in the same intervention a microneurosurgical interfascicular suture was performed. This was a thenar branch of the median nerve. (After the endoscopic surgery had been stopped and the operation was continued by means of the open technique it was possible to see the thenar branch of the median nerve.) One female patient had a sensation of numbness at the radial side of her ring finger even 3 months after the operation. (Here, a sensitive branch of the median nerve was injured).

Two patients (0,8%) underwent revision surgery 4 months after the operation because they still had paresthesiae after the operation. The operations of these two patients were started endoscopically; however due to difficulties during the operation the intervention had to be completed by means of the open technique. Even after revision surgery, no mitigation of the paresthesiae became apparent. (No neurovascular structures were injured).

In the case of nine hands, an iatrogenic injury of the superficial arterial palmar arch occurred. The haemorrhages were stopped by means of bipolar coagulation or a suture. These haemorrhages did not lead to any change in the operative technique. In some cases the haemorrhages were also stopped by means of compression. No postoperative bleeding occurred.

In the case of two patients, uneventful swellings occurred. Likewise, in the case of two patients, wound healing was prolonged but without any wound infection. (the female patient had diabetes mellitus.) In the case of one female patient with insulin-dependent diabetes mellitus, a wound infection and secondary healing occurred.

Altogether 42 patients were operated on by means of the open technique. These were mainly patients with advanced CTS with neurological motor and sensory deficit symptoms. In this group, one patient (2.4%) suffered a nerve lesion. The thenar branch of the median nerve was taken care of surgically during the operation.

In the case of two patients the wound healing was prolonged, one of the patients had diabetes mellitus type II. In this group three patients had swellings. After prolonged immobilisation the swelling was regressive. One female patient suffered a wound infection, which was healed by means of antibiotics and local wound revision. In the case of two patients algodystrophy was diagnosed.

The patients (280) were asked postoperatively after 3 months up to 36 months to name "subjective complaints". Of those operated on 93.5% did not have any paresthesiae or pains; 6.16% of those operated on still had slight paresthesiae, mainly during the night. One patient (0.34% of those operated on) had the same complaints as before the operation (paresthesiae in the area of distribution of the median nerve, known polyarthritis).

Of the 292 patients, 274 underwent a postoperative follow-up electrophysiological examination. Of those who were examined again, 79.11% had a distal latency of less than 4.5 ms. The average distal latency in the group operated on by means of the open technique was 7.83 ms before the operation. After the operation, the average value was 4.42 ms. In the endoscopic group, average distal latency was 6.34 ms. After the operation, this value was 4.31 ms. With similar preoperative average values of distal latency, preoperative and postoperative differences are almost the same in both groups.

Fitness for work, including the resilience of the hand, was possible after 38.2 days in the case of the group operated on by means of the open technique. In the group operated on endoscopically it was possible for the patients to return to work after 24.8 days.

Nineteen patients who had previously been operated on by means of the open technique on the other side and subsequently operated on the other side and subsequently operated on endoscopically said that endoscopic surgery was less painful than the open technique and that they had used their hands earlier and the scar was less sensitive after endoscopic surgery. Patients doing heavy manual work assessed the endoscopic method as superior to the open method, as subjectively the coarse strength of the hand came back faster.

Discussion

The multitude of reports in literature renders it impossible to make a valid statement about the exact rate of complications during the open incision of the flexor retinaculum. According to earlier reports, the risk of complications in the case of open therapy is approximately between 2% and 13.5% [8, 9, 14, 15, 17, 21, 23]. In our patient population, we had a complication rate of 2.4% in the group operated on in open technique. Symptoms which occur following the operation include pains in the area of the operation scar, the frequency of which varies from 4% up to 30% of the patient population, depending on the study. Later prospective randomised reports verify the observation that patients who undergo open surgery suffer pain in the area of the scar more frequently and for a longer period of time than comparable patients operated on by means of the endoscopic technique [4]. In our case material, 25% of the patient population mentioned pains in the area of the operation scar. Postoperative traumatic pains occurred less frequently in endoscopically operated patients and eased off more quickly than in the group operated on by means of the open technique. Nineteen patients (who had previously been operated on by means of the open technique on the other side and then were operated on endoscopically), i.e. about 8% of the patient population, had the possibility of comparing both methods of operation. This small group confirmed that the endoscopic technique caused slight postoperative pain. Therefore the hand operated on endoscopically was used earlier for daily work and thus regained the strength of the closed fist earlier. These observations are in accordance to other reports [1, 6].

In large populations the rate of complications in the case of endoscopic two- and one-opening method is also considered to be low. From 147 cases, Agee described two cases of transitory pains caused by the ulnar nerve and one in completely incised retinaculum [1]. The rate of complications associated with endoscopic methods is lower than that of the open method and is indicated to be 1.66%–5% [2, 6, 10, 14, 22]. In our group, three partial motor dysfunction symptoms of the nerves were registered (1.2%). There was no case of ulnar impairment, no

M. Sudeck occurred. In two cases (0,8 %) we had to perform revision surgery because the first operation had no abolished or alleviated the complaints. Here, no partial severance of the flexor retinaculum was to be seen, nor were there any lesions of neurovascular structures. In the case of one patient (0.4 %) with insulin-dependent diabetes mellitus, a wound infection and secondary healing occurred. In summary, the current rate of complications is 2.4 %.

When the endoscopic operation technique is applied, neurolysis of the median nerve cannot be performed, nor can the carpal canal be inspected. Also synovectomy is not possible. Many authors, including Gelberman et al. [12] and Foulkes et al. [11], reported that a neurolysis of the median nerve is not necessary.

Our results as well as the studies by Chow [6], Brown et al. [4], Agee et al. [1] and Palmer et al. [20] show clearly that the endoscopic procedure is superior to the open technique with regard to postoperative complaints, i. e. the shorter duration of the inability to work and particularly less pain and paresthesia at night.

In our opinion, there is an indication for endoscopic incision of the flexor retinaculum in case of patients without thenar atrophy, with a short period of anamnesis, without any suspected intracarpal masses or an inflammatory process, and who wish to return to work as soon as possible. If there is any restricted extension in the wrist, the closed method of operation is not suitable for technical reasons. In the case of patients with a high distal latency, e. g. of more than 7.5 ms, first the endoscopic technique should be applied; if the carpal tunnel is too narrow so that it is not possible to pass through the operation should without inhibition be continued by means of the open technique. A further indication for the endoscopic operation method is to be considered in the case of older people without any signs of thenar atrophy. Since the great advantage of endoscopic surgery is the early resilience of the hand operated on, it is possible to attain above all in the case of old patients the goal of removing pain and regaining the use of the hand early for necessary self-reliance by applying the closed technique. Those patients who are planned to be operated on endoscopically should be examined thoroughly in advance, and only those for whom this approach is suitable should be selected. The patients should also be informed about an open operation, the risk of injury to neurovascular structures and that it may be necessary to continue the operation by means of the open technique if any difficulties occur during the operation.

In summary, we can conclude that endoscopic incision of the flexor retinaculum certainly is a safe method of operation on the carpal tunnel syndrome if it is performed or supervised by experienced surgeons. Preoperative anatomic preparation exercises are recommended.

The endoscopic incision of the flexor retinaculum must not be performed unless the transverse fibres can be clearly identified. If this should not be possible, we urgently recommend to stop the endoscopic method. Likewise, endoscopic surgery should be stopped and the operation continued by means of the open technique if the carpal canal is too narrow so that it is not possible to pass the trocar through. Then the operation must be completed by means of the proven open technique.

References

1. Agee JM, McCarroll HR Jr, Totosa RD, Berry DAR, Szabo M, Peimer CA (1992) Endoscopic release of the carpal tunnel syndrome: A randomized prospective multicenter study. J Hand Surg Am 17 : 987–995
2. Arner M, Hagberg L, Robsen B (1954) Sensory disturbances after two portal endoscopic carpal tunnel release A: preliminary report. J Hand Surg Am 19 : 548–551
3. Brandt T, Dichgans J, Diener HC (eds) (1988) Therapie und Verlauf neurologischer Erkrankungen. Verlag Kohlhammer Stuttgart, pp XI, 963
4. Brown RA, Gelbermann RH, Seiler JG, Abrahamsson SO, Weiland AJ, Urbaniak JR, Schoenfeld DA, Furcolo D (1993) Carpal tunnel release: A prospective randomized assessment of open and endoscopic methods. J Bone Joint Surg Am 75 : 1265–1275
5. Chow JCY (1989) Endoscopic release of the carpal ligament: a new technique or carpal tunnel syndrome. Arthroscopy 5 : 19–24
6. Chow JCY (1993) The Chow technique of endoscopic release of the carpal ligament for carpal tunnel syndrome; four years of clinical results. Arthroscopy 9 : 9–12
7. de Krom MC, Knipschild PG, Kester AD, Thijs CT, Boekkooij PF, Spaans F (1992) Carpal tunnel syndrome; prevalence in general population. J Clin Epidemiol 45 : 373–376
8. Dellon AL, Mackinnon SE (1988) Surgery of the peripheral nerve. Thieme, New York
9. Erdmann MWH (1994) Endoscopic carpal tunnel decompression. J Hand Surg Br 19 B : 5–13
10. Feinstein PA (1993) Endoscopic carpal tunnel release in a community-based series. J Hand Surg Br 18 : 451–454
11. Foulkes GD, Atkinson RE, Beuchel C, Doyle JR, Singer DI (1994) Outcome following epineurotomy in carpal tunnel syndrome: a prospective, randomized clinical trial. J Hand Surg Am 19 : 539–547
12. Gelberman RH, Pfeffer GB, Galbraith RT, Szabo RM, Rydevik B, Dimick M (1987) Results of treatment of severe carpal-tunnel syndrome without internal neurolysis of the median nerve. J Bone Joint Surg Am 69A : 896–903
13. Kelly CP, Pulisetti D, Jamieson AM (1994) Early experience with endoscopic carpal tunnel release. J Hand Surg Br 19 : 18–21
14. Kröpfl A, Gasperschitz F, Hertz H (1994) Technik, Ergebnisse und Gefahren der endoskopischen Karpaltunnelspaltung: 32. Jahrestagung der Österreichischen Gesellschaft für Plastische, Ästhetische und Rekonstruktive Chirurgie. Baden/Vienna 13–15 Oktober
15. Langloh ND, Linscheid RL (1972) Recurrent and unrelievaed carpal-tunnel syndrome. Clin Orthop 83 : 41–47

16. Mackinnon SE, McCabe S, Murray JF, Szalai JP, Kelly L, Novak C, Kin B, Burke GM (1991) Internal neurolysis fails to improve the results of primary carpal tunnel decompression. J Hand Surg Am 16 : 211 – 218

17. Nathan PA, Meadows KD, Keniston RC (1993) Rehabilitation of carpal tunnel surgery patients using a short surgical incision and an early program of physical therapy. J Hand Surg Am 18 : 1044 – 1050

18. Okutsu I, Ninomiya S, Takatori Y, Ugawa Y (1989) Endoscopic management of carpal Tunnel syndrome. Arthroscopy 5 : 11 – 18

19. Pain KWE, Polyzoidis KS (1983) Carpal tunnel syndrome. Decompression using the Paine retinaculatome. J Neurosurg 59 : 1031 – 1036

20. Palmer DH, Paulson JC, Lane-Larsen CL, Peulen VK, Olson JD (1993) Endoscopic carpal tunnel releasse: a comparison of two techniques with open release. Arthroscopy 9 : 498 – 508

21. Phalen GS (1984) The carpal-tunnel syndrome. J South Carolina Med Assoc 80 : 298 – 301

22. Slattery PG (1994) Endoscopic carpal tunnel rerlease. Use of the modified Chow technique in 215 cases. Med J Aust 160 : 104 – 107

23. Terrono AL, Belsky MR, Feldon PG, Nalebuff EA (1993) Injury to the deep motor branch of the ulnar nerve during carpal tunnel release. J Hand Surg Am 18 : 1038 – 1040

The Dilemma of Treating Carpal Tunnel Syndrome: Open Versus Endoscopic Release

P. Brüser and G. Larkin

Summary

A comparative evaluation of the open and endo-scopic treatment of carpal tunnel syndrome is undertaken using a systematic technology assess-ment which examines safety, efficacy, effectiveness and economical cost-benefit analysis based on cited references. According to this the endoscopic method is encumbered with a higher complication rate and, since several studies also show that the time for return to work is not reduced by shorter incisions, with higher costs.

Furthermore, the results of a controlled study are presented in which different incision lengths (4.5 and 2.5 cm) for the open method are compared and subsequently contrasted with an endoscopic control group [1]. With the exception of weaker grip strength for the long incision in the third postoperative week no significant differences were found.

With the incidence of carpal tunnel syndrome in the general population estimated at about 1% [16], open release of the flexor retinaculum has been regarded not only as one of the most common but also as one of the most effective and safest operations in hand surgery.

Follow-up results focussed primarily on the quality of reinnervation and reduction of pain. The main complications were recurrences, usually due to incomplete division of the flexor retinaculum, and injuries to the palmar cutaneous branch of the median nerve.

This – perhaps deceptive – calm was to change with the introduction of the endoscopic procedure. Follow-up criteria now include, among others, increased patient comfort, scar tenderness and time for return to work. New expressions such as radial and ulnar pillar pain were coined and larger collective statistics were published under the perceptible influence of industrial and commercial know-how.

Any innovation will often initially come under the crossfire of the more conservatively minded, or it may also be taken on board without critical scrutiny for the simple reason that it is modern. Therefore, a systematic technology assessment has been demanded before introducing any new medical technology in order to support advantages und disadvantages as well as indications with hard facts. The golden standard in the present case is open decompression plus, if need be, neurolysis of the median nerve.

A systematic technology assessment comprises:

1. Safety and feasability of the method
2. Efficacy (advantages for the patient in specialized centres)
3. Effectiveness (advantages for the patient in general hospitals)
4. Economical cost-benefit analyses [9]

The first step in the development of a new technology is directed towards proving the safety and feasability of the method. Cadaver studies of endoscopic releases of the flexor retinaculum have shown that incomplete divisions occurred in up to 55.8%, even with experienced operators [8]. In 1994 Rowland and Kleinert [14] reported 38% incomplete releases in experiments using the two-portal technique.

However, more important than this experimentally proven lack of safety of the procedure – the clinical relevance of which has at least not been proven in this form and for which a learning effect must be partly conceded – seems to be the alarmingly high number of grave postoperative complications which have been published [6,10,12,15]. A survey by 30 members of the German Society for Hand Surgery revealed that over a period of 2 years they had to reoperate on 38 patients after a previous endoscopic procedure. Apart from 20 incomplete releases of the flexor retinaculum, among other problems 14 divisions of individual major nerve trunks and digital nerves were found. A survey by the American Society for Surgery of the Hand [11] also revealed a considerably higher number of serious complications following endoscopic than after open release of the flexor retinaculum. In fact, the survey also showed a considerably higher number of serious complications after open release than was so far to be gathered from the literature.

These weighty arguments are first of all directed at the safety and feasability of a procedure. Everyone who knows, albeit just faintly, the dire consequences of a nerve divided in the region of the carpal tunnel or the palm must be conscious of this hazard. It is obviously a misconception that the median nerve is decompressed but not touched during the endoscopic operation. Most hand surgeons would feel more at ease to see the nerve and, if necessary, to touch it than endoscopically to touch it without seeing it – with all the dire consequences this would involve.

Without putting the theoretical advantages of an innovation into question, the arguments forwarded so far are serious enough to view the endoscopic method of operating very critically and in no way can it be described in general as an alternative. The learning curve of the endoscopic method – and this is undisputed even among advocates of endoscopic release – is incomparably steeper than that of the open release.

The decisive question is whether a higher intra-operative complication rate should be accepted for the sake of a possibly only short-term advantage in the postoperative treatment phase.

The second question within the framework of the systematic technology assessment deals with establishing proof of an improvement in treatment results. Only controlled studies can adequately answer this question. A comparison between both operative methods is rendered more difficult, however, by the fact that there is obviously no standardized operative technique for the open form of treatment due to variations regarding incision, type of neurolysis, treatment of the flexor retinaculum und follow-up treatment.

The studies of Agee et al. [1] and Brown et al. [3] are regularly cited as proof for the superiority of the endoscopic form of treatment. The former is a multi-centre study (ten centres) in which the operative procedure for the open treatment is not even described. In the latter study an incision is chosen which reaches from the mid-palm to 3.5 – 4.5 cm proximal to the wrist flexor crease, thus having a total length of about 10 – 11 cm.

This appears to us to be a decisive point of criticism. "Patent comfort" depends primarily upon the length and site of the incision. If the endoscopic method is compared with an open form of treatment which selects an incision reaching beyond the distal wrist crease several centimetres up the distal forearm, then it will inevitably have advantages with regard to scar tenderness, grip strength and subsequently also time for return to work. The cause of most complications with the open treatment of carpal tunnel syndrome is probably an inadequate

incision dividing branches of the palmar cutaneous nerve. The reason for recurrences is usually an incomplete release. Büchler et al. [4] give a conservative estimate for the incidence of re-explorations at 0.5 %.

Whereas the incision lengths are already defined for the endoscopic technique, they are often not even described for the open method. The "standard incision", however, can reach anywhere from the mid-third of the palm to about 3 – 4 cm proximal to the wrist crease [2, 5] and have lengths ranging from about 8 cm to 1.5 – 2 cm for a minimal incision distal to the distal wrist crease [2].

The question therefore arises as to which incision is necessary for open release and with which endoscopic procedure should it be compared in order to make comparable assessments possible. We have compared the results of open carpal tunnel release using a short incision (2.5 cm) with those following a long incision (4.5 cm) in a prospective randomized study (Fig. 1) [13]. Eighty patients were operated upon with a follow-up rate of 100%. These results were then

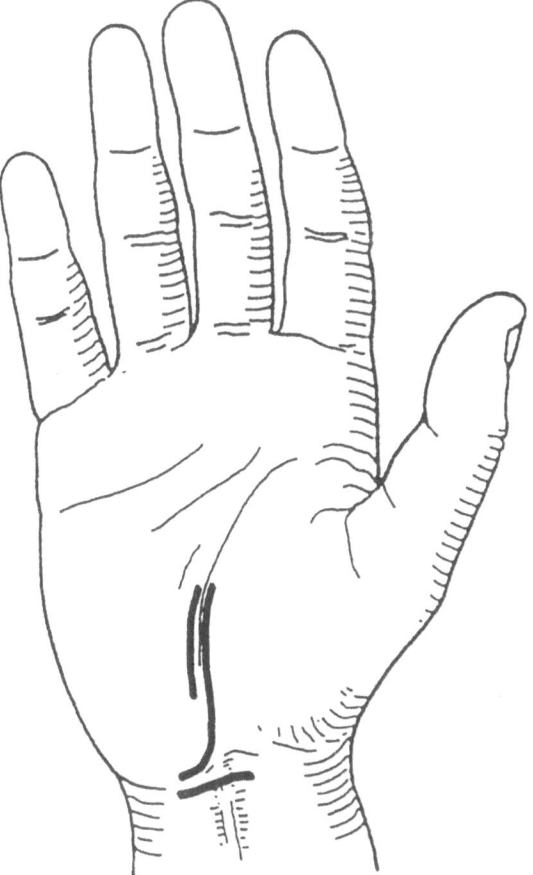

Fig. 1. Long incision (4.5 cm) and short incision (2.5 cm) over the carpus for open release; Agee's endoscopic approach (2 cm) in the wrist flexion crease

Fig. 2.
Significant reduction of night symptoms in the first postoperative week. Patients with long incision were symptom-free after 2 weeks; residual symptoms up to the 6th week in patients with short incision (7.8%) and in endoscopic group (8%)

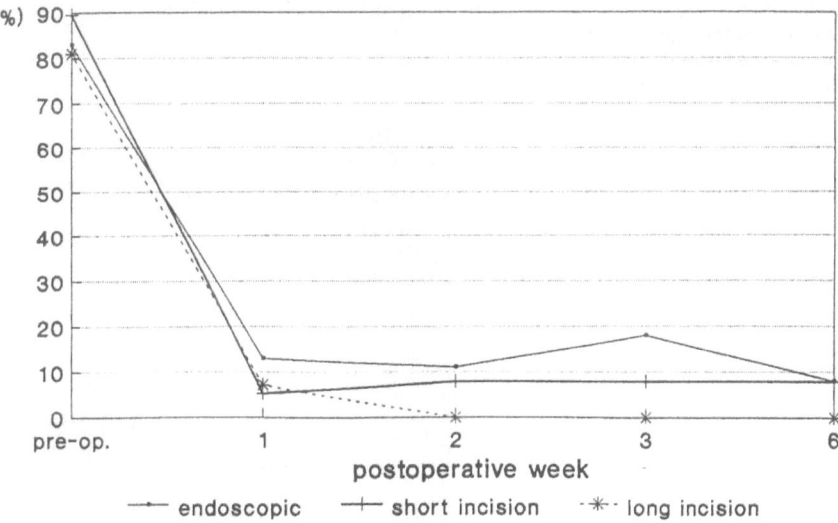

Fig. 3.
Increased scar tenderness after long incision with maximum score difference of 0.6 points as compared with the short incision without statistical significance

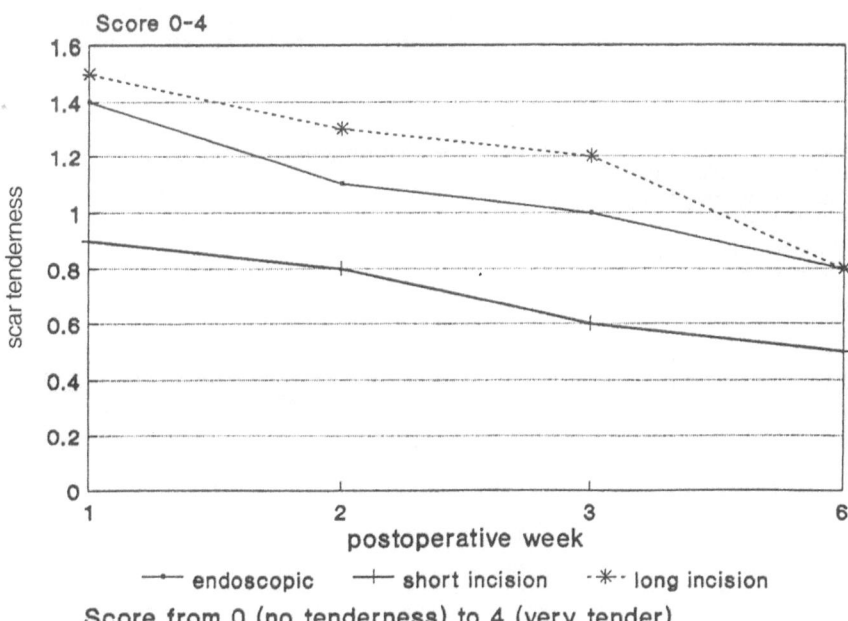

Score from 0 (no tenderness) to 4 (very tender)

compared with those of Agee et al. [1] after endoscopic carpal tunnel release. The variables examined included grip strength, key and pulp pinch strength, sensory function and scar tenderness. Subjective variables such as tingling, numbness and night symptoms were specifically enquired into, and time for return to work was noted. The study design and variables examined were identical with those of Agee et al. [1]. The examinations took place preoperatively, as well as 1, 2, 3 and 6 weeks postoperatively. The endoscopic method demonstrated no advantages over the short incision (Figs. 2, 3). The median time for return to work was 21 days

for the short incision group (Fig. 4). The long incision resulted in a 10% loss of strength only during the first 3 weeks (Fig. 5).

A similar comparison between a minimal incision and an endoscopic procedure, modified after Chow [7], also produced no statistical difference. We must realize, however, that again with "minimal incisions" the danger of incomplete release will possibly increase.

The third aspect of a technology assessment, i.e. the effectiveness of the method in general hospitals or in the hands of surgeons who are not likely to carry out such an operation so often, has not been

Fig. 4.
Median times for return to
work after endoscopic
release and after short and
long incisions

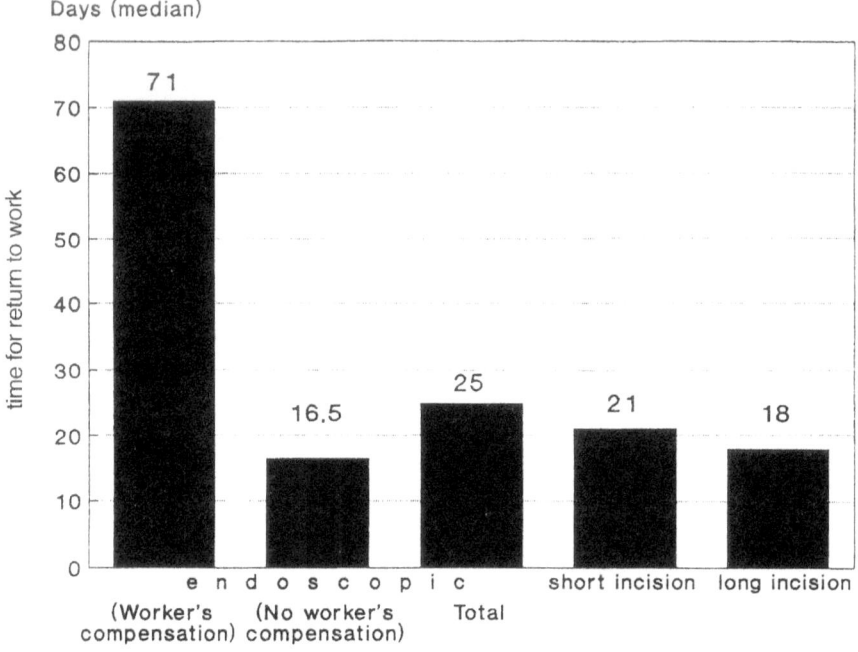

Fig. 5.
Statistically significant difference in grip strength loss
of 10 % for the long incision
in the second and third postoperative week

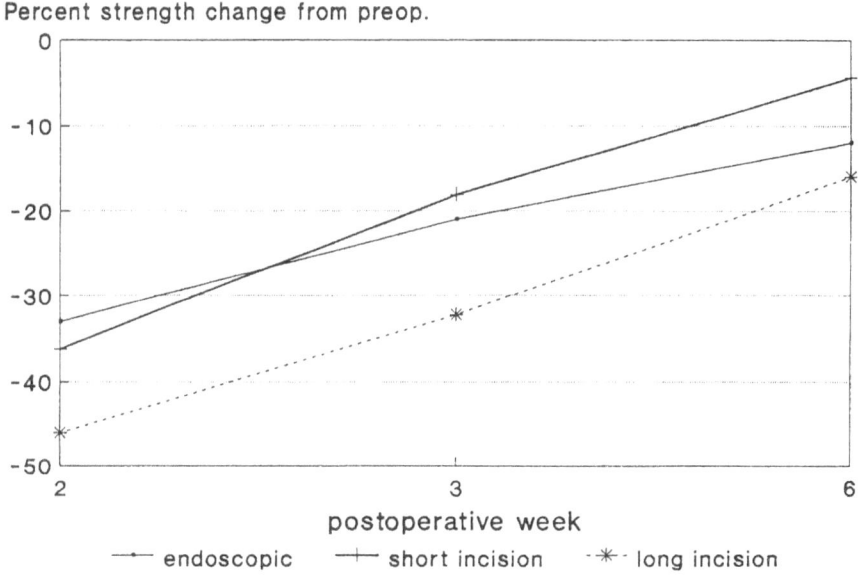

answered to date. Wherever the operation is performed infrequently, this learning curve is dispropor-
tionately prolonged. Personal inclinations or the
compulsion of having to carry out a method declared
the standard technique is at complete variance with
the safety needs of the patient. With at least comparable postoperative results, the simpler method is then
usually the preferable one.

Economic aspects as the fourth criterion of a technology assessment should always be considered
whenever methods are comparable with regard to the
first three parameters. Brown et al. [3] compared both
methods and calculated a difference of U.S. $ 491 in
favour of the open form of treatment. The difference
resulted from the price of the disposable endoscopic
sets and the additional cost of time. The extra costs
for the equipment were not included here.

So if on critical scrutiny the endoscopic technique does not offer any advantages, then its disadvantages should be examined all the more carefully. The steeper learning curve and the higher costs have been mentioned. The endoscopic technique provides less overall visualization, is more difficult and involves – currently at least – higher risks.

References

1. Agee JM, McCarroll HR, Tortosa RD, Berry DA, Szabo RM, Peimer CA (1992) Endoscopic release of the carpal tunnel: a randomized prospective multicenter study. J Hand Surg Am 17:987–995
2. Bromley GS (1994) Minimal-incision open carpal tunnel decompression. J Hand Surg Am 19:119–120
3. Brown RA, Gelberman RH, Seiler JG, Abrahamsson S-O, Weiland AJ, Urbaniak JR, Schoenfeld DA, Furcolo D (1993) Carpal tunnel release: a prospective, randomized assessment of open and endoscopic methods. J Bone Joint Surg Am 75:1265–1275
4. Büchler U, Goth D, Haußmann P, Lanz U, Martini AK, Wulle C (1983) Karpaltunnelsyndrom: Bericht über 56 Nachoperationen. Handchir Mikrochir Plast Chir 15 [Suppl] 3–12
5. Eversmann WW (1993) Entrapment and compression neuropathies. In: Green DP (ed) Operative Hand Surgery, 3rd edn. Churchill Livingstone, New York, pp 1341–1385
6. Feinstein PA (1993) Endoscopic tunnel release in a community-based series. J Hand Surg Am 18:451–454
7. Hallock GG, Lutz DA (1995) Prospective comparison of minimal incision "open" and two-portal endoscopic carpal tunnel release. Plast Reconstr Surg 96:941–947
8. Heest van A, Waters P, Simmons B, Schwarz JT (1995) A cadaveric study of the single-portal endoscopic carpal tunnel release. J Hand Surg Am 20:363–366
9. Jennett B (1986) High technology medicine. Benefits and burdens. Oxford University Press, Oxford
10. Murphy RX, Jennings JF, Wukich DK (1994) Major neurovascular complications of endoscopic carpal tunnel release. J Hand Surg Am 19:114–118
11. Palmer AK, Toivonen DA (1995) Complications of endoscopic and open carpal tunnel release. 50th Annual Meeting of the American Society for Surgery of the Hand, San Francisco, 13–16 September
12. Piza-Katzer H, Laszloffy P, Herczeg E, Balogh B (1996) Komplikationen bei endoskopischen Karpaltunnel-Operationen. Handchir Mikrochir Plast Chir 28:156–159
13. Richter M, Brüser P (1996) Die operative Behandlung des Karpaltunnelsystroms: Ein Vergleich zwischen langer und kurzer Schnittführung sowie endoskopischer Spaltung. Handchir Mikrochir Plast Chir 28:160–166
14. Rowland EB, Kleinert JM (1994) Endoscopic carpal-tunnel release in cadavers. J Bone Joint Surg Am 76:266–268
15. Smet de L, Fabry G (1995) Transection of the motor branch of the ulnar nerve as a complication of two-portal endoscopic carpal tunnel release. A case report. Hand Surg Am 20:18–19
16. Szabo RM (1991) Carpal tunnel syndrome – general. In: Gelberman RH (ed) Operative nerve repair and reconstruction, vol II. Lippincott, Philadelphia, pp 869–888

Part 6
Neuronavigation

Neuronavigation: A CAS System for Neurosurgery

F. Hor, M. Desgeorges, M. Traina, and Y. S. Cordoliani

Summary

We describe our experience with neuronavigation in 140 operated cases. We propose the systematic use of MR for the operative imaging and describe our results concerning the precision of procedures as a function of models of markers. The study of postoperative MR shows best tumoral resection of infiltrating gliomas and a smaller injury for the cerebral parenchyma. The best system of neuronavigation is that associating a good mechanical precision of the guiching system guidance to optical techniques for superimposition of contours in the operative microscope.

Introduction

Neuronavigation is a topic regularly found in congresses in the past 2 – 3 years. It is also a subject of discussion between neurosurgeons who support this technique and those who question its contribution. Some even say they have no need for neuronavigation to operate well. Would it be necessary again to prove the usefulness of the operative microscope, bipolar coagulation, and all advances made by neurosurgery accumulated over the past 30 years? The true problem is to devine what neuronavigation really is, and what we expect from it. This will allow us to specify the system of neuronavigation that is most efficient and most capable of development. It is useless that medical imaging shows us increasingly small and precise lesions if we have no instruments to treat them with efficancy. Surgical precision has to match radiological precision.

Materials and Methods

Since August 1994 we have been using the microscope MKM manufactured by Zeiss. This system of neuronavigation contains a microscope with variable focalization and integrated autofocus. The support of the microscope is a robot system that authorizes and controls six degrees of freedom. A seventh degree can be represented by the autofocus system. This microscope is connected to a stereotactic workstation that comprises radiological data (MR or CT) of the patient. A system of cranial markers matches the radiological volume of the patient's head to real anatomical marks during surgery. In other words, the cutaneous markers represent a virtual stereotactic frame. Two models of external markers exist. They can be glued to the skin; a limited shave is necessary to expose the scalp. They can also be screwed into the cranial bone, which necessitates drilling a small hole in the bone under local anesthesia.

To fully exploit possibilities offered by neuronavigation we prefer to use MRI by an acquisition of 3D SPGR sequence with injection of gadolinium. A total 124 axial linked slices 2 mm thick are thus obtained and transferred onto the stereotactic workstation. Using neuronavigation software the neurosurgeon then locates the lesion marking the contours on each axial slice where it is visible. He can also identify the risk zones by delimiting their contours in the same manner. One thus obtains volumes of interest that are automatically integrated in the image, and that are therefore visible on all plans of reconstruction, sagittal, coronal, perpendicular, and parallel to the trajectory. The neurosurgeon defines this trajectory by two points: the target point and the entry point. He is able to modify this route, i.e. of the optical axis of the microscope, if he is not satisfied by his surgical approach.

Before surgery it is necessary to memorize the radiological markers and to make the autofocus system of the microscope MKM recognize the exact site of the cutaneous markers. This is achieved by optical coding. These are of course the same that have been used for the radiological examination and surgery. The microscope automatically proceeds towards the axis of the trajectory.

During surgery, the neurosurgeon permanently sees in his binocular by a system of superimposition of images the target site or direction if the axis of the microscope has been modified its initial trajectory. He also sees, in superimposition, the distance of his

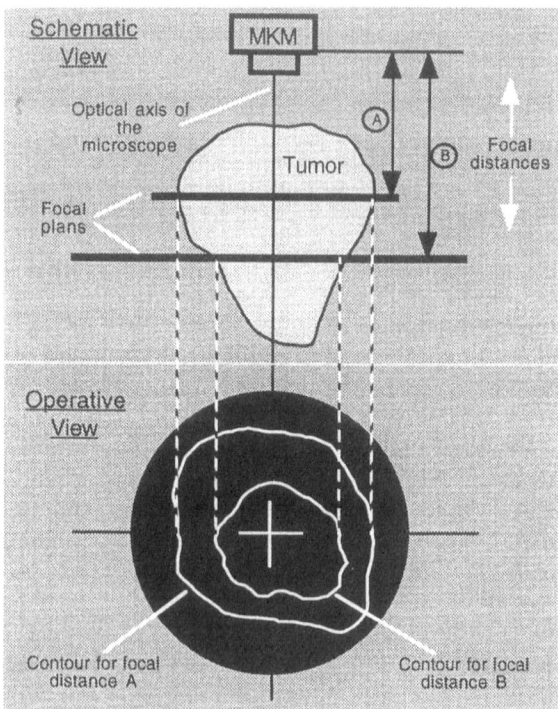

Fig. 1. Schematic and operative views

target from the focal point of the microscope as well as contours of the lesion and possible risk zones whose size varies as a function of the zoom. Visible contours in the binocular are these included in the perpendicular plan to the trajectory passing by the focal point (Fig. 1).

Working with the MKM microscope is very easy with the hand control; two control switches control all displacements. The first authorizes all displacements isocentric to the observed point, and the second allows all movements parallel to the trajectory. A voice control identifies the voice of the neurosurgeon that can thus modify the zoom, focus and lighting, request the autofocus or a slide of the operative field. The microscope is able to place itself automatically on the operative trajectory or on another position that has been recorded. It is therefore a robot with very precise motion control. Finally, the MKM microscope can be driven by an assistant using a hand control panel what allows the surgeon to keep his two hands in the operative field.

During surgery the stereotactic workstation permanently shows in real time the slices axial, coronal, sagittal, and perpendicular or parallel to the optical axis of the microscope. The neurosurgeon is therefore able to make an autofocus on an indeterminate anatomical structure and to look to the stereotactic workstation to know its exact position. It is possible to show a MR or CT slice in the binocular. It should be

noted that the MKM system has developed substantially over the past years. Recent versions have improved the imaging with better definition and a greater rapidity in the processing of images. A head tracking system is also available to mobilize the operating table or the head of the patient. Finally, a program of spine surgery is being currently developed to be able to locate with great accuracy vertebral or medullar lesions and the pedicular screwing.

Results

The correlation between the image volume and the head of the patient is satisfactory: on the average 1.2 ± 0.65 mm for screwed markers and 2.7 ± 1.1 mm for glued markers. It is necessary to know that if the stereotactic workstation announces an overall precision of 2 mm, for example, this does not mean that the contours, target or trajectory will be displaced by 2 m. In fact in our experience the accuracy is excellent for all anatomical structures which are not very mobile (sinus, vasculo-nervous elements near the skull base). The precision remains very satisfactory for parenchymal structures which are more mobile as long as traction is not excessive.

We have performed 140 operations in stereotactic conditions, i.e., using the stereotactic workstation. Most pathologies and locations have been operated using the MKM: extracerebral lesion of the convexity or skull base, deep and superficial cerebral lesions, tumoral or vascular pathologies. Some patients have been operated in a sitting position. The accuracy remains excellent for lesions of the pineal region or extracerebral lesion. A difference of approximately 2 – 4 mm is found for brain stem lesions resulting from the fact that the radiological acquisition used for surgical preplanning is performed in dorsal decubitus.

All craniotomies performed using the MKM system are reduced because they are always centered on the lesion. When the trajectory is near a venous sinus, its contour appearing in the binocular allows safe opening of the dura. The cortical vessels are well visible in sequence 3D SPGR, because of gadolinium, and the neurosurgeon is thus able to define a trajectory that avoids them. We sometimes use two series of different images that are recorded automatically by the stereotactic workstation. The anatomical sequence SPGR can thus be associated: (a) 3D SPGR and CT for the bone structure of skull base (petrous bone mainly), (b) 3D SPGR and FLAIR sequence to clearly distinguish the contours of some infiltrating tumors (oligodendroglioma, for example), and (c) 3D SPGR and functional MR for lesions ocated near or in motor areas. We work on the mixing of images 3D SPGR and SPECT to identify more accurately the zones with necrosis, tumoral infiltration or oedema. Most patients of our series have benefited from a postoperative early MRI, between the second and fifth days after surgery, followed by another MRI, 3 months after surgery to evaluate the quality of the tumoral resection.

On a total of 123 operated tumors, the contribution of the system of MKM neuronavigation is dual: reduction of the size of surgical approach and best preservation of the normal parenchyma, tumoral resections more complete, particularly for infiltrating gliomas.

The quality and duration of patient survival with malignant gliomas is surely linked to the capability of performing surgical resection as complete as possible. In our series out of 42 operated malignant gliomas resection has been judged very satisfactory in 37 cases (88%) and the complementary treatment (radiotherapy and/or chemotherapy) has appeared more efficient. We had to operate twice on 12 patients having glioblastomas previously operated on using the MKM and having undergone a complementary treatment. All these surgical procedures were performed approximately 1 year after the first and have allowed patients to survive significantly longer. Five patients have survived over 18 months with a good quality of survival.

In the case of benign extracerebral tumors such as neurinomas, meningiomas and cholesteatomas the vascular-nervous structures are better protected because the neurosurgeon knows whether he has drawn one or several zones of interest, their site and their distance compared to the focal point where he is operating.

The use of the ultrasonic aspirator for infiltrating tumors and the CO_2 laser coupled to the MKM (OPMILAS) for the meningiomas located on the skull base make tumoral resection easier and faster. An adapter which is fixed on the optical block of the MKM microscope permits connecting an endoscope that is used to control the operative field in lateral vision (70°).

Two lesions of the dorsal and cervical medulla have been operated by placing four markers glued near the median line. In this case the procedure is less accurate: approximately 5–10 mm. The use of the MKM in surgery of the spinal cord is therefore interesting to center the surgical approach and consequently to limit laminectomy in front of the lesion and to perform myelotomy at the appropriate level. The difference of precision is due to the absence of stable fixity and to the position of the patient on the operating table. When the program of spinal surgery is available, the precision of spine and spinal cord surgery should be more satisfactory. The MR acquisition of spinal lesions being performed in dorsal decubitus because of the antenna, and surgery being performed in procubitus, it is necessary to avoid to bend the patient's spine on the operating table.

Discussion

Concerning the precision of neuronavigation procedures, three markers are sufficient. We prefer to systematically place four markers. The use of five markers or more does not really increase the precision of the procedure. Finally, we place the four radiological markers overhead the front and the temporal region at the same side of the lesion, sometimes more in the rear if the lesion is posterior. We place markers around the ear for lesions of the cerebral posterior fossa. It is unnecessary to distribute the markers around the skull as this does not increase precision and makes optical coding more difficult.

Neuronavigation is more a concept than a supplementary neurosurgical instrument. The neurosurgeon is able to anticipate his operation, to reflect on his surgical approach and to mark the contours of the lesion which will help him to find it. This allows him to be more attentive to foreseeable difficulties, to the real form of the lesion. He knows in which direction and to what depth a tumoral prolongation extends. He knows risk zones that appear in superimposition in the perpendicular plan passing by the focal point. He is therefore able to anticipate this risk zone simply using the focus variation (alone the MKM system currently gives this possibility).

Permanently knowing the site of the lesion to reach, the surgical procedure is more serene and more rapid for the neurosurgeon. As it allows greater precision and smaller operative injury, neuronavigation provides better operative results for the patient. The time devoted to preplanning and to the optical coding is therefore recovered in operative rapidity for the surgeon and in safety for the patient.

Since the beginning of neuronavigation a problem exists which has not yet been solved. It is due to the brain shift created by the draining of CSF or tumoral cysts and by the progressive resection of the tumor that is not taken into account by the stereotactic workstation. In our experience deep lesions near the median line vary little anatomically. The greatest anatomical alterations are visible for subcortical lesions when an important mass effect exists. We have found twice a displacement of approximately 1 cm concerning superficial lesions. Each time it involved a patient operated in prone position while the MRI had naturally been made in a supine position. On the other hand, we have not found significant differences when the MR acquisition was made in supine position and when the patient was operated in the same position, head turned laterally to 45°.

It is indispensable to conceive an intraoperative imaging system that will allow an update of the imaging in real time. Except for interventional MR systems the solution will probably be 3 D ultrasonography coupled to echo Doppler.

Conclusion

Neuronavigation represents an anatomical assistance to the operative procedure, and all surgeons who have used this technique consider neuronavigation as a progress for neurosurgery. It can indeed combine the efficiency of microsurgery with the precision of stereotaxy. The real benefit of neuronavigation will have to be appreciated for each pathology over a period of several years so as to make a comparative study of patients operated with and without this technique. The patterns of minimally invasive neurosurgery, image-guided neurosurgery, and neuronavigation today represent the modern techniques of future neurosurgery.

Neuronavigation-Assisted Microsurgery of Intracranial Lesions

C. Schaller, B. Meyer, D. van Roost, and J. Schramm

Summary

We have tested a prototype infrared guided computerized navigation system (SPOCS; Aesculap/ISG) for the microsurgical treatment of intracranial pathology. Forty-eight patients with a total of 53 lesions – 14 (26.4%) <2 cm, 33 (62.3%) 2–4 cm, 6 (11.3%) >4 cm – were included. The navigator was applied for biopsy procedures in three cases, for cyst fenestration/shunting in two cases and for general microsurgery-assisting purposes in the remaining 48 lesions, including transnasal approaches. The accuracy of the intraoperative navigation was in the range of 3 mm or better in 39 patients (81.3%) during the course of the operation and it had to be aborted in seven patients (14.6%) due to technical dropouts and due to other problems in two patients (4.2%). The level of accuracy achieved was not altered by the patient's position except for an unsatisfactory result with the sitting position due to skin movement together with the fiducial markers. The system proved to be helpful for craniotomy planning and for intraoperative guidance in situations of a narrow visual field. It was especially useful for the localization of small lesions in highly eloquent brain regions and for the definition of the margin of large lesions.

Introduction

An increasing number of neurosurgical departments is equipped with so-called neuronavigational devices [1–4, 6, 7, 9, 15, 16, 18]. The expectations to these systems are high, since they should help the neurosurgeon to improve his performance with respect to conventional microsurgical techniques, which could alternatively be performed under stereotactic guidance [17]. However, there is still a lack of precise indications for the use of these devices in a given neurosurgical environment, and their technical and practical limitations still need to be analyzed. In order to be of assistance to the surgeon, these systems should be helpful during each surgical step from the beginning, including surgical planning and location of craniotomy. Furthermore they should provide a reliable intraoperative orientation and assist in the localization of critical structures, definition of the lesion border, and resection control. Altogether this should possibly lead to a minimization of the whole procedure and its associated morbidity. We report on our 1 year experience with such a navigational device.

Patients and Methods

Navigational Device

The system is called SPOCS, which stands for surgical planning and orientation computer system (provided by Aesculap AG, Tuttlingen, Germany), and has an image processing software implemented for 3D reconstruction of axial CT or MR images (ISG, Toronto, Canada). It consists of a UNIX workstation, a digitizer for 3D reconstruction, an infrared camera array with three sensors, and a so-called head follower with four LEDs. The latter has to be attached to the Mayfield clamp. In addition, there are LED-marked pointers of various lengths and shapes, which serve as the tool for intraoperative navigation. The head follower and the pointers have to be connected to a tool box for the integration of information omitted by the LEDs and received by the infrared sensors. A free line of sight between the infrared camera array, which is placed in roughly a 1.5-m distance, and the pointer and the follower is therefore needed for the faultless functioning of the navigator.

In our study we used mainly 128 axial MR images per case with a slice thickness of 2 mm for 3D reconstruction. The data material was transferred to the system via optical disks or by the local glassfiber network. CT images were only used for patients in whom the bony structures were of special interest. The scans are taken with the patient's head being marked with 6 to 12 skin-adhesive fiducial markers, which serve as artificial anatomical landmarks. Preoperative data acquisition, preregistration, and planning can be performed on the day before surgery, whereas for operative registration the patient has to be positioned with his head fixed in the Mayfield clamp and with the whole navigation system in its place for surgery. We used up to 50 surface points on the patient's scalp for improving the eventual error as obtained with skin fiducial markers only, this error being expressed as

"RMS" (root mean square). The position of the pointer tip as perceived by the system is visualized as a hair cross on a 20″ computer screen and can be compared to the surgeon's own impression according to real anatomical landmarks (Fig. 1).

Patients

Apart from selective amygdalohippocampectomies for medically intractable epilepsy, which are the subject of another study, we operated upon 48 patients (22 female, 26 male), aged 7–77 (mean 46) years

Fig. 1. Planning the approach to right-sided trigonal meningioma. A transsulcal approach was chosen after a temporo-occipital craniotomy was performed with the patient in a lateral position. The navigator was helpful for intraoperative guidance to the tumor through the respective sulcus. The hair cross can be used for assessment of the actual accuracy. Here, the tip of the pointer touched the dura, which is reflected by the hair cross in all four images

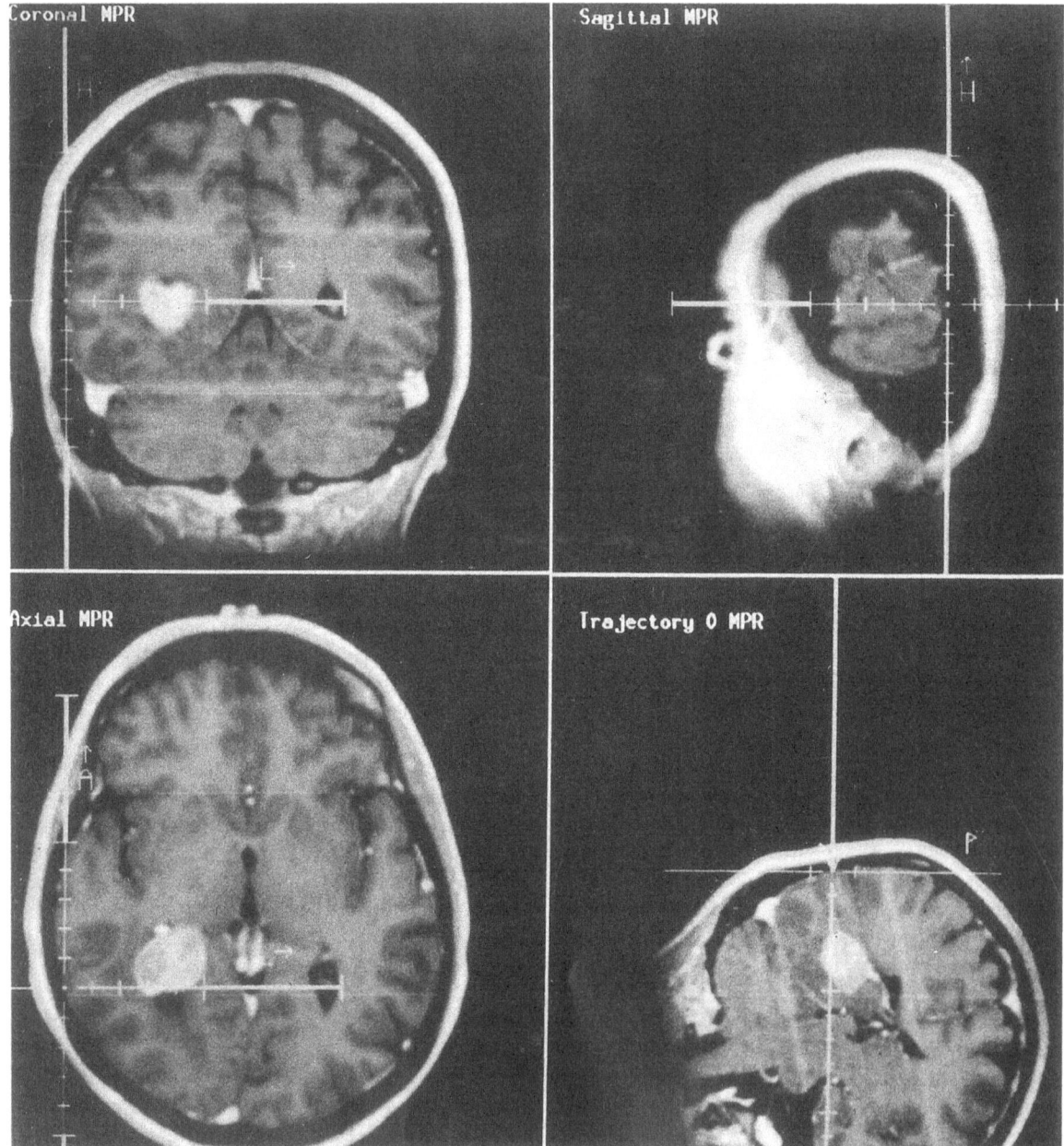

harboring a total of 53 intracranial lesions. Forty-six of the lesions were purely supratentorial, six were located at around the skull base, and one patient suffered from an acoustic neuroma. The pathologies found were as follows: 20 gliomas, seven meningiomas, five metastases, five cysts, four abscesses, four cavernomas/vascular malformations, three chordomas/chondrosarcomas, one gyral abnormality/epileptic focus, one acoustic neuroma, one pituitary adenoma, one multiple sclerosis, and one progressive multifocal leucoencephalopathy. Thirty-six patients were placed in the supine position, five in the lateral decubitus position, three in the prone position, two in the astronaut position, and one each in the park bench and sitting position.

Results

Intraoperative navigation was aborted due to technical problems on seven occasions: there were problems related to the software on three occasions, problems with the pointers in two cases, and the head follower was hit by the microscope rendering registration invalid in another two. All of these problems were encountered early during the series, thereby accounting for a failure rate of 41.7% (five failures) in the first 12 cases, but of only 4.9% (two technical failures) of the consecutive 41 lesions.

In another two patients navigation had to be aborted due to skin movement together with the fiducials in the case of the sitting position and due to shortness of the pointer in the first of the two transnasal approaches. An accuracy level of 3 mm or better during the whole operation or its most important steps was achieved in 39 patients (81.3%). In six of these patients the accuracy decreased with increasing operating time, and therefore resection control could not be performed adequately in two of these. Except for the patient who had to be operated upon in the sitting position no problems related to patient positioning were encountered.

Navigation was rated as extremely useful for the removal of 31 intracranial lesions. This includes seven deep or eloquently located lesions, four approaches through narrow canals with sparing of traversing veins, eight infiltrating tumors with difficult border definition, two freehand biopsies, five cyst fenestrations, four skull base tumors, and one case of purely image guided surgery.

Discussion

There is a diverse range of neuronavigational devices currenty available. The one we used for our present patient series belongs to the optically linked devices with a pen-like pointer as the main tool; hence, it functions as a semi-freehand device. Some systems are guided by ultrasound or magnetic fields or consist of mechanical arms with potentiometers in their joints, while others are microscope-integrated [1–3, 6, 7, 9–11, 13–16, 18]. The accuracy achievable with these systems lies well within the range of 2 mm in the laboratory and within 4 mm in clinical use [2, 3, 9, 13–16, 18, 19].

Problems which remain to be overcome are the intraoperative brain shift when operating according to images obtained days before surgery and merely technical problems such as sufficient instrument design in order to make navigators useful tools in the hand of neurosurgeons. These problems are solved in part by the additional application of intraoperative electrophysiological techniques or by open CT or open MRI [5, 8, 12]. However, this certainly leads to an even more extensive and time-consuming process of surgical planning, and resections and certain indications have to be defined for which the use of a navigator and associated techniques will be of a definite benefit for the patient.

Apart from mainly technical problems encountered early during the series which did not require abortion of the procedures, SPOCS proved to be an accurate tool in 39 patients. It was of particular benefit for the localization of small deep-seated or highly eloquently located lesions such as metastases in the motor strip.

Another excellent indication for intraoperative navigation may be the definition of the border of infiltrating tumors, such as low-grade insular gliomas infiltrating the basal ganglia, and the surgery of skullbase tumors with encasement of vital structures such as the internal carotid artery or cranial nerves. In the case of excision of a cortical epileptic focus, surgery had to be performed solely by image guidance due to absence of visible pathology. Hence, the number of lesions in which navigation was considered to have contributed significantly to the success of the operation or its planning amounted to 31 in our series of 53 lesions; the system applied has been in use for 1 year. Considering a total of approximately 500 elective intracranial procedures apart from surgery of aneurysms or arteriovenous malformations this accounts for roughly 5%–10% of cases in a neurosurgical department covering all aspects of intracranial neurosurgery. Precise planning of the craniotomy is only one advantage of the use of a neuronavigator, but this can lend more confidence especially for bone work with the high speed drill in the posterior fossa close to the transverse and the sigmoid sinus. With improvements in instrument design and an increased speed of image processing, such navigational systems can be used more deliberately and conveniently.

References

1. Apuzzo MLJ (1996) The Richard C. Schneider lecture. New dimensions of neurosurgery in the realm of high technology: possibilities, practicalities, realities. Neurosurgery 38 : 625 – 639

2. Barnett GH, Kormos DW, Steiner CP, Weisenberger J (1993) Intraoperative localization using an armless, frameless stereotactic wand. J Neurosurg 78 : 510 – 514

3. Barnett GH, Kormos DW, Steiner CP, Weisenberger J (1993) Use of a frameless, armless stereotactic wand for brain tumor localization with two-dimensional and three-dimensional neuroimaging. Neurosurgery 33 : 674 – 678

4. Barnett GH, Steiner CP, Weisenbeger J (1995) Intracranial meningioma resection using frameless stereotaxy. J Image Guided Surg 1 : 46 – 52

5. Cedzich C, Taniguchi M, Schäfer S, Schramm J (1996) Somatosensory evoked potential phase reversal and direct motor cortex stimulation during surgery in and around the central region. Neurosurgery 38 : 962 – 970

6. Germano IM (1995) The neurostation system for image-guided frameless stereotaxy. Neurosurgery 37 : 348 – 350

7. Golfinos JG, Fitzpatrick BC, Smith LR, Spetzler RF (1995) Clinical use of a frameless stereotactic arm: results of 325 cases. J Neurosurg 83 : 197 – 205

8. Imai F, Ogura Y, Kiya N, Zhou J, Ninomiya T, Katada K, Sano H, Kanno T (1996) Synthesized surface antomy scanning (SSAS) for surgical planning of brain metastasis at the sensorimotor region: initial experience with 5 patients. Acta Neurochir (Wien) 138 : 290 – 293

9. Kato A, Yoshimine T, Hayakawa T, Tomita Y, Ikeda T, Mitomo M, Harada K, Mogami H (1991) A frameless, armless navigational system for computer-assisted neurosurgery. J Neurosurg 74 : 845 – 849

10. Reinhardt HF, Horstmann GA, Gratzl O (1993) Sonic stereometry in microsurgical procedures for deep-seated brain tumors and vascular malformations. Neurosurgery 32 : 51 – 57

11. Reinhardt HF, Trippel M, Westermann B, Horstmann GA, Gratzl O (1996) Computer assisted brain surgery for small lesions in the central sensorimotor region. Acta Neurochir (Wien) 138 : 200 – 205

12. Rezai A, Hund M, Kronberg E, Zonenshayn M, Cappell J, Ribary U, Kall B, Llinás R, Kelly P (1996) The interactive use of magnetoencephalography in stereotactic image-guided neurosurgery. Neurosurgery 39 : 92 – 102

13. Ryan M, Erisckson R, Levin D, Pellizari C, MacDonald R, Dohrmann G (1996) Frameless stereotaxy with real-time tracking of patient head movement and retrospective patient-image registration. J Neurosurg 85 : 287 – 292

14. Sandeman DR, Patel N, Chandler C, Nelson RJ, Coakham HB, Griffith HB (1994) Advances in image-directed neurosurgery: preliminary experience with the ISG viewing wand compared with the Leksell G frame. Br J Neuosurg 8 : 529 – 544

15. Sipos E, Tebo S, Zinreich S, Long D, Brem H (1996) In vivo accuracy testing and clinical experience with the ISG viewing wand. Neurosurgery 39 : 194 – 204

16. Smith KR, Frank KJ, Bucholz RD (1994) The neurostation – a highly accurate, minimally invasive solution to frameless stereotactic neurosurgery. Comp Med Imaging Graph 18 : 247 – 256

17. van Roost D, Meyer B, Schaller C, Schramm J (1996) Image-guided epilepsy surgery: first experience with a freehand neuronavigation system in selective amygdalohippocampectomy. Annual meeting of the American Association of Neurological Surgeons. Minneapolis, MM, USA, April 1996

18. Watanabe E, Mayanagi Y, Kosug Y, Manaka S, Takakura K (1991) Open surgery assisted by the neuronavigator, a stereotactic, articulated, sensitive arm. Neurosurgery 28 : 792 – 800

19. Zamorano L, Kadi AM, Dong A (1992) Computer-assisted neurosurgery: simulation and automation. Stereotact Funct Neurosurg 59 : 115 – 122

Requirements for Referencing of the MKM Neuronavigation System

J. Kaminsky, G. Arango, T. Brinker, and M. Samii

Summary

The localization method of the MKM neuronavigation system is based on a laser-controlled distance measurement through the microscope optic, which is mounted to a six-axis robotic arm. Due to its optical guidance, it differs from other neuronavigation systems and provides the possibility of continuous navigation during microscopic neurosurgery. The accuracy of the system was studied under experimental conditions.

Two marker systems (titanium bone screws patent no. 1129 F1, 18 mm with exchangeable marker tip, F. L. Fischer, Germany, and 1 × 4 mm microscrews, Leibinger) were used in a fixated cadaver head. The best marker selection/positioning and a CT-protocol to achieve a high 3-D resolution with minimal radiation exposure were evaluated. The accuracy of the MKM system was studied in relation to different geometrical fiducial arrangements.

To achieve CT data with a high resolution and with minimal radiation exposure a CT procedure with 1-mm slices and exclusive scanning of planes of interest was developed. The robotic arm and the optical system of the microscope allow high precision measurements with a deviation below 1–2 mm. The reference points should be spherically arranged around the target volume; linear marker arrangements must be avoided. The studied fiducials provide high precision navigation.

Introduction

Neuronavigation systems are used to optimize the localization and removal of intracerebral lesions with minimal damage to other structures, to reduce complications and hospitalization time and to take over a teaching function [1, 3, 14].

We investigated the accuracy of the MKM neuronavigation system that is based on an operation microscope mounted to a six-axis robotic arm. In contrast to all other navigation tools [1–4, 6–11, 13–15, 17], the position of the focus point (navigation position) is calculated by evaluation of the movements of the robotic arm and of a laser-assisted distance measurement through the optical system. The focus point is indicated by a shining green cross, visible in the microscopic field. Contactless continuous navigation and projection of additional neuronavigation information (distances, targets, contours) is therefore possible even to anatomical structures with limited access.

The special features of the MKM navigation system must be taken into consideration for: (a) selection of the fiducial system, (b) CT-examination procedure, and (c) reference procedure. The present study examines experimentally the influence of these factors on accuracy.

Material and Methods

The MKM neuronavigation system of Zeiss (Oberkochen, Germany) was used. According to its technical specifications, the robotic arm provides an accuracy of ± 0.75 mm. The microscope is equipped with a laser-controlled autofocus function with an accuracy of 0.25 mm. The optical system provides a spatial resolution of 10 μm.

The controlling digital alpha workstation and the neuronavigation software STP is manufactured by Leibinger (Freiburg, Germany).

Fiducials

The fiducials as proposed by F. L. Fischer consist of titanium bone screws patent no. 1129 F1, 18 mm long, on whose tip a small ball is adapted for CT/MRI examination. For the intraoperative reference procedure this ball is replaced by a flat marker tip with a central dot that can be focused.

In a fixated cadaver head six microscrews and three 18-mm long titanium screws were positioned for a suboccipital retrosigmoidal approach. For a transfacial approach three of these titanium screws were fixed to the cranial bone beyond the hair (Fig. 1).

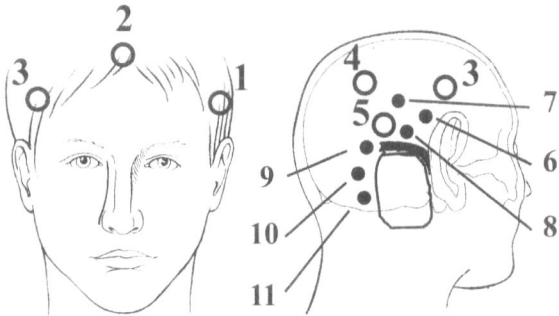

Fig. 1. Positions and numbers of the marker screws for a transfacial and a posterior approach in a fixated cadaver head. Filled circles, Leibinger microscrews; open circles, Zeiss/Fischer titanium bone screws

Table 1. Marker combinations for the reference procedures

Marker type	Number of markers used		
Microscrews	6	7	8
	7	8	9
	7	9	10
	8	9	11
Zeiss/Fischer titanium screws	1	2	3
	3	4	5

With the two different fiducial arrangements 24 and 12 reference procedures with 4 and 2 different fiducial combinations, respectively, were performed following a CT scan (Table 1). The fiducials were compared by evaluation of the reference deviation as indicated by Leibinger software.

In a second experiment following craniotomy three microscrews were implanted to the dorsal wall of the meatus acusticus internus, the posterior semicircular canal and the citellis angle of the same cadaver head before a CT scan was performed. After referencing, the localization deviation (error) between the CT-calculated and the measured position was evaluated for each of the three targets as the mean and standard deviation.

CT Examination Protocol

The Leibinger software requires a CT examination with gantry = 0. After generation of a scout view 1-mm-thick slices were orientated through the region of interest. Three additional slices were orientated through each fiducial. The examination was performed with a scan window of 225 mm offering spatial resolution of 0.44 mm/pixel. In contrast to continuous CT scans, the gaps in the data set had to be indicated manually during date transfer to the workstation.

Reference Procedure

To examine the influence of the different geometrical marker arrangements, reference points were positioned either as a line, triangle or square to a precisely manufactured metal plate with millimeter scaling. Starting from the geometrical center of each figure measurements were obtained at a distance of 0, 25, 50, 100 and 200 mm (Fig. 2). For each of the different reference point arrangements a reference procedure and determination of the five measure points was repeated ten times.

For each experiment, all measured points were documented with their x, y and z coordinates (absolute values). The mean and standard deviation (SD) were calculated. To demonstrate the differences between true and measured position, localization errors were calculated. The localization error [10] is cal-

Fig. 2.
Different geometrical arrangements were used to reference the MKM system to a millimeter-scaled metal plate. Reference point arrangements as a line (×), square (□) and triangle (●) are marked on a metal plate with millimeter scaling. All figures have a circumference of 200 mm. Starting from the geometrical center of the figures five measurement points (o) were marked at 0, 25, 50, 100 and 200 mm. Their and z values were measured in dependence of the reference figure

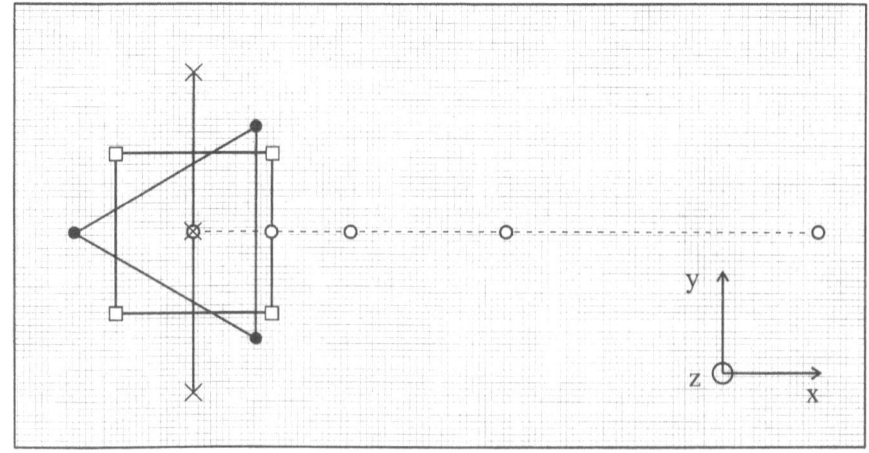

ig. 3.
T artefacts due to an
riginal Zeiss/Fischer fiducial

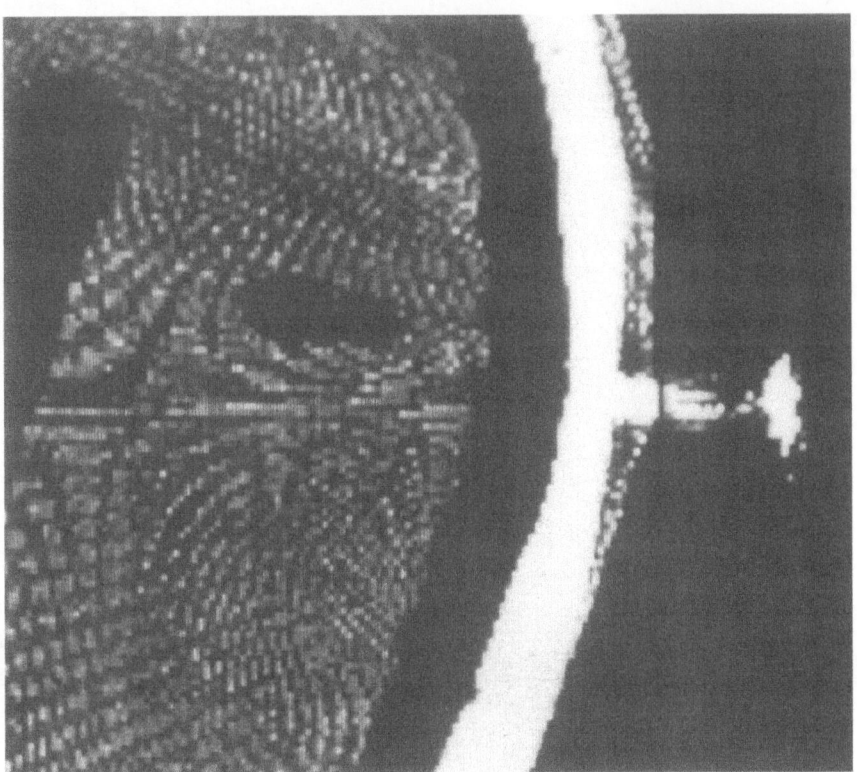

:ulated as the root square sum of the difference be-
ween the theoretical (x_{th}, y_{th}, z_{th}) and the measured
:oordinates, $\sqrt{(x-x_{th})^2 + (y-y_{th})^2 + (z-z_{th})^2}$. All data
amples were checked to be normally distributed and
:ompared to each other by an unpaired Student's t
est. Statistically significant differences were calcula-
ed at $p < 0.01$.

lesults

iducials

Jsing the Leibinger microscrews the software in-
licated a reference deviation of 0.21 ± 0.11 mm
24 measurements, mean ± SD). Using the original
Zeiss/Fischer titanium screw system the deviation
vas 0,29 ± 0.15 mm ($n = 12$). The original fiducials

were sometimes difficult to localize due to CT
artefacts (Fig. 3).

After implantation of the fiducials into the anatomical target structures of the cadaver head, the best
accuracy was achieved with a triangular reference
point arrangement. Comparing the 18-mm-long titanium screws with the microscrews, the latter gave a
slightly better localization accuracy (Table 2).

CT Examination

The CT protocol worked fine and could be used in all
cases. We observed no influence on accuracy by this
discontinuous scanning method. Data transfer from
the CT scanner to the MKM workstation was less convenient and time-consuming as the gaps in the data
set had to be indicated manually.

able 2. Targeting localization error (mm)

1arker	Arrangement	MAI	PSCC	CA	Total
1icroscrew	Triangle ($n = 20$)	0.67 ± 0.20	0.71 ± 0.37	0.54 ± 0.12	0.64 ± 0.26
	line ($n = 10$)	9.19 ± 0.39	7.77 ± 0.36	4.15 ± 0.31	7.04 ± 2.19
'itanium screws	Triangle ($n = 20$)	1.23 ± 0.18	1.25 ± 0.40	1.16 ± 0.29	1.21 ± 0.30

1AI, meatus acusticus internus; PSCC, posterior semicircular canal; CA, citellis angle.

Fig. 4.
The relationship between
geometrical arrangements of
the reference points and the
localization error with increasing distance between
reference and target area

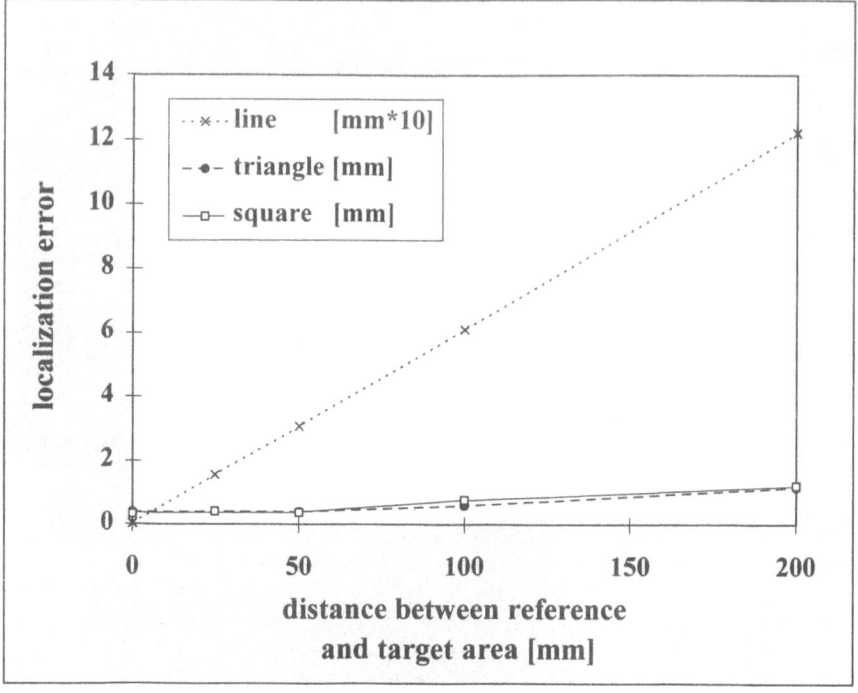

Reference Procedure

Accuracy loss increases with the distance between reference area and target structure. This effect is considerable, leading to deviations higher than 10 cm when a linear fiducial arrangement is chosen. A geometrical reference point arrangement such as a triangle or square leads to a deviation of less than 2 mm even at distances of 200 mm (Fig. 4).

Discussion

According to our experiments the MKM system provides an accuracy in the range of 1–2 mm. Similar values can only be achieved by stereotactic ring systems [10, 12, 16] and infrared neuronavigation systems [2, 6]. This navigation result is based, however, on some technical preconditions, particularly the arrangement of the fiducials for the reference.

From a practical point of view the fiducials should be implanted in the region of the craniotomy as this offers the possibility for easy rereferencing. Repeated referencing procedures become necessary when the patient has to be moved during surgery, which is not uncommon in difficult skull-base approaches. In order to keep the fiducials visible in cases of trans-facial approaches, one cannot implant microscrews in the operative field. In this special case we prefer the 18-mm-long titanium screws of F.L. Fischer which remain visible during surgery if they are implanted in the hairline of the patient.

Our experiments demonstrated that the arrangement of the markers within or nearby the operative field provides high accuracy under the condition that the fiducials used are arranged spherically around and as near as possible to the target volume. It is imperative to avoid under these circumstances a linear arrangement of the fiducials as this causes high intraoperative deviations. It is important to note that such a deviation is not detected by the implemented software, which measures the "reference deviation". This discrepancy between the value obtained by the software of the MKM system and the intraoperative anatomical deviation has been already described for other neuronavigation and image fusion systems [5, 14].

CT Examination Protocol

When a navigation accuracy of less than 1–2 mm is the goal the CT scan must offer a similar or better resolution. Since the markers can only be placed in the cranium at a certain distance from the operative region of interest, a big volume has to be scanned. Using 1-mm-thin slices 80 to 100 slices would be necessary, causing a significant radiation exposure. Therefore we performed a CT-scan allowing gaps between the area of surgical interest and the markers. With this method we observed no influence on accuracy. Data transfer for this procedure is slightly more complicated and time-consuming as the gaps in the data set had to be indicated manually to the work-

ation. An appropriate automatic procedure should
e implemented to the software.

onclusion

he MKM is a precise neuronavigation system which
ffers an accuracy of 1–2 mm if some fundamental
quirements are fulfilled. Without loss of accuracy
e referencing procedure can be adapted to specific
ractical requirements which concerns the preopera-
ve CT scan and the positioning of the fiducials.

eferences

. Barnett GH, Kormos DW, Steiner CP, Weisenberger J (1993)
   Intraoperative localization using an armless, frameless
   stereotactic wand. Technical note. J Neurosurg 78 : 510–514
. Drake JM, Prudencio J, Holowaka S, Rutka JT, Hoffman HJ,
   Humphreys RP (1994) Frameless stereotaxy in children.
   Pediatr Neurosurg 20 : 152–159
. Friets EM, Strohbehn JW, Hatch JF, Roberts DW (1989) A
   frameless stereotaxic operating microscope for neurosurgery. IEEE Trans Biomed Eng 36 : 608–617
. Galloway RL Jr, Maciunas RJ, Edwards CA (1992) Interactive
   image-guided neurosurgery. IEEE Trans Biomed Eng
   39 : 1226–1231
. Hill DL, Hawkes DJ, Gleeson MJ et al. (1994) Accurate frameless registration of MR and CT images of the head: applications in planning surgery and radiation therapy. Radiology
   191 : 447–454
. Horstmann GA, Reinhardt HF (1994) Microstereometry: a
   frameless computerized navigating system for open microsurgery. Comput Med Imaging Graph 18 : 229–233
7. Kato A, Yoshimine T, Hayakawa T et al. (1991) A frameless,
   armless navigational system for computer-assisted neurosurgery. Technical note. J Neurosurg 74 : 845–849
8. Koivukangas J, Louhisalmi Y, Alakuijala J, Oikarinen J (1993)
   Ultrasound-controlled neuronavigator-guided brain surgery. J Neurosurg 79 : 36–42
9. Laborde G, Gilsbach J, Harders A, Klimek L, Moesges R,
   Krybus W (1992) Computer assisted localizer for planning of
   surgery and intra-operative orientation. Acta Neurochir
   (Wien) 119 : 166–170
10. Maciunas RJ, Galloway RL, Latimer JW (1994) The application accuracy of stereotactic frames. Neurosurgery
   35 : 682–694
11. Reinhardt HF, Horstmann GA, Gratzl O (1993) Sonic stereometry in microsurgical procedures for deep-seated brain
   tumors and vascular malformations. Neurosurgery 32 : 51–57
12. Rezai AR, Hund M, Kronberg E et al. (1996) The interactive
   use of magnetoencephalography in stereotactic image-
   guided neurosurgery. Neurosurgery 39 : 92–102
13. Smith KR, Frank KJ, Bucholz RD (1994) The neurostation – a
   highly accurate, minimally invasive solution to frameless
   stereotactic neurosurgery. Comput Med Imaging Graph
   18 : 247–256
14. Tan KK, Grzeszczuk R, Levin DN et al. (1993) A frameless
   stereotactic approach to neurosurgical planning based on
   retrospective patient-image registration. Technical note. J
   Neurosurg 79 : 296–303
15. Watanabe E, Watanabe T, Manaka S, Mayanagi Y, Takakura K
   (1987) Three-dimension digitizer (neuronavigator): new
   equipment for computed tomography-guided stereotaxic
   surgery. Surg Neurol 27 : 543–547
16. Weil SM, van Loveren HR, Tomsick TA, Quallen BL, Tew JMJ
   (1987) Management of inoperable cerebral aneurysms by the
   navigational ballon technique. Neurosurgery 21 : 296–302
17. Zamorano L, Jiang Z, Kadi AM (1994) Computer-assisted
   neurosurgery stystem: Wayne State University hardware and
   software configuration. Comput Med Imaging Graph
   18 : 257–271

Experiences with the Neuronavigation System EasyGuide Neuro

U. Spetzger, M. H. Reinges, G. A. Krombach, V. Rohde, I. Slansky, L. Mayfrank, and J. M. Gilsbach

Summary

This report summarizes our clinical experiences with the neuronavigation system EasyGuide Neuro in a series of 62 cranial neurosurgical procedures. The purpose of this study was to evaluate the accuracy, safety and the routine handling of this system in a wide range of cerebral operations performed from December 1995 to November 1996 in our department.

EasyGuide Neuro permits an exact preoperative planning and is an excellent tool for neurosurgical training. The dynamically displayed visualization of the individual anatomy around the lesion allows a virtual walk to the target area in order to determine the optimal surgical approach.

The navigation system was successfully used for exact trephination planning and for precise localization of the transsulcal or transfissural microsurgical approach to deep-seated lesions, as well as for guided biopsies and catheter insertions.

Introduction

The detailed visualization of the cerebral anatomy by means of modern neuroradiological imaging techniques allows an exact preoperative planning of the surgical strategy. However, it can be difficult to transpose the information of the radiological images to the real anatomy of the surgical field. Neuronavigation systems have partially solved this problem by coupling the spatial image data with the real spatial coordinates of the patient [1, 5, 7, 10 – 12, 14, 15]. Nowadays, neuronavigation systems are able to facilitate surgical procedures due to a precise localization of the target area and to improve intraoperative orientation which provides the opportunity to reduce surgical trauma.

Material and Methods

Navigation System

EasyGuide Neuro is an armless and frameless graphic-interactive navigation system developed for image-guided neurosurgery (Fig. 1). The system consists of the following elements: a UNIX 4.0 workstation with the implemented stereotactic software and a high-resolution monitor. The optical localizing system (two infrared-sensitive cameras) is fixed to the operating table and the probe (pointer with infrared light-emitting diodes) is connected via a cable to the navigation unit.

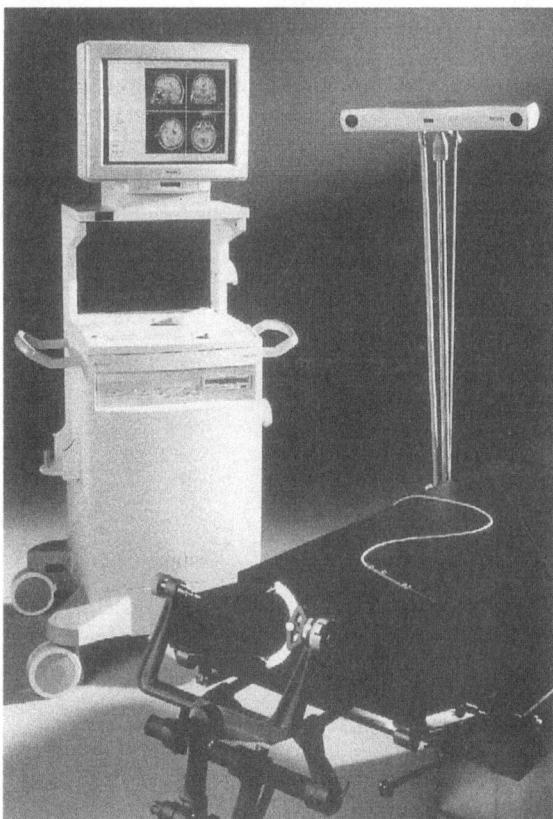

Fig. 1. EasyGuide Neuronavigation system with the workstation, camera system and wire-connected pointer

Data Acquisition

After placement of the external fiducials (multimodal hydrogel skin markers) on the scalp of the patient, all patients of our series were evaluated with MRI using a standardized protocol (T1-weighted, gadolinium enhanced 3-D images with a 3-mm slice thickness). In the present study, data transfer to the navigation system was performed via optical disks.

Intraoperative Application

After induction of general anaesthesia the patient was positioned on the operating table and the skull was fixed in a Mayfield clamp. In patients treated under local anaesthesia, the head was fixed atraumatically with adhaesive tape (Fig. 2). The camera system was fastened to the operating table and the skin markers were touched with the pointer for patient-image registration. This registration procedure is generally performed in a random mode and takes about 5 min. It allows the linkage between the spatial coordinates of the radiological data and the real spatial coordinates of the patients, and on-line graphic-interactive navigation is possible. For accuracy control, the skin markers are consecutively touched again and the navigation system automatically displays the deviation. In our study the patient-image registration procedure was repeated if the deviation of the majority of the fiducials exceeded more than 3 mm. Occasionally, it was necessary to remove such skin markers with an extreme deviation

(more than 6 mm) and discard them from the registration procedure in order to improve the accuracy of registration. Finally, we routinely performed a plausibility check by pointing to anatomical structures such as the tip of the nose or the external auditory meatus, the position of which can easily be confirmed on the monitor screen.

Next, the trajectory and the direction of the surgical approach were determined. The craniotomy was planned and individually tailored in size and shape by projection to the scalp with the pointer. The skin incision was adjusted to these requirements and was marked on the patients head.

After disinfection and draping of the surgical field, a sterilized probe (autoclaved at 121 °C) was used. Sterile covering should be performed with caution because a displacement of the camera position will produce inaccuracy. Therefore, the cloths should not be fastened to the camera stand.

After skin incision and craniotomy has been performed, additional registration markers can be read in. We usually took the small burholes at the trephination margins used for the dura sutures. For intradural navigation, the unsolved problem of brain shift must be taken in to account. An adequate positioning of the patient's head, a small surgical approach and a limited, exactly centered exposure of the cerebral cortex will reduce the brain shift in the beginning. Intracerebral orientation and localization should be performed at an early stage of the operation because the loss of cerebrospinal fluid, tumour debulking or the drainage of cystic lesions considerably reduce the accuracy.

Fig. 2.
Frameless, percutaneous biopsy of a cystic occipital lesion via navigated twist-drill trephination. The head of the patient was fixed with adhesive tapes and the whole procedure was performed with the patient under local anaesthesia

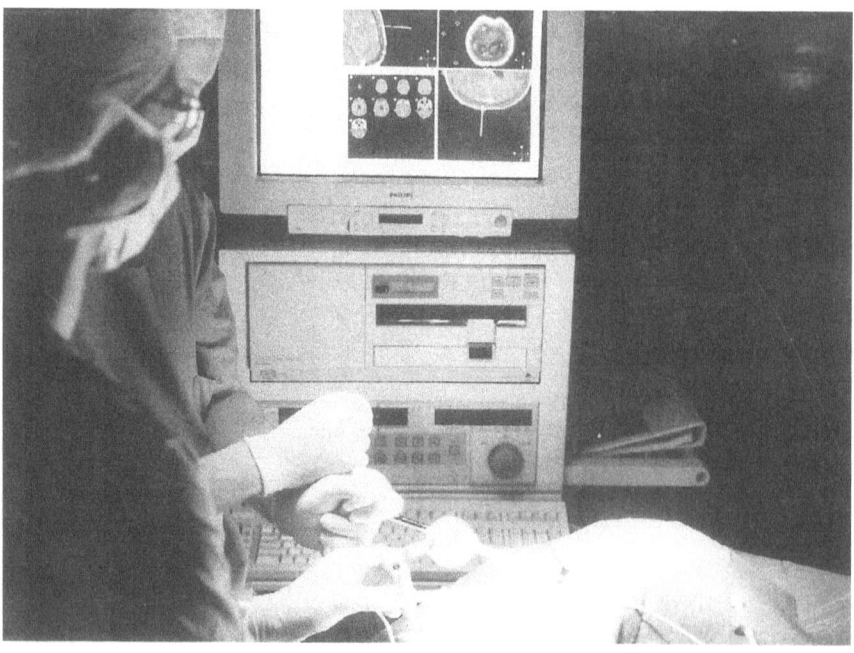

Results

The clinically proven accuracy of the system was 3.9 mm (range 1.6–16.8 mm). In our series we had three cases of intolerable inaccuracy (more than 10 mm) which prevented the use of the system. This error could be explained by an accidental dislodgement of the camera stand. In none of these three cases did the inaccuracy of the system cause any harm to the patient.

The handling of the navigation system was considered comfortable because the user interface on the monitor screen is simple and mainly self-explanatory. The probe or pointer is lightweight and highly flexible due to the thin wire connection, and the remote control when sterilely covered enables intraoperative manipulation of the navigation system even by the surgeon himself. Apart from a dysfunction of a pointer due to a damaged infrared light-emitting diode, we had no malfunction of the hardware or software components of the system. Table 1 and 2 show the whole range of neurosurgical procedures and the different lesions treated using Easy-Guide Neuro from December 1995 to November 1996.

Table 1. Summary of the 62 cases undergoing navigated neurosurgical procedures according to histopathology

Histopathology	n
Glioma	20
Metastasis	9
Cavernoma	6
Pituitary adenoma	6
Miscellaneous	6
Meningioma	4
Vascular lesion	4
Hydrocephalus	3
Lymphoma	2
Cerebral abscess	1
Intracerebral haemorrhage	1

Table 2. Application of EasyGuide Neuro in 62 different cerebral neurosurgical operations

Surgical procedure	n
Microneurosurgery	50
Frameless biopsy	4
Insertion of ventricle catheter	3
Amygdalohippocampectomy	2
Percutaneous puncture	2
Extirpation of a foreign body	1

Discussion

It is important to take into consideration that the precise position of the camera system in relation to the coordinates of the patient's head is the fundamental principle of the navigation process [1, 8, 15]. Impairment of this stable unit by dislodgement of the camera or the head position can cause fatal disturbances. Therefore, a continuous reregistration or tracking would be a decisive improvement which avoids such a dangerous error and makes the whole system safer.

Obviously, maximal accuracy is desirable for all navigation systems; however, only selected neurosurgical procedures require a "pinpoint precision" [2, 3–4, 6, 9, 12, 13]. The clinically proven accuracy of Easy Guide Neuro is below the accuracy of frame based stereotactic systems [6], but an accuracy of 3–4 mm was estimated to be sufficient for the field of neuronavigation. Continuously reduced intradural ac-

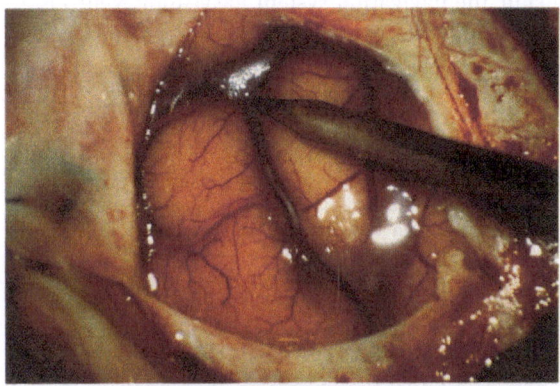

a

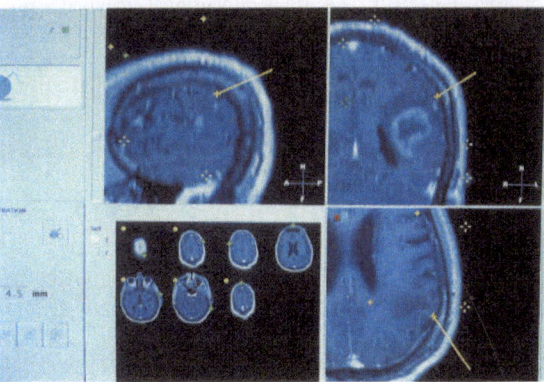

b

Fig. 3. a Intraoperative photography demonstrating the intact superficial temporal cortex in a patient with a malignant glioma. A microsurgical transsulcal approach to the lesion is planned. The tip of the probe points to a cortical vein. **b** Screen display with the tip of the probe at the superficial cortical vein demonstrated in **a**. The navigation system enables the graphical-anatomical interactive localization since the position of the pointer is dynamically indicated in axial, coronal and sagittal MR images on the monitor screen

curacy due to the so-called brain shift is a well-known and at present still unsolved problem of all available navigation systems [1, 5, 7, 12 – 15]. However, intradural navigation especially on the intact cortical surface enables the determination of the optimal sulcus or fissure for the approach of small and subcortical lesions (Fig. 3). The graphic-interactive navigation offers a precise orientation and makes localization and detection much easier. We have the impression that, especially in deep-seated lesions located in eloquent areas, a well-aimed transsulcal and targeted microsurgical approach is essential in reducing surgical morbidity [3, 12].

The virtual elongation of the probe facilitates the planning of the surgical access. It is also useful for planning the insertion of ventricular catheters or the drainage of cystic lesions. The trajection and the length of the catheter can be determined beforehand, and the pathway can be visualized in multiplanar section view. This option is also a valuable asset for neurosurgical training, because ventricular shunt catheter insertion can be simulated virtually. In conclusion, the navigation system EasyGuide Neuro is a valuable tool for image-guided cranial neurosurgery.

References

1. Barnett GH, Kormos DW, Steiner CP, Weisenberger J (1993) Use of a frameless, armless stereotactic wand for brain tumor localization with two-dimensional and three-dimensional neuroimaging. Neurosurgery 33 : 674 – 678
2. Bertalanffy H, Bechtel S, Seeger W (1993) Regional exposure of cerebral convexity lesions. Neurochirurgie 36 : 81 – 86
3. Buchner H, Adams L, Knäpper A, Rügger R, Laborde G, Gilsbach JM et al (1994) Preoperative localization of the central sulcus by dipole source analysis of early somatosensory evoked potentials and three-dimensional magnetic resonance imaging. J Neurosurg 80 : 849 – 856
4. Hammoud MA, Ligon BL, ElSouki R, Shi WM, Schomer DF, Sawaya R (1996) Use of intraoperative ultrasound for localizing tumors and determining the extent of resection: a comparative study with magnetic resonance imaging. J Neurosurg 84 : 737 – 741
5. Kato A, Yoshimine T, Hayawaka T, Tomita Y et al. (1991) A frameless, armless navigational system for computer-assisted surgery. J Neurosurg 74 : 845 – 849
6. Kelly PJ (1992) Stereotactic resection and its limitations in glial neoplasms. Stereotact Funct Neurosurg 59 : 84 – 91
7. Laborde G, Gilsbach J, Harders A, Klimek L, Mösges R, Krybus W (1992) Computer assisted localizer for planning of surgery and intra-operative orientation. Acta Neurochir (Wien) 119 : 166 – 170
8. Laborde G, Klimek L, Harders A, Gilsbach J (1993) Frameless stereotactic drainage of intracranial abscesses. Surg Neurol 40 : 16 – 21
9. Mayfrank L, Bertalanffy H, Spetzger U, Klein HM, Gilsbach JM (1994) Ultrasound-guided craniotomy for minimal invasive exposure of cerebral convexity lesions. Acta Neurochir (Wien) 131 : 270 – 273
10. Olivier A, Germano IM, Cukiert A, Peters T (1994) Frameless stereotaxy for surgery of the epilepsies: preliminary experiences. J Neurosurg 81 : 629 – 633
11. Reinhardt HF, Trippel M, Westermann B, Horstmann GA, Gratzl O (1996) Computer assisted brain surgery for small lesions in the central sensorimotor region. Acta Neurochir (Wien) 138 : 200 – 205
12. Spetzger U, Laborde G, Gilsbach JM (1995) Frameless neuronavigation in modern neurosurgery. Minim Invasive Neurosurg 38 : 326 – 330
13. Sipos EP, Tebo SA, Zinreich SJ, Long DM, Brem H (1996) In vivo accuracy testing and clinical experience with the ISG viewing wand. Neurosurgery 194 – 204
14. Watanabe E, Mayanagi Y, Kosugi Y, Manaka S, Takakura K (1991) Open surgery assisted by the neuronavigator, a stereotactic, articulated, sensitive arm. Neurosurgery 28 : 792 – 799
15. Zamorano L, Kadi M, Dong A (1992) Computer-assisted neurosurgery. Simulation and automation. Stereotact Funct Neurosurg 59 : 115 – 122

Early Clinical Experience with the EasyGuide Neuronavigation System and Measurement of Intraoperative Brain Distortion

N. L. Dorward, O. Alberti, B. Velani, J. Buurman, A. Dijkstra, N. Kitchen, F. A. Gerritsen, and D. G. T. Thomas

Summary

The initial clinical experience with a prototype of the EasyGuide Neuro surgical navigation system is presented together with some measurements of intraoperative brain shift taken with the neuro-navigator. The EasyGuide Neuro is an infrared-based system employing hand-held localisers with light-emitting diodes tracked by a table-mounted camera array. Standard orthogonal reformats which intersect at the pointer tip position are displayed by the workstation in near real-time. We have prospectively studied the 31 consecutive patients undergoing surgery with the neuronavigator during the first 3 months of clinical use. General patient data, system accuracy checks and surgical impact are described. In a subgroup of 16 patients measurements of brain shift were taken at the brain surface, at the deepest point of surgery and at the bone surface offset. The latter was used to correct for registration errors. We found a significantly greater shift of the cortical surface than of deep structures ($p < 0.02$) with this method. These measurements are presented as evidence supporting the validity of navigational guidance in clinical practice.

Introduction

Neuronavigation systems became commercially available in the early 1990s and since then considerable clinical experience has highlighted some of the advantages of interactive image guidance (in particular preoperative planning and perioperative lesion localisation) as well as demonstrating some distinct disadvantages of the mechanical arm arrangement [10]. Of late a variety of alternative systems have become available utilising different technologies for pointer tracking, including sound [2], electromagnetic field [6] and infrared light [13]. Studies of these methods have revealed a significant interference inherent in the operating room with the sound and electromagnetic systems, making the infrared optical systems superior [4]. The EasyGuide Neuro (Philips Medical Systems) is a commercially available neuronavigation system incorporating hand-held pointers with infrared-emitting diodes (LEDs) tracked by two infrared cameras.

A frequently cited criticism of neuronavigation is the occurrence of brain shift. Most stereotactic systems, whether frame-based or frameless, employ preoperative scan images for surgical guidance. Thus, any change in the morphology of the brain during surgery will not be reflected in the images. Intraoperative shift of cerebral structures may occur due to any or all of pressure changes on skull opening, oedema, gravity, brain and lesion resection and CSF withdrawal [3]. These changes have the potential to invalidate the preoperative images and so could make reliance upon a navigation system positively dangerous. Despite this, little work has been published which directly addresses this question. We have performed a study of these intraoperative brain shifts with measurements made with the EasyGuide Neuro and report the preliminary findings.

Methods and Materials

Patients were selected by the surgical consultant in charge of their care making a request for navigational guidance. Entry criteria were the presence of intracranial or very high cervical (above C2) pathology and the ability to undergo MRI scanning. During the study period we did not have the facility to use CT scans for guidance and so all patients underwent imaging with MRI. Adhesive skin fiducials were applied widely across the skull vault, with a cluster over the region of interest, prior to imaging. All patients received intravenous gadolinium enhancement. Two MRI scanners were used (General Electric Vectra 0.5 T, General Electric Signa 1.5 T) and imagining was performed according to a standard protocol (TR 45, TE 15, flip 50, matrix 192 x 192, FOV 25, thickness 180 and TR 42, TE 15, flip 50, matrix 192 x 256, FOV 23, thickness 160).

The EasyGuide Neuro navigation system comprises a table-mounted camera stand incorporating two infrared cameras, a free standing stack of computer hardware with a high-resolution monitor and a selection of straight and bayonet bipolar pointers with attached LEDs. Images were transferred to the EasyVision CT/M workstation (Philips Medical Systems) via ethernet in DICOM format for segmentation and reformatting as desired. The data set was transferred to the EasyGuide Neuro operating system by optical disc. The position of each skin fiducial in the images was selected manually and stored. With the patient positioned and fixed in the Mayfield head clamp the patient to image registration was performed. The system allows the fiducials to be localised with automatic matching of their stored image position. Following registration the system displays the axial, coronal and sagittal reconstructions corresponding to the tip of the hand-held pointer. Navigation is enhanced through facilities for virtual pointer elongation, remote control by the surgeon and a path-planning facility. Following scalp reflection it is recommended that small divots are made in the outer table of the skull at the periphery of the craniotomy site and that the localised position of these is stored. This allows later regregistration should movement of the head relative to the cameras occur during the procedure.

General data relating to the clinical condition, preoperative imaging and the operative procedure were collected according to a predefined protocol and recorded in a dedicated patient booklet. Standard accuracy measurements were similarly collected and recording including the system-calculated root mean square error (RMSE) of registration and the error of localisation of each fiducial.

Measurement of bone, brain and tumour bed position was performed by the placement of the sterile pointer on the appropriate structure by the surgeon and capture of the localisation by the guidance system. The distance of the pointer tip (localised position) from the relevant structure's surface in the preoperative images was measured using the system measurement tool, with the image magnified to fill the available monitor screen. The measurement was repeated for each of the sagittal, coronal and axial planes for the outer bone surface, cortical surface on skull opening, first visualised lesion surface, deepest lesion margin and cortex at completion. The results were recorded as a positive value when the surface lay beyond the pointer tip (reached by pointer elongation) and as a negative value when behind the tip (pointer shortening). Mannitol, lumbar drainage and ventricular tapping are not used routinely in our surgical practice.

Table 1. Pathological indications for Surgery

Pathology	n	%
Glioma	10	33%
Meningioma	7	23%
Epilepsy	4	13%
Metastases	3	10%
Cavernoma	2	6%
Epidermoid	2	3%
Neuroma	1	3%
Cholesterol granuloma	1	3%

Results

In the 3 months following installation of EasyGuide Neuro (June to September 1996) 31 patients underwent surgery with the assistance of the navigation system. The male to female ratio of the patients was 1 : 1.07 and the age range was 15 – 75 years (mean 45.5). Glioma was the most frequent pathological indication for surgery (33%) followed by meningioma (23%); there was a wide range of other diagnoses (Table 1). Duration of surgery ranged from 1 h 10 min to 8 h 50 min (mean 2 h 24 min) and hospital stay was between 5 and 20 days (mean 8.7). The imaging modality in all cases was a T1-weighted MRI volume acquisition with intravenous gadolinium. The number of adhesive skin fiducials used for patient to image registration was between five and ten (mean 8) resulting in a mean RMSE for registration of 4.0 mm (range 1.8 – 7.3 mm). The nearest-marker test for each of the registration fiducials gave results of 0 – 2.4 mm (mean 1.3 mm). There were no intraoperative system failures and no adverse effects attributable to the use of the neuronavigation system.

The brain cortical surface following dural opening was offset by between 0.3 and 15.7 mm (mean 5.8 mm, SD 4.7 mm). In 79% of cases the brain surface measurement following opening confirmed an outward bulging and in 21% there was an infalling (positive values). The offset of the first identified lesion surface was between 0.0 and 3.1 mm (mean 0.9 mm, SD 1.4 mm), with no discernible offset in 63% of the cases. The tumour bed position was offset by between 0.0 and 12.6 mm (mean 2.9 mm, SD 4.5 mm) with outward bulging in 38%, infalling in 19% and no shift in 43% of cases. The bone surface point measurements showed an offset of between 0.0 and 4.8 mm with a mean of 2.2 mm (SD 1.5 mm). The cortex at completion of surgery was offset by between 0.0 and 10.0 mm (mean 3.1 mm, SD 3.5 mm). Correction of each brain shift measurement with the bone surface offset from that case revealed a mean corrected cortex displacement of 6.0 mm (SD 3.8 mm; Table 2), a mean corrected lesion surface displacement of 1.8 mm (SD 2.5 mm), a mean corrected

Table 2.
Brain shift measurements

	Absolute values			Corrected values		
	Mean shift (mm)	Range (mm)	Standard deviation	Mean shift (mm)	Range (mm)	Standard deviation
Bone	2.2	0.0 – 4.8	1.5	–	–	–
Cortex	5.8	0.3 – 15.7	4.7	6.0	1.4 – 12.1	3.8
Lesion surface	0.9	0.0 – 3.1	1.4	1.8	0.0 – 6.7	2.5
Tumour bed	2.9	0.0 – 12.6	4.5	2.6	0.0 – 11.7	3.5
Cortex (Close)	3.1	0.0 – 10.0	3.5	3.3	0.0 – 10.0	3.4

tumour bed displacement of 2.6 mm (SD 3.5 mm) and a mean corrected cortex shift at closure of 3.3 mm (SD 3.4 mm). The difference between the absolute values for cortex shift and deep shift did not reach statistical significance ($p = 0.1$). When each of the offset values were corrected for the bone/registration offset (subtraction of the respective bone value), shift of the cortex was found to be significantly greater than shift at depth ($0.02 > p > 0.05$).

Discussion

Frame-based stereotaxy has become established as an indispensable tool in routine neurosurgical practice through the achievement of high diagnostic yield and low morbidity in large clinical series [1, 9, 12]. This position has not been altered by the realisation that application accuracy of stereotactic frames is in the region of 2.5 mm and not the submillimetric accuracy initially believed [7, 8]. Frame-based stereotactic techniques have been successfully transferred to open procedures without regard to the introduction of postimaging brain distortion. Neuronavigation systems provide unparalleled intraoperative three-dimensional presentation of the preoperative images and at the same time they give anatomical guidance to the surgeon. Considerable clinical experience has now been reported in the literature with arm-based systems and several studies have revealed similar application accuracies for frame-based stereotaxy and image guidance systems [11]. The origins of errors during computed tomography, MRI, mechanical frame accuracy and registration in frameless surgery are now well understood. However, little work has been published which quantifies this postimaging distortion.

This study provides preliminary point-based measurements of cerebral distortion during surgery. The bone surface offset measurements provide an assessment of registration accuracy as this is a rigid and well-defined structure. The mean bone offset was 2.2 mm, providing and application accuracy which compares well with standard stereotactic frames. The measurements of brain shift following dural opening confirmed the expectation that the brain will bulge when there is an underlying tumour. The mean cortical shift of 6.0 mm is not large and will not affect the localisation of surface structures such as the precentral gyrus or a desired sulcus for entry. Of particular interest is the finding that brain shift at depth is significantly lower than at the surface. The tumour surface position showed very little shift and the deepest tumour margin exhibited a mean shift of just 2,5 mm. These findings suggest that reliance upon the guidance of a navigation system for resection of the frequently more dangerous deep portion of lesions is justified.

Conclusions

The Easy Guide Neuro proved to be a simple, effective and reliable clinical tool in a wide range of pathological conditions. The distortion measurements confirm the occurrence of brain shift following cranial opening. This shift is significantly greater at the cortical surface than at depth. The extent of the shift is not sufficient to negate the value of navigation systems.

Acknowledgements. This research and clinical validation work was performed within the context of the EASI project "European Applications for Surgical Interventions", supported by the European Commission under contract HC1012 in their "4th Framework Telematics Applications for Health" RTD programme. The partners in the EASI consortium are Philips Medical Systems Nederland B.V., Philips Research Laboratory Hamburg, the Laboratory for Medical Imaging Research of the Katholieke Universiteit Leuven, the Image Sciences Institute of Utrecht University & University Hospital Utrecht, The National Hospital for Neurology and Neurosurgery in London, and the Image Processing Croup of Radiological Sciences at UMDS of Guy's and St. Thomas' Hospitals in London.

References

1. Apuzzo MLJ, Chandrasoma PR, Cohen D, Zee C-S, Zelman V (1987) Computed imaging stereotaxy: experience and perspective related to 500 procedures applied to brain masses. Neurosurgery 20 : 930 – 937
2. Barnett GH, Kormos DW, Steiner CP, Weisenberger J (1993) Use of a frameless, armless stereotactic wand for brain tumour localization with 2-D and 3-D neuroimaging. Neurosurgery 33 : 674 – 678
3. Bucholz R, Sturm C, Henderson J (1996) Detection of brain shift with an image guided ultrasound device. Acta Neurochir (Wien) 138 : 637
4. Bucholz RD, Smith KR (1993) Comparison of sonic digitizers versus light emitting diode-based localization. In: Maciunas RF (ed) Interactive image-guided neurosurgery. AANS Publications, pp 179 – 200
5. Golfinos JG, Fitzpatrick BC, Smith LR, Spetzler BSE, Spetzler RF (1991) Clinical use of a frameless stereotactic arm: results of 325 cases. J Neurosurg 83 : 197 – 205
6. Kato A, Yoshimine T, Hayakawa T, Tomita Y, Ikeda T, Mitomo M, Harada K, Mogami H (1991) A frameless, armless navigational system for computer-assisted neurosurgery. J Neurosurg 74 : 845 – 849
7. Maciunas RJ, Galloway RL, Latimer J, Cobb C, Zaccharias E, Moore A, Mandava VR (1992) An independent application accuracy evaluation of stereotactic frame systems. Stereotact Funct Neurosurg 58 : 103 – 107
8. Maciunas RJ, Galloway RL, Latimer JW (1994) The application accuracy of stereotactic frames. Neurosurgery 35 : 682 – 694
9. Ostertag CB, Mennel HD, Kiessling M (1980) Stereotactic biopsy of brain tumours. Surg Neurol 14 : 275 – 283
10. Rohling R, Munger P, Hollerbach JM, Peters T (1995) Comparison of accuracy between a mechanical and an optical tracker for image-guided neurosurgery. Image Guided Surg 1 : 30 – 34
11. Sipos EP, Tebo SA, Zinreich SA, Long DM, Brem H (1996) In vivo accuracy testing and clinical experience with the ISG viewing wand. Neurosurgery 39 : 194 – 202
12. Thomas DGT, Nouby RM (1989) Experience in 300 cases of CT-directed stereotactic surgery for lesion biopsy and aspiration of haematoma. Br J Neurosurg 3 : 321 – 325
13. Zamorano LJ, Nolte L, Kadi AM, Jiang Z (1994) Interactive intraoperative localization using an infrared-based system. Stereotact Funct Neurosurg 63 : 84 – 88

Neuronavigation and Neuroendoscopy

A. Nabavi, A. Behnke, B. Petersen, H. Klinge, and H. M. Mehdorn

Summary

In neuroendoscopy orientation and localisation of the endoscope's tip is crucial. Additionally the endoscopic view can be obscured by solid structures or membranes.

To surpass these restrictions we attempted to combine the localisation quality of our neuronavigation system with the visualization capabilities of neuroendoscopy. We designed a handle to attach our endoscope (Aesculap) to our frameless neuronavigation system (ISG Viewing Wand). With this combination the localisation of the endoscope's tip is shown on-line as the crosshair's centre on a computer display. This displays the exact location of the endoscope's tip in stereotactic space in preoperative images. The endoscope's mobility is preserved, and therefore it has a broader range of application than if combined with a rigid stereotactic frame.

We tested the applicability in 10 patients with peri- and intraventricular lesions and loculated hydrocephalus. The assembly is easy. Especially in infant ventricular, cystic malformations this method is of great value. This simple attachment enables us to combine endoscopes with frameless stereotaxy. The precise localisation in stereotactic space as well as flexibility and excellent visualization broaden the applicability of neuroendoscopy and neuroendoscopic-assisted microneurosurgery.

Introduction

Although neuroendoscopy provides an excellent view of the area of interest, a precise localisation in space is lacking. In pathologically altered settings, as in multi-loculated hydrocephalus or tumors, this is a major restriction. To overcome this obstacle stereotactic guidance has been advocated [2, 5, 6].

Since April 1994 we have been applying the frameless stereotactic neuronavigation system ISG viewing wand. The experience we gathered with 300 cases prompted us to extend the application to neuroendoscopy. Therefore we designed a handle to attach rigid endoscopes to this guidance system.

Methods and Materials

The viewing wand has been described elsewhere [5]. In short, images are acquired prior to the operation, three-dimensionally reconstructed and registered to the patient in the OR using a mechanical arm. The tip of a probe attached to this arm is displayed on a computer monitor as the centre of a crosshair in different planes.

In order to attach an endoscope to this arm we designed a handle which is a centrally hollowed cylinder into which endoscopes can be introduced (Fig. 1). The endoscope's length is adjusted to the

Fig. 1.
From *top* to *bottom*: viewing wand probe, the handle and the Aesculap endoscope. The endoscope is introduced into the cylinder. The ruler which is loosely attached to the handle stops the endoscope at the length of the probe. Then the screw on the bottom of the cylinder is carefully tightened

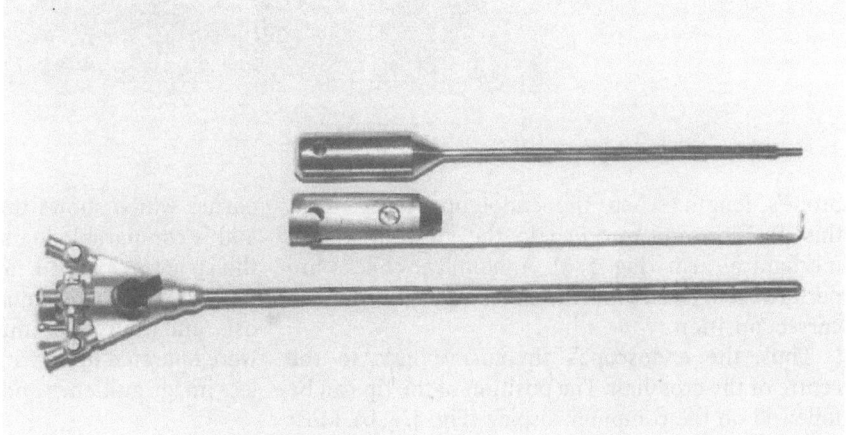

Fig. 2.
The endoscope within the cylinder which is attached to the viewing wand. The ruler is detachable

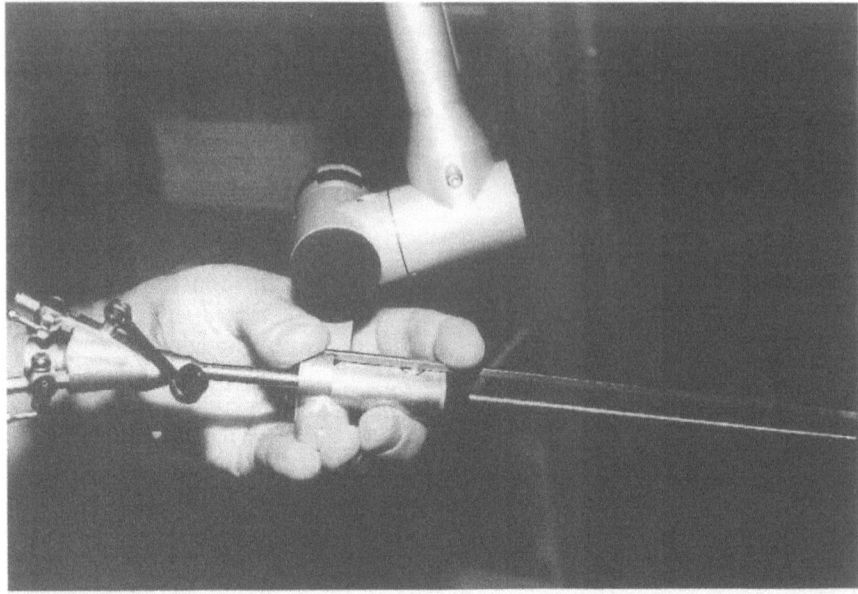

Fig. 3.
The setup in the OR. The viewing wand as well as the endoscope are draped. Two Leyla retractors are attached to the viewing wand, keeping the wand and the endoscope in place. Note the ease in holding the endoscope–arm combination if the arm's counterweight is balanced correctly

probe's length. Then the endoscope is fixed in this position and attached to the viewing wand's mechanical arm (Fig. 2, 3). Anatomical checks are performed to ensure that the endoscope is fixed in the correct position.

Thus, the endoscope's tip corresponds to the centre of the crosshair. The position of the tip can be followed on the computer display (Fig. 4, a, b). Most important is the view displayed in the bottom right corner which shows the trajectory of the approach and is comparable to an ultrasound view, displaying the structures ahead. Ten patients (two with intra-/paraventricular tumours, four with aqueductal stenosis and four with multiloculated hydrocephalus) were operated upon with this combination of frameless image guidance and neuroendoscopy.

Fig. 4 a, b.
Two computer displays showing the image guidance provided in the OR. **a** The endoscope has just been introduced into the lateral ventricle. The in-line view shows the trajectory aiming at the foramen of Monro.
b The endoscope's tip further advanced within the third ventricle aiming at the basilar artery

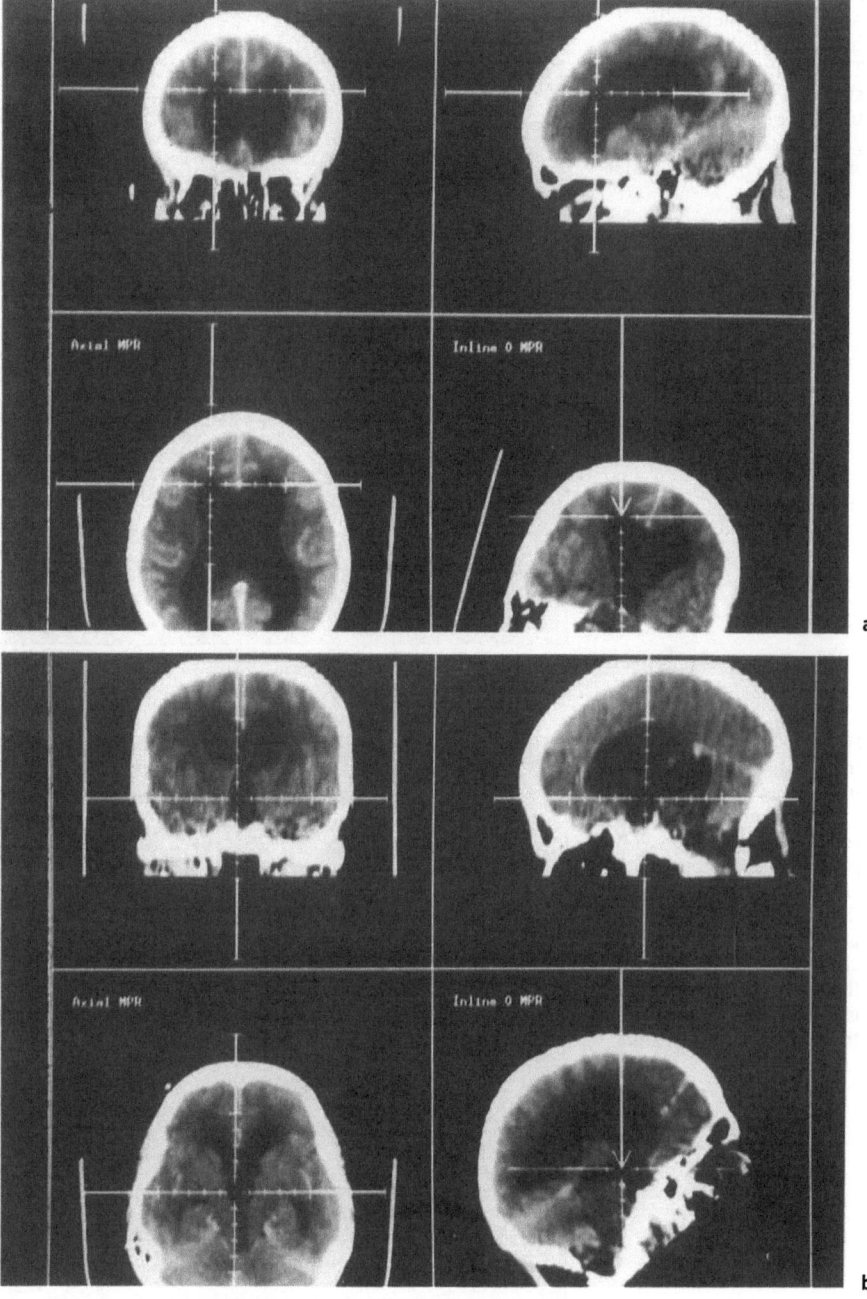

a

b

Results

The preoperatively reconstructed 3D images can be processed on a workstation to aid planning or facilitate teaching (Fig. 5, and 6). In the OR the assembly of arm, handle and endoscope is easy and takes less than 2 min. The precision is within 3 mm in the axis of the endoscope; thus, the deviation is only in the axis of the trajectory with no deviation in other directions. This inaccuracy is corrected by visual control through the endoscope itself.

For the paraventricular tumors (one astrocytoma and one intraventricular meningioma) the wand guided the endoscopic approach and controls the endoscopic-assisted microneurosurgical procedure.

With the multiloculated hydrocephalus, preoperative visualisation was facilitated. In two cases intraventricular contrast medium application allowed the demonstration of noncommunicating cysts. These

Fig. 5.
Aqueductal stenosis with consecutive hydrocephalus. The translucency shows the approach via the right ventricle. The basilar tip and the posterior cerebral arteries are visible at the opposite side

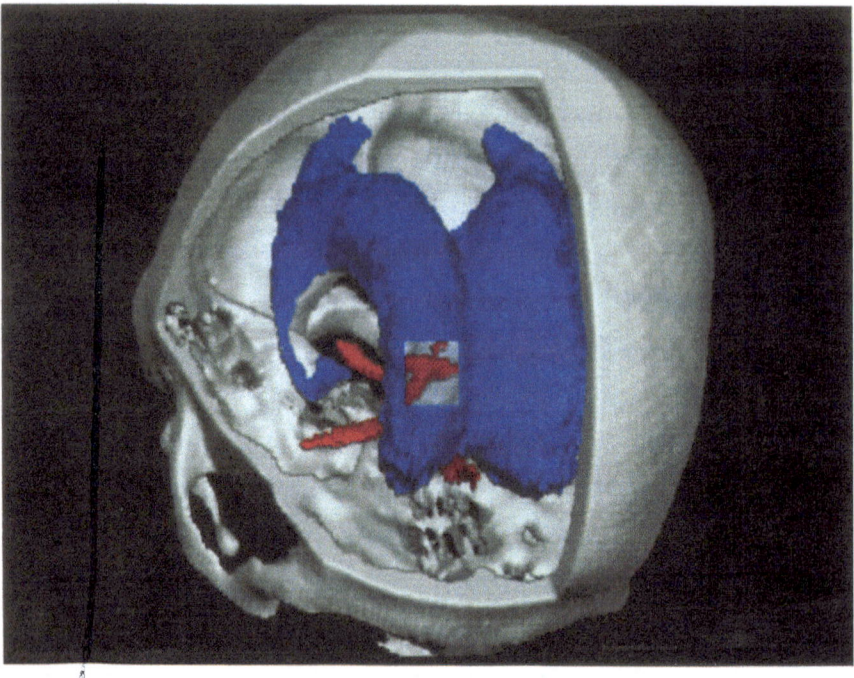

Fig. 6.
In loculated hydrocephalus, as in this child who had suffered a bacterial meningitis, the combination of neuronavigation and endoscopy is effective. The localisation of the endoscope's tip within the pathologically altered ventricular system is crucial, but fluid leakage can render this method ineffective. *Blue,* ventricles; *grey,* cysts; *red,* shunt

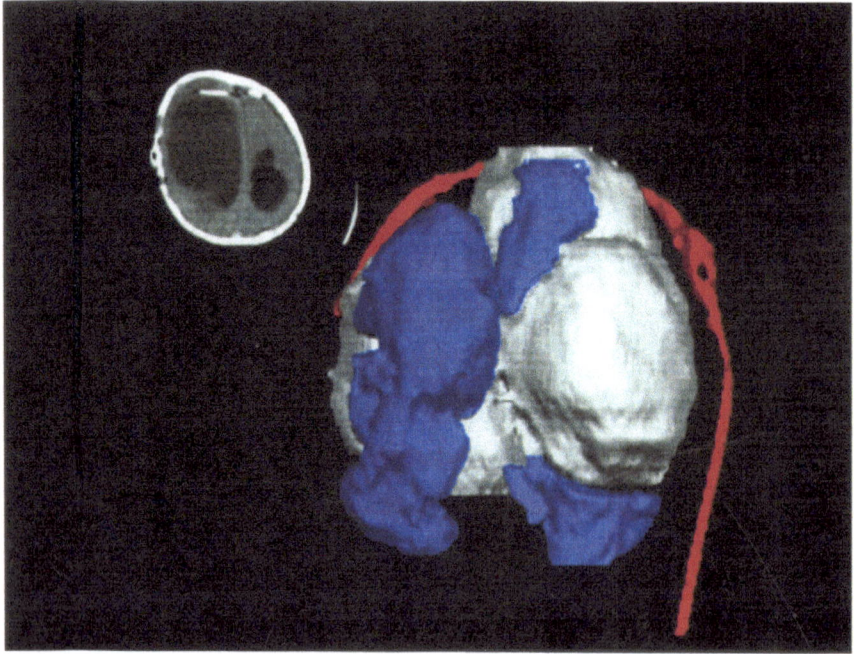

were impressively visible in 3D reconstructions for preoperative planning. In these cases the display of the trajectory proved to be of special value since the septations were not translucent. The image guidance enabled us to image the structures obscured by this veil, thus permitting us to determine the appropriate and safe location for initial fenestration. However, in the case shown in Fig. 6 the cysts collapsed leading to a massive brain shift and rendering the image guidance ineffective after the initial approach. With aqueductal stenosis the combination of neuroendoscopy and neuronavigation was of special advantage in planning the trajectory and visualizing the basilar artery within the prepontine cistern (Fig. 5).

Discussion

Various adjuncts to microneurosurgery have been developed, resulting in advances in various areas, such as the fairly recent frameless stereotaxy [8] and the renaissance of neuroendoscopy [1, 3]. Neuro-endoscopy provides excellent visualisation with the drawback of a very narrow field of view, rendering orientation cumbersome. To overcome this difficulty endoscopes were attached to stereotactic frames [2, 6, 7, 9. 11]. This helped to determine the endoscope's location but confined effectiveness of the system and depleted it of its mobility.

To regain flexibility while preserving localisation in space we devised an attachment for endoscopes to our frameless neuronavigation system. The combination with neuronavigation allowed the visualisation of the endoscope's tip in relation to the surrounding brain as well as a view of the trajectory of the approach.

The system is flexible and does not confine the surgeon to a rigid path, while determining the precise localisation of the endoscope's tip. So far the combination of these two methods has not complicated the procedure. The mechanical attachment of the endoscope is not a hindrance since the arm is moveable. In fact, if balanced well, the arm allows further fixation in addition to the Leyla retractors, which can be attached to the endoscope. For 3rd ventriculostomy frame-based and frameless sterotaxy as well as endoscopy alone [4] are applicable. For cystic lesions frameless stereotactic guidance for neuorendoscopy is a valuable adjunct. The cyst walls can be very rigid and conceal structures, thereby preventing direct visualisation; common anatomic landmarks do not exist. For localisation the wand is ideal; the computer display shows the hidden objects allowing further dissection and a more thorough fenestration. However, in the specific case shown in Fig. 6 leakage of the cyst fluid and CSF led to a massive shift rendering the wand useless. A brain shift to this extent, which occurs rarely, poses a major obstacle to image guidance, minor shifts can be compensated by the excel-lent visual control provided by the endoscope. Nevertheless, larger cysts with considerable internal pressure have to be excluded. So far this approach was used for the ventricular system in normal and pathologically altered states. With further experience and careful selection solid tumors may come into the realm of interactive image-guided neuroendoscopy for minimally invasive tumor removal [10,11].

In summary we believe that the lack of precise localisation in space due to the narrow field of view in neuronavigation can be overcome by combining neuronavigation with neuroendoscopy. Although this combination is demanding on the surgeon it provides excellent visualisation with stereotactic precision.

References

1. Bauer BL, Hellwig D (1994) Minimally invasive endoscopic neurosurgery – a survey. Acta Neurochir Suppl (Wien) 61:1–12
2. Caemaert J, Abdullah J (1993) Diagnostic and therapeutic stereotactic cerebral endoscopy. Acta Neurochir (Wien) 124:11–13
3. Cohen AR, Perneczky A, Rodziewicz GS, Gingold SI (1995) Endoscope-assisted craniotomy: approach to the rostral brain stem. Neurosurgery 36:1128–1130
4. Drake JM (1993) Ventriculostomy for treatment of hydrocephalus. Neurosurg Clin N Am 4:657–666
5. Drake JM, Rutka JT, Hoffman HJ (1994) ISG viewing wand system. Neurosurgery 34:1094–1097
6. Grunert P, Perneczky A, Resch K (1994) Endoscopic procedures through the foramen interventriculare of Monro under stereotactical conditions. Minim Invasive Neurosurg 37(1):2–8
7. Hellwig D, Bauer BL (1991) Endoscopic procedures in stereotactic neurosurgery. Acta Neurochir Suppl (Wien) 52(1):30–32
8. Lewis AI, Keiper GL Jr, Crone KR (1995) Endoscopic treatment of loculated hydrocephalus. J Neurosurg 82(5): 780–785
9. Maciunas RJ (ed) Interactive image-guided neurosurgery. Neurosurgical topics. AANS Publications
10. Manwaring KH (1993) Intraoperative Microendoscopy. In: Maciunas RJ (ed) Interactive image-guided neurosurgery. Neurosurgical topics. AANS Publication, pp 217–232
11. Zamorano L, Chavantes C, Moure F (1994) Endoscopic stereotactic interventions in the treatment of brain lesions. Acta Neurochir Suppl (Wien) 61:92–97

Neuronavigation to the Petrous Bone: A Cadaveric Study

G. Arango, J. Kaminsky, T. Brinker, U. Thorns, and M. Samii

Summary

The technical advantages of the MKM neuronavigation system from Zeiss were tested during cadaveric dissections of the petrous bone. The optical probe, contours and plans which are projected directly on the microscope field enhanced orientation and provided accurate anatomical monitoring of the dissections. These features need practice and understanding of some factors and must be considered advantageous in the performance of petrous bone surgery.

Introduction

Surgeons involved in the operative management of lesions in the area of the petrous bone are challenged by one of the most complex anatomies of the skull base [1, 12]. Neuronavigational aids should provide anatomical monitoring, enhance orientation and increase safety while working around the labyrinth and neurovascular structures embedded in the bone.

The MKM (multi coordinate manipulator) microscope from Zeiss is a high precision neuronavigation system whose probe is based on a laser-assisted distance measurement which defines the coordinates according to the focused point of the microscope's optical system. The location of the "optical probe" (depth of focus) can be observed in the system's monitor with reconstructions of the neuroradiological examinations in axial, sagittal and coronal planes as well as parallel or perpendicular to the direction of the axis of the microscope. Targets or contours of interest can be marked or drawn with the system's software on the preoperative radiological examinations and appear directly on the field of view of the microscope during dissection [6]. The present study examines whether these special features of the MKM microscope are useful during cadaveric dissections of the petrous bone, i.e. opening of the internal acoustic canal (IAC) from the suboccipital route and mastoidectomies with preservation of the labyrinth block and the neurovascular structures.

Methods and Material

Ten fixed cadaver heads were used; in four of them coloured silicones injected into the vascular spaces (Silastic 3110, Dow Corning, Germany). Three fiducials (Leibinger 1.4 mm titanium microscrews) arranged in a 3.5 cm triangle were implanted above the mastoidal area. CT scans with 1-mm slices were made of the area that included the three fiducials and the entire petrous bone. The CT scans were transferred by means of a magnetic tape to the system's software (Leibinger).

In the system's software the fiducials were identified and the coordinates with their respective order recorded. The structures in the petrous bone were marked using contours with maximal eight fold amplification on the monitor on all 1-mm axial slices where the structures appeared.

The facial nerve, lateral semicircular canal (SCC), posterior SCC, anterior SCC and utricle were contoured for the performance of mastoidectomies. For navigations to the IAC, the posterior SCC with its ampulla, common crus and utricle were contoured as described. Additionally, several "plans" were drawn using the "entry and target point" functions to evaluate its possible usefulness.

Dissections

The heads were fixed in a Mayfield head holder and the referencing was performed with autofocus and maximal zoom over the 1-mm fiducial microscrews. To expose the IAC a lateral suboccipital craniectomy was performed. The dura was opened exposing the internal acoustic meatus. Before drilling towards the fundus of the IAC a previously planned entry point was focused with our optical probe on the petrous bone lateral to the meatus in order to determine the area which should be drilled in order to access the fundus without violating the labyrinth. The direction of the optical axis of the microscope was adjusted to the direction of drilling using the parallel reconstructions from the monitor and the guide directly on the microscope's field of view. Subsequently the focus function was used to scan the location of the contours of the labyrinth in the depth. The plan-defined entry point at the meatus and target at the fundus was tested.

For mastoidectomies the procedure consisted in the complete drilling of the mastoid antral and infralabyrinthine cells with skeletonizations of the osseous labyrinth and exposure of the facial nerve from the tympanic portion and fallopian canal. Before drilling deep in the antral cells, the field was scanned in the depth with the focus function in order to assess the position of the contours. The drilling was continued using autofocus functions in the different depths to evaluate the usefulness of the contouring strategies (over cells superimposed on the structures vs over the structures themselves). The value of the lateral SCC facing the middle ear and antral cells as "target", which is normally used as a landmark, was assessed.

Results

The navigational information is projected into the microscopic field (Fig. 1)

Optical Probe

A green cross in the middle of the microscope field indicates the optical probe. Flat surfaces perpendicular to the axis of the microscope such as fiducials or the bone before craniotomy appear correctly navigated once focused. This can be confirmed by following their position in the monitor. If surfaces are presented tilted or tangential to the axis of view the position of focus does not always correspond to the plane (depht of focus). This effect can be minimized by using maximal zoom to provide a smaller field of view and thus a more exact localization of the probe. This feature of the MKM system was found to be

important, e. g. when drilling over a tangential plane as for the exposition of the IAC.

Contours

Contours will appear directly on the field of view of the microscope if the optical probe is focused on the perpendicular plane in which the contours were drawn. Before drilling, the exact position of the contoured structures can be scanned by manually "going in" with the focus function. This feature allowed better orientation and corrections in the direction of drilling. Three-dimensional reconstructions served as valuable means to analyse the drawn contours from different angles of view (Fig. 2).

The use of several contours in a close spatial relationship can be confusing as they will appear superimposed and have the same colour (green). This is disturbing unless exact recognition is evident or the area of interest is the borders rather than the structures themselves.

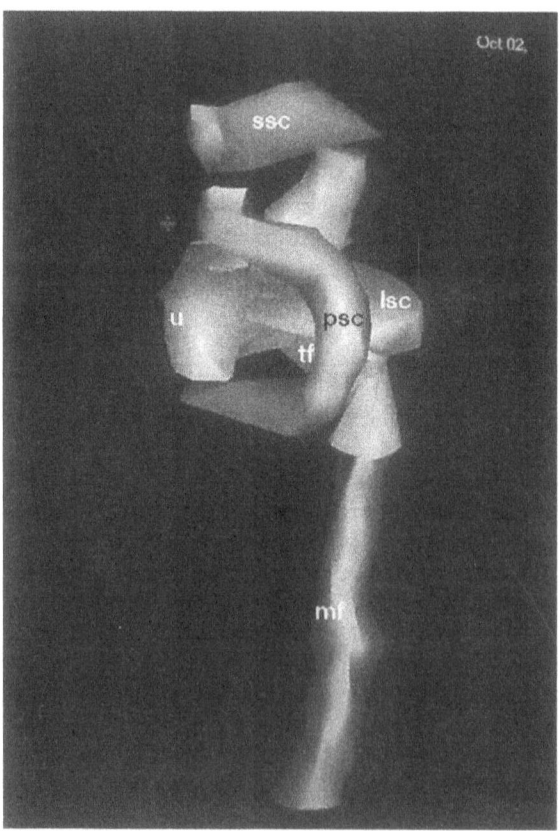

Fig. 2. Three dimensional reconstruction of the contoured labyrinth and the facial nerve in its tympanic and mastoidal portions. This reconstruction can be moved in all directions as to accomodate to the surgical axis. *SSC,* Superior; *lsc,* lateral semicircular canal; *psc,* posterior semicircular canal; *u,* utricle; *tf,* tympanic; *mf,* mastoid portion of facial nerve

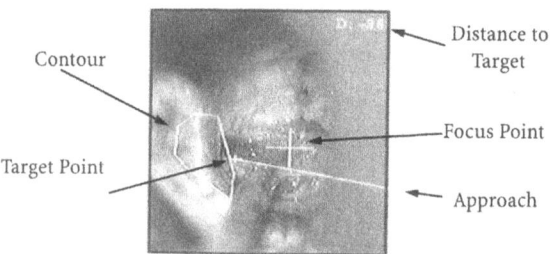

Fig. 1. All of the navigation features of the unique optical system of the MKM can be seen: contours at the area of the labyrinth, a line from the entry point to the fundus which is our target (plan) and the distance from the optical probe (*large cross*) to the target expressed in millimeters in the upper right corner

Entry-Target Function

One entry and target point are called a plan. Multiple plans can be marked in the preoperative CT scan and used at different stages of the dissection procedure. A solid line shows the axis between entry and target points.

The target appears as a small green cross at all times and doesn't need to be in the actual depth of focus. Additionally, the exact distance from the optic probe to the target will be displayed. This feature was at all times highly reliable in our experimental petrous bone surgery.

Monitor

The monitor displays the location of the optical probe and its axis on the preoperative CT. Orientation is aided by reconstructions in the axial, sagittal and coronal planes. The parallel and perpendicular reconstructions provided the highest anatomical correlation between the microscope field of view and the neuroadiological examinations and demand special attention.

Discussion

Previous reports have shown that neuronavigation systems provide high accuracy [5, 7–9, 11, 15, 16]. According to our previous study the MKM system is comparable in this regard [6]. The laser-controlled distance measurement, the projection of the navigational plans and the contours through the microscope are features unique to the MKM neuronavigation system. The optical probe is technically highly accurate; in our experience, its practical accuracy sometimes demands accomodating the surgery to the system. However, as routine develops, the unique technical features can be incorporated with ease.

The contours can be used for orientation around neurovascular structures and the labyrinth. The most useful and reliable features of this navigation system were the plans projected in the microscope with their exact distances toward the targets and the perpendicular neuroradiological reconstruction which provided an interactively correct correlation of the surgical field and the studies.

This reconstruction demands the highest attention; with increasing familiarity its advantage becomes evident over the other reconstructions. Preservation of the labyrinth when exposing the IAC in a patient with preserved hearing can not be overemphasized [3,10,13,14]. Navigational aids that would provide accuracy in the 1- to 2-mm range could be

incorporated to avoid fenestration in certain cases and improve hearing preservation.

Transpetrosal approaches that aim at preservation of function can pose narrow corridors around the labyrinth and may leave unapproached corners that result in recurrences or residual lesions in this area. Neuronavigational aids scan the anatomical area of interest and correlate it to the radiological studies [2, 4] they provide more control in the surgical management of these patients.

A knowledge of the advantages and limitations is essential for the proper use of neuronavigational aids. Their constant use will help to develop navigation strategies suited for each case in clinical practice. To achieve this, their use can be incorporated in the laboratory for training and to gain the desired experience, and they should be used clinically in selected cases. Navigation as a teaching tool with its potential to provide anatomical monitoring improves the quality of the 3-D orientation within the surgical field and incorporates accurate analysis of the neuroradiological study.

References

1. Anson BJ, Donaldson JA (1981) Surgical anatomy of the temporal bone, 3rd edn. Saunders Philadelphia
2. Chakeres DW, Spiegel PK (1983) A sytematic technique for comprehensive evaluation of the temporal bone by computed tomography. Radiology 146: 97–106
3. Domb GH, Chole RA (1980) Anatomical studies of the posterior petrous apex with regard to hearing preservation in acoustic neuroma removal Laryngoscope 90: 1769–1776
4. El Azm M, Samii M, Bini W (1988) Computed tomographic studies in acoustic neurinomas in relation to the vestibular organ and hearing function. In: Frayse B, Lazorthes Y (eds) Neurinomes de l'Acoustique. Fabre, pp 81–89
5. Friets E, Struhbehn JW, Hatch JF, Roberts DW (1989) A frameless stereotaxic operation microscope for neurosurgery. IEEE Trend Act Bio Eng 36(6): 608–617
6. Kaminsky J, Arango G, Brinker T, Samii M (1997) Requirements for referencing of the MKM neuronavigation system. Minimal Invasive Technique for Neurosurgery, ed. Hellwig, Bauer, Springer-Verlag (1998)
7. Galloway RL, Maciunas RJ, Edwards CE (1992) Interactive image-guided neurosurgery. IEEE Trend Act Bio Eng 39(12): 1226–1231
8. Golfinos JG, Fitzpatrick BC, Smith LR, Spetzler RF (1995) Clinical use of a frameless stereotactic arm: results of 325 cases. J Neurosurg 83: 197–205
9. Maciunas RG, Galloway RL, Latimer JW (1994) Application accuracy of stereotactic frames. Neurosurgery 35: 682–695
10. Matula C, Diaz-Day J, Czech T, Koos WT (1995) The retrosigmoid approach to acoustic neurinomas: technical, strategic, and future concepts. Acta Neurochir (Wien) 134(3–4): 139–147
11. Nabavi A, Manthei G, Blömer U (1995) Neuronavigation: Computergestüztes operieren in der Neurochirurgie. Radiologe 35: 573–577
12. Pellet W, Cannoni M, Pech A (1990) Oto-neurosurgery. Springer, Berlin Heidelberg New York, pp 5–40

13. Samii M, Matthies C, Tatagiba M (1991) Intracanalicular acoustic neurinomas. Neurosurgery 29 : 189 – 199
14. Tatagiba M, Samii M, Matthies C, El Azm M, Schönmayr R (1992) The significance for postoperative hearing of preserving the labyrinth in acoustic neurinoma surgery. J Neurosurg 77 : 677 – 684
15. Ungersböck K, Rossler K, Matula C, Koos WT (1996) Rahmenlose Stereotaxie: klinische Erfahrungen mit dem stereotaktisch geführten Operationsmikroskop MKM und dem Neuronavigationssystem Easy guide. Zentralbl Neurochir [Suppl] : 16 – 17

16. Watanabe E, Watanabe T, Manaka S, Mayanagi Y, Takakura K (1987) Three dimentional digitizer "Neuronavigator": new equipment for computed tomography-guided stereotaxic surgery. Surg Neurol 27 : 543 – 547

First Experience with the BrainLab VectorVision Neuronavigation System

H. K. Gumprecht, D. C. Widenka, and C. B. Lumenta

Summary

BrainLab VectorVision is a neuronavigation system that links a free-hand probe, tracked by a passive marker sensor system, to a virtual computer image space on a patient's preoperative CT. Infrared flashes emited by two cameras are reflected by passive marker spheres that are mounted near the head of the patient and on surgical instruments. Using the data of both cameras, the software can calculate the tree-dimensional position of a sphere and therefore the three-dimensional position of the entire tool. During a 6-month period 40 patients with intracerebral mass lesions in different locations and of different sizes were operated on with the guidance of the BrainLab system. In five cases the system could not be used because of technical errors. In the other patients, the BrainLab VectorVision navigation system proved to be a helpful tool in planning and guiding surgery of cerebral mass lesions.

Introduction

Progress in computer and imaging technology provides a variety of techniques that can assist neurosurgeons in localizing the relevant pathology. The stereotactic technique has been used to map image space on to physical space [7, 10]. In addition to frame-based methods, frameless systems are now available [1, 3, 9, 11]. In this paper we present our experience in surgery with 40 patients with different brain mass lesions using the BrainLab VectorVision Neuronavigation system.

The Computer System

One of the new neuronavigation systems is the BrainLab VectorVision, an intraoperative image-guided, frameless, armless localization system. It consists of a method for the registration of image and physical space, an intraoperative localization device, and a computer display of images and provides a real-time feedback of the location of the surgical instrument. A free-hand probe is tracked by a passive marker sensor system to virtual computer image space on a patient's preoperative CT images. Magnetic resonance images can be fused with the CT data. Two infrared cameras are attached at different angles to the tool. The infrared flashes are reflected by passive marker spheres which are mounted near the patients head and on surgical instruments. Using the data of both cameras the software calculates the three-dimensional position of a sphere and therefore the 3-D position of the whole tool. The working principle of the system makes it necessary that an unobstructed view to the tool marker is provided.

Components of the VectorVision

Hardware and Software

We use an alpha computer with a RISC processor, 510-megabyte hard disc storage, and a 64-megabyte memory bank. The VectorVision software runs on windows NT 3.51. The computer is arranged on a transportable trolley; the infrared cameras are mounted on the trolley but can be positioned anywhere in the operating room (Fig. 1).

Fiducials

The skin fiducials consist of a plastic base and two different markers: spherical markers for CT imaging and hemispherical markers with a central cone for intraoperative referencing. The markers can be locked to the plastic sockets using the key type mechanism. The fiducials are simply attached to the patients head with double adhesive tape.

Mayfield Clamp Adapter

The Mayfield clamp adapter is a star-shaped tool which has to be mounted rigidly on the Mayfield

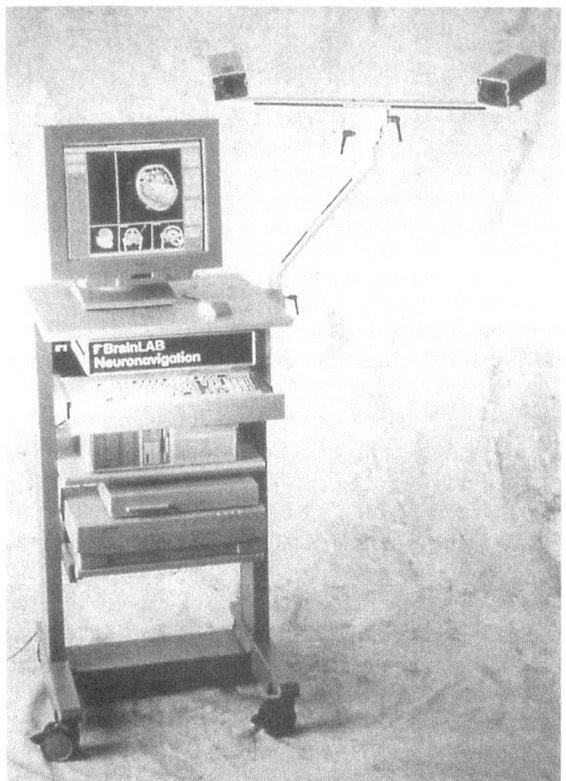

Fig. 1. The computer workstation with the two infrared light-emiting cameras mounted to the left side of the trolley

headrest near the patients head. Three passive markers are attached to this tool. It provides continuous and dynamic patient referencing during the procedure.

VectorVision's Probe

This probe is equipped with two passive reflective markers. It is used for the referencing of the skin fiducials and provides wireless surgical tracking during the operation.

Universal Instrument Adapters

The star-shaped instrument adapters with three passive marker spheres can be attached by the surgeon to the preferred surgical instruments such as the bipolar forceps, suction tube, or even the endoscopes (Fig. 2].

Preoperative Preparation

Preoperatively the surgical approach has to be considered using information from MR and/or CT images. Five fiducials are attached to the patients head using double adhesive tape. The spherical markers are secured to the plastic sockets. A contrast-enhanced computer tomographic scan with 2- to 3-mm slices, depending on the size of the lesion, is performed. The data are archived on a magnetic optical disc. The time elapsed between imaging and surgery might vary from a period of hours to days. To avoid movements of the simply attached fiducials, we prefer to perform preoperative imaging procedure in the awake patient just before surgery.

During initiation of anesthesia the data are transferred to the computer workstation via the optical disc. The lesion is flagged out in colors and reconstructions of the three spheres (axial, coronal, and sagittal) as well as three-dimensional reconstructions

Fig. 2.
The instruments for working with BrainLab Vector-Vision. From *left* to *right*: the pointer, the instrument for referencing the skin markers; a bipolar forceps equipped with a marker tool; the May-field adapter (the rigid reference for the cameras); a tool for mounting on bigger instruments, e. g., endoscopes

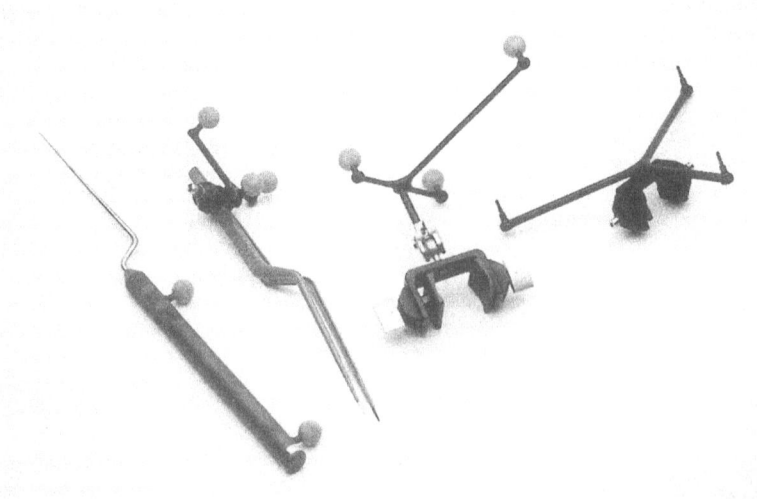

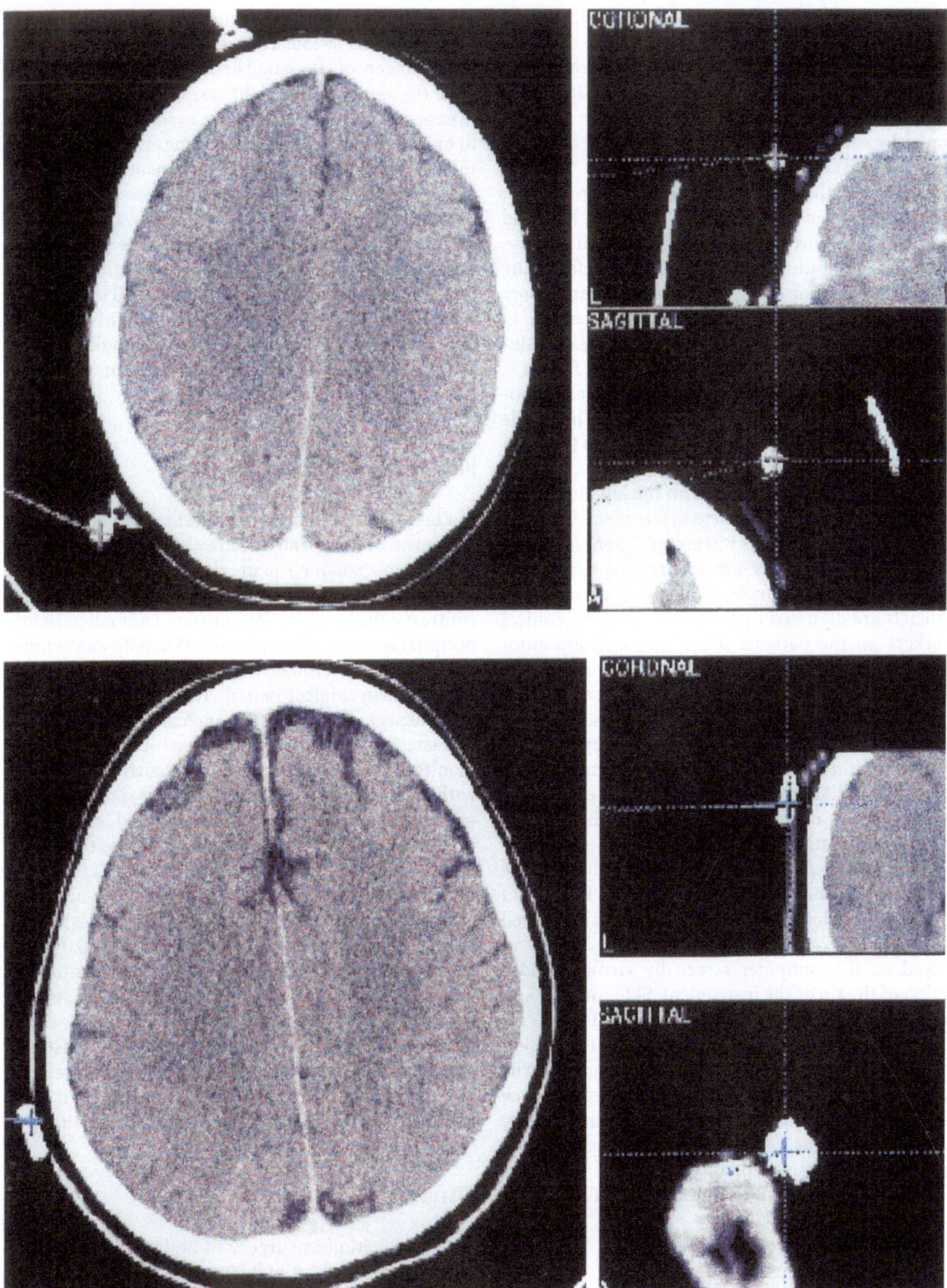

Fig. 3. The accuracy is checked during surgery by pointing on the skin markers (*blue* and *grey cross*). The overall error is within 2 – 4 mm

are performed on the computer screen. Using information about the complex lesion from the different images, a plan for the optimal surgical approach can be developed. If necessary, preoperative MR images can be fused with the CT data. The cameras are calibrated by manipulating a calibration rod with two reflective markers in front of the cameras.

Afterwards the patient is positioned for surgery; the head is fixed very carefully into the Mayfield headrest. Skin movement has to be avoided because displacement of the skin markers before referencing will lead to a major error. The spherical markers are removed with care; the plastic sockets remain in place. The operation site is prepared with sterile solution and the operation field, inclusive of all fiducials, is draped with a large sterile foil. The draping is continued leaving the foil-covered area free. The sterile star-shaped Mayfield adapter is mounted to the Mayfield headrest, in our setting on the left side. The infrared cameras are arranged on the left side of the surgeon so that an unobstructed view even after positioning the microscope is warranted. Then the sterile hemispherical markers are secured to the plastic sockets. Using the special VectorVision probe, the fiducials are digitized by tipping each of the conical markers on the patients scalp. The software automatically registers this data, calculating each coordinate link with the corresponding marker point from the CT data set. Passive markers are attached to the preferred instruments and the tip is registered at the calibration cone of the marker array mounted to the Mayfield clamp. This maneuvre supplies the software with the offset of the tip of the instrument relative to the attached reflective marker.

The optimal surgical approach, exactly over the lesion, can be considered by defining the borders of the brain mass lesion. The depths of the lesion as well as the direction of the operative approach can be displayed on the computer screen by virtual prolongation of the tip of the instrument. Skin incision and craniotomy can be kept quite small. Except for standard skull-base approaches we use straight skin incisions. The lesion is removed microsurgically. The system provides a real-time intraoperative feedback of the position of the working tool. During surgery the accuracy of the system is checked by pointing on anatomical landmarks or on skin markers fixed at positions where they don't move (Fig. 3). The error is within the range of 3–4 mm.

Materials

From May 1996 to October 1996 40 patients with cerebral mass lesions of different sizes and histological natures were operated on in our department using the VectorVision neuronavigational system. The size of the lesions ranged from 1.7 ccm to 55 ccm with a mean of 21 ccm. The location of the lesions were the skull base in eight cases, cortical lesions in 18 cases, subcortical lesions in 12 cases, and brain stem in one case. The histological findings were gliomas in 16 cases, meningiomas in 11 cases, metastatic tumors arteriovenous malformations in two cases, chordoma in one case, a melanoma in one case, and a radiation necrosis in one case. The duration of the procedures ranged from 1.5 h to 4 h and was not prolonged by the use of the neuronavigation system. All lesions but one could be completely resected. This patient deteriorated postoperatively. In five cases the system did not work because of technical errors of unknown causes.

Illustrative Case Report

In Figs. 4 and 5 a small cortical lesion (1.8 ccm) of one patient is demonstrated. This 48-year-old man suffered from a sudden seizure during his daily work. A cranial computer tomographic scan without and with contrast enhancement demonstrated a small parieto-occipital lesion on the right side. When he was admitted to our department he was fully conscious and without neurological deficit. The X-ray of the lung was suspect of a central bronchial carcinoma. The metastatic work-up did not show other metastatic manifestations. Although an alternative treatment with stereotactic LINAC radiosurgery is available in our department the patient was prepared for surgery under neuronavigational guidance because the histological diagnosis was unclear at that time. Figure 4 demonstrates the computer display after craniotomy exactly over the tumor. The cross which indicates the tip of the instrument is still within the bone so that an error of 2 mm is obtained. The situation after tumor removal is shown in Fig. 5. Biopsies from the tumor margin after resection had been without tumor cells. The histological nature of the tumor was a metastasis of a bronchial carcinoma. The surgery was successful and the patient was without neurological deficit. The postoperative MRI demonstrated no residual tumor.

Discussion

Neuronavigation in surgery of brain mass lesions is becoming a field of high interest in neurosurgery. Frame-based stereotactic techniques for surgery of deep-seated brain lesions are well described by Kelly et al. and Weiner et al. [4, 5, 6, 12]. Roberts et al. [9] have given an account of a computer-based system for integration and display of CT data in the micro-

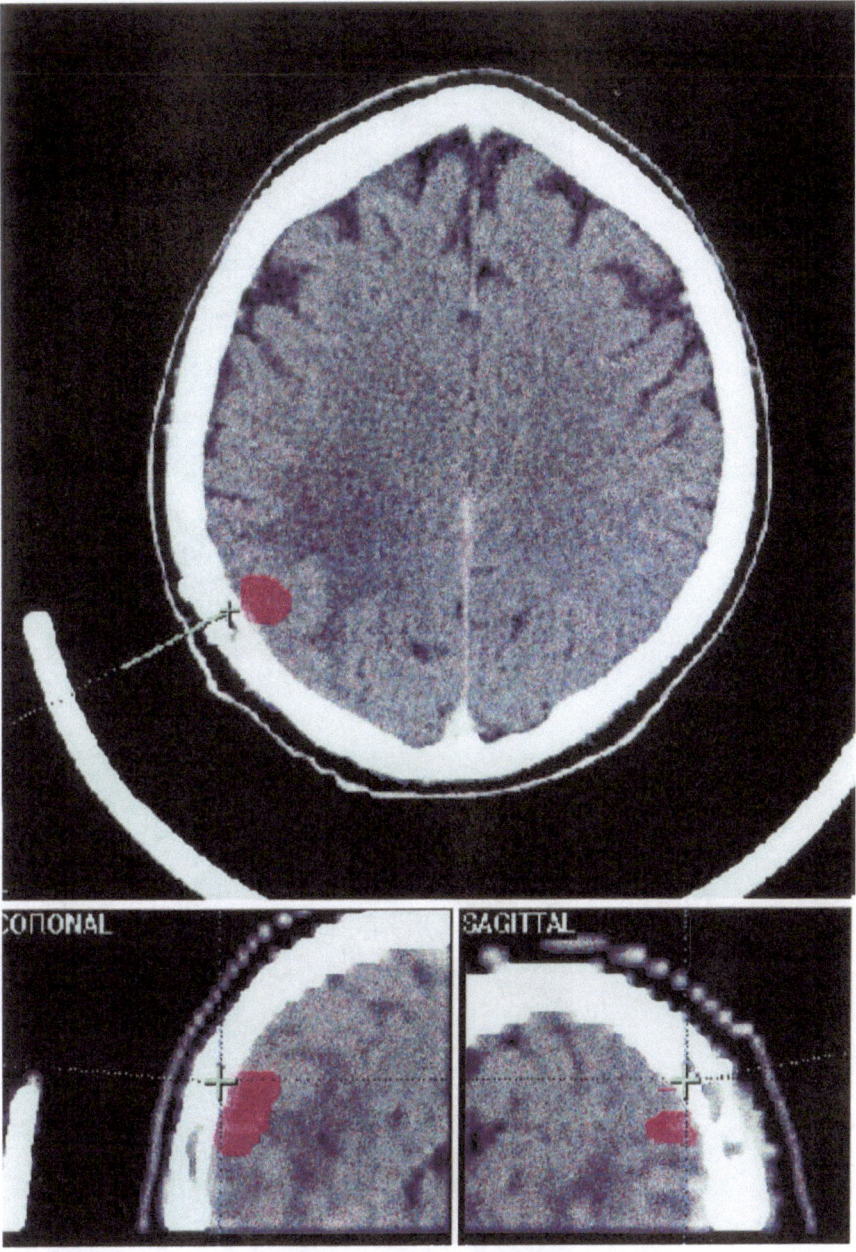

g. 4.
T scan of a patient with a
nall cortical lesion, the
largins are determinated
traoperatively

cope without requiring a stereotactic frame. Several uthors described their developments and experi- nce with frameless neuronavigation systems con- isting of operating arms [8, 11]. A frameless, armless, avigational system has been described by Kato et al. 3]. The Stealth station, one of the new navigation omputer workstations, works by light-emitting diode LED) point sources [1]. The LEDs are detected by hree cameras. Other systems work in a similar ashion. A special probe that is prepared with LEDs is eeded to work with these systems.

The working principle of the BrainLab Vector- Vision is the reflection of infrared flashes by a passive marker sensor system. Two infrared light cameras emit infrared flashes. These flashes are reflected by special spherical markers mounted in a rigid position near the patients head and on surgical instruments.

The system provides wireless marker tools that can be attached to any surgical instrument. This is an advantage over other neuronavigation systems. The surgeon may mount marker tools to his preferred surgical instruments and he does not need to change

Fig. 5.
CT scan of the same patient as shown in Fig. 4 after tumor removal

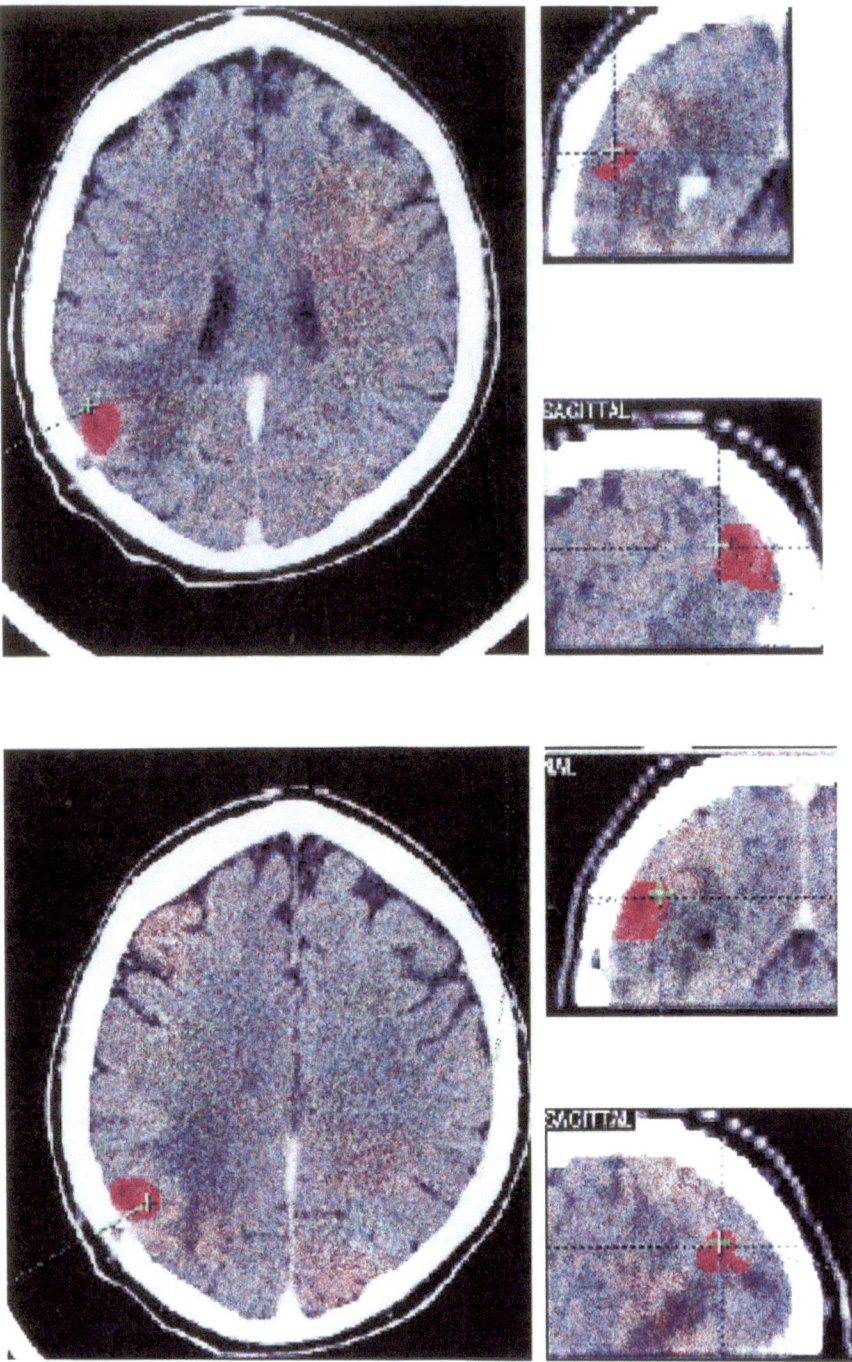

the instrument for intraoperative checking of the position of his working area. The system is easy to combine with endoscopic surgical procedures by attaching the reflecting sensors to the endoscope. Even biopsies or the exact placement of catheters into cystic lesions are possible using the VectorVision. Another advantage is the now available adaptation to the operating microscope.

The additional amount of time needed for preparation, CT scanning, and computer planning is quite small, i.e., around 20 min. It does not prolong anesthesia or surgery. The marker referencing is performed within seconds and the target can be approached accurately and quickly. The size of the skin incision and craniotomy can be minimized which may reduce operative time. The working principle of this system, similar to other systems with camera detection, makes it necessary to provide an unobstructed view to the marker tools. This may limit the applicability of navigation systems working that way. However, with good planning of the arrangement of all of the equipment in the OR this should not be a real problem. VectorVision as well as other systems is based on preoperative images; real-time imaging is not yet possible and the problem of brain shifting is not solved. Ultrasound control and biopsies from the margin after tumor resection are still necessary to verify complete tumor removal. The accuracy of the system was within 2 – 4 mm in our series, which is similar to that of other navigation systems. This error proved to be acceptable in skull-base lesions, small cortical lesions, and subcortical lesions bigger than 1.5 cm in diameter. VectorVision is very helpful in the exact planning of the operative approach of those cerebral mass lesions. For small deep-seated lesions the error is not acceptable because CSF drainage and mannitol administration may lead to displacement of those small tumors. In these cases, the frame-based stereotactic-guided technique is superior and more accurate [2]. All in all, the navigators should assist the surgeon to plan and perform surgery, but they should not replace his anatomical knowledge and skill. Keeping advantages and disadvantages in mind, the BrainLab Vector-Vision neuronavigation system has proved to be a helpful tool in the surgery of brain mass lesions.

Conclusion

- The work with VectorVision is fast and easy.
- The existing instruments can be equipped with reflecting markers.
- The accuracy within 2 – 4 mm is acceptable.
- The system is wireless and without any mechanical restrictions.
- The reflecting markers can be detached from the instruments and separately gas sterilized while the other tools can be autoclave sterilized.
- Combination with endoscopy is possible.
- Combination with the operating microscope is now possible.
- The problem of brain shifting is not solved.

References

1. Germano IM (1995) The neurostation system for image-guide, frameless sterotaxy. Neurosurgery 37 : 348 – 351
2. Gumprecht HK, Lumenta CB (1996) Comparison of stereotactic guided surgery with 3 D computer planning and Brain-Lab VectorVision neuronavigation system. Second Joint Meeting of the Japan Neurosurgical Society and the Deutsche Gesellschaft für Neurochirurgie, Naruto, Japan, October 19 – 21
3. Kato A, Yoshimine T, Hayakawa T, Tomita Y, Ikeda T, Mitomo M, Harada K, Mogami H (1991) A frameless, armless navigational system for computer-assisted neurosurgery. J Neurosurg 74 : 845 – 849
4. Kelly PJ, Alker GJ, Goerss S (1982) Computer-assisted stereotactic laser microsurgery for the treatment of intracranial neoplasms. Neurosurgery 10 : 324 – 331
5. Kelly PJ (1988) Volumetric stereotactic surgical resection in intra-axial brain mass lesions. Mayo Clin Proc 63 : 1186 – 1198
6. Kelly PJ (1988) Volumetric stereotaxis and computer-assisted stereotactic resection of subcortical lesions. In: Lunsford LD (ed) Modern stereotactic neurosurgery. Nijhoff, Boston, pp 169 – 184
7. Kelly PJ (1986) Computer assisted stereotaxis: new approaches for management of intracranial intra-axial tumors. Neurology 36 : 535 – 541
8. Koivukangas J, Louhishalmi Y, Alakuijala J, Oikarinen J (1993) Ultrasound-controlled neuronavigator guided brain surgery. J Neurosurg 79 : 36 – 42
9. Roberts DW, Strohbein JW, Hatch JF, Murray W, Kettenberger H (1986) A frameless stereotactic integration of computerized tomographic imaging and the operating microscope. J Neurosurg 65 : 545 – 549
10. Shelden CH, McCann G, Jaques S, Lutes HR, Frazier RF, Katz R, Kuki R (1980) Development of a computerized micro-stereotaxic method for localization and removal of minute CNS lesions under direct 3-D vision. J Neurosurg 52 : 21 – 27
11. Watanabe E, Watanabe T, Manaka S, Mayanaki Y, Takakura K (1987) Three dimensional digitizer (neuronavigator): new equipment for computed tomography-guided stereotaxic surgery. Surg Neurol 27 : 543 – 547
12. Weiner HL, Kelly PJ (1996) A novel computer-assisted volumetric stereotactic approach for resecting tumors of the posterior parahippocampal gyrus. J Neurosurg 85 : 272 – 277

Anatomically Guided Neuronavigation: First Experience with the SulcusEditor

K. Niemann, U. Spetzger, V. A. Coenen, B. O. Hütter, W. Küker, and D. Graf von Keyserlingk

Summary

Our first clinical experience with the SulcusEditor (SE) in neurosurgical patients with tumors in eloquent regions of the brain are reported. The SE is a software tool for the interactive delineation of the cerebral sulcal pattern in 3 D FLASH MRI data sets. The method has been implemented on RISC workstations under UNIX using an X-Window interface based on the OSF-MOTIF Toolkit. The anatomical definition of sulci is supported by a three-plane display with linked cursor movements. Thus problems in the definition of sulci tangent to the section plane are eliminated. The outcome of the segmentation procedure may then be passed to customary rendering software for 3 D visualization. So far the MRI data of 24 patients with tumors in eloquent regions of the brain have been subjected to the labelling procedure. Digitization of 10–12 sulci was considered sufficient for neurosurgical planning. Logistic requirements – the results of the SE analysis had to be made available to the neurosurgeons preoperatively within a 48-h period after the MR acquisition – could be fulfilled in more than 70% of cases. Labelling takes approximately 4 h, the bidirectional data transfer in a heterogeneous clinical network 2 h. Future developments will comprise improvements of the user interface as well as the intraoperative availability of the results.

Introduction

An important goal of preoperative MR imaging of the brain is to supply the neurosurgeon with relevant information on the exact topography of an intracranial space-occupying lesion. On this basis he or she establishes a 'mental image' of the patient's individual anatomy. In supratentorial tumors, the anticipation of potential hazards for eloquent regions of the cerebral hemispheres encountered during the intended neurosurgical route is crucial. The advent of dedicated medical workstations for neuronavigation which process 3 D MR volume acquisitions has catalytically reinforced preoperative planning strategies. The image slices are no longer understood as static uncorrelated icons but as dynamic 3 D voxel representations which may be interactively roamed and anatomically explored on the computer screen.

However, images contain too much information for humans or machines to process in detail [53]. The wish to segment and classify relevant anatomical structures from an eventually noisy [6,52] routine 3 D MRI data set so as to achieve information reduction is still a challenge for computer vision [57]. The prevailing approach has been the search for predominantly automatically working segmentation algorithms, the ultimate goal being a compromise between the segmentation outcome and the amount of (costly) user interaction necessary. A multitude [1, 2, 5, 6, 11–15, 20–22, 24, 27, 29, 44, 49, 52] of different algorithmic solutions have been suggested (for a review and a taxonomy of MRI segmentation algorithms see [8] and [12]). Only recently do we observe an increased scientific interest in the application of neural networks [1] and active contour models using anatomical pre-knowledge [21, 49] for brain surface and brain structure segmentation. Although scientific progress has been made [8, 12], in clinical routine settings segmentation is still hampered by the serious restrictions [22].

Patient- and MRI-Related Artifacts

Routine MR images in a clinical environment are prone to system- and patient-related sequence, reconstruction-algorithm, field, and noise artifacts [7, 23]. Therefore, the theoretical model assumptions of intensity-based segmentation algorithms are seldomly fulfilled [6, 55]. Furthermore, algorithms for segmentation still lack robustness [22]. A – perhaps supervised – preprocessing of the data is considered mandatory [12, 46, 55].

Re-editing of the Segmentation Outcome

Because of artifacts, depending on the clinical or scientific questions involved and the accuracy needed, the results of (semi-)automated segmentation require more or less time-consuming manual re-editing even in sophisticated environments [13, 14, 22, 24, 26]. With cerebral sulci, problems arise in the segmentation of their deep course as well as in narrow sulci (young patients, brain edema) because of the partial volume phenomenon [11]. Near the surface of the brain where the subarachnoid space of the sulcal clefts is contiguous with the subarachnoid CSF space covering the crowns of the gyri, segmentation problems may be partly superseded by erosion. Active contour models have also been suggested in this area [15, 27] (see however [29]). Segmentation artifacts are more pronounced in the basal parts of the brain, including the clinically important temporobasal and temporomesial regions. This may be partly attributable to the uneven surface of the cranial fossae. In 3 D volume renderings this leads to a pseudopattern of the sulci which has no anatomical equivalent. Colchester et al. conclude that 'segmentation remains one of the time-limiting steps in preoperative planning and preparation for intra-operative guidance' and that 'certain tasks such as tumor segmentation usually still require input from the surgeon' [14]. (On the issue of tumor segmentation cf. also [8, 10, 12]).

From Segmentation to Classification

The segmentation process decomposes an MRI image into regions but in a strict sense does not classify the compartments [12]. It more or less reliably mirrors physicochemical characteristics, i.e., material properties which have an impact on the MR registration. The regions are then classified or labelled, either with or without supervision into compartments such as gray matter, white matter, fat, background, CSF, or lesion. Although these denominations are suggestive one has to keep in mind that the classification process itself is subject to systematic and unsystematic errors. Hence, it is evident that even in an ideal image with a high signal-to-noise ratio (SNR), the class membership of a voxel representing gray matter is not very strongly correlated with its histology proper [45, 49, 51] or its nerve cell/fibre quotient or even its CSF content.

Top-down Approaches Using Anatomical Knowledge

Sulcal clefts will hopefully be represented in the CSF class. At this stage they are not further classified as anatomical entities and bear no anatomical labels. Therefore data-driven bottom-up approaches might at this stage be complemented by knowledge-based top-down approaches [2, 30, 42, 50]. The latter have to cope with the broad range of individual anatomical variation [9, 16, 18, 19, 33, 37, 38, 41, 50], the known problems in establishing and fitting an anatomical atlas [4, 5, 9, 30, 33, 35, 50], as well as different coexisting medical terminologies [3, 15, 16, 18, 19, 32, 37, 41, 50]. In neurosurgical patients with intracranial masses a more or less pronounced shift and distortion of neighboring anatomical structures must be expected so that atlas-oriented approaches are of limited value [21].

Lack of Evaluation

Usually the result of the segmentation outcome is assessed by medical experts. Yet there is no gold standard to compare with [1]. From a methodological view point it is doubtful whether a standard evaluation of segmentation results is logistically feasible [1, 12]. Methods for evaluating the morphological quality of the segmentation outcome – in terms of its value in making clinical decisions – or a comparable measure for time efficiency of the different scientific algorithmic approaches are still not well developed.

In a clinical setting any segmentation or classification approach should in a pragmatic way be problem-oriented and respect the knowledge domain. In neurosurgical visualization, information reduction is necessary so as not to flood the surgeon with irrelevant intricate image details. As regards the surface of the cerebral cortex, we assumed that the detailed interactive anatomical labelling of cortical topography would enhance orientation by high level symbolic anatomical description. However, we did not know whether such an approach was logistically feasible in a clinical setting.

Methods and Material

The premises were the following:

1. We concentrated on patients with superficial and deep tumors in eloquent precentral [25, 31, 40], temporal, and temporobasal regions.
2. It was considered unnecessary to label the complete sulcal pattern of the two hemispheres. Instead, detailed labelling of structures encountered on the route between the trepanation site and the tumor, and the topographical relationships of the mass in a limited area were considered relevant. Other sulci were registered only when they promoted the digitization process.

3. Data transfer and labelling had to be completed within a 48-h interval after MRI acquisition so that the result could be discussed preoperatively.

MRI studies were performed on a Magnetom 1.5 T imaging system (Siemens Medical Systems, Erlangen, Germany) with a standard head coil. A T1-weighted fast low-angle shot (FLASH) gradient echo sequence was used with the following parameters: TR = 40 ms; TE = 5 ms; inversion time = 40 ms; flip angle = 40; 3 D partitions = 128; slice thickness = 1.60 mm; FOV = 300 mm; matrix = 256 × 256. From the VMS scanner, the data were transferred in encrypted form via LAN/WAN or magneto-optical disks to the Department of Anatomy which is 200 m away from the main building of the RWTH Aachen University Hospital.

As a labelling tool we used the SulcusEditor (SE) [34] which is an in-house software development of the Department of Neuroanatomy. It is an offspring of the software modules used for the digitization, interpolation, and triplanar re-editing of digital atlases of the human brain [33] (C++, OSF-Motif 1.2, X11R5) and is installed on a LAN of Sun SPARCstations (Sun Microsystems Inc., Mountain View, California, USA). The functionality of the user interface of the SE comprises a correlated three-plane display (sagittal, axial, and frontal view) with optional linked cursor movements. Editing may be performed in each of the windows. The sulci are interactively defined directly on the screen and labelled in the region of interest starting with the major sulci. In a coarse-to-fine strategy, the sulcal pattern is gradually refined. The images may independently from each other be zoomed and displayed in scrollable windows so as to achieve greater precision in the on-screen digitization. Labelling follows the terminology of von Economo and Koskinas [19] which is concise, 'unsurpassed to date' [56] and pleasantly reprinted in Yasargil's Microneurosurgery [56, pp 70–72]. Among others, the works of Duvernoy [16], Eberstaller [18], Naidich et al. [32], and Ono et al. [37] served as reference.

The 3 D reconstructions and the oblique slices as shown in the figures were subsequently produced using the SunVision 1.2 software (Sun Microsystems Inc., Mountain View, California, USA). In cube mode with texture-mapped rendering a 2 D representation of the volume's texture is mapped around the volume, which may be deliberately scaled and rotated in 3 D. Furthermore, one may nondestructively cut into the volume. Six orthogonal clipping planes and a single oblique plane are available. The oblique plane may be used to simulate the neurosurgical route. In lightbox mode, the volume is displayed in a noncorrelated triplanar display and one oblique plane.

Fig. 1a–c. Case 481. **a** Lightbox mode. The tumor is situated between the inferior end of the superior precentral sulcus (*dark orange*) and the central sulcus (*yellow*). The white lines define the section plane in **b**. **b** Lightbox mode, oblique plane. The approximate location of the trepanation (intraoperative view: **c**) is indicated by an ellipsoid

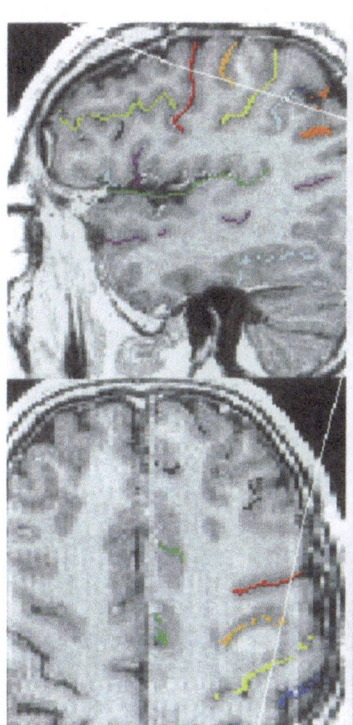
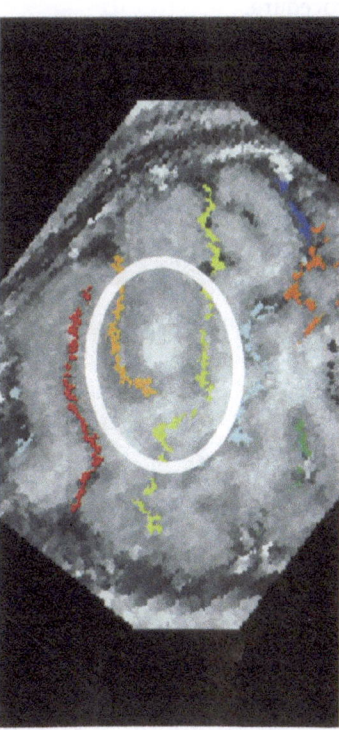
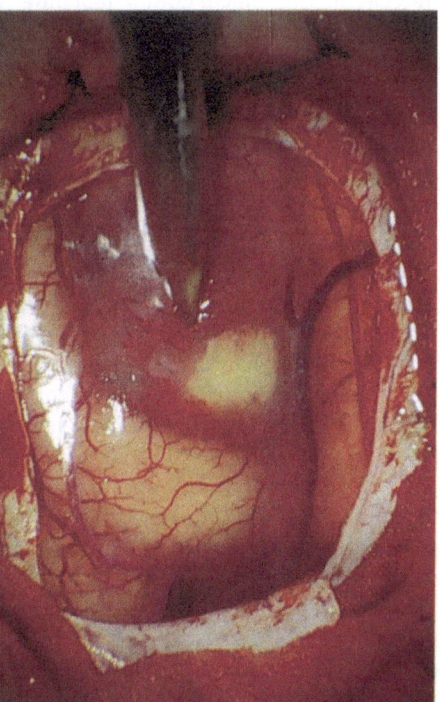

a   b   c

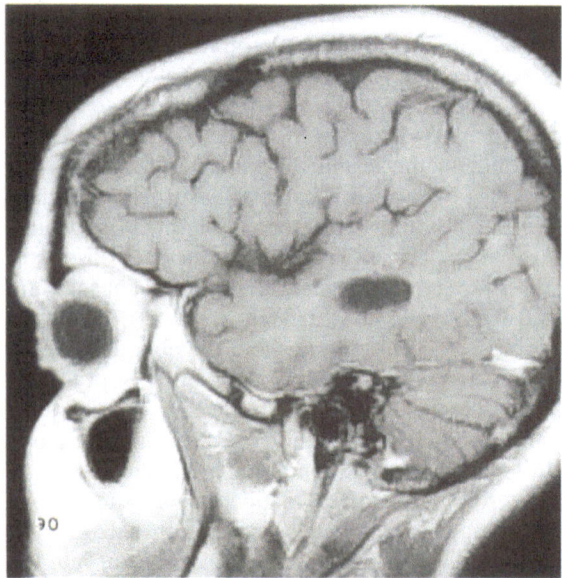

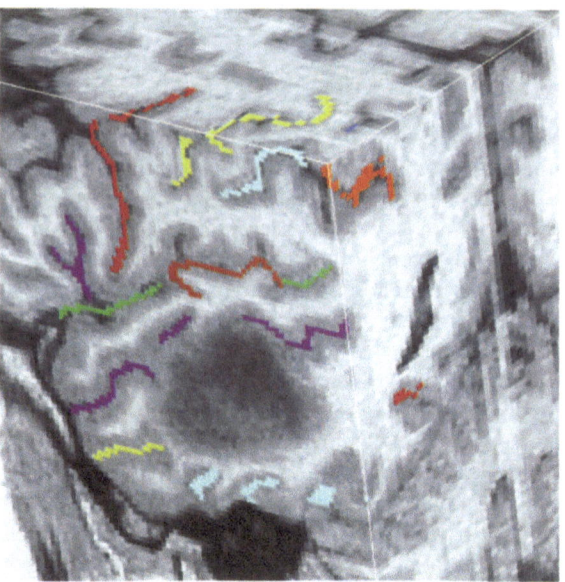

a   b

Fig. 2a, b. Case 540. **a** Post-operative scan (SL = 5 mm) demonstrating lesion site. **b** Reconstruction of astrocytoma from FLASH 3D acquisition (SL = 1.6 mm) in Sun Vision cube mode. *Magenta*, superior temporal sulcus; *yellow*, middle temporal sulcus; *cyan*, inferior temporal sulcus; *red*, collateral sulcus; *brown*, surface of Heschl's gyri

Results

At the time of writing, the MRI data sets of 24 patients underwent the SulcusEditor labelling procedure. Exemplarily we report on three patients in whom a closer analysis of the sulcal pattern was indicated.

Case 481. A 31-year-old male presented with a 3-week history of jacksonian fits starting in the right hand with loss of consciousness, postical weakness confined to the right arm and temporary dysphasia. The neurological status was inapparent apart from an anisocoria with left-sided mydriasis. MRI findings were initially interpreted as a high-grade left-sided astrocytoma in the middle frontal gyrus occipitally bordered by the precentral gyrus. Detailed labelling of the pericentral area revealed that the tumor's precise location was within the precentral sulcus and that it was intercalated between the lower end of the superior precentral sulcus and the central sulcus (Fig. 1a, b). This was in good accordance with intraoperative findings (Fig. 1c). The histological examination of the tumor was in favor of a fibrillary astrocytoma with focal anaplasia (WHO grade III).

Case 540. A 28-year-old female presenting with a 2-year history of headache and photophobia. closer analysis with the Sulcus Editor (Fig. 2) revealed a left hemispheric tumor located in the middle temporal gyrus (T2). It was clearly demarcated by the superior (t1) and middle (t2) temporal sulci. In this case, the

SE findings guided the surgeon in the degree of dissection and debulking of the mass occipitodorsally. Intraoperatively it was confirmed that the tumor (a grade III astrocytoma) did not invade T1 or T3. The postoperative outcome was without any complications apart from a short-lasting temporary dysphasia 2 days after the intervention.

Case 161. A 76-year-old female recently experiencing seizures but no motor deficits. Using the SE we

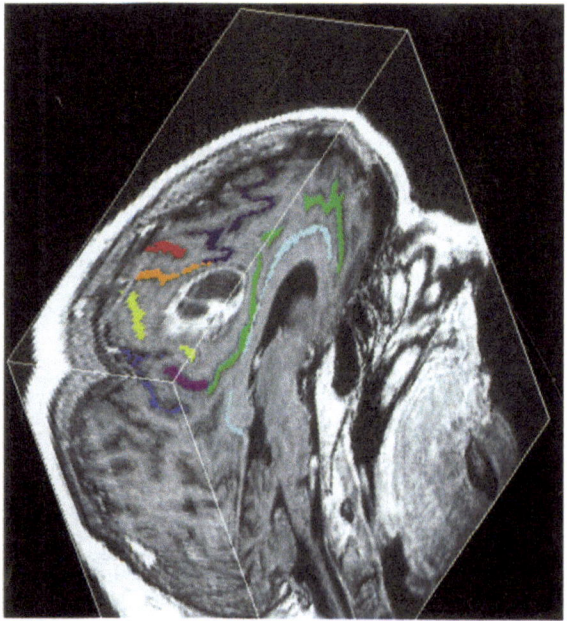

Fig. 3. Case 161. Parasagittal meningioma. Position of mass in cube mode. *Dark orange*, superior precentral sulcus; *red*, inferior precentral sulcus; *yellow*, central sulcus; *green*, cingulate sulcus

could demonstrate the exact position of this para-sagittal meningioma. It compressed the precentral and the cingulate gyri. The tumor was situated between the medial part of the superior precentral sulcus, the upper knee of the central sulcus (Fig. 3b, c), and the cingulate sulcus. Intraoperatively the brain-tumor interface was heterogeneous. Whereas in the basal parts the meningioma was loosely attached to the precentral gyrus, it could not be clearly seperated form the gyrus occipitodorsally.

Time Requirements and Logistics

Approximately 2 h were needed for data transfer from and to the hospital, including organization and the reformatting of the data for the SE. For the detailed labelling of ten to 12 sulci including the texture-mapped 3D reconstruction 4 h were necessary.

Advantage

By the interactive segmentation prior definitions of sulci are protocolled in 3D and may be used for the definition of surrounding sulci. The result of the definitions can be controlled immediately in three orthogonal planes – a quality control step of the editing procedure. The problem of structures tangentially oriented to one section plane (see comment of Gildenberg [26] can be completey resolved (Fig. 4).

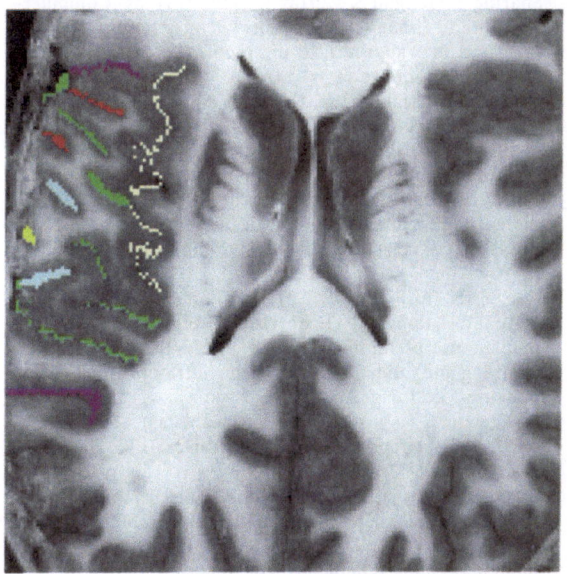

Fig. 4. Visible human male. Axial macroscopic slice 77. *Green,* labelling with SE demonstrates multiple sections of Sylvian fissure. *Yellow* central sulcus; *cyan,* anterior and posterior sub-central sulci; *red,* precentral inferior sulcus

Pitfalls

In six out of the 24 cases the results were not preoperatively available due to lack of information flow and logistic pitfalls in the sequence of events within the clinical routines. The time needed for the digitization is still too long and would be substantially reduced by amendments of the user interface, which are under preparation. In one case, the results could not be demonstrated early enough because of hardware failure of the stand-alone demonstration workstation in the Neurosurgical Department. Some cases which underwent labelling obviously did not meet the selection criteria (e. g., one peri-insular/precentral process, three frontal glioblastomas, two cases of thalamic cavernoma in which additionally superimposition of a thalamic atlas was requested [33]). The common denominator was that they represented the 'complex anatomical case' (see below). A serious drawback of the SE in the eyes of the clinical users is that, although it helps in creating the mental 3D image, it is too detached from the intraoperative situation. Furthermore, ambiguites in color codes and lack of discernibility of some color combinations were criticized.

Discussion

In a comment on a recent paper of Kikinis et al. [26] Bakay, Gildenberg, and Kelly critically assess the benefit of interactive 3D neurosurgical planning. If we consider only the cons put forward we find:

1. Time consumption and time-efficiency are critical factors. This concerns the re-editing of the segmentation outcome, as well as segmentation and rendering time.
2. The greatest benefit is to be expected for the complex anatomical case. The latter occurs infrequently.
3. The fundamental dilemma of using preoperative image-based data to localize intraoperatively (CSF loss, brain shift). In this area, only recently has progress been achieved [14].

Data on the time requirement of the interactive labelling of cerebral sulci [24, 29, 44] in a routine clinical setting are lacking and data on the quality of the segmentation achieved are scarce [29]. This item still has to be quantified with regard to the SE. In (semi-)automated segmentation, Colchester et al. [14] stress the role of manual painting or outlining which was 'used at various stages' and like Kikinis et al. [26] see a necessity for a technician (in the case of tumor segmentation by a surgeon) to correct the segmentation results. Two-dimensional editing tools in

VISLAN are preferred by the users over 3D tools [14]. Gerritsen et al. [22], present 'a rather extensive tool' which can be used directly on scanner data or for the interactive correction of semiautomatic segmentation [20]. From the industrial point of view he criticizes the exaggerated expectations in the degree of automation of segmentation procedures.

In an elegant, clinically oriented morphological study Naidich et al. [32] propose a comprehensive set of anatomical rules for the definition of cerebral sulci in sagittal MRI slices based on visual inspection only. The achieved accuracy is estimated to be approximately 90%. To our knowledge, there is presently no computer vision system with which these knowledge-based rules could be implemented. We regard the SE as a digital complement of the thorough visual inspection suggested by Naidich et al. Although the human visual system [28] is excellent in 2D feature detection, this is not the case for continuity detection in a 3D image series. The SE, which digitally keeps track of 2D definitions, enables this 2D – 3D transition. At the present stage it is considered a modular and versatile cortical topography tool which can easily pass the results obtained to other software (e.g., for rendering, CSF or vessel segmentation, neuronavigation) and also be used for comparative morphometric analysis [35]. It helps to acquire high quality anatomically normative segmentation algorithms [12]. Because of the well-known phenomenon of cortical variation and plasticity [39], however, it cannot be assumed that intra-operative orientation by sulci or gyri alone will be considered sufficient in the future. Neuronavigation systems will have to incorporate functional data from fMRI, PET, MEG, and CEEG as well as the results of the rapidly evolving highly specialized clinical imaging groups [14, 26, 30, 31, 40]. In a heterogeneous clinical environment, these data may stem from different sources. We expect that with the given diversity of commercially available neuronavigation systems the ability of the workstations to cautiously integrate preprocessed or labelled image data might have a positive impact on user acceptance and become a competitive advantage.

Acknowledgements. We are indebted to Prof. Dr. H. Bertalanffy, Prof. Dr. J. M. Gilsbach, Dr. L. Mayfrank, Dr. V. Rohde, and Dr. I. Slansky (Department of Neurosurgery, Technical University (RWTH) Aachen) for participation in the preliminary evaluation and critical discussions. We thank L. Adams, L. van Es, and R. Rüger (Philips Medical Systems, Best, The Netherlands) for the evaluation of integration aspects.

References

1. Amartur SC, Piraino D, Takefuji Y (1992) Optimization neural networks for the segmentation of magnetic resonance images. IEEE Trans Med Imaging 11 : 215 – 220
2. Arata LK, Dhawan AP, Broderick JP, Gaskil-Shipley MF, Levy AV, Volkow ND (1995) Three-dimensional anatomical model-based segmentation of MR brain images through principal axes registration. IEEE Trans Biomed Eng 42 : 1069 – 1078
3. Axer H, Niemann K (1994) Terminology of the thalamus and its representation in a part-whole relation. Methods Inform Med 33 : 488 – 495
4. Bajcsy R, Lieberson R, Reivich M (1983) A computerized system for the elastic matching of deformed radiographic images to idealized atlas images. J Comput Assist Tomogr 7 : 618 – 625
5. Barillot C, Gibaud B, Gee JC, Lemoine D (1995) Segmentation and fusion of multimodality and multi-subjects data for the preparation of neurosurgical procedures. In: Beolchi L, Kuhn MH (eds) Medical imaging. Analysis of Multimodality 2D/3D images. Studies in Technology and informatics 19. IOS Press, Amsterdam, pp 70 – 82
6. Beil W, Ottenberg K, Stiehl HS (1995) Towards automatic segmentation of two-dimensional brain tomograms. In: Beolchi L, Kuhn MH (eds) Medical Imaging. Analysis of Multimodality 2D/3D Images. Studies in Technology and informatics 19. IOS Press, Amsterdam, pp 158 – 174
7. Bellon EM, Haacke EM, Coleman PE, Sacco DC, Steiger DA, Gangarosa RE (1986) MR-artifacts: a review. AJR 147 : 1271 – 1281
8. Bezdek J, Hall L, Clarke L (1993) Review of MR image segmentation techniques using pattern recognition. Med Phys 1033 – 1048
9. Bookstein F (1991) Morphometric tools for landmark data. Cambridge University Press, Cambridge New York
10. Bottomley PA, Hardy CJ, Argersinger RE, Allen-Moore G (1987) A review of ^{1}H nuclear magnetic relaxation in pathology: are T_1 and T_2 diagnostic? Med Phys 14 : 1 – 37
11. Bullmore E, Brammer M, Rouleau G, Everitt B, Simmons A, Sharma T, Frangou S, Murray R, Dunn G (1995) Computerized brain tissue classification of magnetic resonance images: a new approach to the problem of partial volume artifact. Neuroimage 2 : 133 – 147
12. Clarke LP, Velthuizen RP, Camacho MA, Heine JJ, Vaidyanathan M, Hall LO, Thatcher RW, Silbiger ML (1995) MRI segmentation: methods and applications. Magn Reson Imaging 13 : 343 – 368
13. Cline HE, Lorensen WE, Kikinis R, Jolesz F (1990) Three-dimensional segmentation of MR images of the head using probability and connectivity. J Comput Assist Tomogr 14 : 1037 – 1045
14. Colchester ACF, Zhao J, Holton-Tainter KS, Henri CJ, Maitland N, Roberts PTE, Harris CG, Evans RJ (1996) Development and preliminary evaluation of VISLAN, a surgical planning and guidance system using intra-operative video-imaging. Med Image Analysis 1 : 73 – 90
15. Davatzikos C (1996) Using a deformable surface model to obtain a shape representation of the cortex. IEEE Trans Med Imaging 15 : 785 – 795
16. Duvernoy H (1991) The human brain. Surface, three-dimensional sectional anatomy and MRI. Springer, Berlin Heidelberg New York
17. Ebeling U, Huber P, Reulen HJ (1986) Localization of the precentral gyrus in the computed tomogram and its clinical application. J Neurol 233 : 73 – 76
18. Eberstaller O (1890) Das Stirnhirn. Ein Beitrag zur Anatomie der Oberfläche des Grosshirns. Urban & Schwarzenberg, Wien Leipzig

1. Economo von C, Koskinas GN (1925) Die Cytoarchitektonik der Hirnrinde des erwachsenen Menschen. Springer, Berlin Heidelberg New York, pp 28 – 32

2. Fontana F, Dellepiane S, Vernazza G (1995) Interactive segmentation for target outline. In: Beolchi L, Kuhn MH (eds) Medical imaging. Analysis of multimodality 2 D/3 D images. Studies in Technology and informatics 19. IOS Press, Amsterdam, pp 112 – 120

3. Ge Y, Fitzpatrick JM, Dawant BM, Bao J, Kessler RM, Margolin RA (1996) Accurate localization of cortical convolutions in MR brain images. IEEE Trans Med Imaging 15 : 418 – 428

4. Gerritsen FA, van Veelen CWM, Mali WPT, Bart AJM, deBliek HLT, Buurman J, van Eeuwijk AHW, Hartkamp MJ, Lobregt S, Moreira Pereira Ramos L, Polman LJ, vanRijen PC, Visser P (1995) van Eeuwijk Requirements for and experiences with covira algorithms for registration and segmentation. In: Beolchi L, Kuhn MH (eds) Medical imaging. Analysis of multimodality 2 D/3 D images. Studies in Technology and informatics 19. IOS Press, Amsterdam, pp 4 – 27

5. Henkelmann RM, Bronskill MJ (1987) Artifacts in magnetic resonance imaging. Rev Magn Reson Med 2 : 1 – 126

6. Höhne KH, Hanson WA (1992) Interactive 3D segmentation of MRI and CT volumes using morphological operations. J Comput Assist Tomogr 16 : 285 – 294

7. Kido DK, LeMay M, Levinson AW, Benson WE (1980) Computed tomographic localization of the precentral gyrus. Radiology 135 : 373 – 377

8. Kikinis R, Langham Gleason, Moriarty TM, Moore MR, Alexander III E, Stieg PE, Matsumae M, Lorenson WE, Cline HEP, Black PMel, Jolesz FA (1995) Computer-assisted interactive three-dimensional planning for neurosurgical procedures. Neurosurgery 38 : 640 – 651

9. LeGoualher G, Barillot C, Le Briquer L, Gee JC, Bizais Y (1995) 3D detection and representation of cortical sulci. In: Lemke HU et al. (eds) Computer assisted radiology, CAR '95, Springer, Berlin Heidelberg New York, pp 234 – 240

10. Leibovic KN (ed) (1990) Science of vision. Springer, Berlin Heidelberg New York

11. Mangin JF, Frouin V, Bloch I, Regis J, López-Krahe J (1992) Automatic construction of an attributed relational graph representing the cortex topography using homotopic transformations. SPIE Math Methods Med Imaging III 2299 : 110 – 121

12. Mazziotta JC, Toga AW, Evans A, Fox P, Lancaster J (1995) A probabilistic atlas of the human brain: theory and rationale for its development. The international consortium for brain mapping (ICBM). Neuroimage 2 : 89 – 101

13. Mueller WM, Yetkin FZ, Hammeke TA, Morris III GL, Swanson SJ, Reichert K, Cox R, Haughton VM (1995) Functional magentic resonance imaging mapping of the motor cortex in patients with cerebral tumors. Neurosurgery 39 : 515 – 521

14. Naidich TP, Valaranis AG, Kubik S (1995) Anatomic relationships along the low-middle convexity: part I – Normal specimens and magnetic resonance imaging. Neurosurgery 36 : 517 – 532

15. Niemann K, Naujokat C, Pohl G, Wollner C, von Keyserlingk DG (1994) Verification of the Schaltenbrand and Wahren stereotactic atlas. Acta Neurochir (Wien) 129 : 72 – 81

16. Niemann K, van Nieuwenhofen I, Hütter BO, Thron A, Gilsbach J, von Keyserlingk DG (1995) The Sulcus editor: an interactive tool for atlas generation and surgical planning. Hum Brain Mapping [Suppl]: 66

17. Niemann K, van Nieuwenhofen I, Weber R, Matthies G, Theilig A, von Keyserlingk DG (1996) The Talairach proportional grid concept revisited. Neuroimage 3 : S 119

18. Niemann K, van Nieuwenhofen I, Berks G, von Keyserlingk DG (1996) The Schaltenbrand and Wahren stereotaxic atlas: conflicts in a histological database resolved by fuzzy set representation. In: Zimmermann HJ (ed) EUFIT '96. Fourth European congress on intelligent techniques and soft computing. Proc Vol III, pp 2117 – 2122

37. Ono M, Kubik S, Abernathey CD (1990) Atlas of the cerebral sulci. Thieme, Stuttgart

38. Paxinos G (ed) (1990) The Human nervous system. Academic Press, San Diego

39. Rademacher J, Caviness VS Jr, Steinmetz H, Galaburda AM (1993) Topographical variation of the human primary cortices: implications for neuroimaging, brain mapping, and neurobiology. Cereb Cortex 3 : 313 – 329

40. Reinhardt HF, Trippel M, Westermann B, Horstmann GA, Gratzl O (1996) Computer assisted brain surgery for small lesions in the central sensorimotor region. Acta Neurochir (Wien) 138 : 200 – 205

41. Retzius G (1896) Das Menschenhirn. Studien in der Makroskopischen Morphologie. Norstedt & Söner, Stockholm

42. Roland PE, Graufelds C, Wåhlin J, Ingelman L, Andersson A, Ledberg A, Pedersen J, Åkerman S, Dabringhaus A, Zilles K (1994) Human brain atlas: for high-resolution functional and anatomical mapping. Hum Brain Mapping 1 : 173 – 184

43. Rowberg AH, Ramey J (1995) The seven venal sins of user interface design. In: Lemke HU et al. (eds) Computer assisted radiology, CAR '95. Springer, Berlin Heidelberg New York, pp 243 – 240

44. Schiemann T, Bomans M, Tiede U, Hoehne KH (1992) Interactive 3D-segmentation SPIE Visualization in Biomed Comput 1808 : 376 – 383

45. Solsberg MD, Fournier D, Potts DG (1990) MR imaging of the excised human brainstem: a correlative neuroanatomic study. Am J Neuroradiol 11 : 1003 – 1013

46. Soltanian-Zadeh H, Windham JP, Peck DJ, Yagle AE (1992) A comparative analysis of several transformations for enhancement and segmentation of magnetic resonance image scene sequences. IEEE Trans Med Imaging 11 : 302 – 318

47. Sonka M, Tadikonda SK, Collins SM (1996) Knowledge-based interpretation of MR brain images. IEEE Trans Biomed Imaging 15 : 443 – 452

48. Spitzer VM, Whitlock DG, Kilcoyne RF, Scherzinger L, Rubinstein D, Russ P (1995) The visible human male for teaching and reference in radiology. In: Lemke HU et al. (eds) Computer assisted radiology, CAR '95. Springer, Berlin Heidelberg New York, pp 677 – 683

49. Székely G, Kelemen A, Brechbühler C, Gerig G (1996) Segmentation of 2-D and 3-D objects from MRI volume data using constrained elastic deformations of flexible Fourier contour and surface models. Med Image Anal 1 : 19 – 34

50. Talairach J, Tournoux P (1993) Referentially oriented cerebral MRI anatomy. Atlas of stereotaxic anatomical correlations for gray and white matter. Thieme, Stuttgart New York

51. Vandersteen M, Beuls E, Gelan J, Adriaensens P, Vanormelingen L, Palmers Y, Freling G (1994) High field magnetic resonance imaging of normal and pathologic human medulla oblongata. Anat Rec 238 : 277 – 286

52. Vincken KL, Koster ASE, Viergever MA (1995) Probabilistic hyperstack segmentation of MR brain dat. In: Beolchi L, Kuhn MH (eds) Medical imaging. Analysis of multimodality 2 D/3 D images. Studies in Technology and informatics 19. IOS Press, Amsterdam

53. Walters D (1990) Computer vision analysis of boundary images. In: Leibovic KN (ed) Science of vision, pp 365 – 397

54. Weinberg R (1992) Future directions in neurosurgery visualization. In: Apuzzo MLJ (ed) Neurosurgery for the third millenium. Neurosurgical topics. American Association of Neurological Surgeons Publications Commitee, Los Angeles, pp 47 – 63

55. Wells WM III, Grimson WEL, Kikinis R, Jolesz FA (1996) Adaptive segmentation of MRI data. IEEE Trans Med Imaging 15 : 429 – 442

56. Yasargil MG (1994) Microneurosurgery. Vol. IV A. Thieme, Stuttgart

Image-Assisted Surgery of Brain Tumors and Other Intracranial Lesions: A Preliminary Report

G. Broggi, A. Franzini, D. Servello, M. Fornari, S. Giombini, M. Grisoli, and I. Dones

Summary

Three different systems for image-assisted surgery were used for craniotomy in 130 patients affected by intracranial lesions of different types. A mechanically guided system (OAS, Radionics) and two optically guided systems (Easy-Guide, Philips and SMN, Zeiss) were randomly used in this series.

Nine to 11 cutaneous cranial markers were glued on the scalp before neuroimaging examination as fiducial points to allow a computerized reconstruction of a virtual "stereotactic" 3D space containing the lesion of interest. In 20 patients 1.5 T brain MRI sequences were performed with an imaging protocol dedicated for use with Easy-Guide, SMN and OAS; in the other patients CT scans (50 slices, 3 mm thick) was performed.

This is a report of our preliminary experience with these systems for image-assisted surgery with particular regard to mathematical accuracy data and, on the basis of these results, to the reliability of hardware, software, impact on training of OR personnel and duration and quality of surgery. The advantages and limits of these frameless systems are discussed.

Introduction

Traditional brain surgery is limited in the choice of a safe trajectory to reach deep-seated lesions and in the intraoperative control of the boundaries of surgical ablation. In this regard many CT or MRI-coupled systems for intraoperative assistance of neurosurgical procedure were studied and put on the market [1, 3, 7–10]. The aim of these systems is basically to match CT or MR images with the intraoperative field seen under the microscope allowing the continuous use of such instrumentation by the surgeon through monitor viewing or through the microscope.

During the last decade frame-based stereotactic systems have been the only tools available to assist the surgeon in reaching deep and small lesions along a chosen trajectory [2, 4]. Although very accurate, frame-based systems limit movements during surgery, thus impairing intraoperative modifications of surgical strategy and perhaps also limiting any emergency anaesthesiologic procedure during operation. Moreover, these systems usually allow a burr-hole opening of the skull, while a wide craniotomy may be difficult to perform. On the other hand, because of these considerations, stereotactic frames remain mandatory, at present, when very high precision is needed, i.e. for multiple and serial brain biopsies and for the implant of intracranial electrodes.

Frameless systems for image-assisted brain surgery have been used in a consecutive series of 130 patients affected by supratentorial lesions (i.e. tumors or arteriovenous malformations (AVM)). This trial includes both mechanically (OAS; Radionics) and optically (Easy-Guide, Philips and SMN, Zeiss) driven devices. The advantages of the use of these systems are reported.

Patients and Method

In all systems four to 11 skin markers (fiducials) were glued on the patient's head. These fiducials constitute the reference points of a virtual space in which the head is placed. These markers are visible on CT or MR scans and allow the constitution of a virtual stereotactic frameless space. CT and MR are performed the day before or early in the morning of the day of surgery. Images are then transferred to the workstations. The patient is properly positioned in a conventional Mayfield headrest (Mayfield 2000, USA). Before surgery the virtual space described by the markers is rechecked in the operating theatre as the space described by these cutaneous fiducials. CT and MR images are thus matched with the real surgical anatomy.

The first system used (Easy-Guide, Philips, The Netherlands) consists of a UNIX-operated workstation to which CT or MR images are transferred before surgery. CT and MR transverse, coronal, and sagittal sections are reconstructed and shown on the monitor during operation. The equipment is then connected to a 3D camera group fixed to the oper-

Fig. 1.
Representation of the pointer reaching the boundaries of a tumor during surgery of a glial tumor. The Easy-Guide system (Philips) was used here

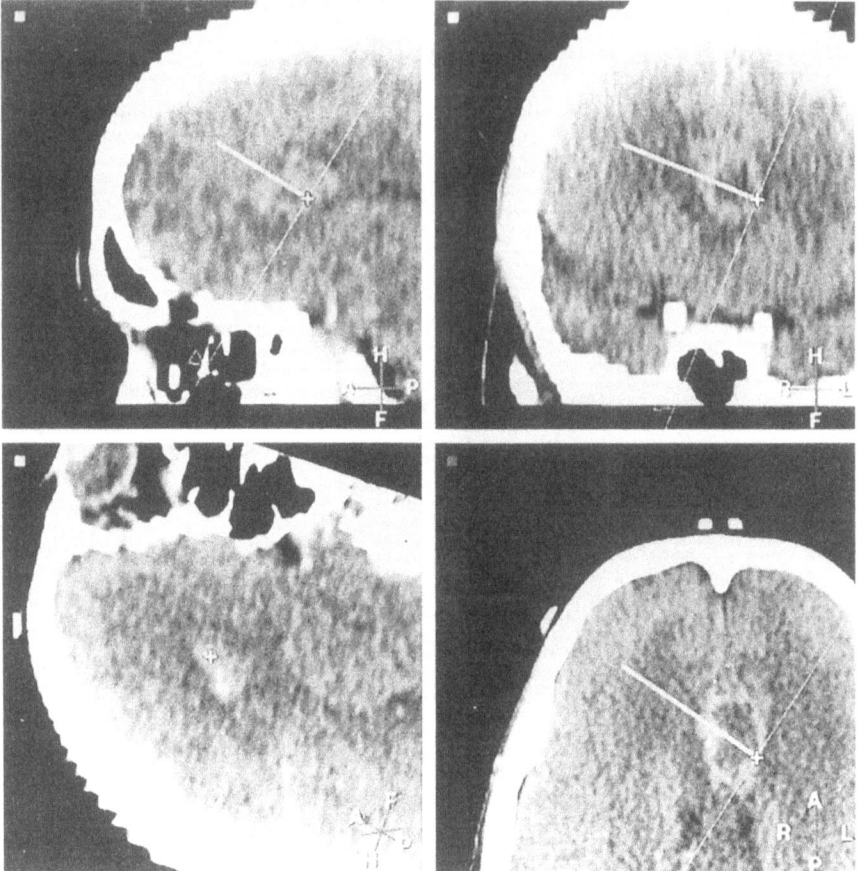

ating table; this reads the movements of a 3-LED pointer handled by the neurosurgeon. Then the system is ready to assist the surgeon showing a segment on the CT/MR images that moves with the pointer in the surgeon's hand. Images of the transverse, coronal and sagittal plane are shown in (Fig. 1).

The second system (OAS, Radionics, USA) is conceptually similar to the first system. In this system the pointer to identify skin markers and to navigate introperatively is fixed to an articulated arm connected to an encoder that sends messages to a UNIX-operated workstation. The arm is fixed to the Mayfield clamp and can be moved by the surgeon during operation to compare its position on the CT or MR images.

With these systems the trajectory to reach the lesion and the size and position of craniotomy can be accurately checked with the pointer. Four checkpoints are then made by drilling the skull. These additional markers are sent to the workstation and their coordinates can be periodically verified during surgery in order to ascertain whether movements of the patient's head have occurred after craniotomy or during the operation.

Moreover, the intracerebral lesion can be virtually delineated before the skin, skull and dura opening by means of an electronic virtual elongation of the pointer for Easy-Guide and with a mechanical elongation of the pointer for OAS. This allows the calculation of the distance of the lesion from the bone and dural surface and a comparison of different angles of trajectory to reach the lesion. During and at the end of surgery the boundaries of tumour ablation or the position of eloquent structures can be verified by the pointer on CT or MR images of the preoperative condition.

The third system (SMN, Zeiss) consists of a workstation connected to a 3D camera group that can alternatively read both a LED pointer or a group of LEDs mounted on the operative microscope.

The boundaries of a lesion and the trajectory to reach it can be previously electronically drawn on neuroimages and reproduced in the microscope finder. Moreover, intraoperative head movements are automatically adjusted by the workstation reading the position of a 3-LED group attached to the Mayfield clamp.

With the assistance of this equipment 130 patients were operated on. Seventy were operated on with

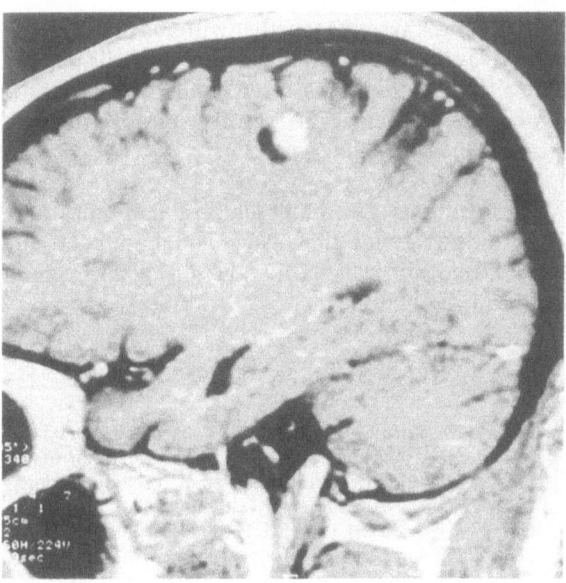

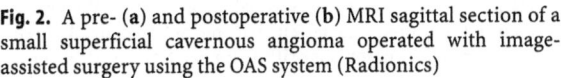

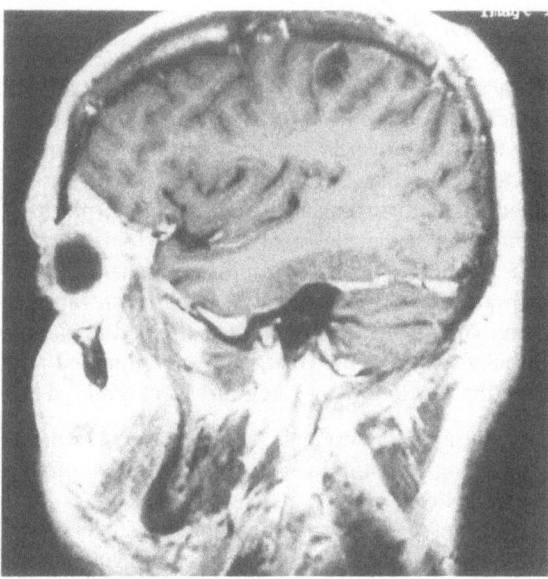

b

Fig. 2. A pre- (**a**) and postoperative (**b**) MRI sagittal section of a small superficial cavernous angioma operated with image-assisted surgery using the OAS system (Radionics)

Easy-Guide assistance, 40 were submitted to operation assisted by OAS and 20 were operated on with SMN. These systems were randomly chosen each time.

Patients were affected by the following diseases: 90 had supratentorial gliomas (60 temporal 20 temporo-occipital, six parietal and four thalamic), five patients had falx meningioma, 30 had arteriovenous malformations (five AVM and 25 cavernous angiomas), one had pituitary adenoma, one was subjected to the ventriculoperitoneal shunting procedure for hydrocephalus and three patients underwent microvascular decompression at the posterior fossa for trigeminal neuralgia. This last surgery was used as an internal control of these systems precision by checking some anatomical landmarks i. e. cranial nerves and arteries. All patients were then submitted to postoperative CT and MR to check the reliability of these systems in helping the surgeon to remove the lesions.

Results

The accuracy of these systems was calculated and expressed (in square millimetres) as the error between the recognition of markers on CT and MR images and the localization of these markers on the patient's head when positioned for surgery. A second correction of this value was made by localizing four checkpoints on the skull before and after craniotomy and during different steps of surgery, calculating the error due to undesired movements of the skin where the fiducials were glued on the head and to movements of the head in the headrest.

The mean accuracy of these systems in our series was 2.2 mm ± 1.0 (mean ± SD) without remarkable differences between the different systems. This accuracy is a mathematical calculation that does not take into account the brain shift due to the opening of the dura, to CSF aspiration and to the removal of a large lesion. Standard size craniotomies were performed and no particular attention was paid to positioning the patient so as to avoid CSF leakage. However, these systems exhibited sufficient accuracy as can be observed in the postoperative CT and MR images (Fig. 2). During operation, the localization of well-recognizable anatomical markers (such as bone or intracranial anatomical landmarks) with the pointer was successfully matched with CT or MR images.

The neuronavigation systems appeared to be very useful in assisting the neurosurgeon in the preoperative planning of the trajectory to the lesion intraoperatively in confirming the identification of the sulcal approach and finally in defining ablation boundaries as they appeared on CT and MR. Furthermore, it must be recalled that image-assisted surgery is related to neuroradiological images that, by definition, only show the visible parts of a biological lesion. The intraoperative findings are thus limited by such a restriction, i. e. the intraoperative accuracy is a radiologically matched result.

Conclusion

The accuracy of optically and mechanically guided frameless neuronavigators is sufficient to aid planning the neurosurgical removal of brain tumours. In the case of tumours in the vicinity of the skull base or "fixed" structures such as meningiomas, pituitary adenomas and cranial nerves, the systems showed their usefulness in confirming the chosen surgical approach and the accuracy was used as an intraoperative control. In the 90 patients affected by neuroepithelial tumours the treatment was improved provided that the following guidelines were adhered to:

- Topographic assessment of eloquent areas in the planning of tumour removal in the dominant hemisphere
- Planning of tumour removal along predetermined trajectories enabling the inclusion or avoidance of eloquent structures when reaching deep lesions
- Gross delimitation of the tumoural boundaries during resection of intrinsic neuroepithelial tumors

The limitations of the employed systems mainly concerns the lack of an on-line control of the volume modifications induced by surgery and by brain movements associated with CSF leakage [5]. The criticism of the use of surface markers which undergo skin movements is neglected in view of the level of accuracy obtained with these systems. Volume resection of a brain tumour remains a radiologically measurable entity while the total "biological" resection cannot be presently checked. Thus an accuracy of 2–8 mm is a highly acceptable standard for these systems in order to be helpful in everyday neurosurgery.

Finally, in our opinion, the tested systems represent a considerable evolution of surgical methodology for brain tumour removal but cannot be considered as an alternative to conventional frame-based stereotactic procedures. The indications for frameless and frame-based procedures are different and overlap just in a few particular applications, for example the removal of a cavernoma or drainage from cystic cavities or abscesses [6]. Further evolution of these systems together with the use of intraoperative optical, ultrasound or radiological controls that allow a continuous reset of the accuracy and reformatting of images will greatly improve the surgical quality of results.

Acknowledgements. We thank Philips, Radionics and Zeiss industries for the instrumental support and Mr Masullo and Miss Cadeddu for their precious technical assistance.

References

1. Barnett GH, Kormos DW, Steiner CP, Weisenberg J (1993) Use of a frameless, armless stereotactic wand for brain tumor localization with two-dimensional and three-dimensional neuroimaging. Neurosurgery 33(4):674–678
2. Gomez H, Barnett GH, Estes ML, Palmer J, Magdinec M (1993) Stereotactic and computed-assisted neurosurgery at the Cleveland Clinic: review of 501 consecutive cases. Cleve Clin J Med 60(5):399–410
3. Heilbrun MP, Koehler S, MacDonald P, Siemionow V, Peters W (1994) Preliminary experience using an optimized three-point transformation algorithm for spatial registration of coordinate systems: a method of non invasive localization using frame-based stereotactic guidance systems. J Neurosurg 81(5):676–682
4. Kitchen ND, Lemieux L, Thomas DG (1993) Accuracy in frame-based and frameless stereotaxy. Stereotact Funct Neurosurg 61(4):195–206
5. Koivukangas J, Louhisalmi Y, Alakuijala J, Oikarinen J (1993) Ultrasound-controlled neuronavigator-guided brain surgery. J Neurosurg 79(1):36–42
6. Laborde G, Klimek L, Harders A, Gilsbach J (1993) Frameless stereotactic drainage of intracranial abscesses. Surg Neurol 40(1):16–21
7. Smith KR, Frank KJ, Bucholz RD (1994) The Neuro-station – a highly accurate, minimally invasive solution to frameless stereotactic neurosurgery. Comput Med Imaging Graph 18(4):247–256
8. Tan KK, Grzeszczuk R, Levin DN, Pelizzari CA, Chen GT, Erickson RK, Johnson D, Dohrmann GJ (1993) A frameless stereotactic approach to neurosurgical planning based on retrospective patient–image registration. Technical note. J Neurosurg 79(2):296–303
9. Zamorano LJ, Nolte L, Kadi AM, Jiang Z (1993) Interactive intraoperative localization using an infrared-based system. Neurol Res 15(5):290–298
10. Zinreich SJ, Tebo SA, Long DM, Brem H, Mattox DE, Loury ME, VanDer Kolk CA, Koch WM, Kennedy DW, Bryan RN (1993) Frameless stereotaxic integration of CT imaging data: accuracy and initial application. Radiology 188(3):735–742

Part 7
MIEN: Prospects for the Future

Intraoperative Diagnostic and Interventional MRI in Neurosurgery: First Experience with an "Open MR" System

F. K. Albert, C. R. Wirtz, V. M. Tronnier, J. Hamer, M. M. Bonsanto, A. Staubert, M. Knauth, and S. Kunze

Summary

In December 1995, an "open MRI" unit was installed inside a radiofrequency cabin adjacent to one of the operating theatres at the Department of Neurosurgery, University Hospital, Heidelberg College of Medicine. Since then, 60 open intracerebral procedures have been performed using this scanner for three main applications of intraoperative diagnostic and interventional MRI (IODIM):

1. Resection control of brain tumours
2. On-line guidance for stereotactic interventions (e.g. biopsy, cyst aspiration, catheter placement
3. Acquisition of actual data for the intraoperative update of neuronavigation

Our first results show that IODIM provides an additional support for the neurosurgeon to improve the efficiency of tumour ablation and of targeting deep-seated structures in order to manipulate them under visual control. Furthermore, it is a reliable method of intraoperatively updating the neuronavigational dataset.

Introduction

For the majority of intracranial lesions magnetic resonance imaging (MRI) is superior to other imaging modalities with regard to sensitivity and spatial resolution. This holds true for preoperative diagnosis and treatment planning, and for postoperative control of success. As the missing link, *intraoperative* real-time application of MRI could provide the neurosurgeon with immediate intra-procedural information both to improve the efficiency and to lower the morbidity of the interventions. Until recently, the intraoperative use of MRI during open brain surgery was impossible due to intrinsic technical problems: the inadequate configuration of the scanners and the strict requirements of safe anaesthesiology and sterile surgery. The introduction of so-called open MRI scanners specifically designed to provide wide access to the patient's various body regions when positioned inside the magnetic field has now completely changed this situation, bringing the neurosurgeon closer to the option of using the MRI scanner similarly to an intraoperative fluoroscopic device.

Since December 1995, 60 open intracerebral procedures have been performed using an open MRI scanner at the Department of Neurosurgery, University of Heidelberg. The main applications were the following:

1. Resection control of brain tumours (especially gliomas)
2. Real-time guidance for stereotactic interventions (e. g. biopsy, cyst aspiration, catheter placement)
3. Acquisition of actual data for the intraoperative update of neuronavigation

The conceptional requirements were realized by the installation of a C-shaped "open" magnet inside a radiofrequency cabin just adjacent to one of the neurosurgical operating rooms (OR). This allows operations under standard microneurosurgical conditions with short-term intraoperative diagnostic MRI and continuous imaging for on-line MR-guided interventional procedures [10]. Here we report our first year results with *intraoperative diagnostic and interventional MRI* (IODIM).

Material and Methods

A C-shaped 0.2 Tesla permanent magnet (Magnetom Open, Siemens AG, Erlangen, Germany) providing a lateral patient access of 240 degrees was installed inside a radiofrequency (RF) cabin immediately adjacent to one of the neurosurgical operating rooms (Fig. 1). For intraoperative use the magnet was supplemented by specially designed sterilisable head coils, a moveable patient's couch with an integrated ceramic head holder, a special mechanism for docking the couch, a local LCD monitor, and an optical fibre light source. The patient's couch served both as an adjustable operating table for routine neurosurgical procedures and as an air-cushion-sup-

Fig. 1.
View from the operating
room to the open MRI scanner (*curved arrow*) inside the
adjacent radiofrequency
cabin. The air-cushion-borne
patient's couch is in docking
position, the sliding door
(*arrowheads*) completely
opened

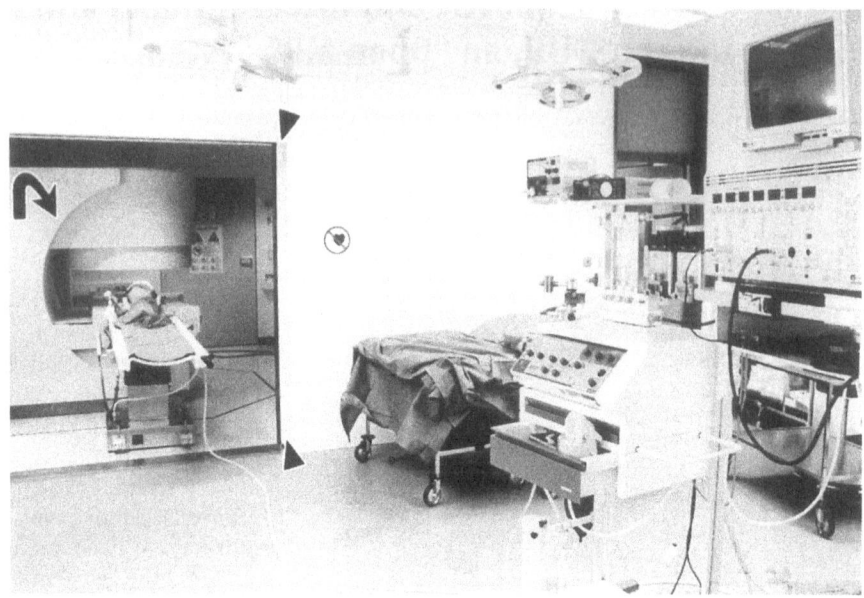

ported vehicle for transporting the patient from the
OR to the magnet (Brandis Company, Weinheim, Germany).

For performing IODIM with the patient under
general anaesthesia, an MR-compatible respirator
(Servo 9000 C, Siemens, Erlangen, Germany) and an
MR-compatible anaesthesia monitor (Maglife, LF
ODAM, Bruker, Wissembourg, France) was used.
MRI-guided *stereotactic* procedures were carried out
either as free-hand on-line controlled manoevres
without head frame, or a ceramic head holder compatible with the Fischer stereotaxy system was
applied combined with a multipurpose head coil.

Resection control cases were done as standard
microneurosurgical procedures outside the RF cabin
without or in combination with one of four different
navigation systems (MKM, Zeiss, Germany; OAS or
OTS, Radionics, USA; SPOCS, Aesculap, Germany;
viewing wand, ISG, Canada). The patient's head was
fixed with a ceramic head holder (Brandis, Weinheim,
Germany) compatible with the Sugita accessories
(Mizuho, Tokyo, Japan). After tumour removal to the
extent that the neurosurgeon had estimated resection
gross total, the patient was transferred to the magnet
under continued anaesthesia with the craniotomy
site still open. For the purpose of safe patient transport a movable table was designed. Its components
consisted of a standard OR table borne by an air
cushion for shock-free transport and of the original
patient's couch of the scanner (Fig. 2). The patients
were both operated on and scanned while remaining
in an unchanged supine or prone position on this
couch. The insertion of the couch for scanning was

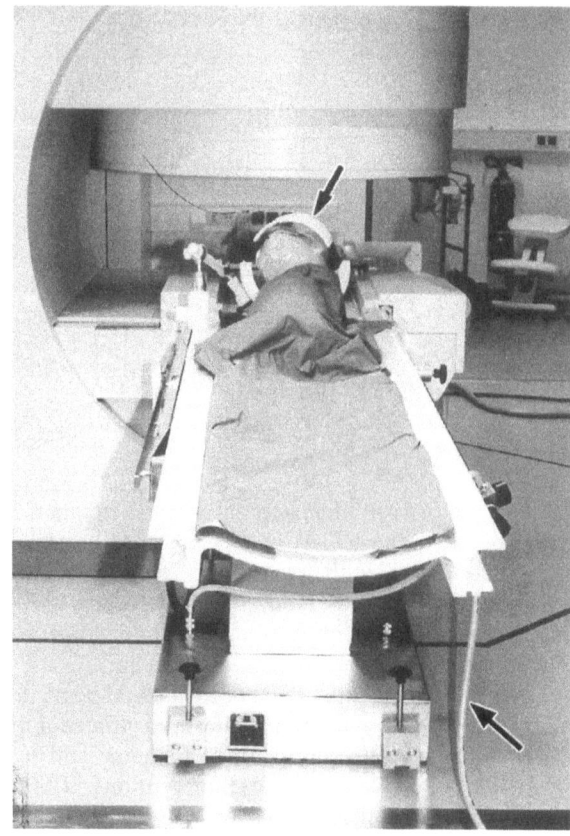

Fig. 2. Movable OR table in docking position before insertion of
the patient's couch into the scanner. *Arrow*, multipurpose head
coil; *arrowhead*, compressed air hosepipe

accomplished by a special docking device at the front of the scanner.

For intraoperative MR resection control (IORC) and for the update of the neuronavigation dataset a standard imaging protocol was used consisting of axial native T1- and T2-weighted series, two sequential axial postcontrast T1-weighted series (0.1 and 0.2 mmol/kg gadolinium-DTPA, Magnevist, Schering AG) and a 3D data set after the second dose of contrast (fast T1, 128 slices, 1.3 mm slice thickness). The latter was applied for navigation update only. For online stereotactic procedures fast sequences (e.g. FISP 2D) were used additionally to T1- or T2-weighted images, displayed on the local monitor inside the RF cabin.

After IORC, the patient was moved back to the OR for closure of the resection site. In the case of residual tumour accessible to further resection, first the updated neuronavigation dataset was transferred to the navigational system via ethernet; then rereferencing was performed based on gadolinium-filled fiducial markers (Howmedica Leibinger Corp., Freiburg, Germany) screwed into the skull prior to intraoperative MR imaging. Final tumour resection was accomplished by navigation guidance, with separate histopathologic examination of the specimen. Prior to the first successful update of interactive navigation in June 1996, the neurosurgeon had to analyse the intraoperative MRI findings by himself and to apply them to the field of operation for removing residual tumour.

Results

From December 1995 to November 1996, 60 IODIM procedures were performed, comprising tumour resection controls, on-line stereotactic biopsies, cyst drainage, and implantation of catheters for brachytherapy. Table 1 shows the numbers of different types of intraoperative diagnostic and interventional procedures done with support of the open MR system. Here we report in particular, on our first results with

Table 1. Different types of diagnostic and interventional intraoperative MRI procedures done with support of the open MR system

Procedure	n
Tumour resection controls	41
On-line stereotactic biopsies	15
Cyst drainage	2
Catheters for brachytherapy	2
Total	60

Table 2. Different types of tumours subjected to intraoperative MRI resection control

Type of tumour	n
High grade glioma	23
Low-grade glioma	11
Pituitary adenoma	3
Metastasis	2
Meningioma (skull base)	1
Radiation-induced necrosis	1
Total	41

Fig. 3.
Left Open MRI scan performed after resection of a recurrent glioblastoma. Residual satellite tumour (*arrow*) inadvertently left behind. This scan was followed by immediate reresection. *Right* Early postoperative MRI on the first day after intraoperatively resection-controlled tumour extirpation, now revealing successful reresection. *Arrow* diffuse enhancement of the parenchyma adjacent to the resection cavity presumably due to persisting gadolinium deposits

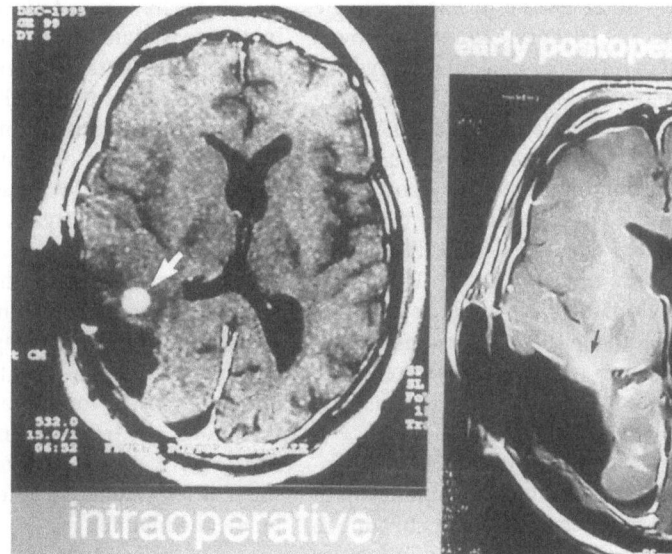

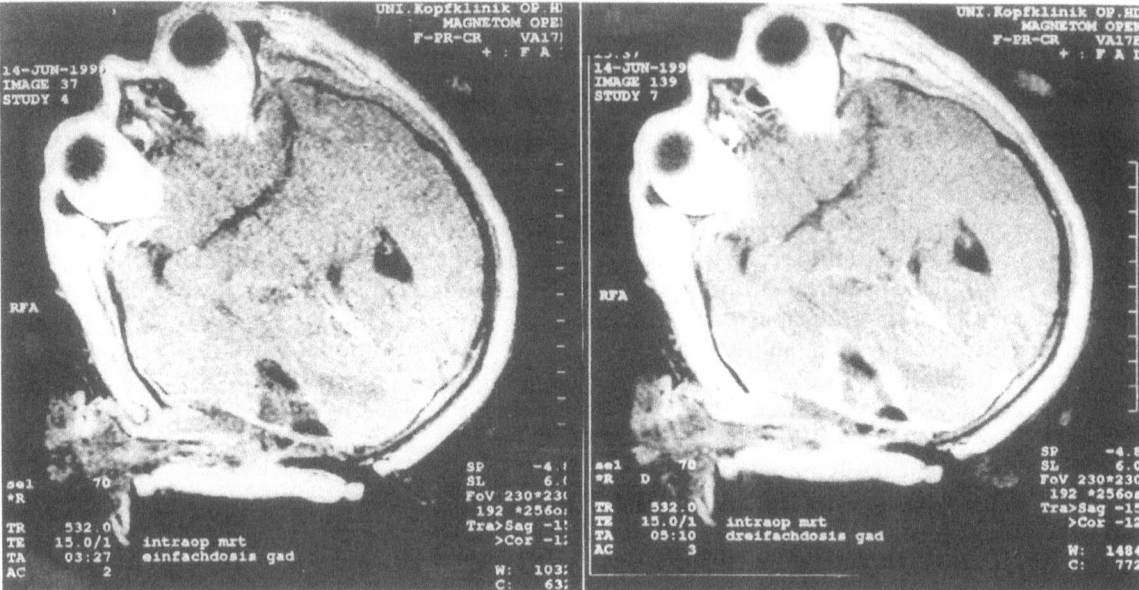

Fig. 4. Intraoperative MRI resection control after glioblastoma extirpation. *Left*, contrast scan immediately after 0.1 mmol/kg gadolinium-DTPA administration, without significant enhancement. *Right*, delayed contrast scan after another 0.2 mmol/kg gadolinium-DTPA was administered showing "nodular enhancement" of the resection margins, highly suspicious of residual tumour. The enhancement fills the outer rims of the initial resection cavity at least partially, suggesting contrast leakage with oozing blood

IORC, which in some cases was followed by immediate extirpation of the MRI-defined area suspected of residual tumour. Most of the resection controls were performed in high-grade and low-grade gliomas (Table 2). Subsequent removal of "residual enhancing tumour" was performed in five cases of glioblastomas (Fig. 3). Tumour was histologically confirmed in three, whereas in two other cases only haemorrhagic materials was found. The latter phenomenon proved to be a critical point of intraoperative imaging, which presumably applies to both MRI and CT. Immediately after tumour resection variable oozing of blood persists at the resection margins carrying contrast medium into the surrounding tissue or depositing it inside blood clots at the surface of the resection wall. This seemingly "nodular enhancement" is then highly suspicious of residual tumour (Fig. 4). This problem was particularly obvious in skull-base tumours such as meningioma and pituitary adenoma. Here, sufficient haemostasis, an indispensible prerequisite to avoid this misleading contrast leakage, is usually much more difficult or even impossible (Fig. 5).

Another interfering factor which we were sometimes confronted with was pronounced enhancement of the adjacent brain parenchyma presumably due to a surgically induced diffuse blood-brain-barrier

(BBB) disruption. Figure 3 (right) shows this phenomenon, which sometimes persisted for at least 24 h as proved by early postoperative MRI the next day. Our explanation is that the gadolinium extravasate had not completely rediffused before the BBB recovered from the surgical trauma and remained then virtually locked out in the parenchyma beyond the restored BBB.

The intraoperative update of the neuronavigation data sets was done in 12 glioma patients operated under image guidance with the MKM system (Zeiss, Germany). Our first experiences were very encouraging. With the gadolinium-filled fiducial markers identified in the intraoperative data set a new registration could be performed. The referencing procedure required only 5–10 min. When we started this completely new method of intraoperative navigation update we had some failures of marker identification in the new data set. Control of this problem was gained with increasing experience and after some modifications of the marker preparation. The mean accuracy for reregistration measured as RMS (root mean square) error was excellent at 0.69 mm. Anatomical check by focusing on bone and the wall of the resection cavity revealed an error of less than 1.5 mm.

On average, for IORC the imaging procedure itself required 58 min. Preparing the patients and transferring them into the magnet and back again took 74 min on average, so that the whole procedure added about 2 h to the total operating time. There was no infectious or traumatising complication resulting from the patient's intraoperative transfer to the magnet or from on-line stereotactic procedures inside the magnet.

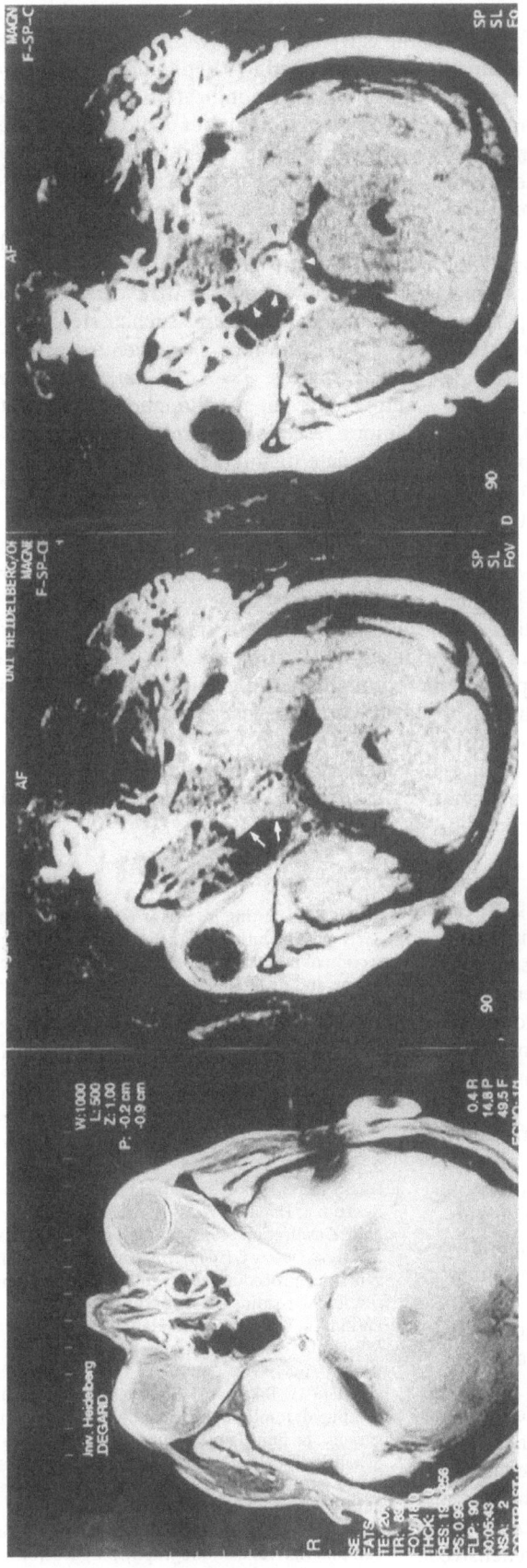

Fig. 5. Pronounced leakage of gadolinium after removal of a recurrent, preirradiated meningioma of the left cavernous sinus and sphenoid cavity. *Left* Preoperative scan. *Centre* Intraoperative resection control after complete tumour removal from the cavernous sinus and the sphenoid cavity (precontrast). The former tumour site is filled with isointense material (*arrows*). *Right* With gadolinium the intraoperative MRI suggests an enhancing tumour in almost the same extent as preoperatively (*arrowheads*), whereas on the next day, the early postoperative MRI confirmed that the tumour had actually been completely removed. (For radiotherapy reasons the amaurotic left eye was also removed)

Conclusion

Our first results show that IODIM provides a promising support for the neurosurgeon to improve the efficiency of tumour ablation and of targeting deep-seated structures in order to manipulate them under visual control. Furthermore, it is a reliable method of intraoperatively updating the neuronavigational data set.

First of all, however, a lot of technical problems had to be solved prior to starting the system, such as the MRI compatibility of the multipurpose patient's couch serving as an OR table, to transport the patient to the open MRI unit and as examination table inside this unit. Other problems included the MRI compatibility of the head fixation device, of the vital function monitoring system, of the respirator, of instruments and of wound retractors used in stereotactic interventions and in open surgery (e.g. for pituitary adenomas).

Another significant source of difficulties which we were confronted with were various new imaging phenomena, obviously related to the immediate surgical trauma, which seriously impeded scan interpretation or even rendered interpretation impossible. These were, for example, the leakage of gadolinium at the resection site secondary to the capillary oozing of blood during examination, or a widespread enhancement encompassing the adjacent brain parenchyma as a result of the temporary blood-brain-barrier disruption.

Both phenomena may either mimic residual tumour or disguise real tumour remnants. These ultraearly enhancement phenomena are unpredictable in appearance and, in our experience, usually do not occur in early postoperative MRI performed 24–48 h after operation [1, 2]. For avoidance, completely new techniques of contrast enhancement are conceivable, such as the application of contrast media prior to the operation with selective uptake by the tumour cells and prolonged activity for many hours or even some days. Particularly in benign skull-base tumours, this might be the only way to gain reliable findings by IORC. One representative of such substances that obviate the necessity of intraoperative application has already been described experimentally [12].

Most of our IORC cases were performed in patients with glioblastomas. This was primarily for two reasons. First, glioblastomas usually show a definite uptake of contrast, which roughly delineates the solid parts of the tumours. Second, since 1989 early postoperative MRI for evaluating the extent of resection has been performed in all glioblastoma patients at our institution in a routine and study-controlled manner, providing us with a comprehensive

knowledge of the various imaging phenomena which the early postoperative situs presents [1, 2]. The fact that we applied IORC to glioblastomas should not be interpreted as an indication that we believe that total resection in this tumour could ever be a realistic goal. Although, in our experience, a certain subgroup of glioblastomas seem to exist in which the enhancing part in MRI represents the biologically decisive compartment responsible for the velocity of postoperative tumour progression.

Up to now, intraoperative MR imaging for on-line guidance in procedures such as stereotactic biopsies or cyst aspirations has been reported only in a few instances [4–6, 8, 10]. Shalit et al. [9] and Okudera et al. [7] applied intraoperative CT for tumour resection control. For this purpose, and for the update of interactive navigation, intraoperative MRI has not previously been reported. There is no doubt that in the realm of high technology [3] IODIM represents a completely new dimension in neurosurgical practice. However, observing the realities of economic requirements and the existence of competing methods of intraoperative procedure guidance [11], IODIM has to be subjected not only to thorough scientific and clinical evaluations, but also to a reasonable cost-effectiveness analysis.

Acknowledgment. Parts of this study were supported by Deutsche Krebshilfe Project 70-1883-Al 3.

References

1. Albert FK, Forsting M, Sartor K, Adams HP, Kunze S (1994) Early postoperative magnetic resonance imaging after resection of malignant glioma: objective evaluation of residual tumor, and its influence on regrowth and prognosis. Neurosurgery 34: 45–61
2. Albert FK, Zenner D, Forsting M (1994) Nutzen des radiologischen Monitoring nach Gliomexstirpation. Klin Neuroradiol 4: 203–219
3. Apuzzo ML (1996) The Richard C. Schneider lectures: new dimensions of neurosurgery in the realm of high technology: possibilities, practicalities, realities. Neurosurgery 38: 625–639
4. Duckwiler G, Lufkin RB, Teresi L, Spickler E, Dion J, Vinuela F, Bentson J, Hanafee W (1989) Head and neck lesions: MR-guided aspiration biopsy. Radiology 170: 519–522
5. Lenz GW, Dewey C (1995) An open MRI system used for interventional procedures: current research and initial clinical results. Proceedings CAR, Berlin, 1180–1187
6. Lufkin RB, Teresi L, Chiu L, Hanafee W (1988) A technique for MR-guided needle placement. AJR Am J Roentgenol 151: 193–196
7. Okudera H, Takemae T, Kobayashi K (1993) intraoperative computed tomographic scanning during transsphenoidal surgery: technical note. Neurosurgery 32: 1041–1043
8. Schenck JF, Jolesz FA, Roemer PB, et al. (1995) Superconduction open-configuration MR imaging system for image-guided therapy. Radiology 195: 805–814
9. Shalit MN, Israeli Y, Matz S, Cohen ML (1979) Intra-operative computerized axial tomography. Surg Neurol 11: 382–384

10. Tronnier VM, Wirtz CR, Knauth M, Lenz G, Pastyr O, Bonsanto MM, Albert FK, Kuth R, Staubert A, Schlegel W, Sartor K, Kunze S (1997) Intraoperative diagnostic and interventional MRI in neurosurgery. Neurosurgery 40 : 891 – 902

11. Woydt M, Krone A, Becker G, Schmidt, K, Roggendorf W, Roosen K (1996) Correlation of intra-operative ultrasound with histopathologic findings after tumour resection in supratentorial gliomas. A method to improve gross total resection. Acta Neurochir (Wien) 138 : 1391 – 1398

12. Zimmer C, Weissleder R, Poss K, Bogdanova A, Wright SC, Enochs WS (1995) MR imaging of phagocytosis in experimental gliomas. Radiology 197 : 533 – 538

Functional Magnetic Resonance Imaging in Clinical Neuroscience: Current State and Future Prospects

D. Auer and L. M. Auer

Summary

F-MRI in recent years has been shown to be an adequate method to study brain activation and carries a number of advantages over PET, especially in terms of spatial and temporal resolution. Even though unsolved methodiological issues exist, valuable clinical application has been demonstrated. Thus, a greater clinical impact than that seen with PET is predicted, mainly because of lack of invasiveness. Hence, repeatability and availability should be possible for a much larger patient group considering the number of MR installations worldwide. Its main applications today consist in visualization of motor and language areas as part of the preoperative evaluation of patients with brain tumors. More complex activation paradigms are under evaluation and may prove helpful to understand the role of those brain regions previously thought to be nonloquent. Other perspectives can be foreseen in the objective assessment of neurological deficits in uncooperative patients pre- and postoperatively and in studies of functional reorganization and possibly plasticity in destructive brain lesions. Finally, the question arises of whether the expected exponential data accumulation from human brain mapping studies can be reintegrated to provide a meaningful understanding of the complexity of human brain function.

Introduction

During the last decade magnetic resonance imaging (MRI) has substantially improved in vivo morphological studies allowing high-resolution three-dimensional display of brain structures and pathologies. Attempts to introduce a 'functional' dimension to MRI previously focused on metabolic studies such as proton or phosphorus magnetic spectroscopy. This approach was, however, limited in its applicability by the spatial as well as temporal resolution. True functional MRI (f-MRI) was first reported by Belliveau in 1991, who produced a map of the human visual cortex derived from dynamic contrast-enhanced MR images [5]. An even more elegant approach to study brain activation is based on oxyhemoglobin as an endogenous contrast, the so-called BOLD (blood oxygenation level dependent) technique, introduced by Ogawa [25, 26]. This technique triggered an enormous interest in brain activation studies with f-MRI during the last few years (for reviews see [17, 22]). The larger number of adequate MR scanners compared to PET scanners and the noninvasiveness of the new method raises the expectation of a much wider application of functional brain imaging in the near future not only in cognitive neuroscience but also as a routine diagnostic tool for neurological and neurosurgical patients. Therefore, the currently available techniques and their possibilities for clinical applications are presented and discussed in this review.

F-MRI: Methodological Aspects

BOLD Contrast: Physiology and MR Technique

Upon neuronal activation, oxygen consumption and blood flow and volume locally increase to meet the higher metabolic demand. The extent and time course of changes in these parameters, however, differ, e.g. a 30–50% rise in blood flow can be accompanied by only a 5% increase in oxygen metabolic rate [12]. These relative differences lead to a net increase in capillary and venous blood oxygenation, which can be measured by a special MR method, the so-called BOLD technique [18, 25, 26]. BOLD contrast is generated by susceptibility or $T2^*$-sensitive sequences: due to the paramagnetic property of deoxyhemoglobin (DeoxHb) as opposed to diamagnetic oxyhemoglobin (OxHb), there is a reciprocal relationship between concentration of DeoxHb and observed signal intensity. A local signal increase of BOLD-fMRI in response to a neuronal stimulus, therefore, reflects a rise in the OxHb to DeoxHb ratio and thus represents a noninvasive, yet indirect measure of neuronal activity. This simplified hypothesis of the BOLD mechanism in fMRI still requires confirmation. Quantitative comparisons with physiological parameters as well as characterization of a probably three-phasic time course of the BOLD response are currently under investigation [11, 18, 23, 26, 29].

To achieve sufficient BOLD sensitivity at clinical field strengths, gradient-recalled echoes with long echo times are used with the conventional fast low-angle shot (FLASH) or echoplanar technique (EPI). There is still some controversy about the respective advantages of both methods. EPI enjoys a wider acceptance due to its multislice capability, the reduction in total acquisition time, and lower sensitivity to effects from large vessels [13, 19, 29].

How to Generate Functional Maps

A typical f-MRI experiment consists of several repetitions of stimulation and activation periods that produce synchronous signal fluctuations locally, i.e., increases during stimulation, decreases during rest (Fig. 1a). Echoplanar sequences allow the study of multiple slices with up to 100 images in a few minutes with a typical spatial resolution of 2–4 mm. To calculate functional maps, a segmentation of activated from nonactivated areas is performed on the basis of the time course of signal intensity, applying various statistical methods. Commonly, split t-test, cross-correlation to an ideal reference waveform, or Fourier analysis [4] are used to pixelwise map statistical

Fig. 1. a Normalized and pooled signal intensity/time profiles from activated regions in normal volunteers repetitively stimulated with a reversing checkerboard. **b** Activation map in a normal volunteer: superimposition of anatomical image with thresholded cross-correlation map (> 0.55). **c** Activation map (as in **b**) in a patient with ruptured parieto-occipital angioma and clinical subtotal right hemianopia

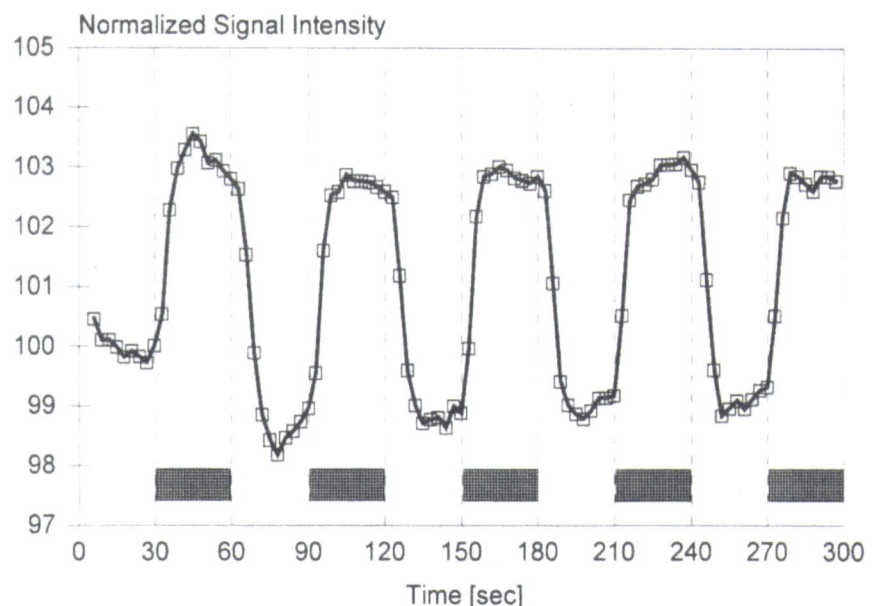

values. The final step is to define an appropriate cut-off and superimpose only thresholded pixels on a high-resolution anatomical MRI; this results in the functional map. Some groups are currently evaluating advanced parametric methods in order to increase statistical power and replace arbitrary thresholds by statistically more meaningful p levels [15, 20]. Since motion artifacts severely affect f-MRI, coregistration of individual images before segmentation is strongly recommended.

Brain Activation Studies: Activation Paradigms

The first tasks explored with f-MRI were visual stimuli and motor activation: delineation of the visual cortex was achieved with light-emitting diodes or reversing flashing checkerboard stimuli presented by use of a projector and mirror devices and alternated with darkness. Contralateral primary and supplementary motor cortices were highlighted by alternation of simple finger tapping tasks with rest periods. Initially, f-MRI studies were designed to reproduce PET data and prove the feasibility of visualizing known functional neuroanatomy. Since these first investigations, a multitude of paradigms have been tested to confirm expected brain activation especially within primary sensory areas (visual, acoustic, sensory cortices), cortical motor areas including cerebellar regions, and classical language areas [17]. In addition, refined technique and careful design of f-MRI experiments enabled detection of differential activation in response to specific stimulus characteristics and cognitive aspects. As a result, complex issues in neuroscience are now amenable to f-MRI. From standpoint of clinical use, the results achieved can be summarized as follows: eloquent brain areas including those with known cognitive function, e. g., internal speech production, [6,9,10,16] have shown consistent activation among subjects and even 'noneloquent' areas in frontal (working memory, mental calculation [7]) and parietal lobes (mental rotation [8]) could be highlighted with carefully designed tasks.

However, some limitations should be considered. Firstly, investigation of overt speech production and motor activities involving the head is limited by motion artifacts. Secondly, control of task performance within the scanner is very restricted since ferromagnetic devices to monitor activity or reaction times cannot be brought into the magnet.

Quantitative analysis of f-MRI results, usually expressed as mean percent signal increase, also need to be cautiously interpreted, especially in the context of patient studies. It is well known that brain activation varies with stimulus characteristics such as intensity or frequency of a sensory stimulus, movement rate, or complexity of mental tasks. However, there is a number of less well studied factors possibly modulating a particular BOLD response. In a preliminary study we have shown that the extent of brain activation is inversely related to motor skill and practice as musicians showed less activation than untrained controls [2]. Thus, differences in premorbid skill and actual performance between, e. g., hemiparetic patients and controls has to be controlled for in terms of movement rate as well as degree of automatization. In addition, for a given task mean percent signal increases have been found to vary not only between individuals, but also intra-individually even in one session. These inconsistencies may be partially explained by variations in mental background activity. Unexpectedly, we found a significantly decreased BOLD response to a standard checkerboard stimulation paradigm when mental calculation was performed in addition to the visual stimulus [3]. Finally, conclusive results of studies on possible masking effects of functional responses in old age, arteriosclerosis, edematous tissue, and centrally acting drugs are missing.

Application in Clinical Neurosciences

Clearly, physiological studies of hemodynamaic coupling and methodological improvements in terms of data acquisition and data analysis are still needed to fully and reliably exploit functional MR data. First clinical studies with f-MRI have, however, shown promising results, and thus main topics for current and future applications are briefly discussed here with an emphasis on neurosurgical patients.

The feasibility to locate sensory and motor brain function allows a better study of neurological deficits (Fig. 1 c). A major advantage is to be expected for tests that rely on subjective judgements, such as perimetric [24] and sensory testing. Also advantageous is the potential to explore deficits in unattentive and possibly also comatous patients. Since there is no restriction on repetitive f-MRI studies, investigation of dynamic changes during recovery of brain function will certainly be useful to explore mechanisms of reorganization and plasticity of the brain. Finally, efforts to analyze complex activation patterns and their disturbances are being made in order to reach a better understanding of psychotic phemomena thought to be related to disordered brain connectivity. The major clinical interest, however, has been focused so far on preoperative brain mapping to covisualize brain lesions and brain function.

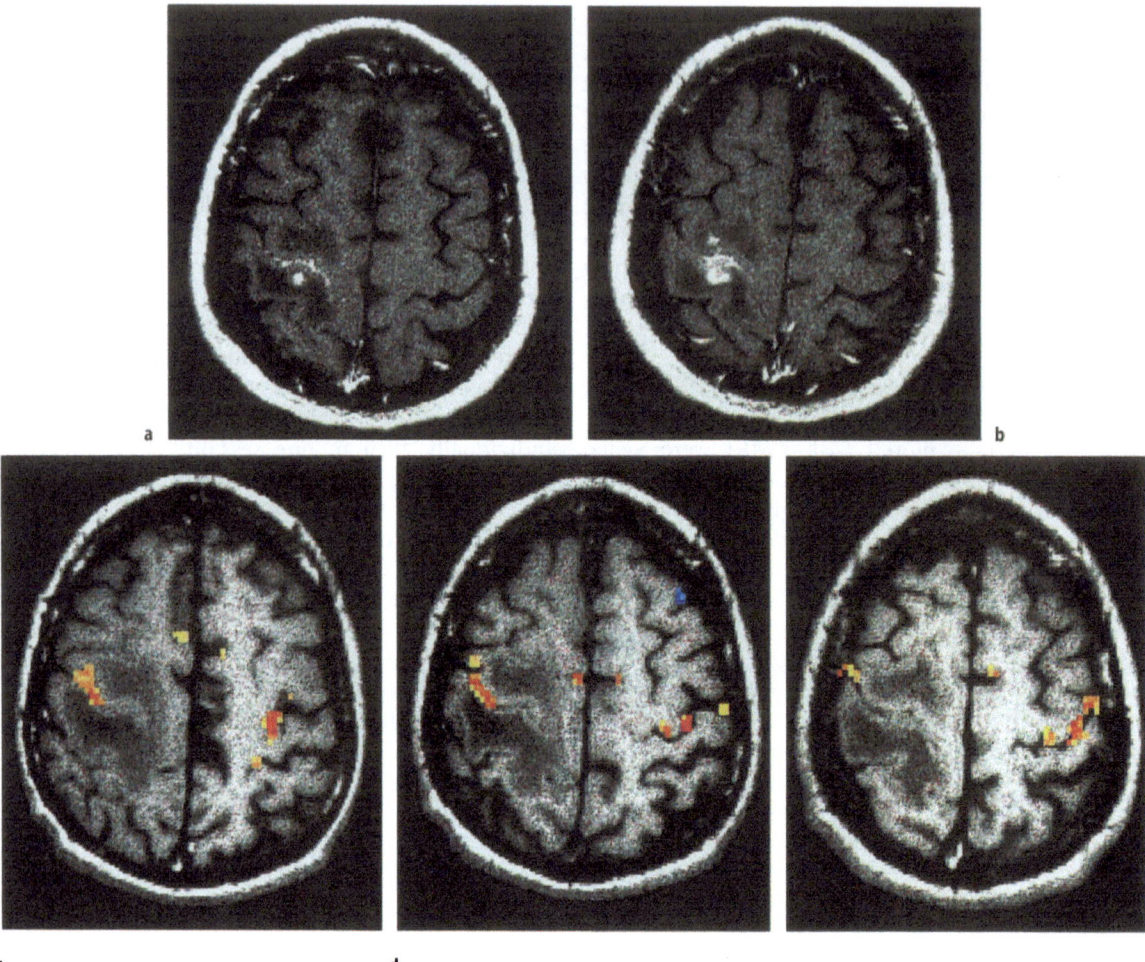

Fig. 2. a, b Contrast-enhanced MRI from patient with biopsy-proven lymphoma involving the central sulcus. **c–e** Consecutive activation maps of a motor task (opening and closing both fists) using relaxed thresholds (CC > 0.4) to reduce false negatives. On affected side, the motor hand area is displaced laterally and downwards compared with the normal side; no activation is detected within the tumor region; biopsy was safely performed

Preoperative Brain Mapping

Visualization of the primary motor cortex (M1) and supplementary motor area (SMA) could reliably be achieved in a number of volunteer studies using simple finger tapping paradigms. Surgical planning in frontal tumors of course requires delineation of the primary motor cortex M1, which may be difficult on conventional tomography if gross anatomical displacement has occurred. Preliminary studies [1, 14, 27, 28, 30] have demonstrated the feasibility of visualizing M1 (Fig. 2) and striate cortex preoperatively in the vicinity of tumors and AVMs [21] with qualitatively satisfactory results showing local topography and displacement. However, with respect to de-fining a reliable nontouch area, activation maps still suffer from a number of limitations: (a) the size of activated areas highly depends on the chosen statistical threshold; (b) geometrical distortion occurs in f-MRI, especially in EPI; (c) contamination from veins cannot be neglected, especially with FLASH; (d) the study of selected functions (e. g., finger movements) shows only parts of the interesting region; and (e) probably most importantly, surgical safety distance to activated area is unclear due to unpredictable local neuronal connections. The aforementioned limitations make clear that current techniques for preoperative brain mapping require careful interpretation that should consider spatial uncertainties. Future methodological developments and a thorough evaluation of large studies are still needed.

Replacement of WADA Test in Epilepsy Surgery?

F-MRI proved capable of defining the dominant hemisphere for language skills with a high correlation to WADA testing in two series [6, 10]. Thus,

Fig. 3 a, b.
Memory and language tasks
in f-MRI. **a** Word association
task highlights left anterior
and to a lesser extent left
posterior language area indicative of left hemispherical
dominance (due to visual
imagery some occipital
activation is also seen).
b Activation map of delayed
recall of word list showing
bilateral hippocampal activation and some temporopolar
activity on the left

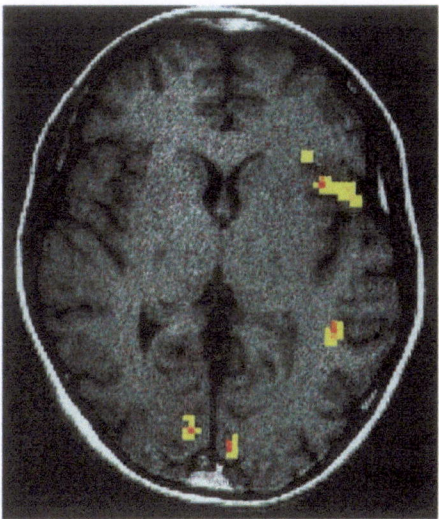

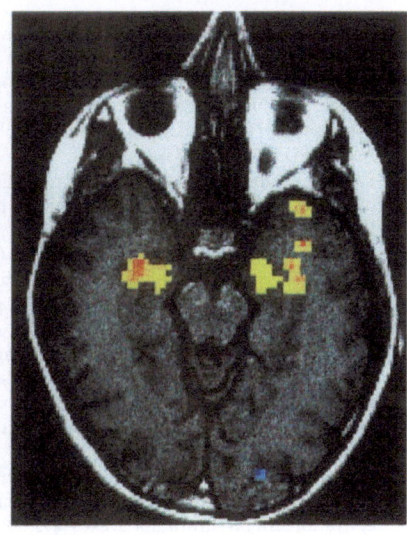

a

b

f-MRI favorably compares with the WADA test, first of all because of the total noninvasiveness of f-MRI and secondly due to the perspective of simply adding functional assessment to routinely performed MR scans of epilepsy patients (Fig. 3 a). Theoretically, f-MRI should be much better than the non-super-selective WADA test for memory lateralization by avoiding possible pitfalls due to variation of anterior and posterior choroidal vascular supply of the hippo-campi. Experimentally, however, paradigm design for hippocampal activation proved difficult and especial-ly lateralization of memory function in relation to the task content has not been thoroughly studied. Initial promising results were obtained using delayed recall of verbal material (word lists) as shown in Fig. 3 b).

Future Application of F-MRI in neurosurgical Patients

Two other areas of f-MRI in neurosurgical patients offer new and intriguing aspects for clinical neuro-science, i. e., the possibility to study the function of so-called noneloquent brain areas. Firstly, routine neurosurgical approaches through these areas may need to be reconsidered if functional activation can reliably be found by complex stimulation paradigms. Secondly, brain mapping of slowly growing tumors and large arachnoid cysts offers the opportunity to document possible plastic changes if function can be shown to be more extensively shifted than can be explained by displacement alone.

References

1. Atlas SW, Howard RS, Maldjian J, Alsop D, Detre JA, Listerud J, D'Esposito M, Judy KD, Zager E, Stecker M (1996) Func-tional magnetic resonance imaging of regional brain activity in patients with intracerebral gliomas: findings and impli-cations for clinical management. Neurosurgery 38 : 329 – 338
2. Auer D, Kraft E, Rupprecht R, Jones RA (1996) Does motor skill influence brain activation in musicians: an f-MRI study. Proceedings of the human brain mapping conference
3. Auer D, Elbel GK, Jones RA (1997) Modulation of BOLD response due to interfering task. Proceedings of the human brain mapping conference, p 94
4. Bandettini PA, Jesmanowicz A, Wong EC, Hydes JS (1993) Processing strategies for time-course data sets in functional MRI of the human brain. Magn Reson Med 30 : 161 – 173
5. Belliveau JW, Kennedy DN, McKinstry RC et al. (1991) Func-tional topographical mapping of the human visual by magnetic resonance imaging. Science 254 : 716 – 719
6. Binder JR, Swanson SJ, Hammeke TA, Morris GL, Mueller WM, Fischer M, Benbadis S, Frost JA, Rao SM, Haughton VM (1996) Determination of language dominance using func-tional MRI: a comparison with the Wada test. Neurology 46 : 978 – 984
7. Burbaud P, Degreze P, Lafon P, Franconi JM, Bouligand B, Bioulac B, Caille JM, Allard M (1995) Lateralization of pre-frontal activation during internal mental calculation: a func-tional magnetic resonance imaging study. J Neurophys 74 : 2194 – 2200
8. Cohen MS, Kosslyn SM, Breiter HC, DiGirolamo GJ, Thomp-son WL, Anderson AK, Brookheimer SY, Rosen BR, Belliveau JW (1996) Changes in cortical activity during mental rotation. A mapping study using functional MRI. Brain 119 : 89 – 100
9. Cuenod CA, Bookheimer SY, Hertz-Pannier L, Zeffiro TA, Theodore WH, Le Bihan D (1995) Functional MRI during word generation, using conventional equipment: a potential tool for language localization in the clinical environment. Neurology 45 : 1821 – 1827
10. Desmond JE, Sum JM, Wagner AD, Demb JB, Shear PK, Glover GH, Gabrieli JDE, Morell MJ (1995) Functional MRI measure-ment of language lateralization in Wada-tested patients. Brain 118 : 1411 – 1419

11. Ernst T, Hennig J (1994) Observation of a fast response in functional MR. Magn Reson Med 32:146–149
12. Fox PT, Raichle ME (1986) Focal physiological uncoupling of cerebral blood flow and oxidative metabolism during somatosensory stimulation in human subjects. Proc Natl Acad Sci USA 83:1140–1144
13. Frahm J, Merboldt K-D, Hanicke W (1993) Functional MRI of human brain activation at high spatial resolution. Magn Reson Med 29:139–144
14. Fried I, Nenov VI, Ojemann SG, Woods RP (1995) Functional MR and PET imaging of rolandic and visual cortices for neurosurgical planning. J Neurosurg 83:854–861
15. Friston KJ, Jezzard P and Turner R (1994) Analysis of functional MRI time-series. Hum Brain Mapping 1:153–171
16. Hinke RM, Hu X, Stillman AE, Kim SG, Merkle H, Salmi R, Ugurbil K (1993) Functional magnetic resonance imaging of Broca's area during internal speech. Neuroreport 4:675–678
17. Kollias SS, Valavanis A, Golay XG, Bösiger P, McKinnon G (1996) Functional magnetic resonance imaging of cortical activation. Int J Neuroradiol 2:450–472
18. Kwong KK, Belliveau JW, Chesler DA, Goldberg IE, Weisskoff RM, Poncelet BP, Kennedy DN, Hoppel BE, Cohen MS, Turner R, Cheng HM, Brady TJ, Rosen BR (1992) Dynamic magnetic resonance imaging of human brain activity during primary sensory stimulation. Proc Natl Acad Sci USA 89:675–679
19. Lai S, Hopkins AL, Haacke EM, Li D, Wasserman BA, Buckley P, Friedman L, Meltzer H, Hedera P, Friedland R (1993) Identification of vascular structures as a major source of signal contrast in high resolution 2D and 3D functional MIT using gradient echoes at 1.5 T. NMR Biomed 7:83–88
20. Lange N (1996) Tutorials in biostatistics. Statistical approaches to human brain mapping by functional magnetic resonance imaging. Stat Med 15:389–428
21. Latchaw RE, Hu X, Ugurbil K, Hall WA, Madison MT, Heros RC (1996) Functional magnetic resonance imaging as a management tool for cerebral arteriovenous malformations. Neurosurgery 37:619–625
22. Le Bihan D, Karni A (1995) Applications of magnetic resonance imaging to the study of human brain function. Curr Opin Neurobiol 5:231–7
23. Menon RS, Ogawa S, Hu X, Strupp JP, Anderson P, Ugurbil K (1995) BOLD based functional MRI at 4 Tesla includes a capillary bed contribution: echo-planar imaging correlates with previous optical imaging using intrinsic signals. Magn Reson Med 33:453–459
24. Miki A, Nakajima T, Fujita M, Takagi M, Abe H (1996) Functional magnetic resonance imaging in homonymous hemianopsia. Am J Ophthal 121:258–266
25. Ogawa S, Lee TM, Kay AR, Tank DW (1990) Brain magnetic resonance imaging with contrast dependent on blood oxygenation. Proc Natl Acad Sci USA 87:9869–9872
26. Ogawa S, Tank DW, Menon R, Ellermann JM, Kim SG, Merkle H, Ugurbil K (1992) Intrinsic signal changes accompanying sensory stimulation: functional brain mapping with magnetic resonance imaging. Proc Natl Acad Sci USA 89:5951–5955
27. Pujol J, Conesa G, Deus J, Vendrell P, Isamat F, Zannoli G, Marti-Vilalta JL, Capdevila A (1996) Presurgical identification of the primary sensorimotor cortex by functional magnetic resonance imaging. J Neurosurg 84:7–13
28. Righini AR, de Divitiis O, Prinster A, Spagnoli D, Appollonio I, Bello L, Scifo P, Tomei G, Villani R, Fazio F, Leonardi M (1996) Functional MRI: primary motor cortex localization in patients with brain tumors. J Comput Assist Tomogr 20:702–708
29. Turner R, Jezzard P, Wen H, Kwong KK, Le Bihan DT, Zeffiro R, Balaban S (1993) Functional mapping of the human visual cortex at 4 and 1.5 T using deoxygenation contrast EPI. Magn Reson Med 29:277–279
30. Yousry TA, Schmid UD, Jassoy AG, Schmidt D, Eisner WE, Reulen HJ, Reiser MF (1995) Topography of the cortical motor hand area: prospective study with functional MR imaging and direct motor mapping at surgery. Radiology 195:23–29

Robots for Neurosurgery?

L. M. Auer

Summary

Previous concerns about the use of robots in surgery do not seem justified according to present experience. Computer technology and methods for image processing are now developing at such a rate that active manipulator arms with robotic function can be considered for neurosurgery of the next generation. The major benefit from this development will be the possibility to operate upon a greater variety of intracranial mass lesions via miniaturized approaches with the consequence of a shorter stay in intensive care and in hospital. Direct and indirect intraoperative 3-D visual control of solid structures will increase safety and reduce complications. The surgeon's capabilities will shift more and more from direct manual action to intelligent planning and development of minimal-risk strategies. Early validation of low-cost solutions instead of ultra-high-tech super-expensive equipment will facilitate wide application and acceptance.

Introduction and Historical Remarks

History of the "automaton"

Ever since the first description of an automatically functioning device as human creation, namely the statues of Venus and Mars by Hero of Alexandria 100 BC, automatic machines have equally fascinated and frightened mankind. Until the late 18th century, inventors were repeatedly accused of sorcery. Some of the most famous robotic puppets in history are the cock of Strasbourg from 1354, the mechanical puppet of Schissler from about 1500, and Baron von Kempelen's chess-playing turk, which was shown around all courts of Europe in the 18th century (this "puppet" never lost chess games and eventually disappeared; the true nature of the phenomenon has thus never been verified). Even in very recent years, similar concerns and anxieties have been expressed over robots as was the case with computers.

History and Development of Robotics

The term "robot" is a creation of the Czech novelist Karel Capek from his 1923 novel R.U.R. – Rossum's Universal Robots; the term "roboti" means slave work. The term "robotics" is a creation of Asimov

Table 1. Neurosurgical robots and robotic applications

Types of robots (company)	Authors	Application
PUMA 200/mark II (Unimation)	Young 1987 [36]	CT-guided stereotaxy
	Doll et al. 1987 [9]	CT-guided stereotaxy
	Drake et al. 1991 [10]	CT-guided microsurgery
AID universal (IMMI)	Benabid et al. 1987 [6]	CT-guided stereotaxy
Austrian frame	Herza and coworkers, unpublished work	US-guided endomicrosurgery
Unnamed	Kelly 1986 [21]	CT-guided microsurgery
ET-01	Jacobi et al. 1990 [17]	Microsurgery
Saarland project	Diewald and coworkers, unpublished work	US-guided endomicrosurgery
MINERVA-Lausanne	Glauser et al. 1989 [12]	CT-guided stereotaxy
Neurobot	Finlay 1993 [11]	CT/MR-guided microsurgery
Tokyo project	Masamune et al. 1995 [24]	MR-guided stereotaxy
Roboscope /LARS	Goradia et al. 1996 [14]	US-guided microsurgery

Table 2.
Nonneurosurgical medical robots

Type of robot (company)	Authors	Application
ROBODOC	Paul et al. 1992 [26]	Orthopedic surgery
LARS	Taylor et al. 1992 [31]	Abdominal surgery
AESOP	Sackier and Wang 1994 [a]	Abdominal surgery
SCARA (IBM 75769	Kavanagh 1994 [20]	Otolaryngology
(W)ELSA, PROBOT	Harris et al. 1995 [17]	Prostatectomy
TISKA	Schurr et al. 1996 [30]	Abdominal surgery

[a] See also product information AESOP 1000, Computer Motion Goleta, CA, 1994.

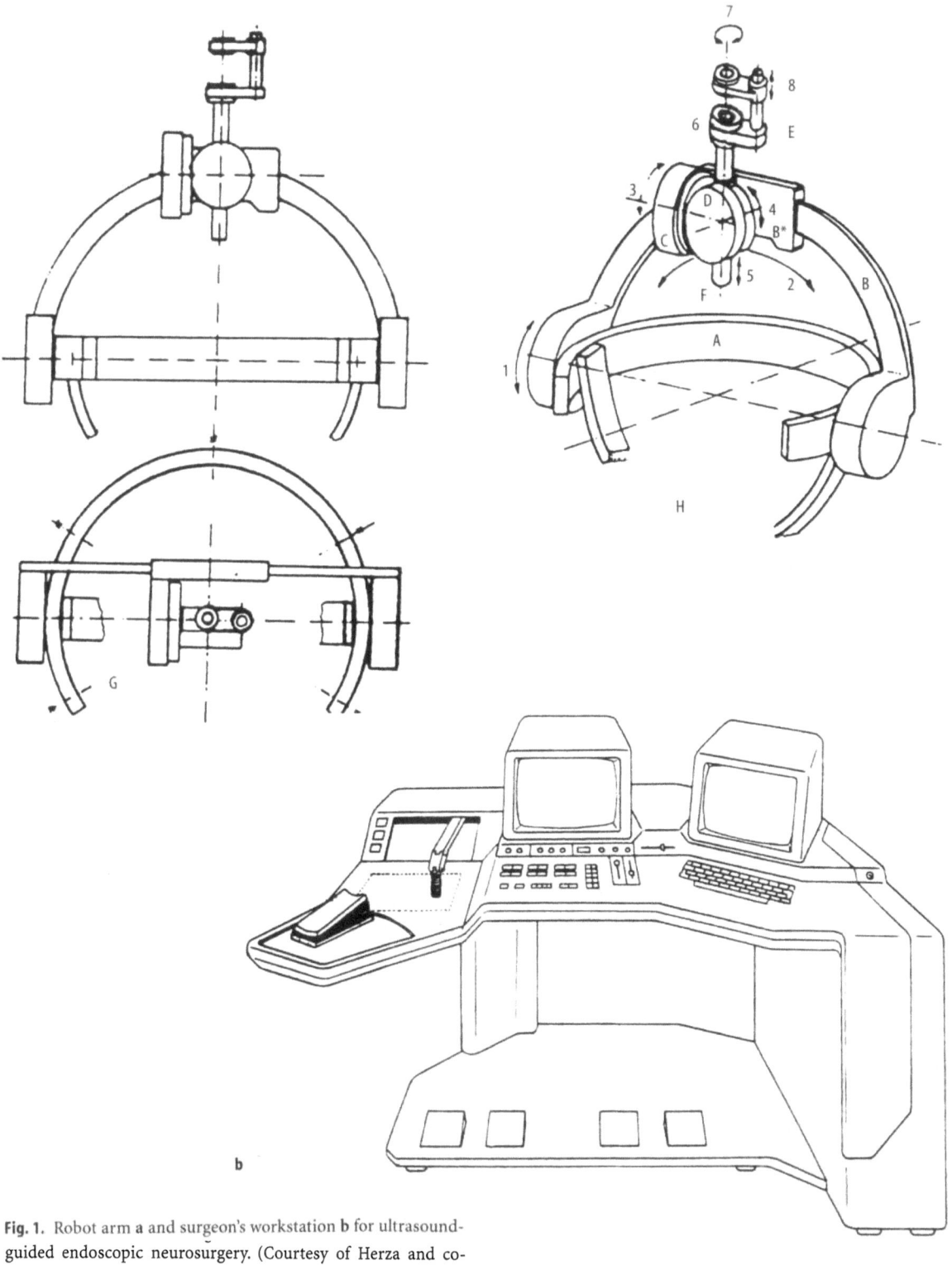

Fig. 1. Robot arm **a** and surgeon's workstation **b** for ultrasound-guided endoscopic neurosurgery. (Courtesy of Herza and co-

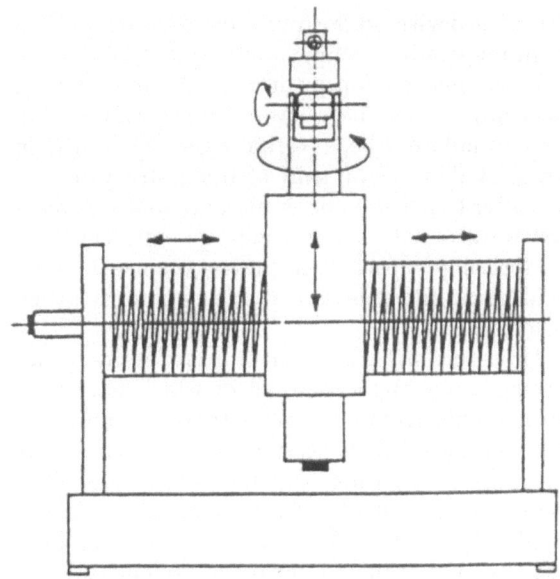
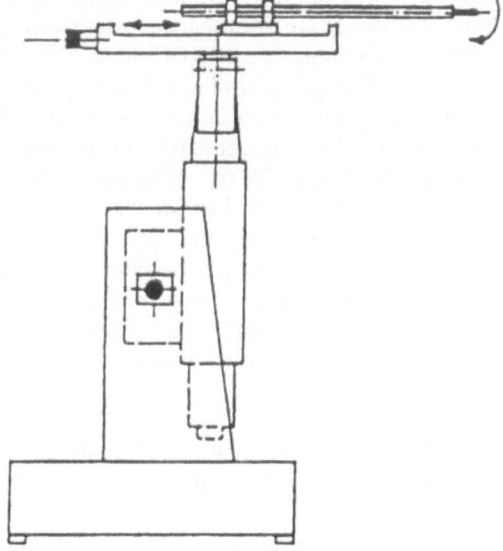

Fig. 2. Design of a fuzzy-logic-driven robot arm for endoscopic and other miniaturized microsurgical procedures. (Courtesy of Diewald and coworkers, University of Saarbrücken)

from 1950. The first industrial robot was designed by Davol in 1961. Since then, robotics have, together with electronics, induced another industrial revolution and the start of deep social changes. Twenty-four more years passed until Kwoh was the first to use a Puma 200 industrial robot for stereotaxy in 1985 [22].

As shown in Tables 1 and 2, several different robotic designs have been planned, developed, and also applied since then for both neurosurgical purposes and other surgical disciplines. In neurosurgery, these first applications have almost exclusively been in stereotaxic procedures [5, 6, 9, 12, 36]. However, some systems have also been used to focus instruments or instrument holders on predetermined targets [10, 11, 13, 17, 21] and noninvasive manipulators have been used, for example, to position an operating microscope. A major disadvantage of all these models has been that intraoperative imaging was not available to correct for brain movements during the operation; the device was still dependent on the spatial information from preoperative images.

Our first project in 1986 included an intraoperative ultrasound imaging system, based on experience with ultrasound-guided endoscopic and other minimally invasive procedures [1, 2], to guide a stereotaxy frame-based robot arm to intracranial targets for endoscopic procedures Herza and coworkers, unpublished work (Fig. 1). A few years later, fuzzy-logic control was the background of a new robot design for endoneurosurgery, again using ultrasound imaging

as intraoperative real-time imaging for actual performance and observation of manipulations (Diewald and coworkers, unpublished work). Moreover, the plan was to coregister and fuse the 3-D ultrasound information with preoperative MRI data sets and with the robot control (Fig. 2).

During the same period, several robotic devices were developed for other surgical disciplines and some of them are even more developed and by far more accepted than all of the neurosurgical models. Some of them are even integrated in clinical routine applications (Table 2). Since 1994, regular meetings are being held on medical robotics [4, 32, 34]. The most successful robotic device is probably the AESOP 1000 third arm device from Computer Motion, which

Fig. 3.
Robot arm LARS as used for comparative in vitro investigation of hand-held endoscopic surgery versus robot-assisted manipulation. From [16]

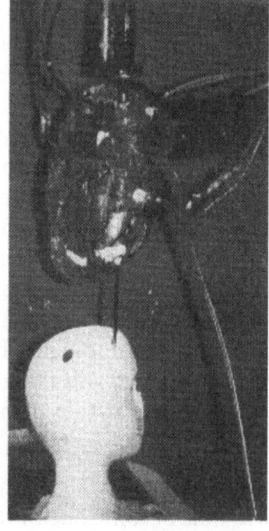

has reportedly been installed in about 500 operating rooms worldwide [29]. Another successful creation is the orthopedic robot "ROBODOC" for robot-assisted hip implantation [4]. In urological surgery, the ultrasound-image-guided robots (W)ELSA and (P)ROBOT have been used in a number of patients for fully automated prostatectomy [8, 15]. Other applications are presently undergoing evaluation for ENT, ophthalmology, and other orthopedic applications such as knee endoscopy and knee-joint replacement [34]. The robot arm LARS which is used for comparative in vitro investigation of hand-held endoscopic surgery versus robot-assisted surgery is shown in Fig. 3.

A Possible Role for Robot-Assisted Devices in the Neurosurgical Operation Room of the Future

The hypothesized helpfulness of robots should not be discussed without considering – as a conditio sine qua non – a combination with real-time orientation the actual topographical circumstances. First neuronavigation systems for computer-assisted neurosurgery have clearly shown the problem of using preoperative, "historical", data sets for intraoperative manipulations. Thus, robotic arms of whichever function can only be discussed in combination with the following three components:

1. Advanced neuronavigation with multimodality image fusion of preoperative and intraoperative imaging using rigid and nonrigid coregistration methods [3, 33] including direct intraoperative intracerebral coregistration by transendoscopic identification of actual landmarks.
2. Instrumentation, i.e., the development of a new generation of efficient microinstruments for the surgical steps to be performed through the minimally invasive approaches. It should be considered that procedures in the target area may not all be minimally invasive with the attack of larger intracranial mass lesions. Among these instruments are combined technologies such as ultrasound morcellation and laser coagulation.
3. Mechatronics, which includes the use of mechanical manipulator arm devices along with electronic steering programs and the development of complex software. The latter considers data from intraoperative imaging and preoperative planning, simulation of surgery, as well as decisions taken by the surgeon during the procedure itself. This development will eventually lead to robotic activities of the manipulator arm for parts of operations.

Reliable and very high quality intraoperative imaging will thus become one of the prerequisites for the use of robotic devices in the future, because any decision for intraoperative manipulations can only depend on real-time information on intracranial morphology. As neurosurgeons have always been aware of and has, in addition, been verified and quantified in recent studies [7, 16], shift of brain structures occurs after trepanation or craniotomy and removal of first intracranial volume compartments. Thus, the "historical" preoperative images are often no longer a valid guide and rules out present neuronavigation systems.

Three imaging systems have been considered for intraoperative application, all of which have been used in clinical routine for years. However, only one of them has been widely accepted as an intraoperative imaging system during the last 15 years, namely ultrasound (US) imaging [1]. The advantages of US are a rapid time sequence for real-time imaging and low cost in comparison with the other modalities. A disadvantage in some instances might be that only part of the brain can be shown with the currently available equipment. On the other hand, it was demonstrated in recent work [7] that preoperative MRI and intraoperative US can be coregistered. This will finally lead to the "refreshing" of preoperative images with the help of intraoperative US. Computerized tomography has also become a candidate since the development of helical scanning techniques [19]. However, radiation, artifacts, and limited accessibility to the operating field have to be considered as disadvantages.

Finally, magnetic resonance imaging has been strongly propagated for intraoperative use in its sandwich-type version. The future will show whether the quality of real-time imaging will become sufficient for surgical decision making. Most probably, however, higher costs factor of 20 – 30 compared with ultrasound imaging will remain a serious handicap for wider application. Also the technical challenge to provide nonmagnetic instrumentation is formidable. Moreover, every centimeter of further separation of the two sandwich-positioned magnets leads to further significant increases in costs of the system.

Reliable and high quality matching or coregistration techniques with high spatial accuracy are still a major challenge for medical imaging specialists; however, these techniques will be available within the next few years. This will enable the surgeon to see any requested planes and views of the patient's brain both during planning and actual surgery. Moreover, morphological information such as the border of a tumor or a neighboring functionally important brain area will be controlled three-dimensionally by the software of the navigation system and the robot arm. Without this development, any robotic arm for neurosurgery would remain as functionless as it would be without efficient instruments.

What is the Actual Use of a Robot Arm?

This section is dedicated to the description of proposed functions for robots, which have to be discussed in view of the technical safety problems. These problems however, have turned out to be less substantial and serious than believed 10 years ago, as can be seen from the successful clinical series in other surgical disciplines mentioned above: with the Robodoc, robot-guided hip surgery has been performed in more than 400 patients without a single complication due to technical problems with the robot [4]; likewise, fully automated prostatectomy has been reported as safe surgery [8]. Thus, it is time for objective consideration of the added value of robots in all fields of surgery including neurosurgery. The most adequate approach is probably the definition of surgical manipulations, where a machine will a priori be safer than the surgeon's hands because of the limitations of the space resolution of human motor activity and above all human eye-hand coordination. The latter has already reached an artificial stage with the introduction of the surgical microscope, where the surgeon works in a virtual orientation space and is at risk of loosing space-orientation in the surrounding area of the operating field. Thus, we loose our "natural" space orientation in the microscopic space. More important than space orientation, however, is the limited spatial survey of the human visual system over solid structures, i.e., The problem that the rear side of an object cannot be controlled simultaneously with the side directed towards the observer. This problem is linked with yet another phenomenon of visuomotor control, i.e., our inability for multifocal attention.

We are becoming aware of these limitations of abilities and methods have been found to overcome these human insufficiencies. Hence, the moment has probably arrived when robotic activities will be widely accepted in neurosurgery as a potential which is not competing with the surgeon's abilities, and when robots will be accepted as a different category to human surgical skill and taken as an augmentation of human skills in general.

The categories and functions of surgical robots so far described are listed below and summarized in the subsequent paragraphs.

Categories of surgical robot activity:

1. Assisting in space-orientation
3. Assisting in surgery [37]
3. Performing surgery [37]
4. Performing radiosurgery

Desirable robot functions:

1. Holding position (steady-hand, third arm)

2. Finding intracerebral location from actual images (real-time navigation)
3. Preventing movements (active constraint)
4. Induced movements
5. Automatic functions (e.g., active constraint)
6. Automatic maneuvers

The Steady-Hand Principle

The need for a holding and guiding device or a manipulator arm has primarily been expressed by surgeons with a wide experience in minimally invasive procedures such as stereotaxic operations, but mainly for endoscopic surgery. This is because all present passive holding and fixation devices are insufficient as soon as movements of the endoscope are required during surgical manipulations. All of these procedures pose the problem of precise target finding from a small approach. In addition, some of them require a third hand to hold, e.g., the endoscope's outer shaft while the surgeon performs transendoscopic surgical manipulations with both hands transendoscopically.

The first common task for a robotic manipulator arm would thus be to function as a reliable third hand. The first approach taken to solve this problem was the development of passive manipulator arms. These, however, turned out to be insufficient for many circumstances of very fine movements during surgery, where the passive arm would give way for a few millimeters and sink into the operating field after activation of its brakes [18, 25, 27].

The first applications of robotic manipulator arms for microsurgery in miniaturized approaches (e.g., transendoscopically) will thus be as an active, i.e., motor-driven third hand, which is moved by aid of remote manipulators such as pedals or hand sticks, or directly by the surgeon. The difference to passive manipulator arms will be that the active manipulator arm will stop precisely in the position requested by the surgeon, a function which has been called "steady-hand principle" [3, 14]. First in vitro experiments for evacuation of hematomas with a convential hand-held endoscope compared with robot-assisted evacuation have shown that robot-assisted endo-neurosurgery is still slower than hand-held manipulation; however, the former is more precise [14].

Real-Time Navigator

A second function will be assistance in space orientation due to the fact that the tip of the surgical instrument is coregistered by real-time intraoperative images and intraoperative, intracerebral landmarks. In addition to these first capabilities, real robotic activities will be helpful.

Active Constraint

This software development will enable the surgeon to outline a surgical field of activity. The program will watch all manipulations and prevent the surgeon from stepping outside this preplanned area and stepping into the no-go area, of e.g., a functionally important surrounding region or any other viable brain tissue outside the actual mass lesion.

Constant-Distance Tool

A further first-generation robotic activity will be to remain at a predefined distance from a border outlined by the surgeon during the planning procedure. The sine qua non importance of real-time intraoperative imaging becomes obvious at this point because this further system will have to consider shifts of tissue during single steps of the operation.

Backtracking Tool

By command of the surgeon, the manipulator arm moves back into a predefined position, e.g., the trepanation hole, exactly along the path of approach or along another track defined by the surgeon.

Training Robot

Another important early function of robot arms will be their role as a user interface for the surgeon's workstation when it is used as a trainer and simulator for education and to simulate operations on actual patients. One example for this application is the simulation of endoscopic operations by coregistration of a robot-arm-guided real endoscope and an actual patient's 3D MR data set, viewed with a so-called virtual endoscopy software. An example of the images obtained with virtual endoscopy software is shown in Fig. 4.

Further steps of development in the more distant future might be fully robotized parts of an operation such as automatic removal of pathology under constant control of the intraoperative imaging system. It has to be appreciated at this point that the clear advantage of a robotic system would be the 3-D intraoperative imaging modality which is able to "see" a structure in three dimensions and guide the robot motions. The human is not able to reliably control the result of a manipulation within solid tissue, while a high-resolution 3-D imaging technique is able to control all morphological changes in quasi real time and to forward this information to a manipulator arm and an effector at its tip.

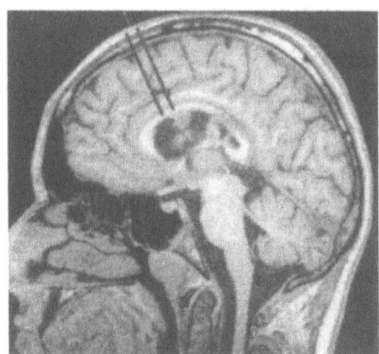

Fig. 4. Simulated endoscopic approach to a ventricular tumor via a frontal trepanation. The MR tomogram shows the planned track for introduction of the endoscope. The sequence of virtual endoscopic views from left to right shows the trepanation with a circular defect of the dura and a view of the underlying subdural space and cortical gyrus (*left*). On the image in the *center* the gyrus is penetrated by the virtual endoscope, which is 6 mm in diameter. As the endoscope is further advanced in the virtual frontal lobe, the lumen of the lateral ventricle is reached (image in the right panel): the structure on the *left side* of this image is the caudate nucleus, while the nodule of tumor in the septum pellucidum is seen on the right side; in the depth of the *center* of the image, the foramen of Monro can be recognized

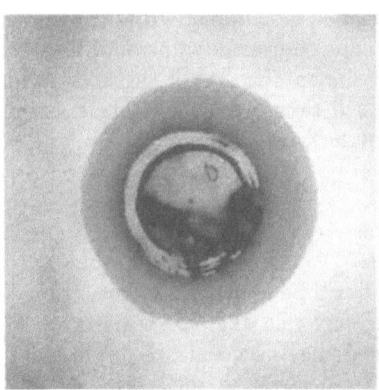
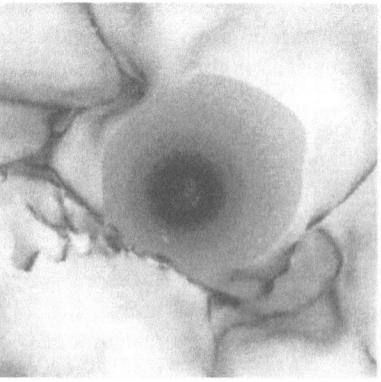
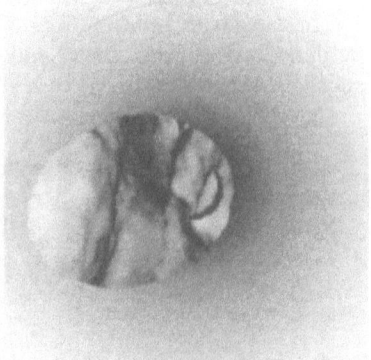

The concept of guiding a manipulator arm with
he aid of real-time intraoperative imaging is con-
idered in a recent project, Roboscope [3]. In this case,
, robot arm compatible for neurosurgical mani-
)ulations [2] is combined with intraoperative ultra-
ound imaging and the development of techniques
or coregistration of preoperative imaging modalities
vith 3-D intraoperative ultrasound imaging.

teferences

1. Auer LM, Van Velthoven V (1990) Intraoperative ultrasound
   imaging. Comparison with CT and MRI. Springer, Berlin
   Heidelberg New York
2. Auer LM, Holzer P, Ascher PW, Heppner F (1988) Endoscopic
   neurosurgery. Acta Neurochir (Wien) 90 : 1 – 14
3. Auer LM (1996) ROBOSCOPE: image-guided manipulator-
   assisted system for minimally invasive endo-neurosurgery
   (Abstract). Proceedings of the SMIT-ROB meeting, Cer-
   nobbio, September 18
4. Bargar WL, Bauer A, DiGioia A, Turner R, Taylor JK, Mc-
   Carthy J, Mears D (1995) Robodoc clinical results – domestic
   multi-center trial, European trial. Proceedings of the MRCAS
   conference Wiley-Liss, New York, pp 308 – 214
5. Benabid AL, Hoffmann D, Lavallee S, Cinquin P, Demongeot J,
   Le Bas JF, Danel F (1991) Is there any future for robots in
   neurosurgery? Advances and technical standards in neurosurgery, Vol 18, Springer, Vienna New York, pp 3 – 45
6. Benabid AL, Cinquin P, Lavalle S, Le Bas JF, Demongeot J, De
   Rougemont J (1987) Computer-driven robot for stereotactic
   surgery connected to CT scan and magnetic resonance
   imaging. Technological design and preliminary results Appl
   Neurophysiol 50 : 153 – 154
7. Bucholz RD et al. (1997) In: Troccaz J, Mösges R (eds) Lecture
   notes in computer science, CVRMed-MRCAS. Springer, Berlin Heidelberg New York, pp 458 – 455
8. Davies BL, Ng WS, Hibberd RD (1993) Prostatic resection;
   an example of safe robotic surgers. In: Robotica, vol 11.
   Cambridge University Press, Cambridge, pp 561 – 566
9. Doll J, Schlegel W, Pstyr O, Sturm V, Maier-Borst W (1987) The
   use of an industrial robot as stereotactic guidance system,
   CAR '79, 374 – 378
0. Drake JM, Joy M, Goldenberg A, Kreindler D (1991) Com-
   puter- and robot-assisted resection of thalamic astrocytomas in children. Neurosurgery 29 : 1
1. Finaly P (1993) A fully active system for assisting in neurosurgery. Industrial robot 20(2) : 28 – 29
2. Glauser D, Flury F, Frankhauser H, Burckhardt CW (1989)
   Configuration of a robot dedicated to stereotactic surgery.
   10th meeting of the world society for stereotactic and
   functional neurosurgery, Maebashi, October
3. Goerss SJ, Kelly PJ, Kall BA (1987) Automated stereotactic
   positioning system. Proceedings of the meeting of the
   american society of stereotactic and functional neurosurgery, Montreal. Appl Neurophysiol 50 : 100 – 106
4. Goradia T, Auer LM, Taylor R (1997) Robot-assisted minimally Invasive Neurosurgical procedures: first experimental
   experience. In: Troccaz J, Grimson E, Mösges R (eds) Lecture
   notes in computer science, CVRMed-MRCAS. Springer, Berlin Heidelberg New York, pp 319 – 322
5. Harris SJ, Mei Q, Arambula-Cosio F, Hibberd RD, Nathan S,
   Wickham JEA, Davies BL. A robotic procedure for
   transurethral resection of the prostate. pp 264 – 271

16. Hill D et al. (1997) In: Troccaz J, Grimson E, Mösges R (eds)
    Lecture notes in Computer Science, CVRMed-MRCAS.
    Springer, Berlin Heidelberg New York, pp 449 – 458
17. Jacobi P, Daniel P, Mugler A, et al. (1990) Diagnosegesteuerte
    Therapierobotertechnik – medizinische und biomedizini-
    sche Aspekte. Z Klin Med 6 : 515 – 519
18. Kanno T, Nonomura K, Katada K (1995) Hyperbaric oxygen
    therapy and CT-fluoroscopy in the treatment of moderate
    hypertensive intracerebral hemorrhage. In: Kanno T (ed)
    Brain '95, Neuron, Nagoya, pp 219 – 225
20. Kavanagh-KT (1994) Applications of image-directed robotics
    in otolaryngologic surgery. Laryngoscope 104/3(I) : 283 – 293
21. Kelly P (1986) Computer-assisted stereotaxis: new ap-
    proaches for the management of intracranial intra-axial
    tumors. Neurology 36 : 535 – 541
22. Kwoh YS, Hou J, Jonckheere EA, Hayati S (Feb 1988) A robot
    with improved absolute positioning accuracy for CT-guided
    stereotactic brain surgery. IEEE Trans Biomed Eng 35(2)
23. Lun-Jou Lo JL, Marsh MW, Vannier VV, Patel Cranofacial
    Computer Assisted Surgical Planning and Simulation,
    WWW-Publication
24. Masamune K, Kobayashi E, Yoshitaka M, Suzuki M, Dohi T,
    Iseki H, Takakura K. Development of a MRI compatible
    needle insertion manipulator for stereotactic neurosurgery.
    pp 165 – 172
25. Mösges R, Schlöndorff G (1988) A new imaging method for
    intraoperative therapy control in skull-base surgery. Neuro-
    surg Rev 11 : 245 – 247
26. Paul HA, Bargar WL, Mittelstaedt B, Taylor et al. (1992)
    Development of a surgical robot for cementless total hip
    arthroplasty. Clin Orthop 285 : 57 – 66
27. Reinhardt H, Meyer H, Amrein E (1988) A computer-assisted
    device for the intraoperative CT-correlated localization of
    brain tumors. Eur Surg Res 20 : 51 – 58
28. Rovetta A, Remo Sala E (MRCAS '95) Robotics and tele-
    robotics applied to a prostatic biopsy on a human patient,
    pp 104 – 110
29. Sackier JM, Wang Y (1994) Robotically assisted laparoscopic
    surgery. Surg Endosc 8 : 63 – 66
30. Schurr MO, Breitwieser H, Melzer A et al. (1996) Experimen-
    tal telemanipulation in endoscopic surgery. Surg Laparosc
    Endosc 6 : 167 – 175
31. Taylor RH, Funda J, LaRose D, Treat M (1992) A telerobotic
    system for augmentation of endoscopic surgery. In:
    IEEE Engineering in medicine and biology society,
    pp 1054 – 1056
32. Taylor RH et al. (1996) Computer integrated surgery, MIT
    Press, Cambridge
33. Thirion J-P. Fast non-rigid matching of 3 D medical images.
    Medical Robotics and Computer Aided Surgery Baltimore
    (USA), Nov. 1995
34. Troccaz J, Grimson E, Mösges R (eds) (1997) Lecture notes in
    computer science. Springer, Berlin Heidelberg New York
35. Watanabe E, Watanabe T, Manaka S, Mayanagi Y, Takakura K
    (1987) Three-dimensional digitizer (neuronavigator): new
    equipment for computed tomography-guided stereotaxic
    surgery. Surg Neurol 27 : 543 – 547
36. Young RF (1987)b Application of robotics in stereotactic
    neurosurgery. Neurolog Res 9 : 123 – 128
37. Pressing et al. (1991)

OR 2015: New Concepts and Technologies in Surgery

V. Urban and R. Schönmayr

Summary

The impact of new technologies developed over recent decades do not adequately meet the integrative for the OR environment. In neurosurgery the rapid improvement of diagnostic procedures provides visualization of pathologies and topographic structures in the submillimeter range. Current surgical procedures should be improved and new surgical techniques should be developed to ensure precision, efficacy, and safety in microanatomical surgery. Many different approaches to minimal invasive neurosurgery are already available. The simultaneous implementation of systems during one operation is limited by today's OR capacity. The study "OR 2015" is presented as an integrative solution to these increasing problems. Virtual reality, microrobotic assistance, navigational aids, and microendoscopic visualization should enable the surgeon to focus on his primary task: safe and effective surgery.

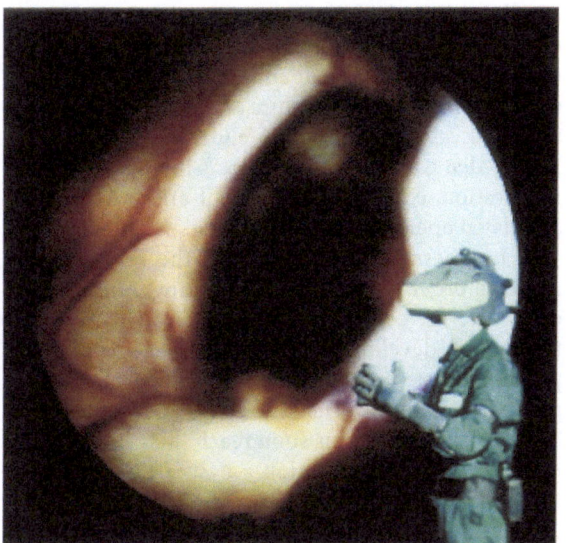

Fig. 1

Operating in the Submillimeter Domain: Endoscopic Neurosurgery in the OR 2015

Who can forget the 1968 science fiction entitled "The Fantastic Voyage"? On a microbe-sized submarine a team of neurosurgeons travel through the circulatory system, the lungs, and finally into the patient brain. There a blood clot is destroyed by laser. And here it was born: the vision of submillimeter surgery within the human organism as microsm (Fig. 1).

Neurosurgical Endoscopy

The microscope is an integral part of today's microsurgery. However, relatively few hospitals worldwide have made the transition to neuroendoscopy. More and more endoscopic procedures are being performed, with the great advantage of inspect the tiniest corners of the central nervous system. Surgical use of the endoscope is still limited, although constantly being improved upon. One important endoscopic advantage, for example, has been eliminating the need for permanent shunt-systems in hydrocephalus.

The University of Marburg's Neurosurgical Clinic has (Dr. Bauer, Dr. Hellwig) reported stereotactical endoscopic evacuation of intracerebral bleeding. The Greifswald University Clinic of Neurosurgery (Dr. Gaab) has presented its first experiences with removal of tumors of the ventricular system. In the Neurosurgical Department at the University of Mainz (Dr. Perneczky, Dr. Hüwel) an endoscopic fenestration technique for the treatment of spinal fluid cysts (syringomelia) has been developed. Dr. Horst-Schmidt-Clinic's Department of Neurosurgery in Wiesbaden (Dr. Schönmayr) has been performing endoscopic surgery since January 1995. In addition to general improvement and development of endoscopic procedures, the main emphasis has been placed on endoscopic cystic removal in newborns and children. Close cooperation with the medical industry has as its goal further developments in complex endoscopic operations of the central nervous system. Such developments are finding increasing acceptance worldwide. Therefore rapid increases in this field are to be expected in the coming years.

Operation Simulation

To take advantage of endoscopic surgical techniques the neurosurgeon needs to plan the optimal route in advance of each operation. However, with procedures dealing in the submillimeter range even today's excellent knowledge of human anatomy is not enough. Here there is the additional challenge of so-called "individual patient anatomy." The length of an optical nerve, or the curvature of a cerebral artery can spell success or failure in an endoscopic procedure. Knowledge gained from diagnosis images is crucial. In addition, computer imagery of entry routes and simulated operations are vital aids.

It is necessary to decide upon the best possible route. In order not to misplace the tiny millimeter of entry into the central nervous system. This requires the further development of interactive preoperative planning systems, along with intraoperative navigational systems as orientation helps during surgery. The navigation systems soon reach their limits, however, if during an operation a cyst is punctured, or there is a charge in the size of a tumor. Therefore there is an urgent need for an intraoperative image projection system. Of greatest help considering the microsurgical operative technical possibilities would be a three-dimensional computer image. Here is a research and development area that neurosurgeons alone cannot cover.

The Computer as "Assistant Surgeon"

In the modern minimally invasive surgery technology thus becoming more and more important. This means that the surgeon is working increasingly before and during operations with the computer. Experience, talent, and intuition is no longer sufficient for the modern surgeon. This step-by-step increase in computer aids has come only gradually. However, often there is a lack of personal training especially in using these new technological advances. Therefore the surgeon is often forced to operate the machines during surgery if he wants to use the obvious advantages which computer imagery can provide.

OR 2015

Modern operating rooms are now full of numerous technical machines. Each of them – ultrasound scalpels, lasers, high-frequency or endoscopic devices, microscopes, computers – must be put into the operation room. With seemingly unlimited numbers of machines, the impression of a technical workshop appears more likely than an operating room. This situation has turned into a frantic free-for-all search for plug-in possibilities. Unfortunately most operation rooms still have the tiled look of the 1960s, with virtually no flexibility to adapt to modern surgical needs. Understandably industrial companies have never thought about an integrated logistics system. The ideal OR 2015 would be a room in which the

Fig. 2.

patient is as usual the center of attention. Directly next to him is the surgeon. Physician-patient contact is thereby restored. At the same time, however, there is a physician-computer-patient interface that would provide the surgeon digital images from the tip of the endoscope and keep an updated overall view of the patient's condition, anatomy, and the navigation route. The third person in the team would be an operational engineer, somewhat as a board engineer on an airplane. His duty is to keep an optimal flow of relevant information available to the surgeon (Fig. 2).

The study of a future OR 2015 also includes more patient centered surroundings. This is especially important because of the trend toward local anesthesia, with operations performed on conscious patients. OR 2015 is equipped as a virtuel-reality room. This means that with the help of computers there is freedom of movement throughout the room. Interior design and other variables could be tailored to fit the situation. New ideas along these lines could be tested and changed according to need.

The second partner in the OR 2015 is the endoscope. Here there are plans for improved equipment necessary for operations in the submillimeter domain. The tip of the endoscope is comparable in size to the head of an ant. What is needed here is complete 3-D optics along with the possibility of an extensive array of sensors and actors. This does not mean robotry but is a form of long-distance tele-operating (Fig. 3).

Reasons

Today's modern microdiagnostics require corresponding surgical techniques. There is a need to offer ever-increasing possibilities of submillimeter surgery. This requires extensive reform of surgical surroundings. Submillimeter surgery is no longer the exclusive domain of neurosurgeons. All surgical disciplines are requested to formulate their needs in this regard.

Horizontal Research

Ideally, an interdisciplinary research and development project of institutional level would be most advantageous. The Fraunhofer Institute in Stuttgart, reponsible for production and automation techniques, is prepared to build the OR of the future. An invitation is thus extended to all physicians, research centers, and commercial companies to take part in this development.

Conclusion

Minimally invasive surgery entails increasing dependence upon submillimeter surgery. The high-tech-standard logistics of today's OR do not meet the necessary integration for microcosmic surgery. The goal is an OR 2015 as testing ground for operating in the range submillimeter. Everyone with developmental ideas along these lines is invited to work as partners in order to produce the ideal working place for all surgical disciplines of the future.

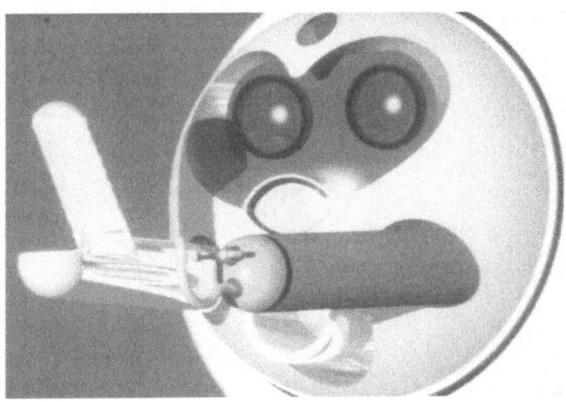

Fig. 3

**Part 8
Neurotransplantation**

Principles of Neural Transplantation in the Central Nervous System

G. Nikkhah and M. Samii

Summary

Transplantation of neuronal and glial cells has developed into an exciting tool to study mechanisms of structural repair and functional recovery in the central nervous system. Today there is convincing evidence that the transplantation approach can significantly ameliorate signs and symptoms in animals models of neurological diseases, e.g. Parkinson's and Huntington's Disease, hypothalamic-pituitary dysfunction, pain, and spinal cord trauma. Under these circumstances the grafted cells may exert their biological activity via a transmitter release, trophic stimulation and/or more specifically, via direct synaptic connections to the host. The transplantation strategy has been introduced into clinical trials for the treatment of Parkinson's disease with grafts of embryonic ventral mesencephalic (VM) tissue in 1987. Surviving dopaminergic neurons and long-term therapeutic benefits have been demonstrated. Many new avenues are now being explored to increase the functional efficacy of transplants and to overcome the use of human embryonic tissue. These include the microtransplantation approach and exploration of new target sides, neurotrophins and other cytoprotective agents, as well as alternative cell resources like precursor cells and genetically modified cells, or direct gene therapy. Therefore, neurotransplantation holds great potential for stimulating further developments in the understanding and treatment of CNS disorders.

Introduction

As clinically practicing neurosurgeons we are confronted daily with patients in whom pathological conditions have induced long-lasting and incapacitating neurological deficits. Examples include spinal cord and head trauma, ischemia and neurodegenerative diseases, but infections and tumor growth as well. State of the art neurosurgical treatment can often improve the clinical condition of the patient or, at least, halt further neurological deterioration, but in many cases permanent deficits have to be dealt with. Intrinsic regenerative potential decreases with age and is only minimal in the adult central nervous system, unlike in the peripheral nervous system. Therefore, there is a great need for novel therapeutic strategies for replacing lost functions in the central nervous system.

Over the past two decades a tremendous scientific effort has disclosed some of the most fascinating insights into brain-repair mechanisms. Neural transplantation is one of the offsprings of this effort and has developed into a neurobiological tool with promising clinical perspectives. The underlying concept is to repair damaged CNS tissue through the implantation of active substitutes that restore and/or replace anatomical and functional losses. Experimental interest has been directed at exploring basic mechanisms of neural degeneration and regeneration, whereas clinical interest has focused on application to patients with neural disorders. This chapter seeks to provide the reader with an overview of the principles involved in neural transplantation, in contrast to some recent publications, which offer more exhaustive and indepth discussion [10, 12].

Historical milestones in Neural Transplantation

The era of neural transplantation began about a century ago when Thompson described his first experimental observations on pieces of adult neocortical tissue taken from adult dogs or cats and then grafted into the neocortical cavities of other adult dogs or cats. He felt he was able to demonstrate transplant survival and concluded that this might become "an interesting field for further research, and I have no doubt that other experimenters will be rewarded by investigating it". In the first half of the twentieth century a number of original studies investigated the experimental circumstances under which neural grafts could survive the transplantation procedure [11]. The most important lesson learned from these studies was the fact that grafts derived from embryonic tissue yielded much better results than grafts taken from neonatal or adult donors both in terms of survival patterns for neuronal and glial graft elements and in their potential to establish afferent and efferent graft-host connections. Consequently, the next major area of scientific investigations was related to the question "Do transplant work?" A first definite answer came in 1979, when two groups independently showed that grafts derived from the embryonic ventral mesencephalon rich in dopaminergic neurons could ameliorate rotational motor asymmetries in a rat model of Parkinson's disease [5, 37]. This was a major scientific breakthrough as it demonstrated that functional deficits could be restored in the adult central nervous system, and it opened up the possibility for the development of novel therapeutic strategies for human neurological diseases. Up until today neurotransplantation has been applied in a large number of animal models of neurological disorders, as summarized in Table 1. Of them all, Parkinson's disease has become the leading area for experimental and clinical investigations related to neural plasticity and regeneration (see also [3, 29]); for this reason, experimental and clinical research in this area will be used frequently in this chapter to exemplify the current state of knowledge and techniques. Finally, clinical transplantation studies have been initiated for patients with Parkinson's disease during the last few years and are beginning to show therapeutically meaningful and long-lasting effects associated with substantial graft survival, both measured by functional neuroimaging (e.g., positron emission tomography or PET) and immunohistochemistry [22, 34].

In order to progress towards a more optimal cell replacement therapeutic concept the major scientific focus has shifted once again, at the turn of the twenty-first century, to a third key question: "How do the transplants work?" Though we still await a better understanding of the underlying mechanisms of neural repair, a number of principles can already be deducted from the current state of knowledge. They are based on the historical development of this field from studies related to structural repair to those concerned with functional recovery and, finally, toward an organotypic reconstruction of parts of the central nervous system. Some of these fundamental principles of neural transplantation are discussed below.

Replacement of Morphological Substrates

The first principle involves the replacement of certain parts or, more specifically, certain cells of the central nervous system, previously destroyed by experimental manipulations or clinical disease processes, through grafted cells. Traditionally, this can be achieved by dissecting the area of the embryonic or fetal brain in which the cells of interest are born and maturing. This is, for example, the ventral mesencephalon for

Table 1. Animal models used in experimental CNS neurotransplantation

Disease	Lesion model	Source of graft tissue
Parkinson's	6-Hydroxydopamine, MPTP	Ventral mesencephalon
Huntington's	Ibotenic or kainic acid	Striatal eminence
Alzheimer's	Fimbria fornix lesions	Septal and basal forebrain
Ischemia	Arterial ligation	Striatal or cortical areas
Hypothalamic pituitary dysfunctions	– Natural mutations like the Brattleboro rat (vasopressin deficiency) or the hypogonadal mouse (growth-hormone deficiency) – Neurohypophysectomy	Hypothalamus
Circadian pacemaker dysfunction	Electrolytic lesions of the suprachiasmatic nucleus	Anterior hypothalamus
Epilepsy	6-Hydroxydopamine kindling Fimbria fornix lesions	Locus coeruleus Hippocampus
Paraparesis	Spinal cord transsection	Spinal cord
Multiple sclerosis	– Natural mutations like the shiverer mouse – X-rays followed by ethidium bromide injections	Oligodendrozytes
Amyotrophic lateral sclerosis	Kainic acid Ventral root avulsion	Spinal cord
Pain	None	Chromaffin cells

dopaminergic neurons, the striatal eminence for GABAergic neurons, and the basal forebrain for cholinergic neurons. One of the reasons for the better graft survival of embryonic and fetal neural tissue is that the axonal projections to the target nuclei are not yet established at the time of dissection, which results in a much smaller trauma to the neuronal cells as compared to dissections at later times, which cause significant axotomies. The tissue is then grafted either as solid blocks or in the form of a cell suspension. Further technical details about transplantation procedures can be found in Dunnett and Björklund

[10]. Most importantly, the choice of the optimal target area for cell implants still represents a major scientific challenge. If a neuronal system with short axonal projections has to be repaired, a homotopic graft placement seems to be the obvious choice, as in the case of the anterior hypothalamic grafts to treat hypothalamic-pituitary dysfunctions. On the contrary, if a neural subsystem with an extensive and long-distance system of axonal projections is damaged, the choice of the target site for implantation is hampered by the fact that growth-inhibiting factors in the adult central nervous system will prevent

Fig. 1 A–F.
Photomicrographs show the intact (**A, C, E**) and the lesioned (**B, D, F**) sides of the hemispheres of a rat with a unilateral 6-hydroxydopamine lesion. The sections are cut in a coronal plane and stained for tyrosine hydroxylase, the marker enzyme for catecholaminergic neurons and neurites. **A** shows the staining pattern of the intact hemisphere at the level of the caudate putamen, whereas **C** illustrates the corresponding distribution of the dopaminergic neurons in the substantia nigra and ventral tegmental area on the normal side. The 6-OHDA-induced lesion produces an almost complete loss of dopaminergic neurons in the substantia nigra region (**D**) and a concomitant denervation of the target areas, as can be seen in **B**. **E** and **F** represent a double TH- and DARPP-32 staining, which demonstrates the close approximation and the intermingling of the dopaminergic dendrites within the substantia nigra pars reticulata in the normal rat (**E**) and the respective changes after 6-OHDA lesion. DARPP-32 is a specific marker for medium spiny GABAergic neurons of the striatum, and shows here the terminals of the striatonigral pathway. CC, corpus callosum; CPU, caudate-putamen unit; CTX, cortex; NA, nucleus accumbens; SNc, substantia nigra pars compacta; SNr, substantia nigra pars reticulata; VTA, ventral tegmental area. Magnifications: A, B, = ×4; C–F = ×10

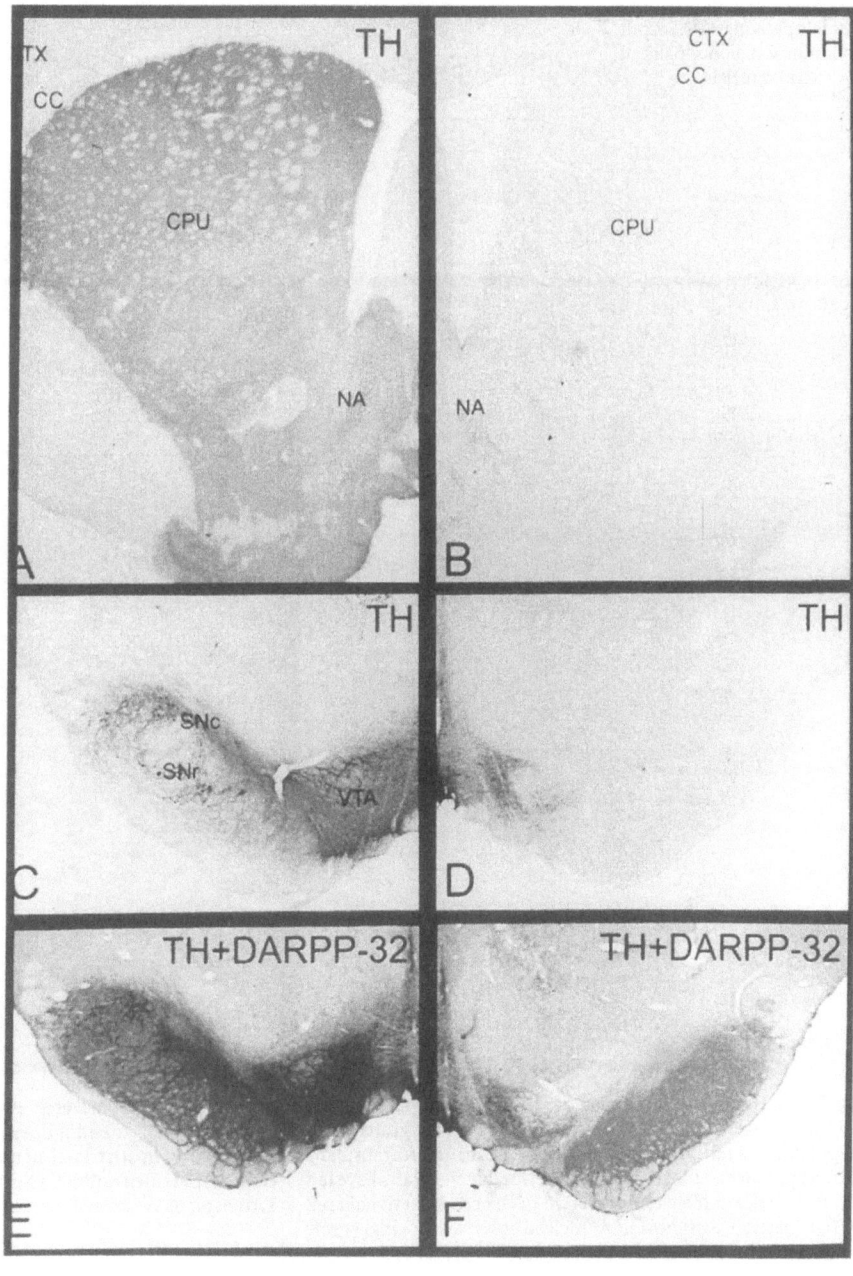

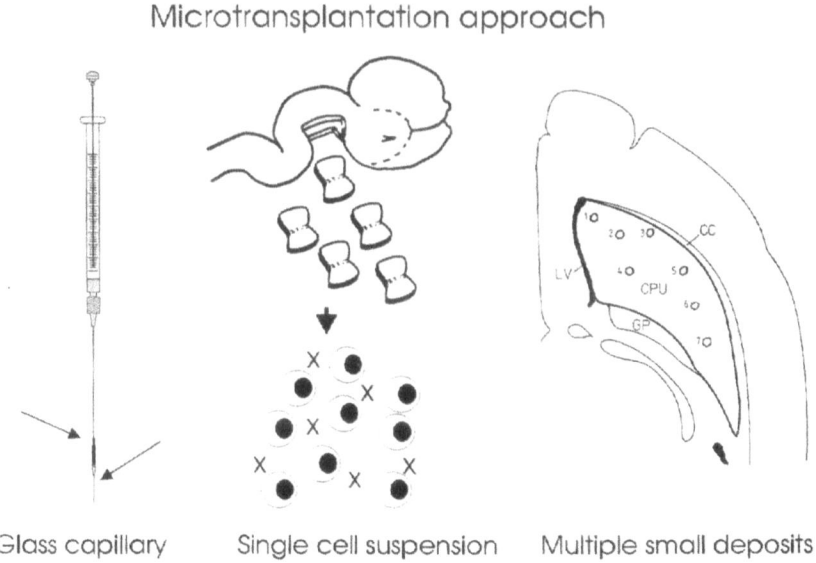

Microtransplantation approach

Fig. 2.
The three modified components of the microtransplantation approach, as used for the implantation of embryonic dopaminergic grafts. (*Left*) The modified microtransplantation instrument consists of a Hamilton microsyringe fitted with a long-shanked glass capillary (OD 50 μm) using a cuff of polyethylene tubing as an adapter. (*Middle*) The preparation of a single cell suspension following the dissection of the ventral mesencephalon with the use of trypsin and DNase (*x*), and *Right* The spreading of small graft deposits over multiple sites in the target area, here shown for the caudate putamen. CC, corpus callosum; GP, globus pallidus; LV, lateral ventricle

Glass capillary   Single cell suspension   Multiple small deposits

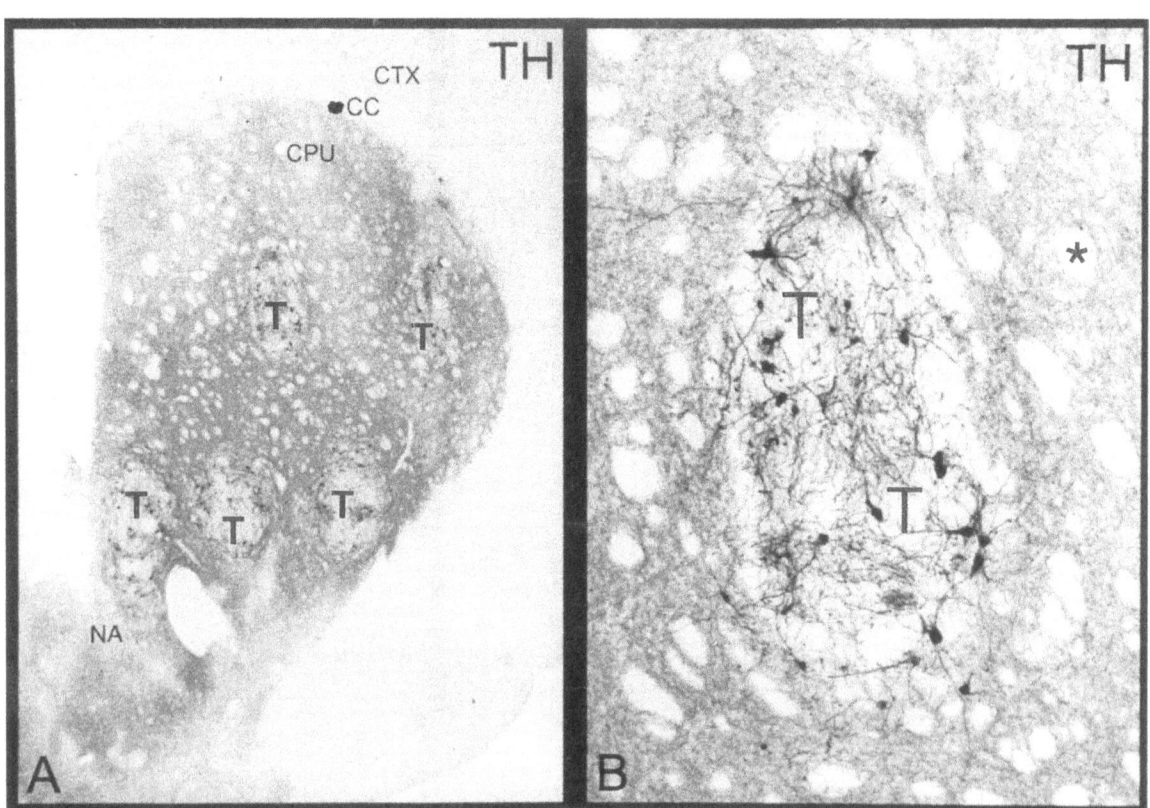

Fig. 3A, B. Higher magnification of a coronal section through a 6-OHDA lesioned and grafted hemisphere. The grafts were implanted as multiple small deposits using the microtransplantation approach as illustrated in Fig. 2. Notably the grafts have led to an almost complete reinnervation of the cross-sectional area of the striatum, as shown in **A**. A close up view of a single graft deposit (**B**) demonstrates multiple surviving dopaminergic neurons, which extend numerous TH-positive neurites into the surrounding host striatal neuropil. CC, corpus callosum; CPU, caudate-putamen unit; CTX, cortex; NA, nucleus accumbens; T, transplant. **A** = × 4, **B** = × 20)

approach was adopted (Fig. 2), which has subsequently been shown to improve dopaminergic graft survival and striatal reinnervation [31, 33]. Using multiple small implants of ventral mesencephalic cell suspensions, an almost complete reinnervation of the caudate putamen can be achieved, as shown in Fig. 3. With this technique a high degree of integration of the transplanted TH-positive neurons and a more extensive fiber outgrowth could be observed (Fig. 4). Microtransplantation techniques have also shown beneficial effects on integration and migration of grafted cells from astrocyte cultures and hippocampal neurons [9, 15].

Restoration of Functional Deficits

Once a morphological substitute has been provided by the transplant, the next question is whether or not a functional influence on the damaged host neuronal circuitry can be exerted from or via the grafted cells. A beautiful example for a graft-induced functional effect is a capacity of anterior hypothalamic transplants to reinstate the overt rhythm in the circadian range in rats, which had been previously eliminated, through electrolytic lesions of the suprachiasmatic nucleus. There are numerous other studies showing comparable transplant-induced functional effects in a variety of animal models, including the ones listed in Table 1, and a comprehensive overview is given by Dunnett and Björklund [12].

On the other hand, it has to be emphasized that the extent of functional restoration that can be achieved with current neural transplantation techniques and strategies is often incomplete as compared to normal behavioral patterns. In animal experiments, drug-induced behavior and simple sensorimotor task are often more likely to be substantially compensated by neural transplants than are complex integrative sensorimotor behaviors such as paw reaching and disengage behavior, or cognitive tests like the water maze and avoidance tasks [13]. As the restoration of lesion-induced functional and behavioral impairments is one of the main goals of neural transplantation, novel concepts have to be developed and evaluated to acquire a more complete level of graft-induced neural repair.

Afferent und Efferent Graft-Host Integration

Some of the current limitations of graft-induced functional recovery have been attributed to the failure of grafted neurons to establish adequate afferent and efferent projection pathways to the host neuronal circuitry. In the nigrostriatal dopaminergic

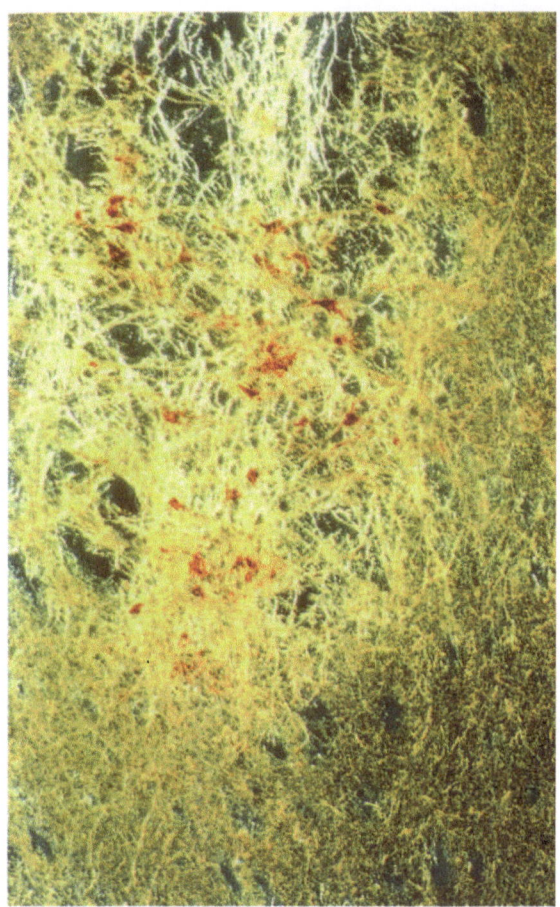

Fig. 4. Darkfield photomicrograph of a TH-stained coronal section through the middle portion of the caudate-putamen unit of a 6-OHDA lesioned and VM grafted rat. It illustrates the extensive neuritic outgrowth and reinnervation of the host striatal neuropil from the grafted dopaminergic neurons, partially avoiding the fiber bundles of the internal capsule. The resulting degree of reinnervation is comparable to the normal side in the close vicinity of the transplant

reconnection over long distance by neural grafts (see also "Afferent and Efferent Graft-Host Integration"). In Parkinson's disease dopaminergic neurons are lost within the substantia nigra and the ventral tegmental area, similar to the experimental situation after a 6-OHDA lesion, which is depicted in Fig. 1D. This leads to a progressive denervation of the corresponding anatomical areas of normal dopaminergic terminal (CPU, Fig. 1B) and dendritic (SNbr, Fig. 1F) innervation. In the "classical" approach dopaminergic grafts have been implanted ectopically into the striatal complex to reinnervate this crucial terminal area. Shortcomings of standard techniques use for intracerebral transplantation have been low graft survival, high variability, considerable implantation trauma, and suboptimal graft integration. In order to overcome these limitations an microtransplantation

Fig. 5.
Three alternative transplantation strategies used to reconstruct the mesotelencephalic dopamine system in a rat model of Parkinson's disease. The dopaminergic graft is highlighted in the *black box*. The principle and likely anatomical connections of the dopaminergic grafts are illustrated for ectopic intrastriatal grafts in adult hosts (*NIS*, nigra in striatum), intranigral grafts in adult (*NIN*, nigra in nigra) and neonatal hosts (*NIN-Neonate*). *Asterisks* indicate presumptive synaptic connections. Ctx, cortex; CPU, caudate-putamen unit; SNpc, substantia nigra pars compacta; SNpr, substantia nigra pars reticulata; STN, subthalamic nucleus

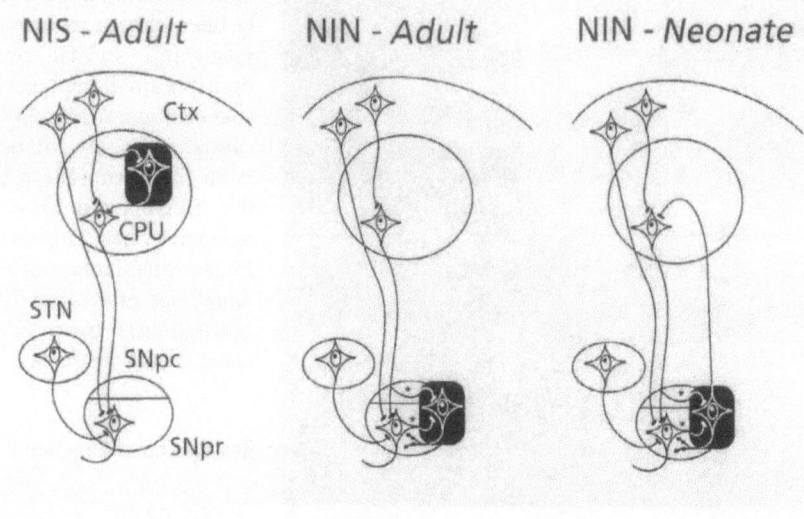

pathway, the cell bodies are localized within the substantia nigra region, whereas the dopaminergic axonal projections and terminal arborizations cover large parts of the di- and telenzephalon. Because of the lack of long-distance axonal growth in the adult central nervous system, nigral transplants have been implanted into their respective target area, i.e., the caudate putamen (see Fig. 5, left third). Whereas some terminal efferent projections can be established in this way, many of the normal physiological afferent inputs and dendritic functions of the dopaminergic neurons are missing [4].

Alternatively, nigral grafts have been implanted homotopically into the adult substantia nigra region, where they can compensate dopamine-agonist-induced motor asymmetry when using the modified microtransplantation approach [28]. It is of interest that this behavioral effect was solely mediated via a local graft-host interaction within the substantia nigra region, as long-distance axon growth towards the caudate putamen could not be observed. When nigral grafts were placed into the neonatal substantia nigra region, such directed long-axon growth and reinnervation of the caudate putamen became evident and was associated with increased behavioral recovery as compared to ectopic intrastriatal grafts in the neonatal host [30, 32]. This indicates that both afferent and efferent control mechanisms are important in determining the extent of graft-induced functional effects.

Homotopically placed grafts of striatal tissue implanted into the ibotenic acid lesioned caudate putamen might receive adequate afferent control, but fiber outgrowth and the establishment of an effective projection system is very limited in the adult central

nervous system, as shown by Wictorin [43]. Similarly, embryonic spinal cord grafts have a restricted growth capacity through white matter tracts in the adult spinal cord [7]. Recent strategies to enhance the afferent and efferent graft-host interactions in the adult CNS include a number of attempts to stimulate long-distance directed axonal outgrowth from fetal nigral grafts through additional manipulations, e.g., by simultaneous intrastriatal and intranigral dopaminergic grafts [26] and by using nigrostriatal bridges of kainic acid in combination with intranigral nigral grafts [44]. This indicates that even in the adult host, target-directed axonal outgrowth can be achieved through the white matter once the appropriate developmental cues are provided. Furthermore, the combined treatment with the antibody IN-1 against myelin-associated neurite growth inhibitory proteins together with local application of neurotrophin-3 after adult spinal cord lesion provided for a more extensive sprouting of corticospinal tract axons in rats [38]. This underlines the potential usefulness of combined treatment modalities in neural restoration to achieve a more complete reconnection of afferent and efferent control pathways, be they graft and/or host derived.

Different Developmental Stages of the Host

Another attempt to increase the outgrowth of neurites from grafted neurons has been made in studies in which the age of the host was modified. Younger hosts, such as neonatal animals or animals which are still in their prenatal development, are generally thought to present a dominance of growth-promoting factors over growth-inhibiting factors in the central nervous system. This is based on the dynamic developmental changes that occur during the stepwise maturation from the embryonic, fetal, and neonatal to the adult stage and include the supporting influence of extracellular matrix and cell adhesion molecules, immature astrocytes, and the reduced expression of inhibiting factors [25, 40]. More extensive long-distance axon growth and target reinnervation has been observed, for example, with embryonic basal forebrain neurons implanted into the neonatal septum [21], striatal precursor cells grafted into the neonatal striatum [35], and also with retinal, cortical and hippocampal grafts into neonatal host animals (for further discussion, see also [30]). Dissected segments of fetal spinal cord implanted into the severed lower thoracic spinal cord of neonatal rats have resulted not only in reformation of reciprocal pathways between the cerebrum, cerebellum, and the spinal cord, but, more importantly, these animals could walk, run and climb again [19], indicative of an almost normal hindlimb-forelimb coordination.

Recently, a modified microtransplantation approach was used to implant fetal dopaminergic cells into the substantia nigra of bilaterally 6-OHDA lesioned rats at postnatal day 3 [30]. The homotopically grafted dopaminergic neurons were shown to reconnect with the host striatum (Fig. 5, right third) and induced behavioral effects on rotation and spontaneous complex behaviors such as locomotor activity and skilled forelimb use [32]. In a follow-up study it was demonstrated that this directed long-distance axon growth can be seen until postnatal day 10 in the rats and that this growth capacity of the grafted neurons is switched off somewhere between postnatal day 10 and 20, the latter phase of time being similar to the nonpermissive adult stage [2]. This time window has also been confirmed with striatal E13.5–14 precursor cell [35] grafted into the striatum of P1, P7 and P12 hosts. The identification of this critical time window for long-distance and directed axonal outgrowth may help to further identify key factors that govern the cellular and molecular events responsible for neuronal plasticity and axonal guidance.

Psychological Influences on Graft Function

Besides anatomical and neurochemical aspects, psychological factors can also influence the development and function of neural grafts. This means that the functional outcome after transplantation may also depend on the training and experiences of the host. It has been shown that the environment as well as conditioning and learning effects can enhance the processes of maturation of transplanted neurons and the efficacy of neural grafts to restore lesion-induced deficits. Enrichment of the environment with climbing frames, shelves, tubes, and branches has appeared to promote the initial rate of fiber outgrowth from cholinergic-rich septal grafts placed into the hippocampus in rats with fimbria-fornix lesions [14]. Associative conditioning to stimuli in a certain environment can lead to priming of stress-induced rotation [41] and the induction of rotation with saline treatment alone [1] in rats with 6-OHDA lesions and nigral grafts. Evidence that transplanted hosts need to "learn how to use their grafts" comes from studies with retinal and striatal transplants (see also [13]). In these experimental studies graft function became only apparent after previous training in the same test or after a conditioned suppression test, which may have helped the animals to correctly interpret the information relayed from the graft.

Clinical Application of Neural Transplantation

Encouraged by experimental results, transplantation of embryonic dopaminergic cells was introduced into the first clinical studies for patients with Parkinson's disease about 10 years ago (overview in [22]). Since then the methodology has been considerably refined from the early open microsurgical implantation of ventral mesencephalic fragments into cavities in the head of the caudate nucleus [24, 27] to the stereotaxic implantation of dissociated dopaminergic cell suspensions distributed over multiple implantation sites in the caudate and putamen nuclei [16, 23]. Worldwide the number of transplanted Parkinson patients has surpassed 200 and, with improved transplantation techniques, meaningful therapeutic benefits could be documented, which were long-lasting and paralleled by significant graft survival, demonstrated both by in in vivo PET studies and postmortem immunohistochemical analysis [22, 34]. These are important observations as they indicate that results derived from animal studies are, at least partly, reproducible in man. For the further application of this method to other disorders of the central nervous system, the experiences in experimental and clinical neural transplantation have identified a number of criteria which need to be considered:

- The disease should have a morphological correlate.
- The lesion should be well characterized and – ideally – only affect one neurochemically defined cell type.
- The severity of symptoms should justify a therapeutic trial.

- Pathogenesis of the disease should not compromise graft survival.
- No better therapeutic alternatives are available.
- Unequivocal experimental evidence of therapeutic benefit exists.

Huntington's disease has become the second area of clinical investigation for neural grafting [39] and transplantation into the damaged spinal cord maybe the next disease to be explored clinically on this basis.

Taken as a whole, the clinical results obtained so far are clearly encouraging. Nevertheless, they also prompt caution: a more widespread use of transplantation as a therapeutic alternative for patients with neurological diseases, such as those listed in Table 1, must await further methodological improvements, a better understanding of graft-host interactions, and, finally, the development of alternative cellular and/or molecular resources.

To provide a continuous forum for discussions about the scientific, clinical, and ethical aspects of neural transplantation a European network NECTAR (Network of European CNS Transplantation and Restoration) has been formed, which has been actively coordinating preclinical and clinical studies related to this topic since 1991, for example, by using common assessment protocols (CAPIT, [20]) and ethical guidelines [6].

Alternative Cell Resources

The past few years have witnessed a tremendous scientific effort toward the establishment of alternative biological resources for the replacement of neuronal and glial cell losses. As can be gathered from Table 2, a wide variety of restorative agents are cur-rently being investigated for their potential to induce morphological repair and functional recovery in the CNS, a development which has been propelled by the exponential advance in cellular and molecular neuroscience. This might also open up the future possibility of substituting for the use of human embryonic material in clinical studies; hopefully it will also enable the design and construction of a neurochemical profile and cytoarchitecture for tailored cell replacement in a specific neurological disease. Today, there is convincing evidence that stem cells do exist in the adult central nervous system which can be purified, modified with ex vivo molecular and gene-transfer techniques, and then be implanted for in vivo neuronal and glial repair [8, 17, 18]. Ultimately, it may even be possible to provide adequate external stimuli to trigger those intrinsic stem cells to develop along a desired cell pathway to replace disease-induced cell losses, as discussed by Brüstle and McKay [8]. Another interesting area of recent scientific progress is the potential use of neural xenografts for clinical application [36]. It can be foreseen that ultimately it may not be a single agent but rather a combination of cellular and molecular restorative components which will be employed in order to achieve more optimal neural repair.

Perspectives for the Future

Neural transplantation has become a valuable tool to explore different strategies for neural repair and to elucidate mechanisms of neuronal and glial plasticity in the central nervous system. It has been possible to demonstrates that grafted neuronal and glial cells survive and integrate into the host brain and are capable of inducing long-term structural repair and

Table 2.
Different experimental approaches to substitute for neuronal and/or glial cell losses in the CNS and to promote neuronal regeneration and axonal outgrowth

Restorative agents	Example
Embryonic cells	Dopaminergic neuroblast
	GABAergic neuroblast
	Cholinergic neuroblast
Cultured neurons	"roller tube" cultures
Neuronal precursor cells	EGF-responsive neurospheres
Genetically modified cells	Fibroblasts, myoblasts
	Immortalized cell lines (e. g. HiB5)
Direct gene/virus transfer	Herpes simplex virus
	Adenovirus
Growth factors	Brain-derived neurotrophic factor (BDNF)
	Glial-derived neurotrophic factor (GDNF)
	Nerve growth factor (NGF)
	Platelet-derived growth factor (PDGF)
Antibodies (IN-1, IN-2)	Myelin-associated inhibitory factors

More detailed information can be found in [10, 42].

functional recovery. Clinical transplantation studies in Parkinson patients are encouraging and underline the great potential of this tool for experimental and clinical neurosciences.

The progress and further success in CNS neurotransplantation will depend on continuous development in a number of different areas, including (a) strategies to improve graft survival and integration, (b) enhanced knowledge about the structure and function of the CNS under normal and pathological conditions, (c) establishment of alternative cellular and molecular resources for restoration of neuronal function, and (d) the role of biotechnological connections. A prerequisite for the further advancement of this field is a fruitful collaboration between basic and clinical scientists to ensure its development into a meaningful therapeutic strategy acceptable to the clinical community and beneficial for its patients.

Acknowledgements. Part of the work presented in this paper was supported by grants from the Deutsche Forschungsgemeinschaft and DAAD.

References

1. Annett LE, Reading PJ, Tharumaratnam D, Abrous DN, Torres EM, Dunnett SB (1993) Conditioning versus priming of dopaminergic grafts by amphetamine. Exp Brain Res 93:46–54
2. Bentlage C, Nikkhah G, Cunningham M, Björklund A (1997) Reformation of the nigrostriatal pathway by fetal dopaminergic micrografts into the substantia nigra is critically influenced by the age of the host (submitted)
3. Björklund A (1992) Dopaminergic transplants in experimental parkinsonsism: cellular mechanisms of graft-induced functional recovery. Curr Opin Neurobiol 2:683–689
4. Björklund A, Dunnett SB, Nikkhah G (1994) Nigral transplants in the rat Parkinson model: functional limitations and strategies to enhance nigrostriatal reconstruction. In: Dunnett SB, Björklund A (eds) Functional neural transplantation. Raven Press, New York, pp 47–70 (Advances in neuroscience, vol 2)
5. Björklund A, Stenevi U (1979) Reconstruction of the nigrostriatal dopamine pathway by intracerebral nigral transplants. Brain Res 177:555–560
6. Boer J (1994) Ethical guidelines for the use of human embryonic or fetal tissue for experimental and clinical neurotransplantation and research. J Neurol 242:1–13
7. Bregman BS (1994) Recovery of function after spinal cord injuries: transplantation strategies. In: Dunnett SB, Björklund A (eds) Functional neural transplantation. Raven Press, New York, pp 489–530 (Advances in neuroscince, vol 2)
8. Brüstle O, McKay RD (1996) Neuronal progenitors as tools for cell replacement in the nervous system. Curr Opin Neurobiol 6:688–695
9. Davies SJA, Field PM, Raisman G (1993) Long fibre growth by axons of embryonic mouse hippocampal neurons microtransplanted into the adult rat fimbria. Eur J Neurosci 5:95–106
10. Dunnett SB, Björklund A (1992) Neural transplantation: a practical approach. IRL, Oxford, pp 1–211 (The practical approach series)
11. Dunnett SB, Björklund A (1994) Introduction. In: Dunnett SB, Björklund A (eds) Functional neural transplantation. Raven Press, New York, pp 1–7 (Advances in neuroscience, vol 2)
12. Dunnett SB, Björklund A (eds) (1994) Functional neural transplantation. Raven Press, New York, pp 1–587 (Advances in neuroscience, vol 2)
13. Dunnett SB, Björklund A (1994) Mechanisms of function of neural grafts in the injured brain. In: Dunnett SB, Björklund A (eds) Functional neural transplantation. Raven Press, New York, pp 531–567 (Advances in neuroscience, vol 2)
14. Dunnett SB, Whishaw IQ, Bunch ST, Fine A (1986) Acetylcholine-rich neuronal grafts in the forebrain of rats: effects of environmental enrichment, neonatal noradrenaline depletion, host transplantation site and regional source of embryonic donor cells on graft size and acetylcholinesterase-positive fiber outgrowth. Brain Res 378:357–373
15. Emmett CJ, Jaques BW, Seeley PJ (1990) Microtransplantation of neural cells into adult rat brain. Neuroscience 38:213–222
16. Freeman TB, Olanow CW, Hauser RA, Nauert GM, Smith DA, Borlongan CV, Sanberg PR, Holt DA, Kordower JH, Vingerhoets FJG, Snow BJ, Calne D, Gauger LL (1995) Bilateral fetal nigral transplantation into the postcommissural putamen in Parkinson's disease. Ann Neurol 38:379–388
17. Gage FH, Ray J, Fisher LJ (1995) Isolation, characterization, and use of stem cells from CNS. Annu Rev Neurosci 18:159–192
18. Groves AK, Barnett SC, Franklin RJM, Crang AJ, Mayer M, Blakemore WF, Noble M (1993) Repair of demyelinated lesions by transplantation of purified O-2A progenitor cells. Nature 362:453–455
19. Iwashita Y, Kawaguchi S, Murata M (1994) Restoration of function by replacement of spinal cord segments in the rat. Nature 367:167–170
20. Langston JW, Widner H, Goetz CG, et al. (1992) Core assessment program for intracerebral transplantations (CAPIT). Mov Disord 7:2–13
21. Leanza G, Nikkhah G, Nilsson OG, Wiley RG, Björklund A (1996) Extensive reinnervation of the hippocampus by embryonic basal forebrain cholinergic neurons grafted into the septum of neonatal rats with selective cholinergic lesions. J Comp Neurol 373:355–370
22. Lindvall O (1994) Neural transplantation in Parkinson's disease. In: Dunnett SB, Björklund A (eds) Functional neural transplantation. Raven Press, New York, pp 103–138 (Advances in neuroscinece, vol 2)
23. Lindvall O, Sawle G, Widner H, Rothwell JC, Björklund A, Brooks D, Brundin P, Frackowiak R, Marsden CD, Gustavii B, Odin P, Rehncrona S (1994) Evidence for long-term survival and function of dopaminergic grafts in progressive Parkinson's diseasse. Ann Neurol 35:172–180
24. Madrazo I, Leon V, Torres C, Del Carmen Aguilera M, Varela G, Alvarez F, Fraga A, Drucker-Colin R, Ostrosky F, Skurovich M, Franco R (1988) Transplantation of fetal substantia nigra and adrenal medulla to the caudate nucleus in two patients with Parkinson's disease. N Engl J Med 318:51
25. McKeon RJ, Schreiber RC, Rudge JS, Silver J (1991) Reduction of neurite outgrowth in a model of glial scarring following CNS injury is correlated with the expression of inhibitory molecules on reactive astrocytes. J Neurosci 11:3398–3411
26. Mendez I, Damaso S, Hong M (1996) Reconstruction of the nigrostriatal pathway by simultaneous intrastriatal and intranigral dopaminergic transplants. J Neurosci 16:7216–7227
27. Molina H, Quinones R, Alvarez L, Galarraga J, Piedra J, Suárez C, Rachid M, Garcia JC, Perry TL, Santana A, Carmenate H, Macias R, Torres O, Rojas MJ, Cordova F, Munoz JL (1991) Transplantation of human fetal mesencephalic tissue in

caudate nucleus as treatment for Parkinson's disease: the Cuban experience. In: Lindvall O, Björklund A, Widner H (eds) Intracerebral transplantation in movement disorders. Elsevier, Amsterdam, pp 99–110 (Restorative neurology, vol 4)

28. Nikkhah G, Bentlage C, Cunningham MG, Björklund A (1994) Intranigral fetal dopamine grafts induce behavioral compensation in the rat Parkinson model. J Neurosci 14 : 3449–3461

29. Nikkhah G, Brandis A (1995) Neurotransplantation bei Morbus Parkinson: Experimentelle und klinische Ergebnisse der funktionellen Rekonstruktion des dopaminergen Systems. Zentralbl Neurochir 4 : 153–160

30. Nikkhah G, Cunningham MG, Cenci MA, McKay R, Björklund A (1995) Dopaminergic microtransplants into the substantia nigra of neonatal rats with bilateral 6-OHDA lesions. I. Evidence for anatomical reconstruction of the nigrostriatal pathway. J Neurosci 15 : 3548–3561

31. Nikkhah G, Cunningham MG, Jödicke A, Knappe U, Björklund A (1994) Improved graft survival and striatal reinnervation by microtransplantation of fetal nigral cell suspensions in the rat Parkinson model. Brain Res 633 : 133–143

32. Nikkhah G, Cunningham MG, McKay R, Björklund A (1995) Dopaminergic microtransplants into the substantia nigra of neonatal rats with bilateral 6-OHDA lesions. II. Transplant-induced behavioral recovery. J Neurosci 15 : 3562a–3570

33. Nikkhah G, Olsson M, Eberhard J, Bentlage C, Cunningham MG, Björklund A (1994) A microtransplantation approach for cell suspension grafting in the rat Parkinson model. A detailed account of the methodology. Neuroscience 63 : 57–72

34. Olanow CW, Kordower JH, Freeman TB (1996) Fetal nigral transplantation as a therapy for Parkinson's disease. Trends Neurosci 19 : 102–109

35. Olsson M, Bentlage C, Cambell K, Wictorin K, Björklund A (1997) Extensive migration and target innervation by striatal precursors after grafting into the neonatal striatum. Neuroscience 79 : 57–78

36. Pakzaban P, Isacson O (1994) Neural xenotransplantation: reconstruction of neuronal circuitry across species barries. Neuroscience 62 : 989–1001

37. Perlow MI, Freed WJ, Hoffer BJ, Seiger Å, Olson L, Wyatt RJ (1979) Brain grafts reduce motor abnormalities produced by destruction of the nigrostriatal dopamine system. Science 204 : 643–647

38. Schnell L, Schneider R, Kolbeck R, Barde Y, Schwab ME (1994) Neurotrophin-3 enhances sprouting of corticospinal tract during development and after adult spinal cord lesion. Nature 367 : 170–173

39. Shannon KM, Kordower JH (1996) Neural transplantation for Huntington's disease: experimental rationale and recommendations for clinical trials. Cell Transplant 5 : 339–352

40. Shewan D, Berry M, Cohen J (1995) Extensive regeneration in vitro by early embryonic neurons on immature and mature CNS tissue. J Neurosci 15 : 2057–2062

41. Snyder-Keller AM, Lund RD (1990) Amphetamine sensitization of stress-induced turning in anmimals given unilateral dopamine transplants in infancy. Brain Res 514 : 143–146

42. Transplantation in the nervous system (1991) Special issue. Trends Neurosci 14 (8)

43. Wictorin K (1992) Anatomy and connectivity of intrastriatal striatal transplants. Prog Neurobiol 38 : 611–639

44. Zhou FC, Chiang YH, Wang Y (1996) Constructing a new nigrostriatal pathway in the parkinsonian model with bridged neural transplantation in substantia nigra. J Neurosci 16 : 6965–6974

Neurosurgical and Restorative Possibilities in the Treatment of Idiopathic Parkinson Syndromes: Neurotransplantation and Neurotrophic Factors

C. Earl, J. Sautter, W. Oertel, and A. Kupsch

Summary

This review focuses on two restorative strategies against Parkinson's disease: (a) intrastriatal transplantation of foetal dopaminergic cells and (b) application of neurotrophic factors.

Neuronal transplantation of dopaminergic foetal ventral mesencephalic cells aims at compensating the striatal dopamine deficit in Parkinson's disease. Clinically, dopaminergic ventral mesencephalic grafts have been demonstrated to ameliorate rigidity, bradykinesia, and efficacy of L-DOPA therapy, for instance, the reduction of L-DOPA induced dyskinesias and on-off fluctuations. Future experimental refinements will concern predominantly the improvement of survival of the transplanted foetal dopaminergic neurones and the use of genetically modified cells.

Neurotrophic factor therapy relates to the prevention and/or restoration of the progressing nigral cell loss in Parkinson's disease. In this regard special emphasis is put on the newly characterised glial cell line derived neurotrophic factor (GDNF). More recent results and perspectives of this concept are discussed.

Introduction

Parkinson's disease (PD) is a slowly progressing neurodegenerative disorder which involves the loss of function of dopaminergic neurones in the A9 (substantia nigra) area [39]. The prevalence of the disease is reported to be 1/1000 in the total population and 1% in the population above 65 years of age. PD is characterised by the loss of melanin-containing neurones in the pars compacta region of the substantia nigra and a reduction in striatal dopamine. Dopaninergic therapy for the treatment of PD (e.g. L-DOPA or dopaminergic agonists in monotherapy or combination therapy) is satisfactory at the early stages of the disease. However, after 3–7 years of L-DOPA treatment the majority (approximately 60%) of PD patients present with side effects, such as dyskinesias and on-off fluctuations (for review see [24, 55]).

Pathophysiology

The pathological hallmark of PD, the degeneration of dopaminergic neurones in the subtantia nigra, leads to a dopamine deficit in the putamen and the caudate nucleus. Consequently, as shown by autoradiographic (glucose uptake) and electrophysiological studies, the output nuclei of the basal ganglia circuit, the nucleus subthalamicus (NST; main neurotransmitter: glutamate) and globus pallidus internus (Gpi; main neurotransmitter: GABA) are overactive (for review see [15a, 62]; see also [18, 19]; see Fig. 1). Using the non-human primate model of PD, the 1-methyl-4-phenyl-1,2,3,6-tetrahydropyridine (MPTP) treated monkey, it has been possible to show that lesioning of the NST leads to a compensation of the behavioural toxin-induced deficit [10, 81].

Thus the flow of information in the basal ganglia in PD patients is considered to be associated with an overactivity of glutaminergic NST neurones and GABAergic GPi neurones which leads to a pathological inhibition of the thalamus (see. Fig. 1) [1, 42] (for review see [28, 56]).

Implantation of Dopaminergic Cells

The neuronal transplantation technique aims at replacing the degenerated striatal dopamine via the implantation of dopaminergic embryonic ventral mesencephalic cells. Following the development of animal models for parkinsonism with selective neurotoxins, such as 6-hydroxydopamine (6-OHDA), it became possible to demonstrate that intrastiatally grafted foetal dopaminergic tissue is able to reverse dopamine deficiency-related disorders in adult animals [11, 63]. In the middle 1980s the discovery of the selective substantia nigra toxin MPTP furthered the understanding of PD and allowed the inclusion of non-human primates in transplantation research (for review see [50]).

Fig. 1.
Simplified scheme showing the motor circuitry in Parkinson's disease. Nigrostriatal dopamine neurones project to the striatum. Dopamine physiologically inhibits D_2 receptors and excites D_1 receptors. Excitatory (glutamate) and inhibitory (GABA) projections between the basal ganglia nuclei are delineated by open and closed arrows, respectively. The projection from the putamen to the Gpe (indirect pathway) contains encephalin as a neuropeptide while the direct pathway (Putamen-Gpi) colocalizes with the neuropeptides dynorphine and substance P. *Dashed lines*, the different neurosurgical possibilities

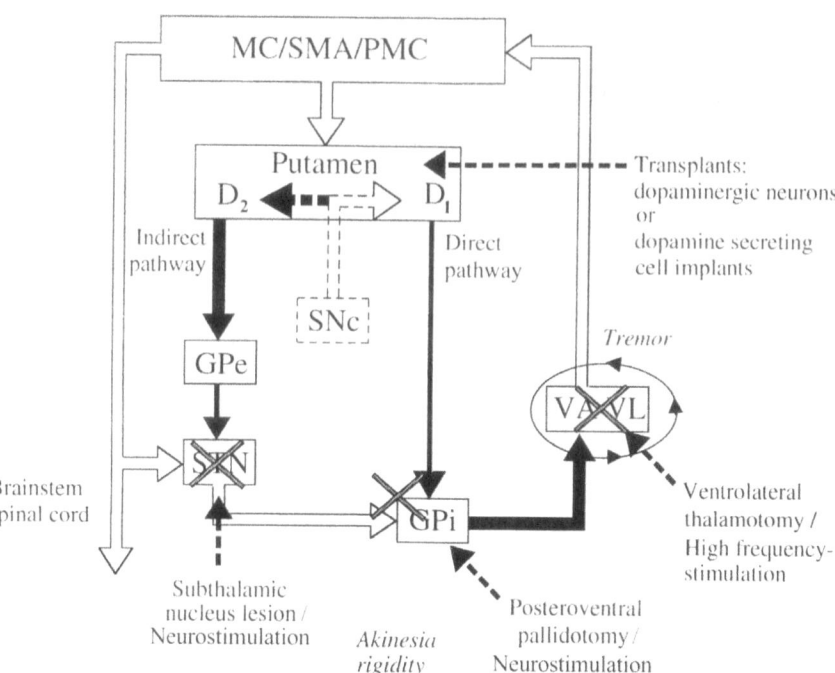

Transplantation Studies in the Rat. The basic knowledge concerning implantation research has been conducted mainly into the rat. Injections of the neurotoxin 6-OHDA into the medial forebrain bundle of the rat result in a stable and highly reproducible dopaminergic-depletion in the ipsilateral striatum leaving the contralateral side unaffected [16]. 6-OHDA treated animals show a spontaneous rotation behaviour which is amplified, and can be quantified, following the systemic administration of dopaminergic agonists such as amphetamine or apomorphine. The recorded turning behaviour allows the effects of implanted dopaminergic nerve cells to be studied. 6-OHDA lesioned animals which have received an intrastriatal implant of dopaminergic foetal cells display a complete compensation in rotational asymmetry, although the compensation is incomplete in other behavioural task (such as in the paw reaching test).

Adrenal Medullary Tissue

From a historical point of view adrenal medullary tissue was the first grafted tissue to be extensively used in the clinic. However, in view of the poor reproducibility rate and the modest clinical benefit (roughly 30% of the patients experience no effect, 30% deteriorate, and 30% benefit transiently for approximately 6–12 months) most centres have abandoned this approach [25, 26]. The cotransplantation of adrenergic tissue and Schwann cells is another

approach and is presently under clinical investigation. The rational behind this approach is the potential production of trophic factors by the Schwann cells and hence might improve the survival rate of the cotransplantated adrenal cells [79]. The first case reports describe an approximately 30% postoperative improvement after 2 years in the parkinsonian symptoms [6,14], however, further carefully designed studies on a small number of patients are warranted to further evaluate this approach.

Foetal Mesencephalic Cells

Implantation studies in animal models of PD have clearly shown the beneficial effects of foetal grafts of dopaminergic mesencephalic tissue which contains the precursor cells of the substantia nigra and the ventral tegmental area [11, 63]. Foetal ventral mesencephalic tissue has a number of advantages over adrenal medullary tissue. For instance, they display a better postoperative survival rate and potential to establish synaptic connections than cells from the adrenal medulla. To date more than 300 PD patients have undergone implantation of human foetal mesencephalic tissue into the striatum at international centres including Belgium, the People's Republic of China, Cuba, the United Kingdom, France, Mexico, Russia, Switzerland, Sweden, the former Czechoslovakia and the United States. However controlled and detailed studies with thoroughly evaluated patients only approximate 30 patients.

Furthermore, results from different centres are often difficult to compare secondary to the divergent evaluation criteria employed. For this reason the Network on European CNS Transplantation and Restoration (NECTAR) in collaboration with scientists in the United States developed the so called CAPIT (core Assessment Protocol for Intracerebial Transplantation) criteria [13]. CAPIT aims at providing guide lines which allow the comparison of different transplanted patients from different centres. In a modified version CAPIT may also be suitable for other surgical interventions in PD patients. It should also be mentioned that NECTAR has provided ethical guidelines for the use of embryonic tissue [12].

In 80 % of the CAPIT-assessed implanted patients one can see a clear therapeutic effect. Both rigidity and bradykinesia are improved primarily on the side contralateral to the implant. There is an amelioration in drug tolerance, reduction in L-DOPA induced dyskinesias, lengthening of the time spent in on, and a reduction in the time spent in off. These effects become apparent 4 – 7 months after the operation. This improvement is presumably due to the outgrowing and maturing dopaminergic fibres within the implanted striatum. To date none of the transplanted patients have been reported to suffer from personality changes [67] although further studies are necessary to investigate this important issue. Using PET with ¹⁸F-fluoro-DOPA it is also possible to show that dopaminergic tissue implanted to the DA-deficient striatum is able to survive for a minimum of 6 years. Furthermore the increase in the PET fluoro-DOPA signals correlates with clinical improvement. In 1995 Kordower et al. [47] published a post-mortem report that described the survival of implanted dopaminergic tissue in the striatum of a patient with PD. The patient had died 18 months after the transplant surgery secondary to an operation unrelated to the transplantation procedure. The report describes that about 200 000 surviving implanted cells were found extensively reinnervating the striatum and representing the first direct evidence that human VM cells survive transplantation in PD patients.

The mechanisms of action of VM grafts have not been completely elucidated. According to present knowledge foetal grafts may be best compared to an "endogenous dopamine pump with a incorporated buffer system". This model maintains that systemically administered L-DOPA and/or locally and/or cellularly produced dopamine may be "buffered" more efficiently in the environment of the transplant. This would explain why L-DOPA therapy causes less dyskinesias after transplantation. Alternatively, therapeutic effects could be partially explained by the release of neurotrophic factors. Thus it has been often speculated that the transient therapeutic effect of adrenal medullary implants may be related to stimulation of surviving dopaminergic nigrostriatal neurones rather than to implantation of catecholaminergic cells.

Future therapeutic grafting strategies embrace the implantation of genetically manipulated cells as vehicle for dopamine replacement, the coapplication of nerve growth factors such as GDNF and/or the application of xenografts. Clinically, a group in Boston (O. Isacson, personal communication) has already xenografted tissue from pig embryos into the striatum of more than seven PD patients; however, detailed accounts are pending. Nevertheless it is evident that xenografting emphasises the numerous open, immunological questions in intracerebral grafting. The open questions in intracerebral transplantation include the following:

1. Which PD patients should be selected for neuronal transplantation. So far it only seems clear that tremor should not be the leading symptom in patient selection.
2. Should transplantation be performed in the early or late stages of the disease?
3. How many cells should be transplanted and to what site (caudate and/or putamen)? At present tissue from four embryos seem to be required per hemisphere.
4. What is the role of non-dopaminergic cells which are cotransplanted?
5. Does immunosuppression increase the survival of the embryonic tissue? What immunosuppressive drugs should be administered? Which cells present antigens? At present it is agreed that tapering off immunosuppression 12 months post grafting does not adversely affect graft function and survival (as assessed by PET studies).
6. Does continued L-DOPA treatment influence graft survival and/or function?

Neurotrophic Factors

During development neurones are abundantly generated in vertebrate nervous system. However, only a proportion of these neurones survive after their axons have reached their target area. This is thought to be the consequence of successful competition of the surviving neurones for a specific target-derived, retrogradely transported neurotrophic factor, present only in limited amounts in the target fields. Nerve growth factor (NGF) must be considered the prototype of such a neurotrophic factor. Apart from NGF the family of the so-called 'neurotrophins' subsumes at present brain-derived neurotrophic factor (BDNF), neurotrophin-3 (NT-3) and neurotrophin-4/5). However, the concept of neurotrophic factors as specific, target-derived molecules, each acting on distinct neuronal types, must be modified because of their high degree of pleiotropism and a considerable overlap in biological activities (for review see [48]). In addition, a number of other non-target-derived molecules exert trophic actions on certain neurones and non neuronal cells. These factors include ciliary neurotrophic factor (CNTF), fibroblast growth factors 1 and 2 (FGF-1/FGF-2), insulin-like growth factors 1 and 2 (IGF-1/IGF-2), muscle-derived differentiation factor (MDF) and members of the transforming growth factor β-superfamily such as GDNF. Several observations in experimental and clinical research suggest the existence of a trophic factor or factors for dopaminergic neurones and are subsequently reviewed.

Glial Cell Line Derived Neurotrophic Factor

GDNF was purified and cloned in 1993, identified by its promotion of survival and morphological differentiation of dopaminergic neurones and increase of their high-affinity dopamine uptake in embryonic midbrain cultures [53]. Furthermore, specific trophic actions have been described for embryonic motor neurones in vitro as well as in vivo [31, 59, 84]. In vivo, it was shown that GDNF acts on transmitter activity in midbrain dopaminergic pathways and on spontaneous and amphetamine-induced motor behaviour in unlesioned rats and induces sprouting of tyrosine hydroxylase-immunoreactive (TH-IR) neurites towards its injection site, suggesting a role for GDNF as a neurotrophic factor for adult dopaminergic neurones in rats [34]. In the 6-hydroxydopamine (6-OHDA) lesion model of the rat neurochemical and behavioural improvements following intranigral administration of GDNF were demonstrated [32]. Following hemitransection of the medial forebrain bundle in rat a significant protection of TH-IR nigral neurones by GDNF was reported [9]. Daily injections of 5 µg (1 µg) GDNF for 14 days above the substantia nigra led to a survival of 84% (82%) of TH-IR nigral cells, whereas in control animals only 50% of these cells survived. In the MPTP mouse model of PD the protective effects of GDNF were investigated in respect to the timepoint of its administration [78]. Striatal injections of GDNF (10 µg on 2 consecutive days) 24 h before MPTP exposure significantly reduced the normally observed TH-IR cell loss of 30% to about 15%. In addition, this effect was accompanied by a protection of density of dopaminergic nerve terminals and dopamine levels in striatum as well as by increased motor behaviour. When GDNF was administered 7 days after MPTP exposure, there was still some recovery of TH-IR cells, increased locomotor behaviour and partial restoration of dopamine levels. However, implantation of encapsulated GDNF-producing cells in rats with unilateral dopamine depletion and parkinsonian symptoms does not decrease apomorphine-induced rotations [54]. Nonetheless, a TH-positive fibre ingrowth into the membranes of the applied capsules was observed.

The most encouraging results stem from a recent investigation employing an animal model with a delayed cell death of nigral dopamine neurones [68, 70]. Animals received an unilateral injection of 6-OHDA into the striatum resulting in nigral degeneration with onset at 1 week and an extensive death of nigral neurones 4 weeks post-lesion. Administration of GDNF for 4 weeks over the substantia nigra at a cumulative dose of 140 µg, starting on the day of lesion, completely prevented nigral cell death and atrophy. A partial protective effect was observed as a consequence of one single injection of 10 µg GDNF at one week post-lesion. The ability of GDNF to induce neurite outgrowth on nigral dopaminergic neurones is further supported by its ability to increase survival, growth and function of intraocular [41, 76] and intrastriatal feotal ventral mesencephalic dopaminergic grafts [66]. These results have been recently confirmed in the non-human MPTP-primate model, where GDNF partially presented both clinical and histological MPTP-induced lesions of the nigrostriated dopamine circuit [86]. In conclusion, GDNF is protective in mechanical and toxic lesions and does represent the most promising candidate for growth factor therapy in PD at present. However, possible adverse effects of GDNF, for instance sprouting or aberrant growth in non-diseased systems, remain to be studied (for review see [87]).

Brain-Derived Neurotrophic Factor

In vitro BDNF revealed distinct effects on survival, morphological differentiation, neuritic growth, protection against MPP$^+$ cytotoxicity and dopamine uptake of foetal mesencephalic dopaminergic neurones [7, 35, 44, 74, 77]. In a preliminary report this neurotrophin was shown to be retrogradely transported to substantia nigra after injection into the striatum indicating that there are functional receptors for BDNF on adult dopaminergic nigrostriatal neurones [82]. Additionally, BDNF has been reported to prevent cell death of axotomized spinal motor neurones in vivo [83]. However, in vivo studies on the neuroprotective function of BDNF on dopaminergic neurones are at present contradictory. The reduction of TH-IR nigral cells in rats after hemitransection of the nigrostriatal forebrain bundle was not prevented by intraventricular infusion of BDNF [45]. Futhermore, no improvement of survival of transplanted foetal ventral mesencephalic neurones by daily intra-striatal injection or chronic intraventricular infusion of BDNF in the 6-OHDA model of the rat was observed [69]. Moreover, mice lacking BDNF showed no affection of survival of midbrain dopaminergic neurones [17b]. In contrast, BDNF-secreting fibroblasts which were implanted near the substantia nigra 7 days before MPP$^+$ infusion markedly increased nigral dopaminergic neuronal survival in the rat [21]. Additionally, it has been reported that intranigral BDNF infusions increased amphetamine-induced rotations and enhanced striatal dopamine meta-bolism, suggesting a presynaptic effect of BDNF on nigrostriatal dopamine system [2]. In summary, experimental evidence concerning the neurotrophic effect of BDNF on nigral dopaminergic neurones is controversial. Thus at present clinical studies with BDNF in PD patients do not seem to be warranted.

Nerve Growth Factor

In vitro no trophic effects of NGF on mesencephalic cultures have been shown so far [43]. In vivo NGF does not protect axotomized nigrostriatal neurones in the adult rat [45]. On the other hand, intraventricular injections of NGF a reported to increase striatal dopamine contents in MPTP-treated mice [23]. NGF might be of interest because of its trophic actions on sympathetic ganglia cells. In vitro catecholaminergic adrenal medullary cells resemble sympathic ganglia cells after application of NGF [4]. An improved survival of transplanted adult adrenal medulla cells after simultaneous, intraventricular application of NGF in 6-OHDA lesioned rats was observed [75]. In non-human primates a trophic support for grafted rhesus adrenal chromaffin cells by cografting excised peripheral nerve as a source for NGF delivery has been reported [46]. Clinically after implantation of autologous adrenal medulla one patient received intraputaminal infusions of NGF followed by a moderate improvement in clinical symptoms 7 months after surgery [58]. No follow-up has been published. Corresponding studies have been initiated in the United States, but with discouraging results (C. Goetz, personal communication). Furthermore, in a few clinical case reports cotransplantation of adrenergic tissue and schwann cells has been communicated, describing on approximately 30% post-operative improvement after two years of the PD symptoms [6, 14]. However, further carefully designed studies on a small number of patients are warrented to evaluate this approach.

Neurotrophin-3 and Neurotrophin-4/5

Apart from BDNF and NGF possible trophic actions of other neurotrophins on dopaminergic neurones remain to be investigated. However, recently NT-3 and NT-4/5 were shown to influence morphological differentiation of rat mesencephalic dopaminergic neurones in vitro [77]. NT-3 mRNA was detected in substantia nigra by in situ hybridisation [22]. Contradictory results exist in respect to whether or not NT-3 promotes survival of mesencephalic dopaminergic neurones in vitro [36, 44]. NT-4/5 elicited a seven-fold increase in the number of cultured dopaminergic neurones as well as an augmentation in dopamine content. In contrast, NT-4/5 had no effect on dopamine uptake capacity [36]. Additionally, 2-week supranigral infusions of NT-4/5 were shown to elevate the turnover of dopamine through both metabolic and release pools and augment the behavioural response to d-amphetamine in rats [3].

Ciliary neurotrophic factor

CNTF exerts protective and neurotrophic functions on embryonic and postnatal lesioned motoneurones in vitro and in vivo. It promotes survival of a wide range of other embryonic neurones in peripheral and central nervous system in vitro (for review see [80]). The most striking features of CNTF until 1995 are the in vivo prevention of lesion-induced degeneration of facial motoneurones after axotomy in newborn rats and the antagonization of neurodegenerative changes in the progressive motor neuronopathy (pmn) genetic mouse model of amyotrophic lateral sclerosis [72,73]. Furthermore, CNTF was shown to rescue nigral, most likely dopaminergic neurones in the hemitransection model of the rat (as assessed by Nissl staining), but it did not protect against axotomy-induced reduction of nigral TH-IR cells [30]. It would be of considerable interest whether or not the TH-negative neurones will be able to synthesis again this rate-limiting enzyme of dopamine synthesis at a later timepoint. In preclinical trials the systemic or intrathecal application of CNTF in rats and sheep led to severe side effects including fever, cachexia, disturbance of the blood brain barrier and others [71], thereby limiting a possible therapeutic clinical application.

Epidermal Growth Factor

In vitro EGF was shown to support survival of both embryonic dopaminergic midbrain and cholinergic forebrain neurones [43]. The trophic actions of EGF required the presence of glial cells proposing an indirect mode of action of EGF on dopaminergic neurones. In this context, it was previously reported that EGF acts also on glial cells promoting their ability to proliferate and to differentiate [33, 52]. In vivo studies revealed that intraventricular administration of EGF in rats 5 weeks after hemitransection of medial forebrain bundle restores about 20% of TH-IR nigral neurones in comparison to vehicle-treated animals [64]. Moreover, intraventricular infusion of EGF also accelerates recovery of striatal dopaminergic parameters, i.e. the dopamine content and TH activity, in the MPTP mouse model [29]. However, corresponding non-human primate studies have not been performed so far. Further investigations should also evaluate possible glial reactions induced by application of EGF.

Fibroblast Growth Factor-2

FGF-2 (previous denotion: basic fibroblast growth factor) exhibits trophic effects on central embryonic dopaminergic and GABA-ergic neurones in culture which appear to be glia mediated [17a]. Transient increases in the amounts of FGF-2 have been described in distinct lesion pradigms of the CNS [20, 27]. The same was suggested to occur in a MPTP-induced lesion of the nigrostriatal dopaminergic system in mice [51]. This may indicate a possible role of FGF-2 in neuronal regeneration, for instance in form of induction of synthesis of NGF or of other trophic molecules in astrocytes [85]. In 6-OHDA-lesioned rats intrastriatal FGF-2 infusions neither prevent striatal dopamine depletions nor diminish behavioural deficits [61]. In the MPTP-mouse model intrastriatal application of FGF-2 via gel foam partially attenuated the toxin-induced damage [60]. However, this effect was only observed if FGF-2 was applied simultaneously or 3 days after intraperitoneal MPTP-injection, whereas a delay of FGF-2 administration vor 7 days after MPTP-injection aborted restoration of transmitters and TH-levels. Apart from these findings, FGF-2 ameliorates rotational behaviour of substantia nigra-transplanted rats with lesions of the nigrostriatal dopaminergic system [57]. Results of ongoing non-human primate studies remain to be awaited.

Insulin-like Growth Factor-1

In vitro IGF-1 was shown to stimulate dopamine uptake of foetal dopaminergic mesencephalic neurones [43]. In addition, the predominant form of IGF-1 in the CNS, des-IGF-1, was very effective in promoting survival of cultured mesencephalic neu-

rones [7]. In vivo, neurotrophic actions of IGF-1 have to our knowledge not been demonstrated. However, IGF-1 gene disruption does not affect the number of mesencephalic dopaminergic neurones [8], and there was no IGF-1 binding in rat substantia nigra or striatum measured by autoradiography [5]. These findings are discouraging in respect to a potential therapeutic effect of IGF-1 in treatment of PD.

Muscle-Derived Differentiation Factor

In vitro MDF induces tyrosine hydroxylase-expression in a variety of CNS neurones, including those of striatum, cerebellum and cortex [37]. Normally, i.e. without MDF, these neurones do not express this enzyme of catecholamine synthesis. Further in vitro studies revealed that MDF enhances TH-mRNA 40-fold in foetal mesencephalic neurones [38]. Preliminary in vivo studies employing infusion of partially isolated MDF reported this molecule to enhance TH activity in dopamine-depleted striata of 6-OHDA-lesioned rats [40]. Furthermore, an increase in striatal dopamine concentrations and a partial compensation of rotational asymmetry were observed. In contrast, dopaminergic parameters were not affected by administration of MDF in control rats suggesting that adult dopaminergic neurones may regain sensitivity towards differentiation factors after lesion.

Transforming Growth Factor-β

GDNF was identified as a member of the TGF-β superfamily. Therefore the potential trophic actions on dopaminergic neurones of other members of this superfamily attracted attention. It has been shown that TGF-β1, TGF-β2, TGF-β3 and activin A exert a survival-promoting activity on cultured dopaminergic neurones of the developing substantia nigra [49, 65]. In addition, TGF-β2 and TGF-β3 mRNAs were detected in developing rat striatum and substantia nigra [65]. However, TGF-β3 did not prevent delayed degeneration of nigral dopaminergic neurones following intrastriatal 6-OHDA lesion [70].

Conclusion

Neurodegeneration in PD is a slow and progressive process that occurs over years before any symptoms of the disease appear. Most of the hitherto utilised animal models, however, lead to a rapid onset of degenerative events. Therefore it is of interest that after intrastriatal 6-OHDA lesion a delayed cell death of nigral neurones may be observed. Although the prevention of this delayed nigral cell death by in vivo application of GDNF features a promising attempt for a possible clinical use of neurotrophic factors in the future, adequate data in respect to the effects of for instance GDNF on non-dopaminergic neurones are not available. The mode of administration represents another major problem in "growth factor therapy" of PD, since the growth factor-proteins possess a relatively high molecular weight preventing them to cross the blood-brain-barrier. An already tested strategy of delivery may be the implantation of genetically modified cells which are able to secrete neurotrophic factors (Lindner et al. 1995). Additionally, one prerequisite for treating PD patients with neurotrophic factors appears to be an early diagnosis of the disease. This is in principle possible with [^{18}F]fluorodopa PET and βCIT-SPECT, although corresponding prospective studies are not available on a larger scale.

In summary, at present neurotrophic growth factors have not yet been sufficiently investigated in respect to a therapeutic application in PD and correspondingly further experiments including non-human primate studies are necessary to study potential beneficial and adverse effects of long-term application of neurotrophic factors before clinical studies should be initiated to a larger extent.

References

1. Albin RL (1995) The pathophysiology of chorea/ballism and parkinsonism. Park Rel Dis 1:3–11
2. Altar CA, Boylan CB, Jackson C, Hershenson S, Miller J, Wiegand SJ, Lindsay RM, Hyman C (1992) Brain-derived neurotrophic factor augments rotational behavior and nigrostriatal dopamine turnover in vivo. Proc Natl Acad Sci USA 89:11347–11351
3. Altar CA, Boylan CB, Fritsche M, Jackson C, Hyman C, Lindsay RM (1994) The neurotrophins NT-4/5 and BDNF augment serotonin, dopamine, and GABAergic systems during behaviorally effective infusions to the substantia nigra. Exp Neurol 130:31–40
4. Anderson DJ, Axel R (1986) A bipotential neuroendocrine precursor whose choice of cell fate is determined by NGF and gluco-corticoids. Cell 47:1079–1090
5. Araujo DM, Lapchak PA, Collier B, Chabot JG, Quirion R (1989) Insulin-like growth factor-1 (somatomedin C) receptors in the rat brain: distribution and interaction with the hippocampal cholinergic system. Brain Res 484:130–139
6. Bakey RAE (1995) Stereotactic intrastriatal cografts of autologous adrenal medulla (AM) and peripheral nerve (PN) improves motor performance in Parkinson's disease, comment. Neurosurgery 37:518–519
7. Beck KD, Knüsel B, Hefti F (1993) The nature of the trophic action of brain-derived neurotrophic factor, des (1-3)-Insulin-like growth factor-1, and basic fibroblast growth factor on mesencephalic dopaminergic neurons developing in culture. Neuroscience 52:855–866
8. Beck KD, Powell-Braxton L, Widmer HR, Valverde J, Hefti F (1995) Igf1 gene disruption results in reduced brain size, CNS hypomyelination and loss of hippocampal granule and striatal parvalbumin-containing cells. Neuron 14:717–730
9. Beck KD, Valverde J, Alexi T, Poulsen K, Moffet B, Vandlen RA, Rosenthal A, Hefti F (1995) Mesencephalic dopaminergic neurons protected by GDNF from axotomy-induced degeneration in the adult brian. Nature 373:339–341
10. Bergman H, Wichmann T, DeLong MR (1990) Reversal of experimental parkinsonism by lesions of the subthalamic nucleus. Science 249:1346–1348
11. Björklund A, Stenevi U (1979) Reconstruction of the nigrostriatal dopamine pathway by intracerebral nigral transplants. Brain Res 177:555–560
12. Boer G, on behalf of NECTAR (1994) Ethical guidelines for the use of human embryonic or fetal tissue for experimental and clinical neurotransplantation and research. J Neurol 242:1–13
13. CAPIT Committee (1991) Core Assessment program for intracerebral transplantations. In: Lindvall O, Björklund A, Widner H (eds) Intracerebral transplantation in movement disorders. Restorative neurology, vol 4, Elsevier, Amsterdam, pp 232–241
14. Date I, Asari S, Ohmoto T (1995) Two-year follow-up study of a patient with Parkinson's disease and severe motor fluctuations treated by co-grafts of adrenal medulla and peripheral nerve into bilateral caudate nuclei: case report. Neurosurgery 37:515–519
15a. DeLong MR (1990) Primate models of movement disorders of basal ganglia origin. Trends Neurosci 13:281–285
15b. Dogali M, Fazzini E, Kolodny E, Eidelberg D, Sterio D, Devensky O, Beric A (1995) Stereotactic ventral pallidotomy for Parkinson's disease. Neurology 45:753–761
16. Earl CD, Reum T, Xie, J-X, Sauter J, Kupsch A, Oertel WH, Morgenstern R (1996) Foetal nigral cell suspension grafts influence dopamine release in the non-grafted side in the 6-hydroxydopamine rat model of Parkinson's disease: in vivo voltammetric data. Exp Brain Res 109:179–184
17a. Engele J, Bohn MC (1991) The neurotrophic effects of fibroblast growth factor in vitro are mediated by mesencephalic microglia. J Neurosci 11:3070–3078
17b. Ernfors P, Lee KF, Jaenisch R (1994) Mice lacking brain-derived neurotrophic factor develop with sensory deficits. Nature 368:147–150
18. Filion M, Tremblay L (1991) Abnormal spontaneous activity of globus pallidus neurons in monkeys with MPTP-induced parkinsonism. Brain Res 547:142–151
19. Filion M, Tremblay L, Bedard P (1991) Effects of dopamine agonists on the spontaneous activity of globus pallidus neurons in monkeys with MPTP-induced parkinsonism. Brain Res 547:152–161
20. Frautschy SA, Walicke PA, Baird A (1991) Localization of basic fibroblast growth factor and its mRNA after CNS injury. Brain Res 553:291–299
21. Frim DM, Uhler TA, Galpern WR, Beal MF, Breakefield XO, Isacson O (1994) Implanted fibroblasts genetically engineered to produce brain-derived neurotrophic factor prevent 1-methyl-4-phenylpyridinium toxicity to dopaminergic neurons in the rat. Proc Natl Acad Sci USA 91:5104–5108
22. Gall CM, Gold SJ, Isackson PJ, Seroogy KB (1992) Brain-derived neurotrophic factor and neurotrophin-3 mRNAs are expressed in ventral midbrain regions containing dopaminergic regions. Mol Cell Neurosci 3:56–63
23. Garcia E, Rios C, Sotelo J (1992) Ventricular injection of nerve growth factor increases dopamine content in the striata of MPTP-treated mice. Neurochem Res 17:979–982
24. German Parkinson Study Group (1995) Pharmakotherapie des Morbus Parkinson. Neuropsychiatrie 9 [Suppl 1] S 45–S 58

25. Goeth CG, Tanner CM, Penn RD, Stebbins III GT, Gilley DW, Shannon KM, Klawans HL, Comella CL, Wilson RS, Witt T (1990) Adrenal medullary transplants to the striatum of patients with advanced Parkinson's disease: 1-year motor and psychomotor data. Neurology 40 : 273–276

26. Goetz CG, Stebbins GT, Klawans HL, Koller WC, Grossman RF, Bakey RAE, Penn RD (1991) United Parkinson Foundation neurotransplantation registry on adrenal medullary transplants: presurgical and 1-year and 2-year follow-up. Neurology 41 : 1719–1722

27. Gomez-Pinilla F, Lee JWK, Cotman CW (1992) Basic FGF in the adult brain: cellular distribution and response to entorhinal lesion and fimbria-fornix transection. J Neurosci 12 : 345–355

28. Guaridi J, Luquin MR, Herrero MT, Obeso JA (1993) The subthalamic nucleus: a possible target for stereotaxic surgery in Parkinson's disease. Movement Dis 8 : 421–429

29. Hadjiconstantinou M, Fitkin JG, Dalia A, Neff NH (1991) Epidermal growth factor enhances striatal dopaminergic parameters in the 1-methyl-4-phenyl-1,2,3,6-tetrahydropyridine-treated mouse. J Neurochem 57 : 479–482

30. Hagg T, Varon S (1993) Ciliary neurotrophic factor (CNTF) prevents axotomy-induced degeneration of adult rat substantia nigra dopaminergic neurons. Proc Natl Acad Sci USA 90 : 6315–6319

31. Henderson CE, Phillips HS, Pollock RA, Davies AM, Lemeulle C, Armanini M, Simpson LC, Moffet B, Vandlen RA, Koliatsos VE, Rosenthal A (1994) GDNF: a potent survival factor for motoneurons present in peripheral nerve and muscle. Science 266 : 1062–1066

32. Hoffer BJ, Hoffmann A, Bowenkamp K, Huettl P, Hudson J, Martin D, Lin LFH, Gerhardt GA (1994) Glial cell line-derived neurotrophic factor reverses toxin-induced injury to midbrain dopaminergic neurons in vivo. Neurosci Lett 182 : 107–111

33. Honegger P, Guentert-Lauber B (1983) Epidermal growth factor (EGF) stimulation of cultured brain cells. I. Enhancement of the developmental increase in glial enzymatic activity. Dev Brain Res 11 : 245–251

34. Hudson J, Granholm AC, Gerhardt GA, Henry MA, Hoffmann A, Biddle P, Leela NS, Mackerlova L, Lile JD, Collins F, Hoffer BJ (1995) Glial cell line-derived neurotrophic factor augments midbrain dopaminergic circuits in vivo. Brain Res Bull 36 : 425–432

35. Hyman C, Hofer M, Barde YA, Juhasz M, Yancopoulos GD, Squinto SP, Lindsay RM (1991) BDNF is a neutrophic factor for dopaminergic neurons of the substantia nigra. Nature 350 : 230–232

36. Hyman C, Juhasz M, Jackson C, Wright P, Ip NY, Lindsay RM (1994) Overlapping and distinct actions of the neurotrophins BDNF, NT-3, and NT-4/5 on cultured dopaminergic and GABAergic neurons of the ventral mesencephalon. J Neurosci 14 : 335–347

37. Iacovitti L (1991) Effects of a novel differentiation factor on the development of catecholamine traits in noncatecholamine neurons from various regions of the rat brain: studies in tissue culture. J Neurosci 11 : 2403–2409

38. Iacovitti L, Evinger MJ, Stull ND (1992) Muscle-derived differentiation factor increases expression of the tyrosine hydroxylase gene and enzyme activity in cultured dopamine neurons from the rat midbrain. Mol Brain Res 16 : 215–222

39. Jellinger KA (1991) Pathology of Parkinson's disease. Changes other than the nigrostraital pathway. Mol Chem Neuropathol 14 : 153–197

40. Jin BK, Schneider JS, Du YY, Iacovitti L (1991) MDF, a muscle factor, produces partial motor recovery in 6-hydroxydopamine lesioned rats by increasing tyrosine hydroxylase activity and catechol levels. Soc Neurosci Abstr 18 : 1296

41. Johansson M, Friedemann M, Hoffer B, Strömberg I (1995) Effects of glial cell lined-derived neurotrophic factor on developing and mature ventral mesencephalic grafts in oculo. Exp Neurol 134 : 25–34

42. Klockgether T, Turski L (1993) Toward an understanding of the role of glutamate in experimental parkinsonism: agonist-sensitive sites in the basal ganglia. Ann Neurol 34 : 585–593

43. Knüsel B, Michel PP, Schwaber JS, Hefti F (1990) Selective and nonselective stimulation of central and cholinergic and dopaminergic development in vitro by nerve growth factor, basic fibroblast growth factor, epidermal growth factor, insulin and the insulin-like growth factors I and II. J Neurosci 10 : 558–570

44. Knüsel B, Winslow JW, Rosenthal A, Burton LE, Seid DP, Nikolics K, Hefti F (1991) Promotion of central cholinergic and dopaminergic neuron differentiation by brain-derived neurotrophic factor but not neurotrophin 3. Proc Natl Acad Sci USA 88 : 961–965

45. Knüsel B, Beck KD, Winslow JW, Rosenthal A, Burton LE, Widmer HR, Nikolics K, Hefti F (1992) Brain-derived neurotrophic factor administration protects basal forebrain cholinergic but not nigral dopaminergic neurons from degenerative changes after axotomy in the adult rat brain. J Neurosci 12 : 4391–4402

46. Kordower JH, Fiandaca MS, Notter MFD, Hansen JT, Gash DM (1990) NGF-like trophic support from peripheral nerve for grafted rhesus adrenal chromaffin cells. J Neurosurg 73 : 413–428

47. Kordower JH, Freeman TB, Snow BJ, Vingerhoets FJG, Mufsan EJ, Sanberg PR, Hauser RA, Smith DA, Nauert GM, Perl DP, Olanow CW (1995) Neuropathological evidence of graft survival and striatal innervation after the transplantation of fetal mesencephalic tissue in a patient with Parkinson's disease. N Eng J Med 332 : 1118–1124

48. Korsching S (1993) The neurotrophic factor concept: a reexamination. J Neurosci 13 : 2739–2748

49. Krieglstein K, Suter-Crazzolara C, Fischer WH, Unsicker K (1995) TGF-beta superfamily members promote survival of midbrain dopaminergic neurons and protect them against toxicity. EMBO J 14 : 736–742

50. Kupsch A, Oertel WH (1994) Neural transplantation, trophic factors and Parkinson's disease. Life Sci 55 : 2083–2095

51. Leonard S, Luthman D, Logel J, Luthman J, Antle C, Freedman R, Hoffer B (1993) Acidic and basic fibroblast growth factor mRNAs are increased in striatum following MPTP-induced dopamine neurofiber lesion: assay by quantitative PCR. Mol Brain Res 18 : 275–284

52. Leutz A, Schachner M (1981) Epidermal growth factor stimulates DNA-synthesis of astrocytes in primary cerebellar cultures. Cell Tiss Res 220 : 393–404

53. Lin LFH, Doherty J, Lile J, Bektesh S, Collins F (1993) GDNF: a glial cell line-derived neurotrophic factor for midbrain dopaminergic neurons. Science 260 : 1130–1132

54. Lindner MD, Winn SR, Baetge EE, Hammang JP, Gentile FT, Doherty E, McDermott PE, Frydel B, Ullman MD, Schallert T, Emerich DF (1995) Implantation of encapsulated catecholamine and GDNF-producing cells in rats with unilateral dopamine depletions and parkinsonian symptoms. Exp Neurol 132 : 62–76

55. Marsden CD (1994) Parkinson's disease. J Neurol Neurosurg Psychiat 57 : 672–681

56. Marsden CD, Obeso JA (1994) The functions of the basal ganglia and the paradox of stereotaxic surgery in Parkinson's disease. Brain 117 : 877–897

57. Matsuda S, Saito H, Nishiyama N (1992) Basic fibroblast growth factor ameliorates rotational behavior of substantia nigra-transplanted rats with lesions of the dopaminergic nigrostriatal neurons. Jpn J Pharmacol 59 : 365–370

58. Olson L, Baklund EO, Ebendal T, Freedman R, Hamberger B, Hansson P, Hoffer B, Lindblom U, Meyerson B, Strömberg I, Sydow O, Seiger A (1991) Intraputaminal infusion of nerve growth factor to support adrenal medullary autografts in Parkinson's disease. Arch Neurol 48:373–381

59. Oppenheim RW, Houenou LJ, Johnson JE, Lin LFH, Li L, Lo AC, Newsome AL, Prevette DM, Wang S (1995) Developing motoneurons rescued from programmed and axotomy-induced cell death by GDNF. Nature 373:344–346

60. Otto D, Unsicker K (1990) Basic FGF reverses chemical and morphological deficites in the nigrostriatal system of MPTP-treated mice. J Neurosci 10:1912–1921

61. Otto D, Unsicker K (1992) Effects of FGF-2 on dopaminergic neurons. Neurosci Facts 3:82–83

62. Parent A, Coté P-Y, Lavoie B (1995) Chemical anatomy of primate basal ganglia. Prog Neurobiol 46:131–197

63. Perlow MJ, Freed WJ, Hoffer BJ, Seiger A, Olson L, Wyatt RJ (1979) Brain grafts reduce motor abnormalities produced by destruction of nigrostriatal dopamine system. Science 204:643–647

64. Pezzoli G, Zecchinelli A, Ricciard S, Burke RE, Fahn S, Scarlato G, Carenzi A (1991) Intraventricular infusion of epidermal growth factor restores dopaminergic pathways in hemiparkinsonian rats. Mov Disord 6:281–287

65. Poulsen KT, Armanini MP, Klein RD, Hynes MA, Phillips HS, Rosenthal A (1994) TGF beta 2 and TGF beta 3 are potent survival factors for midbrain dopaminergic neurons. Neuron 13:1245–1252

66. Rosenblad C, Martinez-Serrano A, Björklund A (1996) Clial cell-line derived neurotrophic factor increases survival, growth and function of intrastriatal fetal nigral dopaminergic grafts. Neuroscience 75:979–985

67. Sass KJ, Buchanan CP, Westerveld M, Marek KL, Fan A, Robbins A, Naftolin F, Vollmer TL, Leranth C, Roth RH, Price LH, Bunney BS, Elsworth JD, Hoffer PB, Redmond DE, Spencer DD (1995) General cognitive ability following unilateral and bilateral fetal ventral mesencephalic tissue transplantation for treatment of Parkinson's disease. Arch Neurol 52:680–686

68. Sauer H and Oertel WH (1994) Progressive degeneration of nigrostriatal dopamine neurons following intrastriatal terminal lesions with 6-hydroxydopamine: a combined retrograde tracing and immunocytochemical study in the rat. Neuroscience 59:401–415

69. Sauer H, Fischer W, Nikkah G, Wiegand P, Brundin P, Lindsay RM, Björklund A (1993) Brain-derived neurotrophic factor enhances function rather than survival of intrastriatal ventral mesencephalic grafts. Brain Res 626:37–44

70. Sauer H, Rosenblad C, Björklund A (1995) GDNF but not TGFβ3 prevents delayed degeneration of nigral dopaminergic neurons following striatal 6-hydroxydopamine-lesion. Proc Natl Acad Sci USA 92:8935–8939

71. Sendtner M (1995) Neurotrophic factors for motoneurons. J Neurol 242:S2–S1

72. Sendtner M, Kreutzberg GW, Thoenen H (1990) Ciliary neurotrophic factor prevents the degeneration of motor neurons after axotomy. Nature 345:440–441

73. Sendtner M, Schmalbruch H, Stöckli KA, Caroll P, Kreutzberg GW, Thoenen H (1992) Ciliary neurotrophic factor prevents degeneration of motor neurons in mouse mutant progressive motor neuronopathy. Nature 358:502–504

74. Spina MB, Squinto SP, Miller J, Lindsay RM, Hyman C (1992) Brain-derived neurotrophic factor protects dopamine neurons against 6-hydroxydopamine and N-methyl-4-phenylpyridinium ion toxicity: involvement of the glutathione system. J Neurochem 59:99–106

75. Strömberg I, Herrera-Marschitz M, Ungerstedt U, Ebendal T, Olson L (1985) Chronic implants of chromaffin tissue into the dopamine-denervated striatum. Effects on NGF on graft survival, fiber growth and rotational behaviour. Exp Brain Res 60:335–349

76. Strömberg I, Björklund L, Johansson M, Tomac A, Collins F, Olson L, Hoffer GB, Humpel C (1993) Glial cell line-derived neurotrophic factor is expressed in the developing but not adult striatum and stimulates developing dopamine neurons in vivo. Exp. Neurol 124:401–412

77. Studer L, Spenger C, Seiler RW, Altar A, Lindsay RM, Hyman C (1995) Comparison of the effects of the neurotrophins on the morphological structure of dopaminergic neurons in cultures of rat substantia nigra. Eur J Neurosci 7:223–233

78. Tomac A, Lindquist E, Lin LFH, Ögren SO, Young D, Hoffer BJ, Olson L (1995) Protection and repair of the nigrostriatal dopaminergic system by GDNF in vivo. Nature 373:335–339

79. Unsicker K (1993) The trophic cocktail made by adrenal chromaffin cells. Exp Neurol 123:167–173

80. Unsicker K, Grothe C, Westermann R, Wewetzer K (1992) Cytokines in neural regeneration. Curr Opin Neurobiol 2:671–678

81. Wichmann T, Bergman H, DeLong MR (1994) The primate subthalamic nucleus. III Changes in motor behaviour and neuronal activity in the internal pallidum induced by subthalamic inactivation in the MPTP model of parkinsonism. J Neurophysiol 72:521–530

82. Wiegand SJ, Alexander C, Lindsay RM, DiStefano PS (1991) Soc Neurosci Abstr 17:1121

83. Yan Q, Elliot J, Snider WD (1992) Brain-derived neurotrophic factor rescues spinal motor neurons from axotomy-induced cell death. Nature 360:753–755

84. Yan Q, Matheson C, Lopez QT (1995) In vivo neurotrophic effects of GDNF on neonatal and adult facial motor neurons. Nature 373:341–344

85. Yoshida K, Gage F (1991) Fibroblast growth factors stimulate nerve growth factor synthesis and secretion by astrocytes. Brain Res 538:118–126

86. Gash DM, Zhang Z, Ovadia A, Cass WA, Yi A, Simmerman L, Russel D, Martin D, Lapchak PA, Collins F, Hoffer BJ, Gerhardt GA (1996) Functional recovery in parkinsonian monkeys treated with GDNF. Nature 380:252–255

87. Lapchak PA, Gash DM, Jiao S., Miller P, Hilt A (1997) Glial cell line-derived neurotrophic factor: a novel therapeutic approach to treat motor dysfunction in Parkinson's disease. Exp Neurol 144:29–34

Preliminary Clinical Experience with Neural Transplantation: Two Case Reports

O. Pogarell, T. Eichhorn, A. Kupsch, W.H. Oertel, P. Brundin, P. Hagell, P. Odin, K. Pietz, H. Widner, and O. Lindvall

Summary

Parkinson's disease is a chronic degenerative neurological disorder with progressive loss of nigrostriatal dopaminergic neurons and subsequent striatal dopamine deficiency. The treatment of choice is levodopa or dopamine-agonist replacement. However, long-term therapy is associated with severe side effects such as motor fluctuations and abnormal involuntary movements. Clinical trials with transplantation of human embryonic mesencephalic tissue into the striatum have provided evidence for long-term (up to 6 years) improvement in akinetic-rigid and axial symptoms.

We report on preliminary clinical results in two men (patient 1 aged 54, patient 2 aged 44 years) with Parkinson's disease, Hoehn and Yahr stage V and III/IV, respectively, during defined off (12 h after drug withdrawal), who underwent stereotactic neurosurgery with implants of human embryonic tissue into putamen and caudate nucleus. Both patients have been evaluated monthly according to the Core Assessment Program for Intracerebral Transplantation for 12 and 8 months prior to surgery, respectively. In addition, F-dopa PET scans, structural and functional MR imaging, neuropsychological assessments, and personality scales were performed.

Patient 1 was operated bilaterally in one surgical session with implants of tissue from a total of seven donors at seven sites (five putamen, two caudate) per side. Patient 2 received tissue from four donors into the left striatum, and in a second surgical session 6 months later tissue from five donors into the right striatum (seven sites each, five putamen, two caudate). Both, embryonic tissue and patients were pretreated with lazaroids (tirilazad-mesylate) prior to transplantation, and lazaroid administration has been continued for 3 days after surgery. Immunosuppressive treatment for approximately 1 year from surgery was started 2 days before surgery and comprised cyclosporine (serum level 100 µg/l), azathioprine (100 mg/day) and corticosteroids (7.5 mg prednisolone/day).

Preliminary data of the clinical follow-up showed a marked to moderate improvement in overall motor performance and motor fluctuations that allowed a gradual reduction of levodopa medication. There were no adverse events or side effects due to neurosurgery. Patient 2 experienced a slight increase in physiological tremor, presumably due to immunosuppression.

Introduction

Parkinson's disease is a chronic degenerative disorder characterized by progressive loss of dopaminergic neurons in the substantia nigra (pars compacta) with subsequent striatal (putamen and caudate nucleus) dopamine deficiency. The treatment of choice is long-term substitution with dopamimetic drugs such as levodopa/dopadecarboxylase inhibitors (DCI) or (synthetic) dopamine agonists [13,14].

Long-term treatment with dopaminergic drugs, however, is associated with increasingly severe side effects, namely motor fluctuations and abnormal involuntary movements, leading to a dramatic impairment in the patient's quality of life. These complications usually occur within 3–5 years after initiation of drug therapy and increase by approximately 10% per treatment year [13, 17]. The underlying pathophysiological mechanisms remain to be elucidate; presumably there is a combination of both, disease and treatment-related central changes in patients with advanced Parkinson's disease [4,16,19].

Management of Parkinson's disease in this stage of the disease is increasingly difficult. The first step can be a change in current pharmacotherapy with reduced interdose intervals, a decrease in daily levodopa doses, and an increase in dopamine agonists (with long half-life). Additional strategies include the use of controlled-release levodopa preparations, amantadines, and/or continuously administered dopamimetics (e.g., subcutaneously administered apomorphine). However, despite these modifications in pharmacotherapy levodopa-associated complications are still a challenge to therapists [2] and there is a need for a more effective, highly tolerable therapy to improve not only parkinsonian symptoms but also long-term complications.

A new approach is the intracerebral transplantation of fetal mesencephalic neurons into the striatum of patients with Parkinson's disease. This procedure restores the dopaminergic innervation of the striatum by healthy dopaminergic neurons and promotes improvement in parkinsonian signs and symptoms [15].

Several clinical studies and case reports have shown the feasibility, safety, and efficacy of this method, but the number of treated patients is still limited, and the results from various centers differ due to different study protocols, so that fetal nigral transplantation must be regarded as experimental therapy [8, 9]. Positron emission tomography studies with (18F)-fluorodopa have shown survival of donor tissue in the striotum of transplanted patients for at least 5 years [11, 12] and a correlation of postoperative striatal fluorodopa uptake with clinical improvement [18]. An accidental postmortem analysis showed extensive striatal reinnervation 18 months after transplantation [7].

To facilitate the comparison of data from various centers and to evaluate different surgical protocols, an international cooperation of neuroscientists has developed standardized clinical guidelines for the assessment of transplanted patients [3, 10].

As a part of a multicenter program (founded by the BIOMED II program of the European Community) we investigated and clinically followed up two German patients with advanced Parkinson's disease, who were enrolled into a transplant program according to the ethical guidelines of the Network on European CNS Transplantation and Restoration [1] and Core Assessment Protocol for Intracerebral Transplantation [3]. Imaging studies (fluorodopa PET) were performed in London, UK. The neurosurgical procedures of transplanting the fetal cells took place in Lund, Sweden. Preliminary case reports with data of clinical follow-up 8 and 6 months postoperatively are presented.

Patients and Methods

Patient 1 was a 53-year-old white man. He first complained of bradykinetic-rigid symptoms predominantly on the lift side in 1981. After diagnosis of Parkinson's disease treatment was started with levodopa/DCI in combination with amantadine, with an excellent initial response. However after several years he developed severe on-off fluctuations with peak of dose dyskineasias and both biphasic and off dystonia (foot dystonia and torticollis). Except for an increase in the daily levodopa dose and a modification of dose intervals no further improvement in drug treatment was possible since the patient tolerated neither treatment with controlled release levodopa preparations nor dopamine agonist treatment. He entered the transplantation program in 1994 with a stabilized levodopa medication of 1400 mg per day. Prior to surgery the patient was in Hoehn and Yahr stage V (unable to walk unassisted) in defined off condition (12 h off dopaminergic drugs) and III–IV during best on.

Patient 2 was a 42-year-old white man, suffering from Parkinson's disease for more than 15 years. Initial symptoms were slight bradykinesia and rigidity in the upper limbs, predominantly on the right side. Diagnosis was confirmed, and levodopa medication was started in 1983, with an excellent response for the first 3–4 years. He was extremely sensitive to levodopa and developed marked motor fluctuations and abnormal involuntary movements under a low levodopa dose of 225 mg daily and despite an optimization of drug therapy with a combination of levodopa, dopamine agonists (pergolide), monoamine oxidase B inhibitors (deprenyl) and anticholinergics (biperiden). He was enrolled into the transplantation program in January 1995. He was in Hoehn and Yahr stage III–IV in defined off and in stage II in best on.

After enrollment into the transplantation program the patients were examined for approximately 1 year prior to surgery, while drug therapy was kept stable. The following tests were performed according to standardized guidelines:

- Monthly clinical evaluations beginning approximately 1 year prior to surgery according to CAPIT (core assessment program for intracerebral transplantation) comprising:
  - UPDRS I, II, III (defined off, best on, off), IV, patient diary
  - Hoehn and Yahr scale
  - Schwab and England scale
  - Timed motor tests: pronation-supination, hand-arm movements, finger dexterity, foot tapping, stand-walk-sit test
- Neuropsychological and neuropsychiatric assessments
- Imaging studies:
  - Fluorodopa PET, to assess the presynaptic dopaminergic system and to evaluate reinnervation of the striatum by the grafted cells postoperatively
- IBZM SPECT, to prove an intact postsynaptic striatal dopamine receptor status
- MRI, to exclude, for example, intracerebral lesions, tumor before stereotaxy

Neurosurgery

The patients were operated on in Lund, Sweden in an established neurosurgical procedure and according to a standardized protocol (review: [20]). Patient 1 was operated on bilaterally in one surgical session. He received fetal tissue from a total of seven donors at seven sites per striatum (five sites in putamen and two sites in caudate nucleus). Patient 2 had two surgical sessions within 6 months. In the first operation tissue from four donors was implanted into the left striatum at seven sites (five putamen, two caudate nucleus). In the second session 6 months later tissue from 5 donors was implanted into the right striatum. Before grafting the fetal tissue was pretreated with lazaroids (tirilazad-mesylate) and the patients received lazaroid-infusions 1 day before and for 3 days after surgery. Long-term immunosuppression (for approximately 1 year from surgery) was started perioperatively with cyclosporine (serum level 100 µg/l), azathioprine (100 mg/day) and prednisolone (7.5 mg/day).

Results

Since this is an ongoing study, only preliminary clinical data can be presented. A detailed statistical analysis is not possible due to the short follow-up period, with only 4 and 3 postoperative evaluations.

Patient 1 was in Hoehn and Yahr stage V in off before surgery. He was not able to stand or walk unassisted. Eight months after the bilateral transplantation the Hoehn and Yahr stage was III in off, i. e., the patient is able to walk and stand without help at any time. Before transplantation the mean UPDRS III sum score of patient 1 in defined of (i. e., 12 h after last medication) was 60.73 and decreased to 25.17 in best on after a single levodopa test. Postoperatively (four evaluations) the mean sum scores in defined off and in best on were 47 and 16 respectively.

Patient 2 improved from Hoehn and Yahr stage III–IV in off to Hoehn and Yahr stage II 6 months after the first operation, at the time of the second, right-sided transplantation. He had a mean UPDRS III sum score of 43.67 and 22.9 preoperatively (defined off and best on after levodopa test, respectively), 19 and 7.6 postoperatively (three evaluations after the first surgical session). The motor improvement in patient 2 after the first grafting was predominantly contralateral to the side of the operation.

According to a patient's diary, motor fluctuations decreased postoperatively with improvement in intensity and duration of off periods during waking day in both patients.

Levodopa medication could be reduced in both patients: in Patient 1 by 200 mg (dose reduction from 1400 to 1200 mg per day), in patient 2 by 75 mg from 225 to 150 mg per day.

Subjectively both patients were aware of an improvement of motor function, without side effects, cognitive or personality changes. Patient 1 reported a mild and patient 2 a marked overall improvement.

With respect to an overall clinical impression there was a mild to moderate improvement in the disease, especially due to improved motor performance and less severe motor fluctuations. There were no complications during or after operation. The tolerability of immunosuppression was good except for a mild transient postural hand tremor in patient 2, presumably due to immunosuppressive therapy with cyclosporine A and/or corticosteroids.

Discussion

Our preliminary results 8 and 6 months after transplantation indicate a mild to moderate clinical improvement in motor performance during both on and off periods. The patients reported an improvement in motor fluctuations, especially intensity and duration of off periods. The fetal nigral grafting allowed a gradual reduction in levodopa medication, which may also have contributed to the improvement in motor fluctuations.

Perioperatively both the patients and the donor tissue were pretreated with lazaroids (tirilazademesylate). Lazaroids are inhibitors of free radical formation and lipid peroxidation [5] and have been shown to improve the survival of cultured mesencephalic neurons [6]. Therefore pretreatment with lazaroids may enhance the viability of grafted neurons and improve the reinnervation of the striatum. These patients are part of a study to assess the effect of lazaroids in fetal nigral grafting in vivo.

Finally, the patients did not experience any serious side effects due either to the operation to immunosuppression (except for the above-mentioned postural tremor in patient 2). Further follow-up is necessary to assess long-term efficacy and tolerability. In conclusion, neural transplantation in Parkinson's disease seems to be a safe and tolerable method and leads to a moderate to marked improvement as early as 6/8 months postoperatively.

References

1. Boer G, on behalf of NECTAR (1994) Ethical guidelines for the use of human embryonic or fetal tissue for experimental and clinical neurotransplantation and research. J Neurol 242:1–13

2. Calne DB (1993) Treatment of Parkinson's disease. N Engl J Med 329:1021–1027

3. CAPIT Committee (1991) Core assessment program for intracerebral transplantations. In: Lindvall O Björklund A, Widner H (eds) Intracerebral transplantation in movement disorders. Restorative neurology, vol 4. Amsterdam, pp 232–241

4. Chase TN, Engber TM, Mouradian MM (1996) Contribution of dopaminergic and glutamatergic mechanisms to the pathogenesis of motor response complications in Parkinson's disease. Adv Neurol 69:497–501

5. Clark WM, Hazel JS, Coull BM (1995) Lazaroids. CNS pharmacology and current research. Drugs 50:971–983

6. Frodl EM, Nakao N, Brundin P (1994) Lazaroids improve the survival of cultured rat embryonic mesencephalic neurons. Neuroreport 5:2393–2396

7. Kordower JH, Freeman TB, Snow BJ et al (1995) Neuropathological evidence of graft survival and striatal innervation after the transplantation of fetal mesencephalic tissue in a patient with Parkinson's disease. N Engl J Med 332:1118–1124

8. Kupsch A, Oertel WH (1996) Operative Therapiestrategien gegen das idiopathische Parkinsonsyndrom. Psycho 22:37–47

9. Kupsch A, Oertel WH (1997) Neue operative Ansätze in der Behandlung des idiopathischen Parkinson-Syndroms: Neurostimulatio und Transplantation. Akt Neurologie 24:49–55

10. Langston JW, Widner H, Brooks D, Fahn S, Freeman T, Goetz CG, Watts R (1991) Core assessment program for intracerebral transplantation (CAPIT). In: Lindvall O, Björklund A, Widner H (eds) Intracerebral transplantation in movement disorders. Elsevier, Amsterdam, pp 227–241

11. Lindvall O, Brundin P, Widner H, Rehncrona S, Gustavii B, Frackowiak R, Leenders KL, Sawle G, Rothwell JC, Marsden CD, Björklund A (1990) Grafts of fetal dopamine neurons survive and improve motor function in Parkinson's disease. Science 242:574–577

12. Lindvall O, Sawle G, Widner H et al. (1994) Evidence for long term survival and function of dopaminergic grafts in progressive Parkinson's disease. Ann Neurol 35:172–180

13. Marsden CD, Parkes JD (1977) Success and problems of long-term levodopa therapy in Parkinson's disease. Lancet 1:345–349

14. Oertel WH, Quinn N (1996) Parkinsonism. In: Brandt T, Diener HC, Caplan LR, Kennard C, Dichgans J (eds) Neurological disorders: course and treatment. Academic, San Diego, pp 715–771

15. Olanow CW, Kordower JH, Freeman TB (1996) Fetal nigral transplantation as a therapy for Parkinson's disease. Trends Neurosci 19:102–109

16. Papa SM, Engber TM, Kask AM, Chase TN (1994) Motor fluctuations in levodopa treated parkinsonian rats: relation to lesion extent and treatment duration. Brain Res 662:69–74

17. Poewe WH (1994) Clinical aspects of motor fluctuations in Parkinson's disease. Neurology 1994; 44 [Suppl 6]: s6–s9

18. Remy P, Samson Y, Hantraye P et al. (1995) Clinical correlates of (18F) fluorodopa uptake in five grafted parkinsonian patients. Ann Neurol 38:580–588

19. Sage JI, Mark MH (1994) Basic mechanisms of motor fluctuations. Neurology 44 [Suppl 6]: s10–s14

20. Widner H, Rehncrona S (1993) Transplantation and surgical treatment of parkinsonian syndromes. Curr Opin Neurol Neurosurg 6:344–349

**Part 9
Molecular Neurosurgery**

Molecular Cytogenetics of Malignant Gliomas

J. Schlegel, G. Stumm, and H.-D. Mennel

Introduction

Glial tumours are the most common intracranial neoplasms in adults. According to the WHO they are classified histologically with different grades of malignancy and include differentiated gliomas, i.e. astrocytomas, oligodendrogliomas and ependymomas (WHO Grade II), anaplastic variants (WHO Grade III) and glioblastoma multiforme (GBM, WHO Grade IV), the most malignant variant. GBM represents a particularly pressing problem. Despite considerable advancements in the clinical management of malignant gliomas, the prognosis remains poor. Consequently, novel therapeutic approaches based on the biological properties of malignant gliomas are required.

Genetic alterations have been implicated in the neoplastic transformation of cancer cells. Increased mutation rates in clonal cell populations arising during tumour progression may be indicative of a malignant phenotype. It has been shown that the activation of dominant acting genes (oncogenes) and loss of function of recessive genes (tumour suppressor genes) accumulates during carcinogenesis of human malignant tumours. The association of genetic changes with different grades of malignancy has also been demonstrated in glial tumour progression. The identification of genetic alterations as contributors to the transition from normal growth of cells to their neoplastic transformation has fueled the hope that interference with the gene products might open new approaches to rational, targeted tumour therapies. However, only a few genetic alterations in humans and experimental models have been characterized in detail. Loss of heterozygosity of chromosome 17p involving the $p53$ tumour suppressor gene appears at early stages of glial tumourigenesis and can be detected in almost the same proportion of tumour cells in all grades of malignancy. Alterations of genes involved in cell cycle regulation such as cyclin-dependent kinase 4 ($CDK4$), the p16 cdk4 inhibitor ($CDKN2$) and the retinoblastoma (RB) gene product can be detected in anaplastic gliomas and seem to be the prerequisite for malignant conversion. Amplification of the gene for the epidermal growth factor receptor and allelic loss of chromosome 10 seem to be consistently associated with GBM. It seems likely that improved screening methods will uncover other genetic alterations associated with tumour progression in glial tumors. Molecular cytogenetic methods have this potential and are introduced in this article.

Molecular Cytogenetic Methods

In situ hybridization is based on findings by Gall and Pardue [5] and has been used to localize certain DNA sequences in tissue sections, nuclei and chromosomes. This technique also allows localization of single genes on defined chromosomal regions and is therefore capable of detecting structural rearrangements. For many years radioactively labelled probes were used for in situ hybridization. This prerequisite for the method, however, may hamper its applicability in routine diagnostics. Radioactive techniques require special laboratory equipment and they are expensive and time-consuming. Moreover, they are hazardous to the laboratory personnel and the environment. The introduction of nonisotopic labelling methods has overcome these problems. Modern molecular catogenetic techniques are based on fluorescence in situ hybridization (FISH). There are now three major variants used for the investigation of tumour genomes: (i) interphase cytogenetics, (ii) chromosome painting and (iii) comparative genomic hybridization (CGH).

For interphase cytogenetics, cloned genomic or cDNA fragments are utilized [1, 2]. Using different fluorochromes it is possible to detect two or more genetic loci simultaneously. The major advantage of interphase cytogenetics is the investigation of tumour cells *in situ*. Therefore, it is possible to detect regional differences in individual tumours. In principle, it allows single cell diagnostics and consequently is applicable to small samples, e.g. stereotactic biopsies. It does not need prior cell culture, which could lead to selection of clonal outgrowth. However, its applicability is hampered by its requirement of known and cloned sequences. The assessment of hybridization dots in single cells may be difficult due to microscopy artifacts, e.g. dots outside of microscopic layer.

For chromosome painting, preparation of metaphase spreads of tumour cells is performed by standard protocols. Chromosomes are heat-denatured and probe hybridization is performed in the same manner as described for interphase nuclei [1, 2]. Chromosome painting is one major advancement of classical cytogenetics. Karyotyping of tumour genomes is limited by extrachromosomal elements and marker chromosomes. Using cloned DNA sequences or specific probes for chromosomes or chromosomal segments, e.g. telomeres or centromeres, assignment of marker chromosomes or amplified genes in double minutes can be made. However, the disadvantages are the same as for classical cytogenetics.

Comparative genomic *in situ* hybridization (CGH) is a novel reverse chromosome painting technique which allows a molecular cytogenetic analysis of complex genomic alterations [3, 6, 7]. It is based differential fluorescence displayed among chromosomes in a normal metaphase spread, following their hybridization with normal and tumour DNAs that have been labelled with different fluorochromes. The ratio of the fluorescence intensities reflects the relative copy number of the tumour DNA compared with the normal DNA. Tumour DNA amplifications and deletions show distinct signals, revealing the chromosomal positions of the target sequences associated with specific alterations.

Use of complete tumour DNA as probe allows identification and chromosomal mapping of unknown DNA sequences altered in tumour cell genomes. It has been shown by several investigators that the results obtained by CGH are reproducible and comparable to the results of cytogenetic investigations [12]. Therefore, CGH has the potential to detect complex genetic changes in glial tumours. CGH does have limitations, however. The proportion of non-neoplastic cells in tumour samples can be quite high. As a result, the estimation of deletions and chromosomal polysomy can be difficult. In contrast, the DNA ratio in regions containing gene amplification depends on the copy number of amplified DNA sequences which frequently reach 50–100 copies [11]. Thus, contaminating non-neoplastic DNA has no significant effect on the signal intensity of regions containing gene amplification. Also, CGH will not detect allelic loss that arises via mitotic recombination, in which two copies of the same allele remain, as occurs with chromosome 17p.

Investigations of Human Gliomas Using Molecular Cytogenetics

Since its introduction, interphase cytogenetics has been used in several studies for the investigation of chromosomal alterations of brain tumours *in situ*. Probes used for human gliomas include the genes for growth factors and their receptors, genes involved in cell cycle regulation and the *p53* gene. The results of these reports are in accordance with the data obtained by molecular and cytogenetic methods. The method is, however, limited to those genes and chromosomal segments which have also been investigated by molecular methods. Novel genetic data have not been discovered by the use of interphase cytogenetics. Reports on chromosome painting of human glial tumours are rare. There are several reasons for this. The resolution of translocations or

Fig. 1.
Summary of molecular cytogenetics of human malignant gliomas investigated by CGH. Gains and losses are calculated for individual chromosomes as percentages ($n = 95$). Gains and amplifications are depicted as positive values, losses as negative values

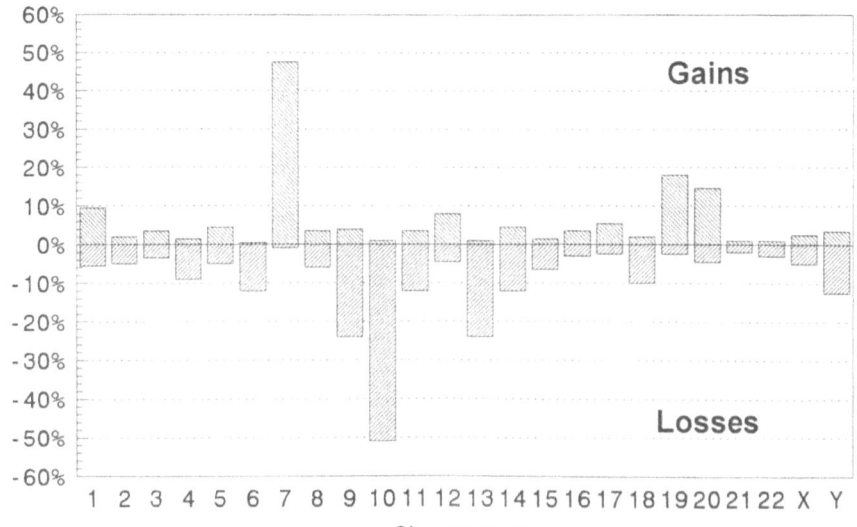

complex chromosomal rearrangements is a major issue in fluorescence in situ hybridization of tumour metaphases. However, these alterations are rare in glial tumours. In addition, chromosome preparation is dependent on prior cell culture of tumour cells. It is well recognized that human malignant gliomas exhibit culture artifacts due to clonal growth or active genetic changes, e.g. the most common genetic alteration in GBM, the amplification of the *EGFR* gene is lost during primary passages of cell culture.

To date, some 100 malignant gliomas have been analysed by CGH [4,6,8,9,11,12]. The results are summarized in Fig. 1. The data obtained by CGH are in accordance with the results of cytogenetic investigations. Using CGH, complex genomic alterations in malignant gliomas have been demonstrated. One of the most interesting results was the demonstration of complementary changes in genes involved in cell cycle regulation. These data are corroborated by recent molecular genetic investigations, emphasizing the importance of this class of genes in malignant transformation in glial tumourigenesis. In conclusion, CGH is a comprehensive and rapid approach for the analysis of complex genomic alterations in glial tumors. Therefore, it may serve as a supplemental tool to define subgroups of human GBM with different clinical behaviour on the basis of genetic changes. It could also help to screen low-grade gliomas for genetic alterations involved in tumour progression. Finally, CGH data indicate that the combined use of cytogenetic and molecular methods for the analysis of these tumours can provide important information regarding the molecular mechanisms involved in glioma development.

Recent Developments

Recently, direct labelling by fluorescence-coupled nucleoside triphosphates has advanced FISH applications. However, the intensity of such probes is usually 15% of that produced by an equivalent probe detected by secondary reagent. This is not problematic when sensitive digital imaging devices are used. However, it could be of concern when results are being recorded by conventional photomicrosocpy. The introduction of combined labelling with different fluorochromes has led to multicolour FISH applications. Using whole chromosome probes it has become possible to visualize each individual chromosome with a single hybridization. Although used for chromosome painting of different tumour types, so far no interphase cytogenetic data have been published using this approach for brain tumours. One of the major advancements in CGH has been the introduction of PCR using degenerate oligonucleotide primers (DOP-PCR), thus allowing amplification of

the entire genome [13]. In principle, it is possible to amplify sufficient DNA for CGH even from a single tumour cell. Therefore, paraffin-fixed material or probes provided by stereotactic biopsies are suitable for DOP-PCR and CGH.

References

1. Cremer T, Lichter P, Borden J, Ward DC, Manuelidies L (1988) Detection of chromosome aberrations in metaphase and interphase tumor cells by in situ hybridization using chromosome-specific library probes. Hum Genet 80:235–246
2. Cremer T, Tesin D, Hopman AHN, Manuelidis L (1988) Rapid interphase and metaphase assessment of specific chromosomal changes in neuroectodermal tumor cells by in situ hybridziation with chemically modified DNA probes. Exp Cell Res 176:199–220
3. duManoir S, Speicher MR, Joos S, Schröck E, Popp S, Döhner H, Kovacs G, Robert-Nicoud M, Lichter P, Cremer T (1993) Detection of complete and partial chromosome gains and losses by comparative genomic in situ hybridization. Hum Genet 90:590–610
4. Fischer U, Wullich B, Sattler HP, Göttert E, Zang KD, Meese E (1994) DNA amplifications on chromosome 7, 9 and 12 in glioblastoma detected by reverse chromosome painting. Eur J Cancer 30:1124–1127
5. Gall JG, Pardue ML (1969) Formation and detection of RNA-DNA hybrid molecules in cytological preparations. Proc Natl Acad Sci USA 63:378–383
6. Joos S, Scherthan H, Speicher MR, Schlegel J, Cremer T, Lichter P (1993) Detection of amplified genomic sequences by reverse chromosome painting using genomic tumor DNA as probe. Hum Genet 90:584–589
7. Kallioniemi A, Kallioniemi OP, Sudar D, Rutovitz D, Gray JW, Waldman F, Pinkel D (1992) Comparative genomic hybridization for molecular cytogenetic analysis of solid tumors. Science 258:818–821
8. Kim DH, Mohapatra G, Bollen A, Waldman FM, Feuerstein BG (1995) Chromosomal abnormalities in glioblastoma multiforme tumors and glioma cell lines detected by comparative genomic hybridization. Int J Cancer 60:812–819
9. Mohapatra G, Kim DH, Feuerstein BG (1995) Detection of multiple gains and losses of genetic material in glioma cell lines by CGH. Genes Chromosom Cancer 13:86–93
10. Scherthan H, Cremer T (1994) Methodology of non-isotopic in situ-hybridization in embedded tissue sections. In: Adolph KW (ed) Methods in molecular genetics, vol 5. Academic Press, San Diego, pp 223–238
11. Schlegel J, Scherthan H, Arens N, Stumm G, Kiessling M (1996) Detection of complex genetic alterations in human glioblastoma multiforme using comparative genomic hybridization (CGH). J Neuropathol Exp Neurol 55:81–87
12. Schröck E, Thiel G, Lozanova T, duManoir S, Meffert MC, Jauch A, Speicher MR, Nürnberg P, Vogel S, Jänisch W, Donis-Keller H, Ried T, Witkowski R, Cremer T (1994) Comparative genomic hybridization of human malignant gliomas reveals multiple amplification sites and non-random chromosomal gains and losses. Am J Pathol 144:1203–1218
13. Speicher MR, duManoir S, Schröck E, Holtgreve-Grez H, Schoell B, Lengauer C, Cremer T, Ried T (1993) Molecular cytogenetic analysis of formalin-fixed, paraffin-embedded solid tumors by comparative genomic hybridization after universal DNA-amplification. Hum Mol Genet 2:1907–1914
14. Weber RG, Sommer C, Albert FK, Kiessling M, Cremer T (1996) Clinically distinct subgroups of GBM studied by CGH. Lab Invest 74:108–119

A Role for Hypoxia in Glioma Biology

K. H. Plate

Summary

Glioblastomas are rapidly growing tumours with large hypoxic areas featuring necrosis. These hypoxic tumour compartments are most likely responsible for the relative resistance of glioblastomas to radiotherapy. In vivo studies suggest that perinecrotic palisading represents the most hypoxic tumour compartment in glioblastomas. Current evidence suggests that hypoxia has several deleterious effects for glioblastoma patients, since it leads to growth factor expression in glioma cells, induces tumour angiogenesis, vascular permeability and genetic instability, and selects for apoptosis-resistant cells. In addition, hypoxia impairs significantly the efficacy of the anti-oedema drug dexamethasone. Taken together, these findings suggest that on a physiological basis, hypoxic tumour cells are the most malignant cell clones in glioblastoma and thus represent a challenging target for novel therapies, for example by targeting heterologous gene expression to hypoxic tumour cells.

Introduction

Glioblastomas are not only the most common (incidence of 2.5 per 100 000 population per year) but are also the most malignant human brain tumours (the 5-year survival rate in 8581 patients examined was 0.3%) [3]. The single most effective treatment is radiation therapy, which extends the mean survival time to 9–10 months, compared with 4 months in patients treated with cytoreductive surgery alone [21]. It is well established, that the efficacy of radiation therapy is dependent on the oxygen tension in the tumour. Current evidence suggests, that the DNA-damaging effect of radiation is directly related to the cellular oxygen tension. Hypoxic tumours are much more resistant to radiation than well-oxygenated tumours [12]. A recent survey of 103 patients with cervical cancer identified hypoxia as an independent prognostic marker of patient survival [11]. Glioblastomas are among the most

hypoxic tumours in humans, as shown by intratumoural oxygen measurements [25]. This observation is reflected on a morphological basis by the presence of large intratumoural necrotic areas, which are a hallmark of glioblastomas and distinguishing them from anaplastic astrocytomas [1]. A series of recent cell and molecular biology studies has significantly extended the current knowledge of the role of hypoxia in tumour cells in vivo. These studies have shown that: (a) hypoxia is a potent inducer of gene expression in tumour cells, (b) hypoxia is the major trigger for new blood vessel growth in malignant tumours, (c) hypoxia induces genetic instability in tumour cells, (d) hypoxia selects for apoptosis-resistant and thus malignant cell clones, and (e) hypoxia impaires the efficacy of drugs used in brain tumour therapy.

These findings not only suggest that hypoxia may be directly involved in tumour progression but in addition identifies hypoxia as a promising target for glioma therapy. In the present manuscript, some recent findings of the role of hypoxia in tumour cell biology are summarized, and some strategies to target hypoxic glioma cells are discussed.

Hypoxia Induces Gene Expression

Cells responds to changes in the microenvironment such as acidosis, hypoglycaemia or changes in oxygen tension by down- or upregulation of certain genes. Hypoxia for example upregulates the transcription of several genes [6]. Most of these genes (e.g. those encoding growth factors and glycolytic enzymes) are hypoxia-inducible in a large variety of cell types, including tumour cells. Some aspects of the mechanism of hypoxia-inducible gene expression have recently been discovered: a crucial transcription factor involved in hypoxic upregulation of several genes is hypoxia-inducible factor-1 (HIF-1) [9]. HIF-1 consists of two subunits, HIF-1α and HIF-1β/aryl hydrocarbon nuclear translocator (ARNT). Mice carrying a targeted disruption of the ARNT gene failed to upregulate genes which normally respond to hypoxia, indicating that ARNT is necessary for a proper physiological response to hypoxia [16]. A downstream target of HIF-1 is VEGF; mice with a heterozygous deletion of VEGF die in utero and have a phenotype similar to ARNT-deficient mice [2, 7]. Although the number of hypoxia-inducible genes is still increasing, current evidence suggests that malignant gliomas take advantage of hypoxia by the upregulation of at least two growth factors, PDGF-BB and VEGF. PDGF-BB being expressed in glioma cells supports glioblastoma growth by autocrine and paracrine cell regulatory mechanisms [26]. VEGF, the major tumour angiogenesis and vascular permeability factor, supports glioma growth via a paracrine effect on endothelial cells ([18]; see below).

Hypoxia Triggers Tumour Angiogenesis and Tumour Oedema

Glioblastomas are among the most hypoxic tumours in humans [25]. Necroses with "palisading" tumour cells and adjacent microvascular proliferations are a reflection of tumor hypoxia and are histopathological hallmarks of glioblastomas. The endothelial-cell-specific mitogen VEGF also has permeability-inducing properties in vivo. VEGF is highly expressed in perinecrotic palisading cells but is downregulated in tumour cells adjacent to vessels, suggesting oxygen-dependent gene expression [17, 22]. Recent studies have shown that: (a) the binding site for HIF-1, termed hypoxia-responsive element (HRE), is necessary for the hypoxic induction of the VEGF gene in glioma cells and (b) that mRNA stabilization sites in the 3′ UTR of the VEGF gene restrict hypoxic gene expression to the perinecrotic palisading cells in situ [5]. VEGF, secreted by the hypoxic tumour cell compartment (e.g. perinecrotic palisading cells), is distributed throughout the tumour by diffusion and binds to VEGF receptors (VEGFR) on vascular endothelial cells, where it enhances endothelial VEGFR expression in an autocatalytic fashion [13]. VEGF binding to vascular endothelial cells leads to a cascade of events, including migration and proliferation of endothelial cells, and induction of fenestrae and vascular permeability in tumour vessels [20]. Thus, the hypoxic upregulation of VEGF most likely is responsible for angiogenesis and oedema observed in malignant gliomas [Fig. 1].

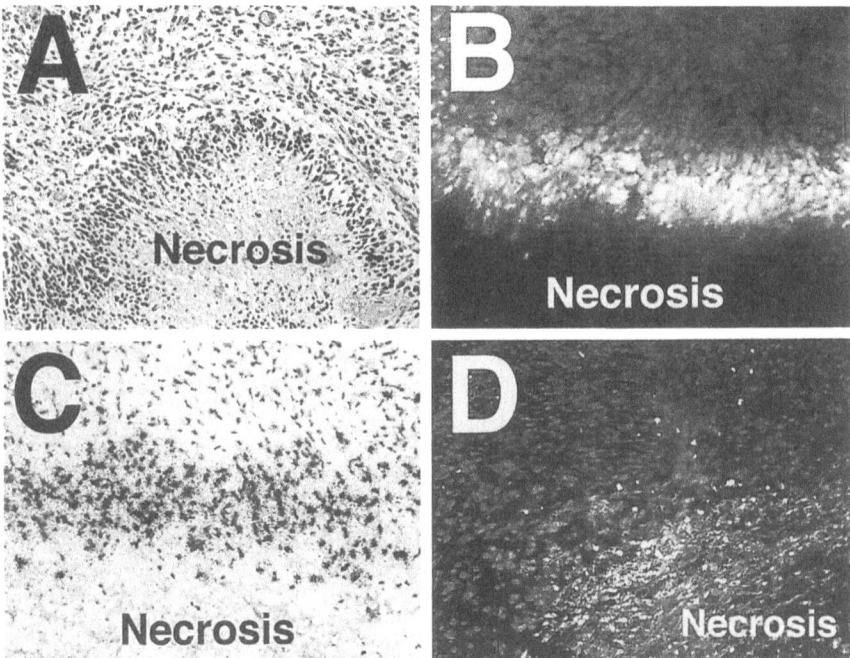

Fig. 1. (A) Hematoxylin-eosin-stained section of a human glioblastoma biopsy specimen demonstrating three distinct tumour compartments: (1) viable tumour cells (*upper part*), (2) tumour necrosis (*lower part*) and (3) tumour cells palisading at the interface between necrotic and viable cells. B Palisading cells (*middle*) represent in fact hypoxic tumour cells in vivo, as shown in this GS9L rat glioma stained for hypoxia-generated nitroimidazole (EF5) adducts with the monoclonal 3.5-Cy 3 antibody. C High expression of VEGF mRNA in hypoxic palisading cells in a human glioblastoma in situ, suggesting that hypoxia is the driving force of tumour angiogenesis and tumour oedema initiated by VEGF. In situ hybridization of VEGF using a ^{35}S-labelled VEGF antisense cRNA probe. D Detection of apoptotic palisading cells in a GS9L rat glioma. In situ end-nick labelling technique for DNA-strand breaks. Note the colocalization of hypoxia (**B**), hypoxia-induced gene expression (**C**), and apoptotic cell death (**D**) in the same tumour compartment

Hypoxia Induces Genetic Instability in Tumour Cells

When tumours develop, they often become more malignant with time, a processed termed tumour progression. Glioblastomas arise either de novo (primary glioblastoma) or by progression from a low-grade astrocytoma (secondary glioblastoma). Epidermal growth factor receptor overexpression and p53 mutations have been found to be specifically associated with primary and secondary glioblastomas, respectively [26]. However, apart from the p53 tumour suppressor gene and some cell-cycle-regulated genes, little is known about the biological significance of the genetic alterations described in glioblastomas [15]. Considering these findings, it is interesting to consider that genetic alterations found in tumour cells are not necessarily primary events but instead may be a consequence of tumour hypoxia. Thus, Reynolds and colleagues have observed a five fold increase in mutations in hypoxic compared with normoxic tumour cells. The authors concluded that "the microenvironment of an incipient developing tumor might itself contribute to genomic instability and mutagenesis, leading to tumor progression and an evolution of the malignant phenotype" [19].

Hypoxia Selects for Apoptosis-Resistant (p53-Deficient) Tumour Cell Clones

Hypoxia may not only induce mutations in tumour cells, but may also confer a growth advantage to apoptosis-resistant cells and may thus select for malignant cell clones [14]. Graeber and colleagues have shown that when p53/- and p53/+ cells are mixed in vitro, several rounds of hypoxia lead to accumulation of p53/- cells. In tumours grown in vivo, areas with a high level of apoptosis corresponded with hypoxic tumour areas; in addition, the apoptotic index was much higher in p53/+ than in p53/- tumours [8]. The hypoxic perinecrotic palisading cells in glioblastomas may play a major role in tumour cell apoptosis, since the apoptotic index is higher in these cells than in other parts of the tumour and the apoptosis gene Fas/APO-1 is highly expressed (24). These findings suggest that tumour-cell apoptosis induced in hypoxic areas leads to selection of p53/- cell clones. This hypothesis is consistent with the observation that p53/- cells have a growth advantage in vivo and accumulate during glioma progression [23].

Hypoxia Impairs the Efficacy of Dexamethasone

Hypoxia induces VEGF expression in malignant gliomas, which subsequently leads to an increase in vascular permeability and to peritumoural oedema (see above). The most effective therapy for peritumoural oedema is the parental application of the synthetic glucocorticoid dexamethasone. The mechanism of dexamethasone-induced oedema suppression was a mystery for many years, but a recent study has shown that dexamethasone downregulates VEGF in glioma cells [10]. Interestingly, the inhibitory effect of dexamethasone on VEGF expression in glioma cells is dependent on oxygen. A 50% downregulation of the VEGF gene by dexamethasone was observed in glioma cells cultured under normoxic conditions, compared to 15% downregulation in cells cultured under hypoxic conditions (our own unpublished observations). Thus, the effect of dexamethasone on VEGF expression in glioma cells in vivo may be directly correlated to the intracellular oxygen tension. These findings suggest, that hypoxic glioma cells in vivo express large amounts of VEGF, irrespective of the presence or absence of dexamethasone.

Hypoxia May Be Used for Targeting Tumour Therapy

The above-mentioned findings show that tumor hypoxia has several deleterious effects for glioblastoma patients and suggest that hypoxia itself is an appropriate target for tumor therapy. Novel therapeutic approaches utilizing these findings include directing gene expression to hypoxic tumor cells. The targeting of heterologous gene expression to hypoxic tumor cells by the hypoxia-responsive element has already been demonstrated [4]. Putative therapeutic approaches may include the targeted expression of suicide genes, such as the herpes-simplex thymidine kinase gene, to hypoxic tumor cells.

Acknowledgements. The help of Dr. Marcia R. Machein, Simone Erhardt and Richard Haas for producing Fig. 1 is gratefully acknowledged. I thank Drs. Georg Breier, Annette Damert and Werner Risau for discussion. The provision of EF5 and the 3.5-Cy3 antibody by Dr. Cameron J. Koch is gratefully acknowledged. The research was supported by grants from the Deutsche Krebshilfe (Dr. Mildred-Scheel Stiftung) and Bundesministerium für Bildung und Forschung.

References

1. Burger PC, Green SB (1987) Patient age, histological features, and length of survival in patients with glioblastoma multiforme. Cancer 59:1617–1625

2. Carmeliet P, Ferreira V, Breier G, Pollefeyt S, Kieckens L, Gertsenstein M, Fahrig M, Vandenhoeck A, Harpal K, Eberhardt C, Declercq C, Pawling J, Moons L, Collen D, Risau W, Nagy A (1996) Abnormal blood vessel development and lethality in embryos lacking a single VEGF allele. Nature 380:435–439

3. Central Brain Tumor Registry of the United States (1995) First annual report CBTR, Chicago

4. Dachs GU, Patterson AV, Firth JC, Ratcliffe PJ, Stuart Townsend KM, Stratford IJ, Harris AL (1997) Targeting gene expression to hypoxic tumor cells. Nat Med 3:515–520

5. Damert A, Machein M, Breier G, Fujita MQ, Hanahan D, Risau W, Plate KH (1997) Two distinct mechanisms contribute to up-regulation of vascular endothelial growth factor in the palisading cells or a rat glioma. Cancer Res (in press)

6. Fandrey J (1995) Hypoxia-inducible gene expression. Resp Physiol 101:1–10

7. Ferrara N, Carver-Moore K, Chen H, Dowd D, Lu L, Sue O'Shea K, Powell-Braxton L, Hillan KJ, Moore MW (1996) Heterozygous embryonic lethality induced by targeted inactivation of the VEGF gene. Nature 380:439–442

8. Graeber TG, Osmanian C, Jacks T, Housman DE, Koch CJ, Lowe SW, Giaccia AJ (1996) Hypoxia-mediated selection of cells with apoptotic potential in solid tumours. Nature 379:88–91

9. Guillemin K, Krasnow MA (1997) The hypoxic response: huffing and HIFing. Cell 89:9–12

10. Heiss JD, Papavassiliou E, Merill MJ, Niemann L, Knightly JJ, Walbridge S, Edwards NA, Oldfield EH (1996) Mechanism of dexamethasone suppression of brain tumor-associated vascular permeability in rats. J Clin Invest 98:1400–1408

11. Höckel M, Schlenger K, Aral B, Mitze M, Schaffer U, Vaupel P (1996) Association between tumor hypoxia and malignant progression in advanced cancer of the uterine cervix. Cancer Res 56:4509–4517

12. Kinsella TJ, Bloomer WD (1981) New therapeutic strategies in radiation surgery. JAMA 245:1669–1674

13. Kremer C, Breier G, Risau W, Plate KH (1997) Up-regulation of flk-1/VEGF receptor-2 by its ligand in a cerebral slice culture system. Cancer Res (in press)

14. Kinzler KW, Vogelstein B (1996) Life (and death) in a malignant tumour. Nature 379:19–20

15. Louis DN (1997) A molecular genetic model of astrocytoma histopathology. Brain Pathol 7:755–764

16. Maltepe E, Schmidt JV, Baunoch D, Bradfield CA, Simon MC (1997) Abnormal angiogenesis and response to glucose and oxygen deprivation in mice lacking the protein ARNT. Nature 386:403–439

17. Plate KH, Breier G, Weich HA, Risau W (1992) Vascular endothelial growth factor is a potential tumour angiogenesis factor in human gliomas in vivo. Nature 359:843–845

18. Plate KH, Risau W (1995) Angiogenesis in malignant gliomas. Glia 15:339–347

19. Reynolds TY, Rockwell S, Glazer PM (1996) Genetic instability by the tumor microenvironment. Cancer Res 56:5754–5757

20. Roberts WG, Palade GE (1997) Neovasculature induced by vascular endothelial growth factor is fenestrated. Cancer Res 57:765–772

21. Salcman M (1980) Survival in glioblastoma: historical perspective. Neurosurgery 7:435–439

22. Shweiki D, Itin A, Soffer D, Keshet E (1992) Vascular endothelial growth factor induced by hypoxia may mediate hypoxia-initiated angiogenesis. Nature 359:843–845

23. Sidransky D, Mikkelsen T, Schwechheimer K, Rosenblum ML, Cavanee W, Vogelstein B (1992) Clonal expansion of p53 mutant cells is associated with brain tumour progression. Nature 355:846–847

24. Tachibana O, Lampe J, Kleihues P, Ohgaki H (1996) Preferential expression of Fas/APO-1 (CD 95) and apoptotic cell death in perinecrotic cells of glioblastoma multiforme. Acta Neuropathol 92:431–434

25. Vaupel PW (1993) Oxygenation of solid tumors. In: Teicher BA (ed) Drug resistance in oncology. Decker, New York, pp 53–85

26. Watanabe K, Tachibana O, Sata K, Yonekawa Y, Kleihues P, Ohgaki H (1996) Overexpression of the EGF receptor and p53 mutations are mutually exclusive in the evolution of primary and secondary glioblastomas. Brain Pathol 6:217–223

27. Westermark B, Heldin C-H, Nistér M (1995) Platelet-derived growth factor in human glioma. Glia 15:257–263

Glycosphingolipids: Diagnostic and Therapeutic Relevance in Human Gliomas

R. Becker, J. Rohlfs, R. Jennemann, H. Wiegandt, H.-D. Mennel, and B. L. Bauer

Summary

Glyycosphingolipids seem to be ideal markers of cellular proliferation and differentiation. The purpose of this study was to evaluate the correlation between distinct glycosphingolipid (GSL)-component profiles of human gliomas with survival time and histopathological malignancy grading.

Three basic GSL-component patterns of human astrocytic tumors have been designated as glioma GSL types I, II, and III and a correlation between these GSL types and histological malignancy grades was demonstrated in earlier studies. In this study GSL-component patterns, survival time, and histopathological malignancy grade of 40 human gliomas were analyzed. GSL type I was present in 18 patients (17 glioblastoma multiforme GBM) WHO IV and 1 anaplastic astrocytoma (AA) WHO III). GSL type II was expressed by 11 patients (seven GBM, three AA, one low-grade astrocytoma (LGA) WHO II). Ten patients presented with GSL type III (three GBM, four AA, three LGA).

The median survival times of patients with GSL types I–III were 56, 51, and 248 weeks, respectively. The Kaplan Meier survival curves differed significantly ($p = 0.0231$, log rank test). However, survival time correlated better to the WHO Grades IV–II (median 51, 118, 396 weeks; Kaplan Meier survival curves differed significantly: $p = 0.0034$, log-rank test). Furthermore, in a given malignancy grade (WHO Grade IV or Grade III), survival time seems to correlate with the different GSL types. In the literature there is one similar report about the ganglioside patterns of 84 human gliomas, confirming these results. In conclusion, the analysis of GSL-type expression might give useful additional information about the proliferative properties of human gliomas in a given malignancy grade. In particular the early prediction of secondary malignancy of low-grade gliomas might be possible with additional GSL-component analysis. The therapeutic relevance of GSL with respect to the concept of adjuvant active immunization therapy is also discussed in this paper.

Introduction

The development of new concepts for therapy of brain tumors is one of the major fields in neurosurgical research. Especially in the most common tumors, the human astrocytomas, various approaches with the exception of surgical resection and radiation therapy have not improved outcome. For this reason, many efforts have been made to learn more about the proliferative properties of gliomas. Several different markers that correlate with cellular proliferation, differentiation, or dedifferentiation have been established in addition to the traditional histopathological categorization and grading of gliomas [2, 3, 10, 22]. To be best addressed examine its use as a target in new diagnostic or therapeutic approaches, the ideal marker of proliferation or differentiation/dedifferentiation should be located on the cell surface. For two reasons glycosphingolipids (GSL) represent such an ideal marker:

1. In a given animal species, the GSL-component pattern reflects the histological cell type, state of cellular growth, maturation, and differentiation [12]. Dedifferentiation is accompanied by a reduction in structural complexity of GSL components.
2. There is evidence that GSL are of importance for a number of functions at the cell surface and intercellular membranes, i.e., cell attachment, receptor function, and synaptic transmission [for review see 27, 35].

Seifert first described ganglioside expression in human brain tumors in 1965 [28]. In the late 1980s we started with the systematic determination of the GSL-component variability among different human gliomas. In 1990 we had already described three distinct GSL-component patterns in human astrocytic tumors that were designated glioma GSL types I–III [16]. A correlation between these glioma GSL types and histological tumor types was demonstrated at least to some extent [17]. Other authors reported about similar observations concerning the diagnostic usefulness of gangliosides [30] and furthermore were able to correlate 1b gangliosides and 6′-LM1 of primary brain tumors with survival [31]. In a further

Table 1. GSL-component distribution in normal brain tissue and gliomas with the GSL types I – III (modified from [16])

Probe	Normal	GSL Type III	GSL type II	GSLType I
Neutral GSL (%)				
GlcCer	–	–	24.1	13.8
GalCer	98.5	93.8	10.7	25.8
CDH	1.5	5.1	47.6	50.2
CTH	–	0.4	7.3	5.6
Gb4Cer	–	0.6	7.5	2.1
nLc4Cer	–	0.05	2.7	2.7
Sulfatide	+	+	–	–
Gangliosides (%)				
Glac 1	4.6	10.4	34.6	49.8
Glac 2	5.0	26.0	27.8	50.2
Gtri 1	3.2	7.9	27.3	< 2.0
Gtri 2	4.8	17.7	10.2	< 2.0
Gtet 1	27.2	18.8	–	–
Gtet 2a	22.2	13.0	–	–
Gtet 2b	19.2	6.7	–	–
Gtet 3b	13.5	4.4		

study these authors reported about the neutral glycolipid composition of human brain tumors [29]. The results were similar to the findings with the reported GSL types I–III. However, a categorization that takes gangliosides as well as neutral GSL into consideration was not applied.

The established categorization of GSL patterns is briefly presented here and its prognostic relevance compared to other GSL categorizations and to the WHO malignancy grading. The detailed analysis of the correlation between GSL types I, II, and III and clinical outcome and other categorizations will soon be presented elsewhere. Furthermore, we present a short review on the concept of adjuvant active immunization of glioma patients with tumor-specific ganglioside vaccines.

Materials and Methods

Tissue Samples

All tissue samples were obtained after surgical removal in the local Neurosurgical Department. Tumors were processed during the operation and immediately there after (intraoperatively) smear preparations [23], and rapid diagnosis with frozen sections were carried out. Postoperatively, routine histology from paraffin-embedded sections was determined. These tests were augmented with immunohistochemical determinations using a panel of antibodies against intermediary filament proteins and other neural differentiation markers. Diagnosis and grading in all cases were performed by one of the

authors (HDM) according to the system of the WHO [21]. Tumors analyzed were gliomas of WHO Grades II, III, and IV (glioblastoma multiforme).

GSL-Component Analysis

All glycosphingolipid-component analyses were performed by RJ by tissue extraction with mixtures of chloroform–methanol–water, separation into neutral and acidic fractions, and further characterization by HPTLC (high-performance thin-layer chromatography) as described earlier in detail [16].

Glycosphingolipid Component Profiles of Normal Brain and Human Intracranial Gliomas

A qualitative analysis and semiquantitative GSL-distribution analysis in the presence of known standards were performed by using a Shimadzu Dual Wavelength TLC Scanner CS 9000, measuring at 440 nm for the orcinol-induced and at 580 nm for the resorcinol-induced color [25]. Yields of orcinol-sulfuric-acid and resorcinol-hydrochloric-acid stain were found to be proportional to the number of monosaccharide and sialic acid residues respectively, present in the glycolipid compounds. Consequently, GSL quantities were calculated by adjustment of the staining values to the amount of monosaccharides and sialic acid present [17]. Three basic types of GSL-component profiles (GSL types I–III) have been established for human intracranial gliomas. Table 1 presents the different components of the neutral and acidic GSL fractions of normal brain and GSL types

I–III, as described below. For detailed information see [16,17]. All tumors could be classified according to the GSL types I–III.

GSL Type I

The neutral components consisted mainly of glucosyl- and dihexosylceramide as well as smaller amounts of trihexosylceramide, and globo- and neolactotetraosylceramide. Galactosylceramide and sulfatide were absent. The acidic components, the gangliosides, of GSL type I were Glac1 and Glac2, trace amounts of sialo- nLc4Cer, and minor amounts of Gtri components. Gtet gangliosides were always completely absent.

GSL Type II

Neutral glycolipids were similar to type I, consisting mainly of CDH, CTH, CTetH, and low proportions of CMH. Sulfatide again was absent. Gangliosides were Glac1, Glac2, Gtri1, and Gtri2.

GSL Type III

This GSL type presented as an overlap of the components in normal brain tissue and GSL type II. The major neutral GSL was CMH, as galactosylcerebroside. CDH, CTH, and CTetH appeared in minor amounts. In the acidic fraction sulfatide appeared as a major component. Glac1, Glac2, Gtri1, and Gtri2 were found as in GSL-type II and the major brain gangliosides Gtet2a, 2b, and 3b were also present.

Statistical Analyses

Survival time was determined as weeks between primary surgery and death or time-point of data acquisition in patients still alive. All deaths were related either directly to the tumor or to complications secondary to the tumor. Kaplan-Meier survival curves were generated for the GSL types I–III and WHO Grades 2–4. The significance of the overall differences in the survival curves were tested by the log rank test [20].

The common occurrence of GSL types and histopathological malignancy grades according to the WHO classification was tested by the Cohen's Kappa coefficient. The dependence of survival time on the determined GSL type in patients with gliomas WHO Grade IV was tested by the Mann–Whitney test (U test).

Immunization

To enhance the immunogenicity of glioma-specific gangliosides (Gtri1, Gtri2) a new method was established by RJ [18]. By ozonolysis and reductamination the immune stimulatory protein keyhole limpet hemocyanin (KLH) was chemically coupled to the gangliosides. The resulting ganglioside conjugates are currently being tested in animal studies [19] and a phase I clinical trial on the effects of active adjuvant immunotherapy with Gtri conjugates in glioma patients is in progress.

Results

Patient Population (Classification and Grading of Tumors According to WHO Criteria)

Forty patients were included in this series. All patients were treated in the local neurosurgical department. Four of these patients presented with a low-grade astrocytoma, WHO Grade II; eight patients had an anaplastic astrocytoma, WHO Grade III; and 28 had a malignant astrocytoma (glioblastoma multiforme), WHO Grade IV. Age, sex and WHO Grade in this patient population are summarized in Table 2.

GSL-Type Expression

In 18 gliomas the GSL type I was defined. One of these was an anaplastic astrocytoma (WHO Grade III). The

Table 2. Patient Population (classification and grading according to WHO)

Diagnoses and WHO grading	No. of patients	Age (years)		Sex	
		Mean	Range	Male (%)	Female (%)
Low-grade astrocytoma (WHO II)	4	43.5	33–49	50	50
Anaplastic astrocytoma (WHO III)	8	46	22–69	50	50
Glioblastoma multiforme (WHO IV)	28	59	39–75	57	43

Table 3. Distribution of GSL types and WHO Grades

	GSL type I	GSL type II	GSL type III	Total
WHO Grade II	0	1	3	4
WHO Grade III	1	3	4	8
WHO Grade IV	17	7	3	27
Total	18	11	10	39

$n = 39$; One tumor of the WHO Grade IV presented as GSL type I–II.

histopathological diagnosis was glioblastoma multiforme (WHO Grade IV) for the remaining 17 tumors. In the group with GSL type II, glioblastomas represented again the major subgroup with seven from a total of 11 patients. Three patients presented with an anaplastic astrocytoma (WHO Grade III) and one patient presented with a low-grade astrocytoma (WHO Grade II). In GSL type III the glioblastomas were still present but represented the minor part with three of ten tumors. Four were anaplastic astrocytomas (WHO Grade III) and three low-grade astrocytomas (WHO Grade II). One patient with a glioblastoma multiforme, WHO Grade IV presented with a mixed GSL type I–II. This tumor was excluded from further analysis of the results. The distribution of WHO grades and corresponding GSL types is presented in Table 3.

Correlation of GSL Types and WHO Grading

There was not a very strong correlation between GSL type and histology, as it was supposed earlier [16, 17]. However, by calculating the Cohen's Kappa coefficient the common occurrence of GSL type I and WHO Grade IV was demonstrated ($\kappa > 0.45$). For GSL types II and III a correlation with any one WHO Grade could not be proven.

Survival Analyses

The survival time was calculated for the different subgroups (the GSL types and WHO malignancy grades). Patients with GSL type I glioma had a median survival time of 56 weeks (range 4–185 weeks, mean 63 weeks); GSL type II had a median survival time of 51 weeks (range 9–396 weeks, mean 118 weeks); and GSL type III 248 weeks had a median survival time of (range 6–477 weeks, mean 227 weeks).

Patients with a glioma of WHO Grade IV had a median survival time of 51 weeks (SD 10 weeks), those with a WHO Grade III glioma, a median of 118.5 weeks (SD 28.6 weeks), and those with a WHO Grade II glioma a median survival time of 396 weeks (SD 53 weeks).

Kaplan-Meier survival curves were calculated for the GSL types and for the WHO grades (Fig. 1). The differences in the survival curves were tested for statistical significance according to the log rank test. As shown in Fig. 1 the survival curves differed significantly ($p = 0.023$ for GSL types I–III and $p = 0.0034$ for the WHO Grades II–IV).

Of substantial interest was whether the different GSL types in a given histological malignancy grade provide additional prognostic information or not. Therefore, the survival times in the different malignancy grades were calculated according to the GSL patterns.

Patients with glioblastoma multiforme GSL type I had a median survival time of 51 weeks ($n = 17$) and those with GSL type II, a median survival time of 46 weeks ($n = 7$). The patients with GSL type III had a median survival time of 26 weeks ($n = 3$). However, all three patients were multimorbid, with a GSL-type III mean age of above 65 years, which has a major influence on survival time.

The eight patients with anaplastic astrocytomas GSL type I had a survival time of 60 weeks (one patient), with GSL type II, 77 weeks ($n = 3$), and with GSL type III more than 218 weeks ($n = 4$). With only four patients suffering from a low-grade glioma (WHO Grade II) it was impossible to achieve consistent data about the survival time in the different GSL-types.

Immunization

The first animal experiments yielded promising results. It was possible to enhance the immunogenicity of ganglioside vaccines by chemical coupling with KLH and substitution with monophosphoryl-lipid-A (MPL-A) [18]. The animals showed significantly enhanced antiganglioside antibody titers compared with controls.

Nude mice that developed anti-Gfpt1 antibody titers after immunization with different ganglioside-macroprotein conjugates against the bronchial-carcinoma-specific ganglioside Gfpt1 developed no tumors after subcutaneous inoculation of bronchial

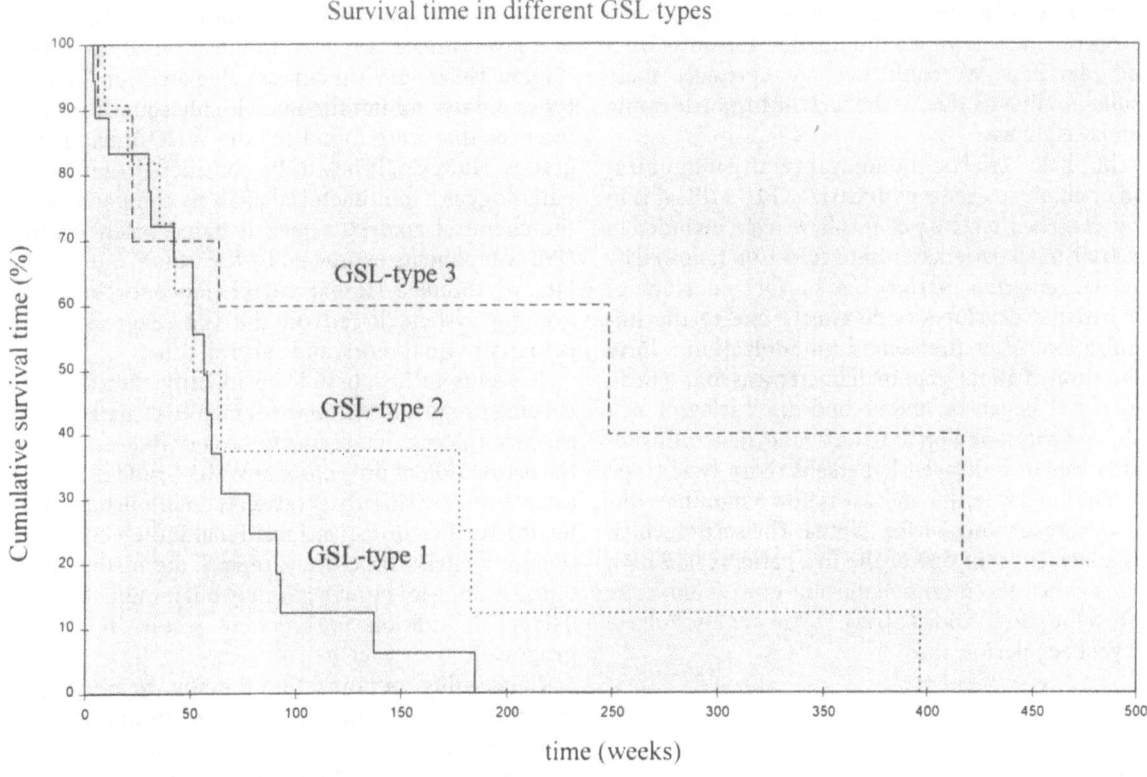

Survival time in different GSL types

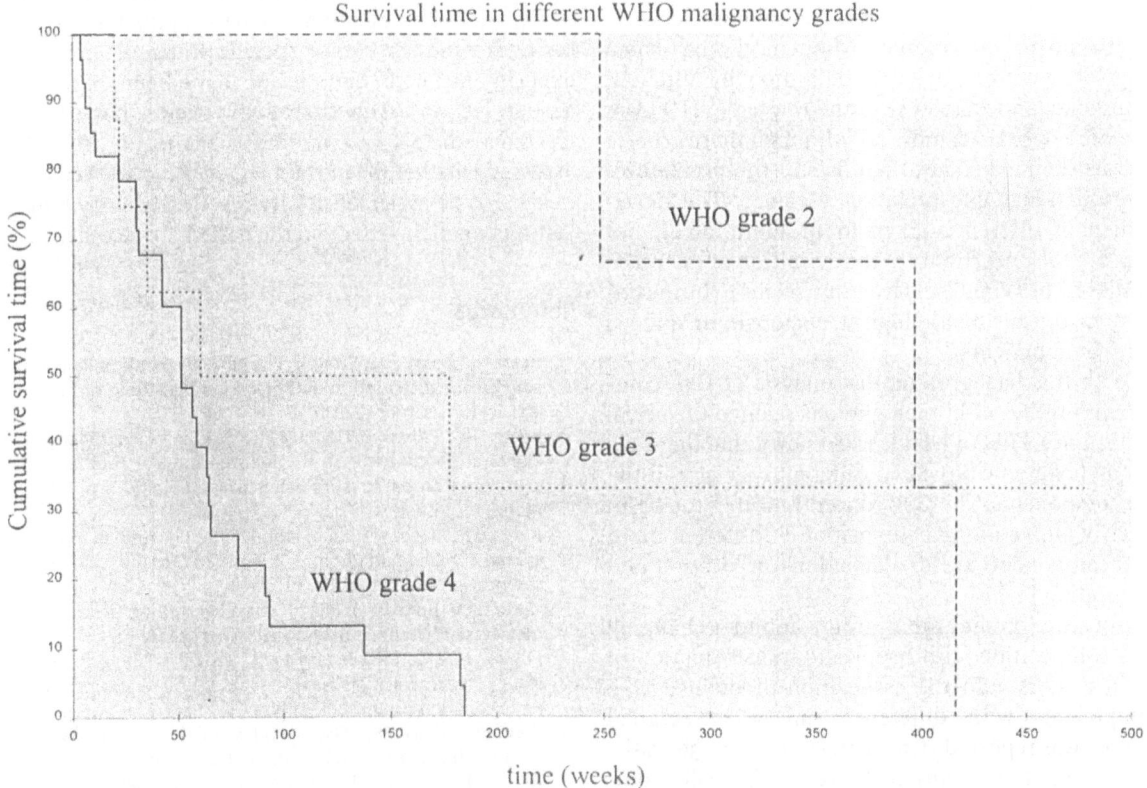

Survival time in different WHO malignancy grades

Fig. 1. Kaplan Meier survival curves for patients with gliomas of
GSL types I – III (**a**) and WHO malignancy Grades II – IV (**b**)

carcinoma cell. In contrast there was tumor growth in all control animals which did not develop anti-Gfpt1 antibody titers. We could not yet reproduce these results in gliomas due to the lack of tumor-forming glioma cell lines.

The phase I trial on the adjuvant active immunization of glioma patients with Gtri2–KLH–MPL-A is in progress. The first five patients have been included in this trial. In all patients a tumor resection followed by external radiation therapy was performed. None of the patients developed side effects due to the immunization. After the second immunization a local induration of about 3 cm in diameter was found in all patients. It began on the second day (delay of 24–48 h) and continued up to 10 days, and, hence, this reaction was of a delayed hypersensitivity type (type IV). The longest follow-up case is now 6 months without local recurrence of the glioma. The serum which was analyzed from one of the five patients had high titers against the immunoadjuvant but no antibody against the ganglioside (Gtri2). Further analyses have not yet been performed.

Discussion

The discussion about the histopathological criteria for diagnosis and grading of human gliomas has a long history, from the first definition by Virchow, Connheim, and Maaß [4] to the revised WHO classification of brain tumors [1]. Although these criteria are well defined today, there is still the problem of interindividual interpretation variance. The development of different markers for proliferation did not improve the uncertainty in the definition of the WHO Grades III or IV, and survival time seems to be in contrast to the histopathological diagnosis in quite a number of patients.

With this background, the analysis of GSL-component profiles of human gliomas seemed of special importance. First of all it is well known that there is a strong taxonomy of GSL in different species and tissue components. The GSL concentration in the brain is fairly high and the composition of different brain GSL components varies, depending on location and maturation [35].

In earlier studies other groups found a changing GSL composition with neoplastic transformation of brain tumors, and the association of defined GSL components with different primary intracranial tumors was reported [1, 6–9, 11, 13, 26, 32, 34, 36–38].

However, the definition of three distinct GSL-component profiles, taking neutral and acidic GSL into account, was a new approach [16]. Initially, these GSL types I–III seemed to correlate well with the histopathological malignancy grade. This correlation is not as close as it was believed. However, the GSL-type of a given tumor is related to the survival time. The Kaplan Meier survival curves calculated for the GSL types did not exhibit the same highly significant differences that were found for the WHO malignancy grades. However, it has to be considered that histopathology is a multifactorial analysis compared with the chemical analysis which is based solely on the GSL components expressed by the tumor. Sung et al. [30, 31] found a similar correlation between their grading system, based on the GSL expression in primary brain tumors, and survival time.

It seems to be possible to identify tumors with varying prognosis within the same WHO malignancy grade, a finding that seems to be very interesting for the estimation of prognosis in WHO Grade II and III astrocytomas. This observation is an important matter for further investigations. Finally the GSL-component analysis should not replace the histopathological grading of human gliomas, but it might be very useful in addition to precisely assess individual prognosis.

Concerning immunization therapy we are still in the initial stages. The expression of tumor-specific GSL components on the cell surface is a well-known phenomenon reported for a variety of tumors [5, 14, 15, 24, 33]. In human gliomas Gtri-gangliosides might be addressed as tumor-specific antigens. Animal experiments with the vaccine have been reported earlier [18, 19]. The first five patients have been treated with this vaccine and it has to be carefully assessed whether the immune response, observed as a delayed-type hypersensitivity reaction has any long-lasting beneficial effect on the patients' outcome.

References

1. Berra B, Gaini SM, Riboni L (1985) Correlation between ganglioside distribution and histological grading of human astrocytomas. Int J Cancer 36 : 363–366
2. Burger PC, Shibata T, Kleihues P (1986) The use of the monoclonal antibody Ki-67 in the identification of proliferating cells: application to surgical neuropathology. Am J Surg Pathol 10 : 611–617
3. Burger PC (1990) Classification, grading, and pattern of spread of malignant gliomas. In: Apuzzo MLJ (ed) Malignant cerebral glioma. AANS, Park Ridge
4. Connheim J, Maas H (1877) Zur Theorie der Geschwulstmetastasen. Virchows Arch Pathol Anat 70 : 161–171
5. Fukuda MN, Bothner B, Lloyd KO, Rettig WJ, Tiller PR, Dell A (1986) Structures of glycosphingolipids isolated from human embryonal carcinoma cells. J Biol Chem 261 : 5145–5153
6. Fredman P (1988) Gangliosides in human malignant gliomas. In: Ledeen RW, Hogan EL, Tettamanti G, Yates AJ, Yu RK (eds) New trends in ganglioside research: neurochemical and neuroregenerative aspects. Liviana, Padova, pp 151–161
7. Fredman P, von Holst H, Collins VP, Granholm L, Svennerholm L (1988) Sialyllactotetraosylceramide, a ganglioside marker for human malignant gliomas. J Neurochem 50 : 912–919

8. Fredman P, Dumanski J, Davidsson P, Svennerholm L, Collins P (1990) Expression of ganglioside GD3 in human meningiomas is associated with monosomy of chromosome 22. J Neurochem 55 : 1838 – 1840

9. Fredman P, von Holst H, Collins VP, Dellheden B, Svennerholm L (1993) Expression of ganglioside GD3 and 3'-sioLM1 in autopsy brains from patients with malignant tumors. J Neurochem 60 : 99 – 105

10. Gerdes J, Schwab U, Lemke H, Stein H (1983) Production of a mouse monoclonal antibody reactive with a human nuclear antigen associated with cell proliferation. Int J Cancer 31 : 13 – 20

11. Gottfries J, Fredman P, Mansson JE, Collins VP, von Holst H, Armstrong DD et al. (1990) Determination of gangliosides in six human primary medulloblastomas. J Neurochem 55 : 1322 – 1326

12. Hakomori S (1981) Glycosphingolipids in cellular interaction, differentiation and oncogenesis. Annu Rev Biochem 50 : 733 – 764

13. Hakomori S (1985) Aberrant glycosylation in cancer cell membranes as focused on glycolipids: overview and perspectives. Cancer Res 45 : 2405 – 2414

14. Hakomori S (1996) Tumor malignancy defined by aberrant glycosylation and sphingo(glyco)lipid metabolis. Cancer Res 56 : 5309 – 5318

15. Hoon DSB, Okun E, Neuwirth H, Morton DL, Irie RF (1993) Aberrant expression of gangliosides in human renal cell carcinomas. J Urol 150 : 2013 – 2018

16. Jennemann R, Rodden F, Bauer BL, Mennel HD (1990) Glycosphingolipids of human gliomas. Cancer Res 50 : 7444 – 7449

17. Jennemann R, Mennel HD, Bauer BL, Wiegandt H (1994) Glycosphingolipid component profiles of human gliomas correlate with histological tumour types: analysis of interindividual and tumour-regional distribution. Acta Neurochir (Wien) 126 : 170 – 178

18. Jennemann R, Gnewuch C, Boßlet S, Bauer BL, Wiegandt H (1994) Specific immunization using keyhole limpet hemocyanin–ganglioside conjugates. J Biochem 115 : 1047 – 1052

19. Jennemann R, Bauer BL, Schmidt R, Elsässer HP, Wiegandt H (1996) Effects of monophosphoryllipid-A on the immunization of mice with keyjole limpet hemocyanin- and muramyl-dipeptide-ganglioside Gfpt 1 conjugates. J Biochem (Tokyo) 119 : 378 – 384

20. Kaplan EL, Meier P (1958) Nonparametric estimation from incomplete observations. J Am Stat Assoc 53 : 457 – 481

21. Kleihues P, Burger PC, Scheithauer BW (1993) The new WHO classification of brain tumours. Brain Pathol 3(3) : 255 – 268

22. Martin H (1994) The importance of the AgNOR analysis in malignant tumors. Zentralbl Pathol 140 : 15 – 22

23. Mennel HD, Rossberg C, Lorenz H, Schneider H, Hellwig D (1989) Reliability of simple cytological methods in brain tumor biopsy diagnosis. Neurochirurgia 32 : 129 – 134

24. Miyake M, Hashimoto K, Ito M, Ogawa O, Arai E, Hitomi S, Kannagi R (1990) The abnormal occurence and the differentiation-dependent distribution of N-acetyl and N-glycolyl species of the ganglioside GM2 in human germ cell tumors. Cancer 65 : 499 – 505

25. Ohsawa T, Nakane-Hikichi K, Nagai Y (1984) Alterations of ceramide monohexoside in human diploid fetal lung fibroblasts during cell aging. Biochem Biophys Acta 782 : 79 – 83

26. Pausch G, Jennemann R, Mennel HD, Bauer BL, Wiegandt H (1992) Gangliosides in meningeomas: correlation of Glac2 to intermediary filament. Acta Neurochir (Wien) 117 : 166 – 171

27. Rodden FA, Wiegandt H, Bauer BL (1991) Gangliosides: the relevance of current research to neurosurgery. J Neurosurg 74 : 606 – 619

28. Seifert H, Uhlenbruck G (1965) Über Ganglioside in Hirntumoren. Naturwissenschaften 52 : 190 – 191

29. Singh LP, Pearl DK, Franklin TK, Spring PM, Scheithauer BW, Coons SW, Johnson PC, Pfeiffer SE, Li J, Knott JC, et al. (1994) Neutral glycolipid composition of primary brain tumors. Mol Chem Neuropathol 21(2 – 3): 241 – 257

30. Sung CC, Pearl DK, Coons SW, Scheithauer BW, Johnson PC, Yates AJ (1994) Gangliosides as diagnostic markers of human astrocytomas and primitive neuroectodermal tumors. Cancer 74(11) : 3010 – 3022

31. Sung CC, Pearl DK, Coons SW, Scheithauer BW, Johnson PC, Zheng M, Yates AJ (1995) Correlation of ganglioside patterns of primary brain tumors with survival. Cancer 75 : 851 – 859

32. Svennerholm L, Bostrom K, Fredman P, Mansson JE, Rosengren B, Rynmark BM (1989) Human brain gangliosides: developmental changes from early fetal stage to advanced age. Biochem Biophys Acta 1005 : 109 – 117

33. Tai T, Paulson JC, Cahan LD, Irie RF (1983) Ganglioside GM2 as a human tumor antigen (OFA-I-1). Proc Natl Acad Sci USA 80 : 5392 – 5396

34. Traylor TD, Hogan EL (1980) Gangliosides of human cerebral astrocytomas. J Neurochem 34 : 126 – 131

35. Wiegandt H (1985) Gangliosides. In: Wiegandt H (ed) New comprehensive Biochemistry, vol 10. Elsevier, Amsterdam, pp 199 – 260

36. Wikstrand CJ, He X, Fuller GN, Bigner SH, Fredman P, Svennerholm L et al. (1991) Occurence of lacto series gangliosides 3'-isoLM1 and 3',6'-isoLD1 in human gliomas in vitro and in vivo. J Neuropathol Exp Neurol 50 : 756 – 769

37. Yates AJ, Thompson DK, Boesel CP, Albrightson C, Hart RW (1979) Lipid composition of human neural tumors. J Lipid Res 20 : 428 – 436

38. Yates AJ (1986) Gangliosides in the nervous system during development and regeneration. Neurochem Pathol 5 : 309 – 329

Subject Index

Errata

The Title of the contribution by H. W. S. Schroeder and M. R. Gaab should read as follows on page VIII of the contents and in the running head on pages 102 and 104:

Endoscopic Management of **Intracranial** Arachnoid Cysts

Dieter Hellwig · Bernhard L. Bauer (Eds.),
Minimally Invasive Techniques for Neurosurgery
© Springer-Verlag Berlin Heidelberg 1998